AF251536

Handbook of
Energy Harvesting
Power Supplies
and Applications

Handbook of Energy Harvesting Power Supplies and Applications

edited by

Peter Spies | Loreto Mateu | Markus Pollak

PAN STANFORD PUBLISHING

Published by

Pan Stanford Publishing Pte. Ltd.
Penthouse Level, Suntec Tower 3
8 Temasek Boulevard
Singapore 038988

Email: editorial@panstanford.com
Web: www.panstanford.com

British Library Cataloguing-in-Publication Data
A catalogue record for this book is available from the British Library.

Handbook of Energy Harvesting Power Supplies and Applications

Copyright © 2015 Pan Stanford Publishing Pte. Ltd.

ISBN 978-981-4241-86-1 (Hardcover)
ISBN 978-981-4303-06-4 (eBook)

Printed in the USA

Contents

3 Piezoelectric Transducers   **79**

Bernhard Brunner, Matthias Kurch, and William Kaal

Loreto Mateu and Peter Spies

Preface

The power consumption of microelectronic circuits and systems is decreasing by successive development of circuit and semiconductor technology. On the other hand, the efficiency of energy transducers such as solar cells, thermoelectric, and inductive generators is increasing by means of material and system improvements. Thus, energy transducers are able to use ambient energy to power small electronic devices such as sensors, microcontrollers, and wireless transceivers. The technology has come to be known as "energy scavenging" or "energy harvesting," the systems with these power supplies are often called "energy-autarkic" or "self-powered systems."

On the one hand, energy harvesting power supplies replace batteries in conventional applications such as consumer products, household appliances, measurement and monitoring applications, and home automation systems. If the battery cannot be replaced completely, at least the length of time before the next recharge can be extended. By eliminating batteries, a significant reduction of waste and battery replacement effort is achieved.

On the other hand, new applications such as wireless sensors in remote or inaccessible areas become possible with energy harvesting. Examples are medical implants, integrated sensors in machinery, engines or plants or rotating equipment. Furthermore, unlimited operation and standby time are possible with energy harvesting.

The growing research into and development of wireless sensor networks are closely linked to energy harvesting. The full benefits of wireless sensor networks cannot be achieved with wires for power supply or battery replacement maintenance. Especially, with an increasing number of nodes in a mesh network, self-powered electronics are mandatory.

At present, several professional applications have established themselves in this domain, mainly in the area of building and home automation, consumer products and condition monitoring. In contrast, a huge new field of applications for energy harvesting, especially for powering wireless sensor nodes are addressed in research and development projects.

Contributions to this book have been made by the leading facilities of applied research in Germany, the Fraunhofer Gesellschaft and the Hahn-Schickard-Gesellschaft, which are both application-oriented research and development providers. They work in publicly funded projects and also conduct research and development for industrial companies around the world.

Thus, this book deals with the basics of energy harvesting technology with a focus on application-oriented implementation. Each chapter addresses a special core technology of energy harvesting including the different transducer principles and related materials, power management, storage devices, and system design. The final chapter introduces different applications of energy harvesting and related system architectures and application devices and discusses relevant converter types.

Chapter 1

System Design

Loreto Mateu and Peter Spies

*Fraunhofer Institute for Integrated Circuits IIS, Nordostpark 93,
90411 Nuremberg, Germany*
loreto.mateu@iis.fraunhofer.de

This chapter deals with the topic of designing an energy harvesting (also called energy-scavenging) system that is composed by an energy harvesting power supply and a low-power load. In such systems, the energy is collected from the environment employing a transducer that transforms the ambient energy into electrical energy for supplying energy autarkic electronic devices.

1.1 Introduction

A self-powered system based on energy harvesting is composed of several blocks (see Fig. 1.1) and each of them has a dedicated section in this chapter. The blocks are:

- Energy transducer (also called energy harvesting generator). It is used to convert the input ambient energy into electrical energy. The environmental energy sources available for conversion may be heat (thermoelectric

Handbook of Energy Harvesting Power Supplies and Applications
Edited by Peter Spies, Loreto Mateu, and Markus Pollak
Copyright © 2015 Pan Stanford Publishing Pte. Ltd.
ISBN 978-981-4241-86-1 (Hardcover), 978-981-4303-06-4 (eBook)
www.panstanford.com

modules), light (photovoltaic cells), radiation (rectifying antennas), and vibration (piezoelectric, electro-magnetic, electro-static transducers).

- Rectifier and storage capacitor. Some energy transducers do not provide DC power, and in this case it is necessary to rectify the current and accumulate it into a capacitor.
- Voltage regulator. It is necessary to adapt the voltage level to the requirements of the powered device or the optional storage element.
- Optional energy storage element. A battery or capacitor, depending on the requirements of the application, is employed as energy storage element. In some applications the powered device can be completely switched off during certain intervals and a battery is not necessary, while in others a permanent supply is mandatory. Furthermore, the energy storage element is required to provide pulse currents for radio transceivers that work in the burst mode; these cannot be generated by the energy transducer itself due to their large internal resistance. In any case, this battery will have a lower weight, volume, and capacity than a battery that is expected to supply power to an electronic device without an energy harvesting generator. Whether a capacitor can be used instead of a battery depends on the requirements of the application.
- Electronic load. It has typically different power consumption modes allowing to operate the device most of the time in a low-power consumption mode. It works in active mode only during limited time periods to decrease its total energy consumption.

1.2 Input Energy

The application of the energy harvesting system determines which energy sources are available in the environment to power it. The main environmental energy sources employed to supply power to, for example, wireless sensor networks (WSNs) are solar, mechanical, and thermal energy. Self-powered devices have normally reduced

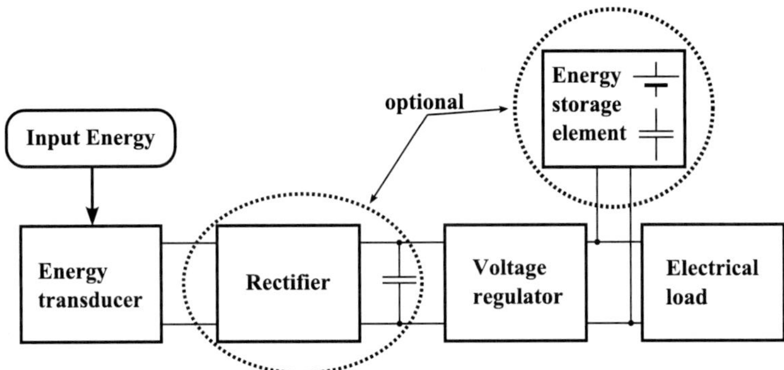

Figure 1.1   Schema of a generic self-powered device.

dimensions (volume around 10 cm^3 or lower) since their most frequent applications are as nodes in a WSN or wearable devices. The dimensions of the energy harvesting power supply are a constraint for the amount of electrical energy that is generated. That is why an accurate comparison of energy harvesting systems can be done only in terms of power per unit of volume (power density) or power per unit of area.

Roundy[4] summarizes in Table 1.1 a comparison between different energy harvesting sources (unshaded part) and energy storage elements (shaded part) in terms of power density (power per unit of volume). Power density for energy harvesting sources under the same input conditions remains constant with time while it does not for energy storage techniques due to leakage currents as it is shown in Fig. 1.2. That is why energy harvesting becomes an alternative to energy storage techniques in long operation time applications where it is not possible to replace or recharge the energy storage element.

Raju[5] gives an estimation of the available power per unit of area for different energy harvesting sources and scenarios. For the case of vibrations and temperature difference, Table 1.2 distinguishes between human and industry energy sources. "Human" refers to the use of the human body as input energy. Therefore, the temperature gradient existing between the human body and the environment and the vibrations associated to the human movement

Table 1.1 Comparison of energy scavenging sources and energy storage elements in terms of power density taking into consideration lifetime

	Power density (μW/cm^2) 1 Year lifetime	Power density (μW/cm^2) 10 Years lifetime	Source of information Source of information
Solar (Outdoors)	15,000—direct sun 150—cloudy day	15,000—direct sun 150—cloudy day	Commonly available
Vibrations	200	200	Roundy et al.[1]
Acoustic noise	0.003 @ 75 Db 0.96 @ 100 Db	0.003 @ 75 Db 0.96 @ 100 Db	Theory
Daily temp. variation	10	10	Theory
Temperature gradient	15 @ 10°C gradient	15 @ 10°C gradient	Stordeur et al.[2]
Shoe inserts	330	330	Starner 1996, Shenck et al.[3]
Batteries (non-recharg. Lithium)	45	3.5	Commonly available
Batteries (rechargeable Lithium)	7	0	Commonly available
Hydrocarbon fuel (micro heat engine)	333	33	Mehra et al.mehra2000six
Fuel cells (methanol)	280	28	Commonly available
Nuclear isotopes (uranium)	6×10^6	6×10^5	Commonly available

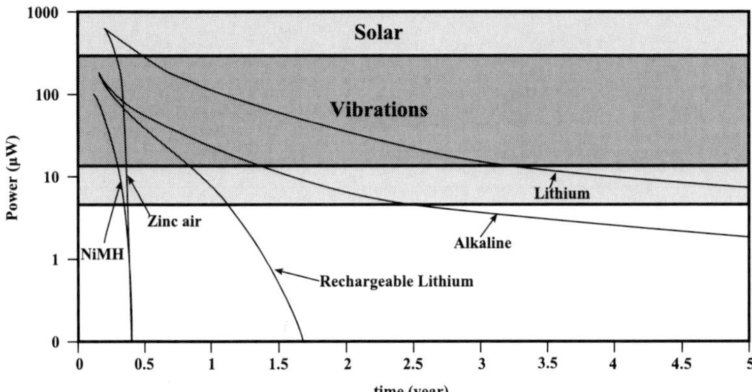

Figure 1.2 Power density versus life time for solar cells, vibrations and batteries.

Table 1.2 Comparison of harvested power per cm^2 for different energy sources and scenarios

Energy source	Harvested power
Vibration/Motion	
Human	$4\ \mu W/cm^2$
Industry	$100\ \mu W/cm^2$
Temperature Difference	
Human	$25\ \mu W/cm^2$
Industry	$1–10\ mW/cm^2$
Light	
Indoor	$10\ \mu W/cm^2$
Outdoor	$10\ mW/cm^2$
RF	
GSM	$0.1\ \mu W/cm^2$
WiFi	$1\ \mu W/cm^2$

can be the input energy of an energy harvesting power supply. In the case of industry, excess heat and machine vibrations are employed as energy source.

Light is an environmental energy source available to power electronic devices. A photovoltaic system generates electricity by the conversion of light into electricity. Photovoltaic systems are found

Table 1.3 Solar power measurements taken under different light conditions

Conditions	Outside, midday	4 inches from 60 W bulb	15 inches from 60 W bulb	Office lighting
Power (μW/cm^3)	14000	5000	567	6.5

from the megawatt to the milliwatt range producing electricity for a wide number of applications: from lighting to wristwatches.

Outdoors, the solar radiation is the energy source for PV system. This solar radiation varies over the earth's surface due to the weather conditions and the location (longitude and latitude). For each location, there exists an optimum inclination angle and orientation of the PV solar cells in order to obtain the maximum radiation over the surface of the solar cell.[6] Table 1.3 shows the power density measured under different light conditions with a silicon solar cell and Table 1.2 presents the power per unit of area for an outdoor and indoor light source. The power per unit of area is three orders of magnitude bigger in the outdoor than in the indoor case.[4] From the measurements displayed in Table 1.3, it is deduced that in the case of indoor light, the power density decreases with the inverse of the square of the distance between the solar cell to the light source.

The principle behind kinetic energy harvesting is the displacement of a moving part or the mechanical deformation of a structure inside the energy harvesting device. This displacement or deformation can be transformed to electrical energy by three different methods: inductive, electrostatic and piezoelectric conversion.

Vibrations are the input energy for the transducers that convert the displacement of a moving part into electrical energy. Vibrations are characterized by their peak acceleration and the corresponding frequency. With these data, an estimation of the electrical energy that can be generated employing vibrations is possible.[7] Table 1.4 gives a list of peak accelerations and frequencies for different industry vibration sources and from this data, it is deduced that vibrations of industry machines have associated accelerations between 60 and 125 Hz.

Table 1.4 List of vibration sources with their respective peak acceleration and frequency

Vibration source	Peak acc. (m/s^2)	Frequency of peak (Hz)
Base of 5 HP 3-axis machine tool with 36″ bed	10	70
Kitchen blender casing	6.4	121
Clothes dryer	3.5	121
Door frame just after door closes	3	125
Small microwave oven	2.25	121
HVAC vents in office building	0.2–1.5	60
Wooden deck with people walking	1.3	385
Breadmaker	1.03	121
External windows (size 2 ft × 3 ft) next to a busy street	0.7	100
Notebook computer while CD is being read	0.6	75
Washing machine	0.5	109
Second story floor of a wood frame office building	0.2	100
Refrigerator	0.1	240

There is also the possibility of employing the human body as a vibration source. Vibrations associated with the human body have accelerations with frequencies under 10 Hz.[8] T. von Büren et al.[9] present a comparison of simulations done with vibrational generators employing measured acceleration data from walking motion at different locations of the human body. Walking is one of the human activities that have more energy associated.[10,11] Mateu et al.[12,13] are also presenting a simulation study for the case of a non-linear model of the electrodynamic generator employing measured acceleration data from different human activities and locations at the human body.

Jansen[14] employs the term human power as short for human powered energy systems in consumer products. The Personal Energy System (PES) research group of the Delft University of Technology distinguishes between active and passive energy harvesting methods when the input energy is provided by the human body. The active powering of electronic devices takes place when the user of the electronic product has to do a specific work in order to power the product that he otherwise would not have done.

The passive powering of electronic devices takes place when the user does not have to do any activity different to the normal tasks associated with the product. In this case, the energy is harvested from the user's everyday actions (walking, breathing, body heat, blood pressure, finger motion, etc.).

The option to parasitically harvest energy from everyday human activity (passive power) implies that an unobtrusive technique has to be adopted. Starner presented human power as possible source for wearable computers.[10] He analyzed power generation from breathing, body heat, blood transport, arm motion, typing, and walking and provides the power dissipated by the human body during several activities. A more recent study appears in reference[15] where the state of the art of passive human power to power body-worn mobile electronics is explained.

Heat can be used as input energy for energy harvesting power supplies where a temperature gradient and heat flow is present. The maximum efficiency for converting the harvested heat into electricity is given by Carnot efficiency:[10]

$$\eta_{\text{Carnot}} = \frac{T_{\text{Hot}} - T_{\text{Cold}}}{T_{\text{Hot}}} \tag{1.1}$$

where T_{Hot} is the high temperature and T_{Cold} is the low temperature of the temperature gradient.

Thermal energy is characterized by the temperature gradient and the heat flow. It is converted into electrical energy with thermogenerators that are fundamentally based on the Seebeck effect. This kind of energy is present, e.g., in machinery (industrial) and in the human body (see Table 1.5). The temperature gradient is mostly obtained between the heat source and the room temperature.

Starner et al.[16] makes an analytical study of thermoregulation in humans.

Leonov et al.[17] have experimental data of skin temperature and heat flow of humans dependence on air temperature for different locations on the forearm. Leonov et al.[18] present the thermal circuit of a thermogenerator on the skin that is composed by three different thermal resistances, the body, the thermogenerator and the ambient air.

Table 1.5 Characteristics of typical energy harvesting power supplies

Energy source	Characteristics	Efficiency	Harvested power
Light	Outdoor	10-25%	100 mW/cm^2
	Indoor		100 μW/cm^2
Thermal	Human	$\sim$0.1%	60 μW/cm^2
	Industrial	$\sim$3%	10 mW/cm^2
Vibration	$\sim$ Hz-human	25-50%	40 μW/cm^2
	$\sim$ kHz-machines		800 μW/cm^2
Radio frequency (RF)	GSM 900 MHz	$\sim$50%	0.1 μW/cm^2
	WiFi 2.4 GHz		0.001 μW/cm^2

1.3 Energy Transducer

The energy transducer is used to convert the available energy into electrical energy. The selection of the energy transducer depends on the kind of available energy for the application under consideration. Therefore, thermoelectric cells are employed for thermal energy and photovoltaic cells for light. For mechanical energy, three different transducers are considered: piezoelectric, electro-dynamic, and electro-static.

The location of the transducer determines the amount of input energy that is available for the energy harvesting power supply and, therefore, the output power obtained for supplying the electronic load. Consequently, to find the location that provides the higher amount of input energy for the relevant application is of special interest.

When the input energy are vibrations, it is necessary to measure the acceleration at the different possible locations of the energy harvesting transducer in order to determine the amplitude and the frequency range of the vibrations.

The principle behind kinetic energy harvesting is the displacement of a moving part or the mechanical deformation of some structure inside the energy harvesting device. This displacement or deformation can be converted to electrical energy by three different methods: electro-magnetic, electro-static, and piezoelectric conversion. Each of these transducers can convert kinetic energy into electrical energy with two different methods: inertial and non-inertial.

Inertial transducers are based on a spring-mass system. In this case, the proof mass vibrates or suffers a displacement due to the kinetic energy applied. The transducer converts the relative displacement of the mass referred to the housing, which causes an inertial force, into electrical energy. Thus, this type of transducer is called an inertial converter. Mitcheson et al. have classified inertial converters as a function of the force opposing the displacement of the proof mass.[19] These converters resonate at one discrete frequency and many of them are designed to resonate at the frequency of the mechanical input source since at this frequency (resonance frequency), the energy obtained is maximum. However, as the converters are miniaturized to integrate them on MEMS devices, the resonance frequency increases, and it becomes much higher than characteristic frequencies of many everyday mechanical stimuli.

For non-inertial converters, an external element applies a pressure that is transformed into elastic energy, causing a deformation that is converted to electrical energy by the converter. In this case, there is no proof mass and the obtained energy depends on mechanical constraints or geometric dimensions.[20] The following paragraphs give an overview of piezoelectric, electro-dynamic and electro-static transducers.

Piezoelectric materials are employed as sensors, actuators, or energy harvesting transducers due to their properties. The piezoelectric effect was discovered by Jacques and Pierre Curie in 1880. Curie's brothers found that certain materials, when subjected to mechanical strain, suffered an electrical polarization that was proportional to the applied strain. Metallizing the piezoelectric materials and connecting electrodes provides a voltage associated with the charge when the electrodes are not short-circuited. The piezoelectric effect can be employed for the conversion of mechanical energy into electrical energy. Table 1.6 shows a summary table with some energy harvesting generators that employ piezoelectrics as transducers. Detailed information about these transducers is provided in Chapter 3.

Electro-dynamic generators, also called voltage damped resonant generators (VDRG), are based on Faraday's law. The principle of

Table 1.6 Summary table of piezoelectric inertial generators

Design Author	Mechanical excitation	Output power	Dimensions
S. Roundy et al.[21] Design 1	$a = 2.25$ m/s^2 $f = 85$ Hz	207 μW @ 10 V	1 cm^3
S. Roundy et al.[21] Design 2	$a = 2.25$ m/s^2 $f = 60$ Hz	335 μW @ 12 V	1 cm^3
S. Roundy et al.[21] Design 3	$a = 2.25$ m/s^2 $f = 40$ Hz	1700 μW @ 12 V	4.8 cm^3
H. Hu[22]	$a = 1$ m/s^2 $f = 50$ Hz	246 μW/cm^3 @ 18.5 V	—

these electromagnetic induction microgenerators is the generation of a current induced on a coil by a moving magnet relative to the coil. The relative motion produces a change in the electromagnetic flux through the coil that causes an electromotive force (EMF) in the coil, following Faraday's law. This induced EMF will generate a current related to the electrical load of the coil that in turn generates a force due to the electromagnetic field and this force will interact with the motion. This flux variation can be realized with a moving magnet whose flux is linked with a fixed coil or with a fixed magnet whose flux is linked with a moving coil. The first configuration is preferred to the second one because the electrical wires are fixed. As the relevant magnitude here is the magnetic flux, the length of the coil is directly proportional to the obtained electric field and therefore, to the generated energy. This means that big transducers with large area coils will perform better than smaller ones, unless a large acceleration is involved with the small-scale generators. Table 1.7 shows a summary table of electro-dynamic generators. The analysis of these transducers is given in Chapter 4.

Electro-static generators, also known as Coulomb-damped resonant generators (CDRGs), are based on electrostatic damping. The implementation of electro-static generators is done using a capacitor with one plate moving against the electric field. If the charge on the capacitor is maintained constant while the capacitance decreases by reducing the overlap area of the plates or increasing the distance between them, the voltage will increase. If the voltage on the capacitor is maintained constant while the

Table 1.7   Summary table of electro-dynamic generators

Design author	Mechanical excitation	Output power	Dimensions
Williams et al.[23]	$f = 4$ kHz Amplitude $= 300$ nm	$0.3\,\mu$W	mm^3
Li et al.[24]	$f = 64$ Hz Amplitude $= 1000\ \mu$m	$10\ \mu$W @ 2 V	1 cm^3
Ching et al.[25]	$f = 104$ Hz Amplitude $= 190\ \mu$m	$5\ \mu$W	—
Amirtharajah et al.[26]	$f = 2$ Hz Amplitude $= 2$ cm	$400\ \mu$W @ 180 mV	—
Yuen et al.[27]	$f = 80$ Hz Amplitude $= 250\ \mu$m	$120\ \mu$W @ 900 μV	2.3 cm^3

Table 1.8   Summary table of electro-static generators

Design author	Mechanical excitation	Output power	Dimensions
Meninger et al.[28]	$f = 2.52$ kHz	$8\ \mu$W	0.075 cm^3
Sterken et al.[29]	$f = 1, 200$ Hz Amplitude $= 20\ \mu$m	$100\ \mu$W @ 2 V	–
Miyazaki et al.[30]	$f = 45$ Hz Amplitude $= 1\ \mu$m	120 nW	–

capacitance decreases, the charge will decrease. The mechanical energy converted into electrical energy is greater when the voltage on the capacitor is constant than when the charge on the capacitor is constant. However, the voltage source needed to place an initial charge on the capacitor plates has a smaller value, if the charge across the capacitor is constrained. A way to increase the electrical energy for the charge constrained method is adding a second capacitor in parallel with the variable capacitor. The disadvantage of this solution is that the value of the initial voltage source has to be increased. The energy conversion principle of electro-static generators is explained in more detail in Chapter 5. Table 1.8 shows the results obtained with some electro-static generators.

A comparison table between piezoelectric, electro-dynamic, and electro-static transducers for the mechanical to electrical energy conversion with their advantages and disadvantages is given by Roundy[4] and Jia[31] and has been put together in Table 1.9.

Table 1.9 Comparison of vibration transducers

Type	Energy density equation	Practical maximum for energy density (mJ/cm^3)	Theoretical maximum for energy density (mJ/cm^3)	Advantages	Disadvantages
Piezoelectric	$U = \frac{\sigma_y^2 k^2}{2Y}$	17.7	355	No external voltage source required, Voltages of 2 to 10 V, No mechanical stops, Compatible with MEMS, Highest energy density	High output impedance, Depolarization, Charge leakage, Brittleness in PZT, Poor coupling in PVDF
Electro-static	$U = \frac{\epsilon E^2}{2}$	4	44	Easier to integrate in MEMS, Voltage of 2 to 10 V	External voltage source needed, Mechanical stops needed
Electro-dynamic	$U = \frac{B^2}{2\mu_0}$	4	400	No external voltage source, No mechanical stops	Maximum output voltage of 0.1 to 0.2 V, Difficult to integrate with MEMS

Table 1.10   Summary table of thermogenerators

Author	Output power	ΔT (K)	Absolute temperature
Stordeur et al.[2]	20 μ W @ 4 V	20	Room temperature to $120°C$
Stordeur et al.[33]	15 μW/cm^2	10	—
Stevens[34]	—	10	—
Seiko[35,36]	1.5 μ W@1.5 V	1–3	—
ThermoLife[37]	28 μW @ 2.6 V	$5K$	30°C
Leonov et al.[18]	250 μW 20 μW/cm^2@0.9 V	—	Room temperature

The variables that appear in the energy density equation for piezoelectric transducers are the yield strength of the material σ_y, the piezoelectric coupling coefficient k, and Young's modulus Y. For the case of the electro-static transducer, ϵ is the dielectric constant and E is the electrical field between the plates. In the case of the electro-dynamic transducer, B is the magnetic field and μ_0 is the magnetic permeability.

Thermogenerators basically consist of one or more thermocouples, each of them composed of a p-type and a n-type semiconductor connected electrically in series and thermally in parallel. The TEG is based mainly on the Seebeck effect and produces an electrical voltage proportional to the temperature difference and to the number of thermocouples since the electrical connection allows to add the voltage obtained from each thermocouple.[32] Table 1.10 and Chapter 6 provide a detailed analysis of this transducer.

Light is another environmental energy source available to power electronic devices. A photovoltaic system provides electrical energy by the conversion of light employing solar cells as transducers. The employment of photovoltaics in portable products is a valid option under the appropriate circumstances. Chapter 7 explains this technology in further detail.

1.4 Rectifier

Piezoelectric, electro-static and electro-dynamic energy harvesting power supplies produce an AC output power. In order to power an electronic load, a rectification of the output power is necessary in those cases. The rectifier can be integrated with the power management unit, such as in the case of electro-magnetic generators,[38,39] and some piezoelectric generators.[40,41]

The rectification of the AC signal for piezoelectric transducers can also be done employing a voltage or a current multiplier.[42] It is also possible to choose between synchronous or asynchronous rectifiers and between half-wave and full-wave rectifiers. This topic is studied in detail in Chapter 9.

1.5 Power Management Unit

State-of-the-art TEGs produce, for instance, 50 mV open-circuit voltage per Kelvin thermal gradient. Typical piezoelectric modules can generate several volts depending on material and displacement. Electronic circuits such as sensors, microcontrollers or wireless transceivers, which are most often used with energy harvesting power supplies work with a supply voltage range between 1.8 and 5 V. Moreover, they need a very constant and well-regulated supply voltage to maximize their performance. Especially, peaks or oscillations generated from kinetic energy transducers will degrade their operation in terms of noise figure, accuracy, or resolution because of parasitic paths between the supply rails and the signal rails. The property to suppress such noise on the supply rail is called power supply rejection ratio (PSRR). This is the ratio between the gain from the input to the output and the gain from the supply rail to the output of the electrical circuit, e.g., an amplifier.

To close the gap between the outputs of the energy transducers and requirement of constant and decoupled supply rails at fixed voltage levels, different power management circuits are used. For employing low voltages, so-called up-converters or boost regulators are required. These blocks are important for thermogenerators when only small thermal gradients are present or when only a small number of thermo-couples are used to achieve a small system size or price. Furthermore, up-converters are helpful with solar cells for the same reasons. These converters are also used to discharge batteries below the required voltage of the circuit to supply. Usually, for a 3.3 V system, the battery is used down to 3.4 V, considering a 0.1 V drop of the power management, namely voltage regulator. Using an up-converter or a combination of an up- and a down-converter (buck-boost), you can discharge the battery down to the minimum battery voltage utilizing the total battery charge. The higher voltage for the application is produced with the up-converter. Care must be taken of the efficiency of these up-converters, which is often significantly lower than the efficiency of down-converters. Furthermore, the efficiency is often maximal only for a certain load current range. Leaving that range in the application, the efficiency will drop and the losses of the power management itself are increased.

Especially with thermogenerators, so-called start-up circuits are used, which enable an operation of the voltage converter down to several millivolts. Problem is the threshold voltage of semiconductors, which is presently at about 0.3 V. This would usually mean that you cannot power a circuit with a voltage below that range, because you are not able to switch any transistor. There are several techniques used in start-up circuits to cope with that problem. They are introduced and explained in detail in Chapter 8. These circuits are often only used during the start-up of the whole system when no battery is present. After a transient time, the voltage converter supplies itself from its own generated higher voltage via a feedback loop and the start-up circuit is disabled. The downside of such start-up circuits is often a very bad efficiency, thus it make sense to disable them during normal operation of the system.

Besides up- and down-conversion, another important of task of the power management is the impedance matching. As the power theory states, a power source will deliver the maximum power to a load if the impedance of the source and the load are equal. Especially, TEGs and solar cells alter their internal resistance due to aging and temperature. To match the internal resistance of the transducer as the source and the power management as the load, so-called maximum power point trackers are used. These systems were formerly used in large photovoltaic plants and are now adapted to energy harvesting systems by a significant reduction in power consumption and performance. The circuits are simply regulation loops of switching voltage regulators, which measure the output power of these regulators. They change the duty cycle of the switching regulator and monitor the output power to arrive at an optimum. An adjustment of the switching frequency of the regulator is equal to a change of its input resistance, which is given by the input inductance and the switching frequency. Thus, the maximum power point tracker adjusts the input resistance of the power management to achieve the maximum output power of the energy transducer.

Regarding piezoelectric transducer, the power management is used to extract more energy from the material by using switched inductors. These inductors build a resonance circuit with the internal capacitor of the piezoelectric material. The non-linear techniques are also explained in Chapter 9.

Another task of the power management is the charge regulation and protection of the energy storage elements in the energy harvesting system. This can be done by charge regulators if large load or charge currents are present, which might damage a battery or a capacitor if not carefully adjusted. Most often, in energy harvesting system, the currents are so small that only simple voltage regulators are needed for energy storage elements. Battery monitoring in terms of calculation the remaining charge in a battery can also be done in the power management unit. Here counting the charge flowing into and out of the battery can be used as well as simple voltage measurement.

1.6 Load Device

The electrical power obtained from an energy harvesting system is very small (1 μW/cm^3 to 100 mW/cm^3) and that is why only low-power loads can be supplied with energy harvesting generators. A typical electronic load consists of a sensor, a microcontroller, and a wireless transceiver (see Fig. 1.3). Table 1.11 shows some current consumption values for those components differentiating between ultra-low-power and conventional components.[43]

During transmission, the power consumption is approximately between 50 to 100 mW depending on the transmission range.[44] The consumed power is much higher in almost all the cases than the available harvested power. The energy present in the environment that can potentially be harvested is mostly discontinuous in nature.

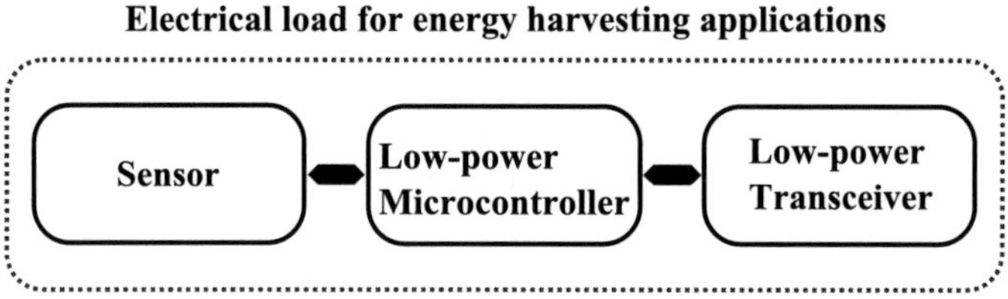

Electrical load for energy harvesting applications

Sensor

Low-power Microcontroller

Low-power Transceiver

Figure 1.3 Block diagram of a general load: sensor, microcontroller, and wireless transceiver.

Table 1.11 Characteristics of low-power loads

Device	Ultra-low-power	Conventional
Microcontroller	160 μA/MHz	500 μA/MHz
Sensor	120 μA	> 1 mA
Transceiver	3 mA	15 mA
Transceiver	120 μA	70 mA

Consequently, there must be an element to store the energy at periods with high ambient power to guarantee an operation at times with low ambient power. This storage element can be a capacitor, or a secondary battery.

So, the harvested energy is accumulated in a storage element and the sensors, microcontroller, and RF transceiver can work in a low-power or standby mode or are totally powered off until there is enough energy accumulated to sense, process, and transmit the data. The next subsection goes deeper into this topic.

1.6.1 *Continuous and Discontinuous Load Operation*

The discontinuous nature of ambient energy sources has consequences on the way in which electronic devices powered by energy harvesting power supplies are operated. In principle, we can distinguish between two situations in which an energy storage element is necessary:[45]

(1) The average power consumption of the electronic device is lower than the average power provided by the energy

transducer. In this case, the electronic device may operate continuously.

(2) The power consumption of the device is higher than the power provided by the energy transducer. The operation must be discontinuous, and the time between operations depends on the stored energy provided by the transducer.

The electronic device is able to operate only when there is enough energy in the energy storage element. Figure 1.4 shows the two cases. The energy storage element is necessary to provide energy during the moments where transiently the provided power is lower than the power consumed by the load. A special case would be if the device is operated exclusively during the time when there is electrical power generated, and the power consumption is at all times smaller than the generated electrical power (Fig. 1.5). In this case, the energy storage element is not necessary, although voltage regulation is.

In the general case of discontinuous operation (Fig. 1.4b), energy is a more relevant magnitude than power for energy harvesting systems because the electrical energy generated determines when an operation can be done and also the time between operations of the load. The power requirements of the load in active mode will determine the energy storage element to select and the power profile of the load. The power consumption of the load in active mode is fixed by its electrical components and the supply voltage. Moreover, the components also determine the enable times needed to enter different power consumption modes.

The eZ430-RF2500[46] is a state-of-the-art wireless transceiver system that combines a MSP430 microcontroller with a CC2500 2.4 GHz wireless transceiver. The ultra-low power consumption of both components is ideal for its utilization in energy harvesting applications. Figure 1.6 shows the current profile of the eZ430-RF2500 as a transmitter and Fig. 1.7 shows the current profile of the acceleration sensor when it does four measurements and the current profile of the eZ430-RF2500 that transmits the data employing a piezoelectric energy harvesting supply.

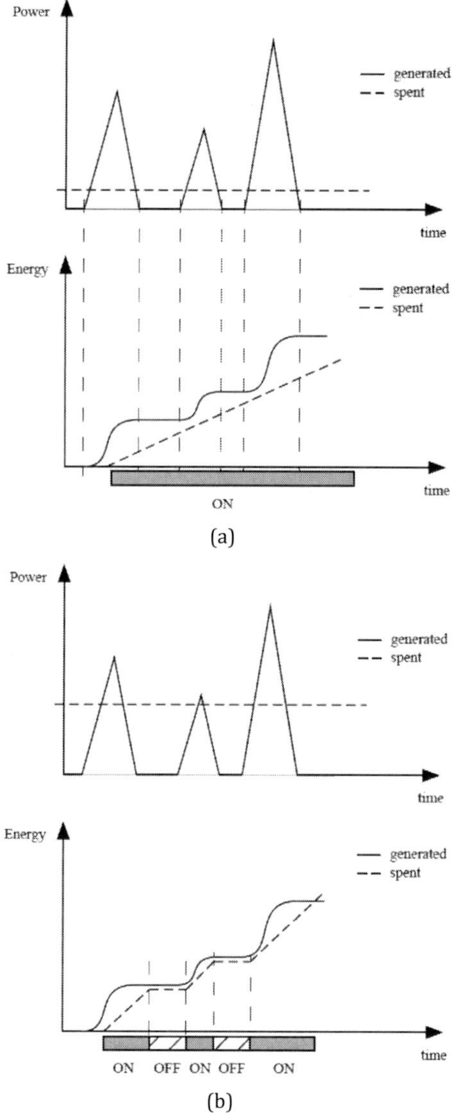

Figure 1.4 Case of continuous (a) and discontinuous (b) load operation. In the case of discontinuous operation, the device must be turned off until enough energy is collected in the storage element.[45]

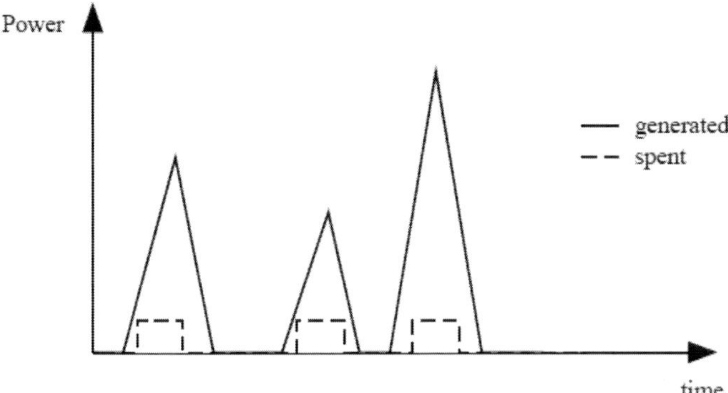

Figure 1.5 Generated and spent power when the device operation is only at times when there is energy generation.

The active time needed by a wireless transceiver for sending a certain amount of data is calculated with Eq. (1.2):

$$T_{\text{active}} = \cfrac{1}{\text{data rate} \times \frac{1\,\text{byte}}{8\text{bits}} \times \cfrac{1}{\left\lceil \frac{D}{n} \right\rceil \text{packet length}}} \tag{1.2}$$

where data rate is the transmission speed in Kbps, D are the data bytes to transmit, n are the data bytes of one packet and packet length is the number of bytes that are transmitted.

Only a minimum current is required in standby mode of most RF transceivers since nearly all the blocks are turned off. In the synthesizer mode, only the blocks associated with the synthesizer (like the crystal and PLL) are turned on. During the transmit and receive mode, all the blocks that are necessary for a transmission and reception are turned on.

When the application requires to transmit data several times per second, Eq. (1.3) is employed to calculate the average current needed by the wireless transceiver.

$$\langle I \rangle = \frac{\left(I_{\text{sleep}} T_{\text{sleep}} + I_{Tx} T_{Tx}\right)}{T_{\text{sending}}} \tag{1.3}$$

where I_{sleep} is the current consumed by the transceiver in sleep mode, T_{sleep} is the time that the transceiver is in sleep mode, I_{Tx} is the current consumed during a transmission and T_{TX} is the time required to send the data and T_{sending} is the period of the

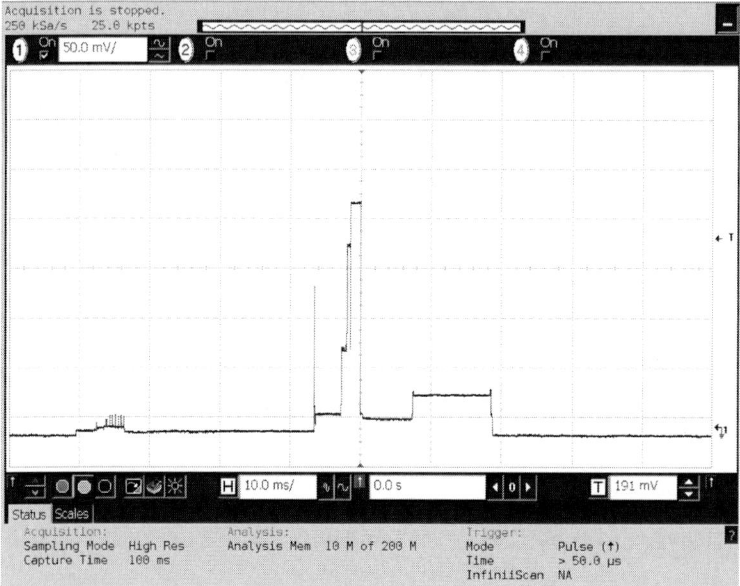

Figure 1.6 Current profile of the eZ430-RF2500 as a transmitter.

transmissions. Thus,

$$T_{\text{sleep}} = T_{\text{sending}} - T_{TX} \tag{1.4}$$

The previous calculus can also be done fixing the average current necessary to power the wireless transceiver and generated by the power converter of the energy harvesting power supply and obtaining the period between transmissions. This case is more realistic that the previous one for energy harvesting applications since it does not imply a redesign of the transducer nor the power management unit to increase or decrease the value of $\langle I \rangle$.

1.6.2 *Low-Power Sensors*

Important parameters to take into account in the selection of sensors for energy harvesting applications are the current consumption both in active and power down modes and the enable response time. The mean power provides the information about the energy consumed by the sensor. The minimum voltage supply and current

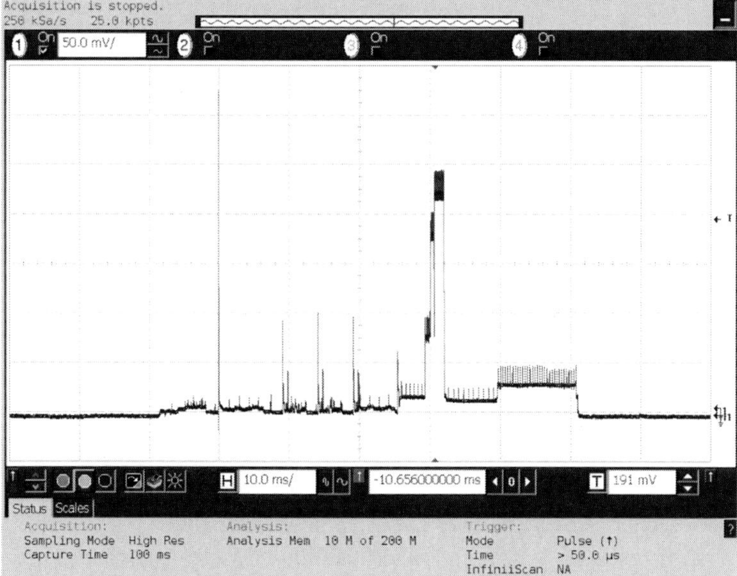

Figure 1.7 Current profile of the eZ430-RF2500 as a transmitter in combination with an accelerometer sensor.

consumption in active mode are fixed parameters that provide the power consumption in active mode. Nevertheless, the time that the sensor is in power down mode changes the total energy required by the sensor. Sensors with low current consumption values in power down mode and low enable response time are the best suited for energy harvesting applications. The enable response time is the time needed to obtain valid data once the low power down mode has been left. Thus, this time extends the time that the sensor will be in active mode.

Sensors can provide analog, digital or both kinds of output. When the output data is available through an I^2C/SPI interface, it has a direct interface for its connection with a microcontroller.

The sensitivity of a sensor is the amount of change in the output signal for the change in the measured parameter. For the case of analog passive sensors, the output signal is expressed in volts and for the case of a digital passive sensors, the output signal is expressed in number of bits. The sensitivity of a sensor will be between the

Table 1.12 Characteristics of body sensors

Signal	Sensitivity	Number of samples per time	Data rate
Heart rate	8 bits	10 samples/min	80 bits/min
Blood pressure	16 bits	1 sample/min	32 bits/min
Temperature	16 bits	1 sample/min	16 bits/min
Blood oxygen	16 bits	1 sample/min	16 bits/min

minimum and maximum values given by the manufacturer on the datasheet for a certain temperature, usually 25°C. Thus, a calibration of the sensor is necessary in order to obtain accurate results. The sensitivity changes versus temperature and this variation is expressed in %/°C.

A conditioning circuit is required to obtain a voltage response from active sensors. Conditioning circuits for resistive, capacitive and electromagnetic sensors are explained by Pallàs et al.[47]

The resolution is the smallest change of the measured quantity that can be detectable by the sensor.

The bandwidth response of a sensor is expressed in Hertz and it is the maximum frequency at which the sensor can make measurements. The data rate is expressed as well in Hertz and corresponds to the frequency at which the measured data is captured.

In wearable applications, sensors measure vital parameters as heart rate, blood pressure, temperature or oxygen in the blood. Table 1.12 presents the required sensitivity, number of samples per time and data rate for some body sensors. Yeatman[48] reports that a total power consumption for a load of 1 µW is a realistic value for body sensors.

Torfs et al.[49,50] adapted the design of a pulse oximeter to have a low-power consumption device that uses an average power of only 62 µW when a measurement is done every 15 s. Figure 1.8 shows the percentage of power consumption of all the components that compose the pulse oximeter device.

The load with lower power consumption is composed by a sensor, an analog to digital converter (ADC) and a transmitter. Yates et al.[51] expose that the power consumption of an ADC with a data rate of 1 Kbps would be 104 nW and that the power consumption of

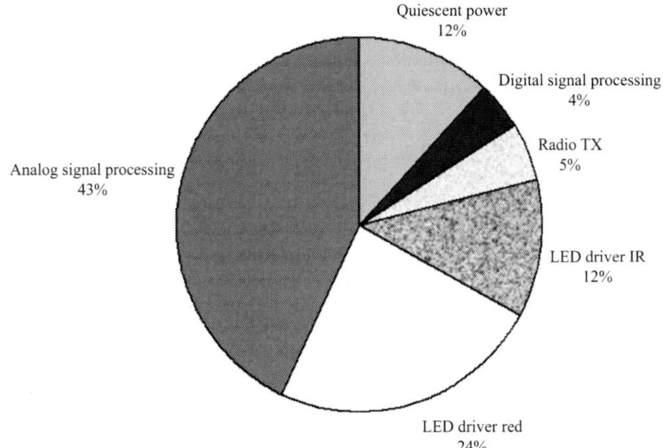

Figure 1.8 Power consumption of the different components of the pulse oximeter device.

a transmitter with the same date rate would be 300 nW. The ADC is operated with a duty cycle of 0.26% and if the temperature sensor MAX6613, which has a power consumption of 20 μW, is operated also with the same duty cycle it has an average power consumption of 5.2 nW. Therefore, a total power consumption of 456 nW for a 1 kbps is required for the sensor, ADC and the transmitter.

1.6.3 *Low-Power Microcontrollers and Transceivers*

Low-power microcontrollers have different operation working modes associated with different current consumptions. In the active mode, the current consumption is the highest and all the clocks are active, whereas in the low-power modes the CPU and some of the internal clocks are disabled. Figure 1.9 shows a generic block diagram of a low-power microcontroller.

Data rate, preamble cycles, and packet length determine the active time needed for the transmission of the data in transceivers. Figure 1.10 shows the typical current profile of a wireless transceiver. In an energy harvesting application, the transceiver is in standby mode most of the time to keep the average power consumption at a minimum level. When the sensed data has to be transmitted, first, some time is needed to enable the synthesizer

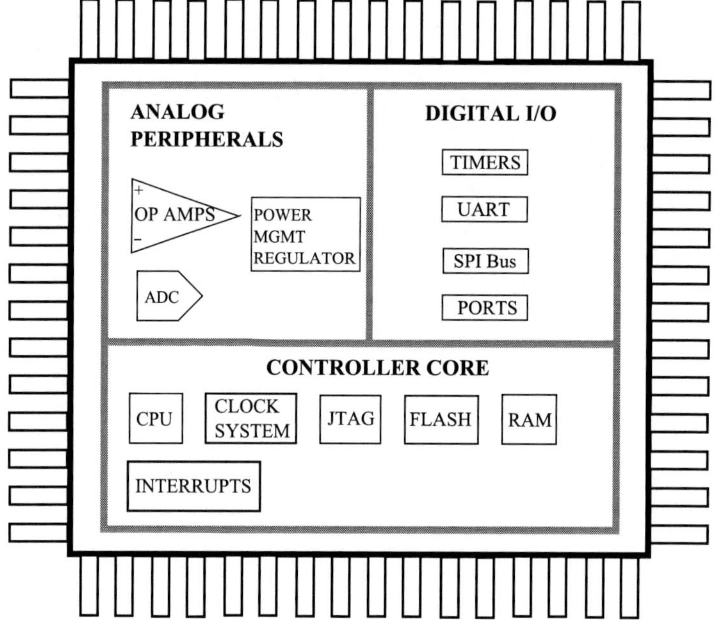

Figure 1.9 Block diagram of a low-power microcontroller.

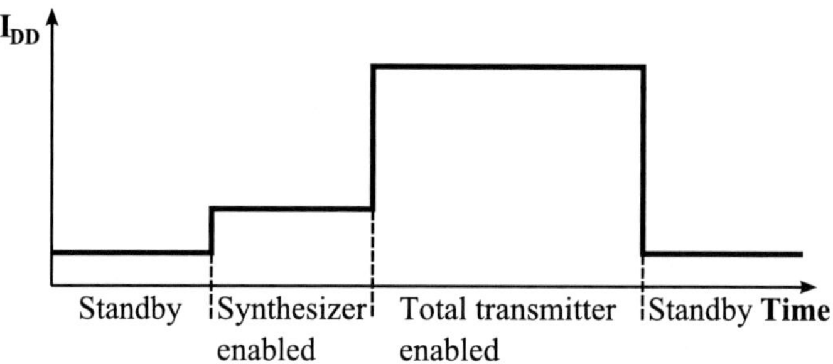

Figure 1.10 Current profile of a transceiver.

and afterward the data is sent. Each of the different modes has its associated current consumption value.

Figure 1.11 shows a block diagram of a generic low-power RF transceiver.

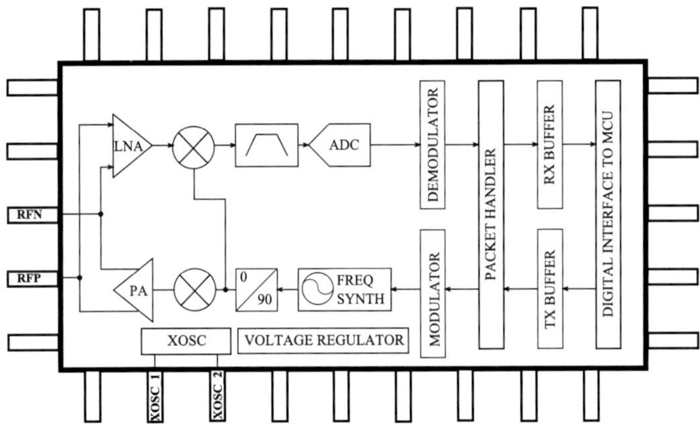

Figure 1.11   Block diagram of a low-power RF transceiver.

1.7 Energy Storage Element

Energy harvesting transducers such as thermogenerators and piezoelements provide only small amounts of electrical power. Moreover, the size and thus the price of the transducer are always related to the power output. Additionally, energy harvesting transducers exhibit large internal resistances, not able to provide large currents without a drop in their output voltage. Finally, typical application scenarios such as the human body or buildings exhibit only small ambient energy sources. In contrast to that, common electronic devices used in energy harvesting systems, especially wireless transmitters, operate in burst mode, transmitting data only during a small period of time and thus require pulse currents during these transmission bursts. Also, microcontrollers are often operated between a full-performance mode, active mode, and a low-power, sleep, or stand-by mode, leading to a pulse current profile of the typical application device. Finally, the application itself, such as sensor data measurement of temperature, moisture, heart rate, etc., is done only during short periods of time, because the interesting physical parameters do not change that frequently to ask for a permanent measurement.

To match the low power output of the transducers with the pulse current requirement of the application, energy storage elements are always needed in energy harvesting systems. These might be rechargeable batteries or capacitors, each having its advantages and downsides.

Regarding volume and weight related to energy content, which is more precisely called gravimetric and volumetric energy density, batteries are superior compared to capacitors. The downside of batteries is their aging depending on application temperature and number of charge and discharge cycles. With that aging comes a reduced maximum capacity and an increased internal resistance, leading to larger voltage drop. Leakage current of both types have to be considered very carefully, since energy harvesting systems collected minimum currents, which might be in the range of these leakage currents. These leakages are of course temperature dependent. Problem with capacitors is their linear decrease of output voltage during discharge. This means, a fraction of the energy cannot be used, because of the minimum supply voltage of the application. A solution that should be investigated very carefully is using boost or step-up converters, because of their own power consumption. In contrast to capacitors, batteries have a flat voltage profile between 80% and 20% of their capacity, making it more easy to use the most of their energy without special means. Moreover, their operating voltage range does not go below a certain voltage such as 3 or 2 V. All these values depend on the technology and manufacturer chosen. The temperature range is another parameter that helps decide one of the two alternatives of energy storage. Batteries usually work from -20 to $+50°$C, which is a range for charging starting at $0°$C. Below that level, the capacity droops significantly. Capacitors are superior regarding this issue. Batteries as well as capacitors will be explained in Chapter 11.

1.8 Combination of Several Input Energies

There are a lot of application environments where several ambient energy sources are available for powering electronic circuits. If the power budget of the electronic consumer is critical and price or board space is not an issue, a simultaneous operation of several energy transducers makes sense. Especially, when using light, there is often also a thermal gradient introduced, which can be employed for additional electrical power generation. At the human body environment, motion, and heat could be used with a combination of piezoelectric films and thin-film thermogenerators. Machines and large motors exhibit heat besides vibrations; thus, a combination of both principles will be promising to increase the available power. Also, the building environment offers the use of a combination of energy transducers such as solar cells and thermogenerators. Furthermore, mobile applications, where the environment is changed by movement such as human beings, animals, and vehicles, can be stuffed with different kinds of energy transducers. Such combination of energy transducers can guarantee a self-powered operation regardless of the situation.

New research approaches try to combine several kinetic energy transducer principles in on system.

Owing to the different nature of the outputs from the different transducers, there will be always a dedicated power management required for each transducer. Thermogenerators exhibit large currents at small voltages, whereas piezoelectric transducers generate larger alternating voltages at small currents compared to TEGs. Electro-dynamic and electro-static converters also produce alternating currents but at a smaller voltage level.

1.9 Energy Neutral Operation

The amount of available power in a system that employs an energy harvesting power supply is limited and it is not constant with time.

It is desirable to have equal amounts of the energy harvested by the generator and the energy consumed by the load to assure an energy neutral operation, in other words, to guarantee that there will always be sufficient energy available to supply the load. This is the same as stating that the average power in the time interval generated and spent must be the same for energy neutral operation.

In Section 1.6.1, the two operation modes of an electronic load powered by an energy harvesting generator are explained: continuous and intermittent. An energy storage element is not necessary if the power consumption of the electronic device is

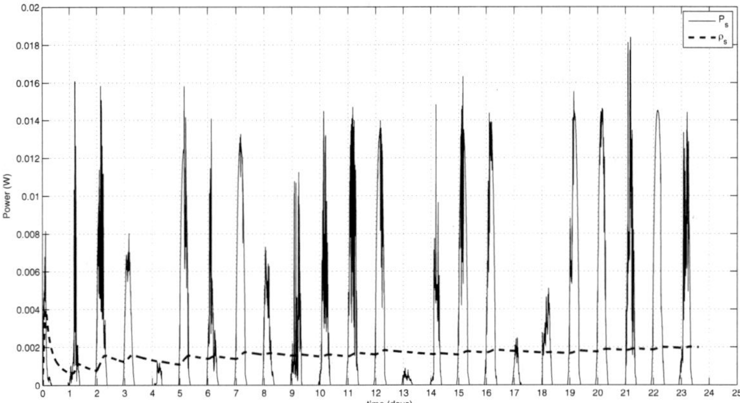

Figure 1.12 Power P_s delivered by a portable panel to an energy storage element as a function of time as a function of time and calculation of its mean power ρ_s.

always lower than the power generated by the energy harvesting generator and is only operated when there is power generated. For the rest of the cases, an energy storage element is necessary, which could be, for example, a battery.

The application and the harvested and consumed energy will determine the operating mode to select. The objective of this section is to present a way to calculate the initial charge of the storage element before starting operation as well as the maximum quantity of energy that is necessary to store. The method is based on the works made by Kansal et al.[52,53]. This technique consists of a model to characterize environmental sources and electronic loads that allows to determine the size of the energy storage element employed as a function of the power consumption profile of the load to assure energy neutral operation.

First of all, it is necessary to define the energy delivered by the transducer to the energy storage element and the energy consumed by the load in a mathematical way. Figure 1.12 shows the power P_s delivered by a portable solar cell with an open circuit voltage of 1.89 V and a short-circuit current of 12.6 mA at a location 42.78°N, 73.85°N[54] to an energy storage element as a function of time. The mean power ρ_s is defined as:

$$\rho_s = \frac{1}{T} \int_T P_s(t)dt, \qquad (1.5)$$

where T is the interval of time considered for the calculations that gives an error of $\pm\Delta$.

The difference between the last maximum and minimum of the mean power delivered by the energy harvesting transducer to the energy storage device is called Δ this means that the error to calculate the energy neutral operation point is $\pm\Delta$. The value of Δ in Fig. 1.12 after 10 days of measurements is 109 μW. This value is reduced to 54 μW after 23 days of measurements. This method is feasible for periodical or quasi-periodical energy harvesting sources where the value of the maximum and minimum mean power provided by the transducer to the storage element converges but not for non-periodical input energies. Kansal et al. developed this method for solar panels in outdoors applications where there is power available only under sunlight and the cycles of converted power caused by day and night alternation are quasi-periodic.[52,53] Nevertheless, this behavior is not limited to solar energy since mechanical energy sources or thermal energy sources can also be periodical or quasi-periodical.[13]

If it is assumed that $T_{\text{lows}-i}$ is the i-th contiguous time duration for which $P_s(t) \leq \rho_s(t)$, then σ_d is defined as the maximum deficit of energy of the energy harvesting transducer (see Fig. 1.13).

$$\sigma_d = \max_i \left\{ \int_{T_{\text{lows}-i}} \rho_s - P_s(t)dt \right\} \qquad (1.6)$$

If it is assumed that $T_{\text{highs}-i}$ is the i-th contiguous time duration for which $P_s(t) \geq \rho_s(t)$, then σ_e is defined as the maximum excess of energy of the energy harvesting transducer (see Fig. 1.13).

$$\sigma_e = \max_i \left\{ \int_{T_{\text{highs}-i}} P_s(t) - \rho_s dt \right\} \qquad (1.7)$$

The energy harvested by the transducer and delivered to the energy storage element will be inside a certain margin limited by E_{smin} and E_{smax}:

$$E_{\text{smin}} \leq \int_T P_s(t)dt \leq E_{\text{smax}} \qquad \forall\, t \qquad (1.8)$$

where E_{smin} is the lower limit and E_{smax} is the upper limit of the energy delivered by the energy harvesting transducer to the energy

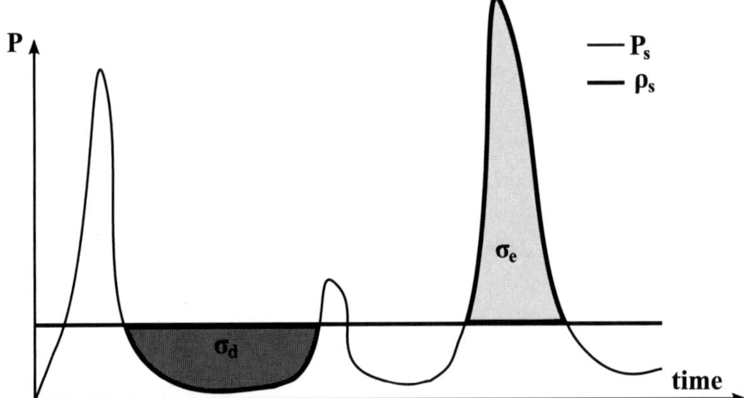

Figure 1.13 Power delivered by the energy harvesting transducer to the energy storage element as a function of time.[13]

storage element. $E_{\text{smin}}(T)$ and $E_{\text{smax}}(T)$ are piecewise functions defined as

$$E_{\text{smin}}(T) = \begin{cases} \rho_s T - \sigma_2 \frac{T}{T_{lows-i}} & \forall T \leq T_{\text{lows}-i}, \\ \rho_s T - \sigma_2 & \forall T \geq T_{\text{lows}-i}. \end{cases} \tag{1.9}$$

$$E_{\text{smax}}(T) = \begin{cases} \rho_s T + \sigma_1 \frac{T}{T_{highs}} & \forall T \leq T_{\text{highs}-i}, \\ \rho_s T + \sigma_1 & \forall t \geq T_{\text{highs}-i}. \end{cases} \tag{1.10}$$

Figure 1.14 shows the power consumption of an electronic load P_l as a function of time. ρ_l is the average power consumption of the

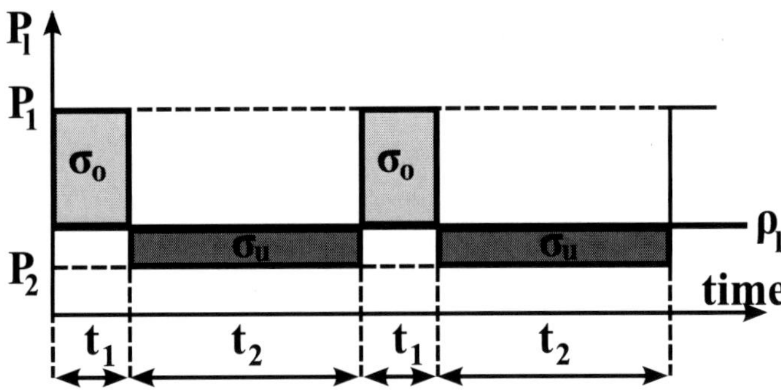

Figure 1.14 Power consumption of the load as a function of time.[13]

load. P_1 is the power consumption of the load in the highest power consumption mode (e.g., transmission mode in a communication module) and it takes place during a time interval t_1. P_2 is the power consumption of the load in the lowest power consumption mode (e.g., standby mode in a communication module) and it takes place during a time interval t_2. Therefore, the electronic load consumption can be defined as a function of the parameters $(\rho_l, \sigma_3, \sigma_4)$.

$$\rho_l = \frac{1}{T} \int_T P_l(t)dt \tag{1.11}$$

If it is assumed that $T_{highs-i}$ is the i-th contiguous time duration for which $P_l(t) \geq \rho_l(t)$, then σ_o is defined as the maximum over-consumption of energy made by the load (see Fig. 1.14).

$$\sigma_o = max_i \left\{ \int_{T_{highl-i}} P_l(t) - \rho_l dt \right\} \tag{1.12}$$

If it is assumed that T_{lowl-i} is the i-th contiguous time duration for which $P_l(t) \geq \rho_l(t)$, then σ_u is defined as the maximum under-consumption of energy made by the load (see Fig. 1.14).

$$\sigma_u = max_i \left\{ \int_{T_{lowl-i}} \rho_l - P_l(t)dt \right\} \tag{1.13}$$

The lower and upper limits of the energy consumed by the electronic load are:

$$E_{lmin}(T) = \begin{cases} \rho_l T - \sigma_4 \frac{T}{T_{lowl-i}} & \forall\, T \leq T_{lowl-i}, \\ \rho_l T - \sigma_4 & \forall\, T \geq T_{lowl-i}. \end{cases} \tag{1.14}$$

$$E_{lmax}(T) = \begin{cases} \rho_l T + \sigma_3 \frac{T}{T_{highl-i}} & \forall\, T \leq T_{highl-i}, \\ \rho_l T + \sigma_3 & \forall\, T \geq T_{highl-i}. \end{cases} \tag{1.15}$$

1.9.1 *General Conditions for Energy Neutral Operation*

When a wearable device, e.g., a node of a WSN, employs an energy harvesting system to be powered, the objective is to eliminate the need to replace or recharge its battery. Thus, it is necessary to assure energy neutral operation or in other words, to assure that the battery will contain always the energy required by the electronic device. Therefore, the energy storage element is defined by two parameters that are its initial charge when it is connected to the energy harvesting system, B_0, and the amount of energy that can be stored, B. The values of both parameters are calculated in this section.

In order to achieve energy neutral operation, the total energy in the system, ΣE, has to be always greater than zero since the energy from the harvesting transducer, E_s, plus the initial energy stored in the battery, B_0 has to be greater than the energy consumed by the electronic load, E_l. Moreover, it is desirable in an energy harvesting system to assure that no energy is wasted or that the battery is damaged due to overcharge. These two conditions can be expressed as:

$$\Sigma E \geq 0 \tag{1.16}$$

$$\Sigma E \leq B. \tag{1.17}$$

The available energy in the system is equal to the initial energy stored in the battery plus the energy harvested by the transducer minus the energy consumed by the load and minus the leakage energy due to the energy storage element.

$$\Sigma E = B_0 + E_s - E_l - \int_T P_{\text{leak}} T \, dt, \tag{1.18}$$

where P_{leak} is the leakage power of the energy storage element.

The value of B_0 can be calculated using the condition expressed by Eq. (1.16). This condition is evaluated when the worst case takes place, that is, when the provided energy from the harvesting transducer to the energy storage device is minimum, $E_{s\min}$, and the energy consumed by the electronic load is maximum, $E_{l\max}$.

$$B_0 + E_{s\min} - E_{l\max} - \int_T P_{\text{leak}} T \, dt \geq 0 \tag{1.19}$$

The above condition can be expressed as:

$$B_0 + E_{s\min} - E_{l\max} - \rho_{\text{leak}} T \geq 0, \tag{1.20}$$

where ρ_{leak} is the mean leakage power of the energy storage element.

In a similar way, the value of B can be calculated using the condition expressed by Eq. (1.17). This condition is evaluated when the worst case takes place, that is, when the generated energy by the harvesting transducer is maximum, $E_{s\max}$, and the energy consumed by the electronic load is minimum, E_{lmin}. In this case, the energy storage element will have at its maximum capacity.

$$B_0 + E_{smax} - E_{lmin} - \int_T P_{leak}\, T\, dt \leq B \qquad (1.21)$$

The above condition can be expressed as:

$$B_0 + E_{smax} - E_{lmin} - \rho_{leak}\, T \leq B \qquad (1.22)$$

If the value of T tends to infinity for the two conditions expressed by Eqs. (1.20) and (1.22), the following is obtained:

$$\rho_s - \rho_l - \rho_{leak} \geq 0 \qquad (1.23)$$

$$\rho_s - \rho_l - \rho_{leak} \leq 0 \qquad (1.24)$$

Equations (1.23) and (1.24) can be simplified as:

$$\rho_s - \rho_l - \rho_{leak} = 0 \qquad (1.25)$$

Substituting in Eq. (1.20), the value given for E_{smin} and E_{lmax} by Eqs. (1.9) and (1.15), respectively, and taking into consideration the previous expression, the following is obtained:

$$B_0 - \sigma_2 \frac{T}{T_{lows}} - \left(\sigma_3 \frac{T}{T_{highl}}\right) \geq 0$$
$$\forall\, T \leq T_{lows} \text{ and } \forall\, T \leq T_{highl} \qquad (1.26)$$

$$B_0 - \sigma_2 \frac{T}{T_{lows}} - (\sigma_3) \geq 0$$
$$\forall\, T \leq T_{lows} \text{ and } \forall\, T \geq T_{highl} \qquad (1.27)$$

$$B_0 - \sigma_2 - \left(\sigma_3 \frac{T}{T_{highl}}\right) \geq 0$$
$$\forall\, T \geq T_{lows} \text{ and } \forall T \leq T_{highl} \qquad (1.28)$$

$$B_0 - \sigma_2 - (\sigma_3) \geq 0$$
$$\forall T \geq T_{lows} \text{ and } \forall T \geq T_{highl} \qquad (1.29)$$

There are two conditions that give as a result the worst scenario and therefore, the minimum value of B_0. One of these conditions occurs when T is equal to T_{lows} and this value is greater than T_{highl}.

The second condition occurs when T is equal to T_{highl} and this value is greater than T_{lows}. For both cases, the same expression is obtained:

$$B_0 \geq \sigma_2 + \sigma_3$$
$$\forall\, T = T_{\mathrm{lows}} \text{ and } \forall\, T \geq T_{\mathrm{highl}} \tag{1.30}$$
$$\forall\, T = T_{\mathrm{highl}} \text{ and } \forall\, T \geq T_{\mathrm{lows}}$$

When the conditions given by Eqs. (1.30) and (1.25) are accomplished, the energy harvesting system can operate forever.

The conditions to avoid overcharging the energy storage element are obtained substituting in Eq. (1.22), the value given for E_{smax} and E_{lmin} by Eqs. (1.10) and (1.14), respectively, a piecewise expression is obtained:

$$B_0 + \sigma_1 \frac{T}{T_{\mathrm{highs}}} - \left(-\sigma_4 \frac{T}{T_{\mathrm{lowl}}}\right) \leq B$$
$$\forall\, T \leq T_{\mathrm{highs}} \text{ and } \forall\, T \leq T_{\mathrm{lowl}} \tag{1.31}$$

$$B_0 + \sigma_1 \frac{T}{T_{\mathrm{highs}}} - (-\sigma_4) \leq B$$
$$\forall\, T \leq T_{\mathrm{highs}} \text{ and } \forall\, T \geq T_{\mathrm{lowl}} \tag{1.32}$$

$$B_0 + \sigma_1 - \left(-\sigma_4 \frac{T}{T_{\mathrm{lowl}}}\right) \leq B$$
$$\forall\, T \geq T_{\mathrm{highs}} \text{ and } \forall\, T \leq T_{\mathrm{lowl}} \tag{1.33}$$

$$B_0 - \sigma_2 - (\sigma_3) \leq B$$
$$\forall\, T \geq T_{\mathrm{highs}} \text{ and } \forall\, T \geq T_{\mathrm{lowl}} \tag{1.34}$$

The maximum amount of stored energy B will occur when the maximum delivered energy by the transducer and the minimum spent energy by the load are coincident in time. In this case, the following expression is obtained:

$$B_0 + \sigma_1 + \sigma_4 \leq B \tag{1.35}$$

Combining Eqs. (1.30) and (1.35), it is deduced that

$$\sigma_1 + \sigma_2 + \sigma_3 + \sigma_4 \leq B \tag{1.36}$$

When this condition is accomplished, no waste energy is produced from the energy harvesting transducer since all the energy generated can be stored in the energy buffer.

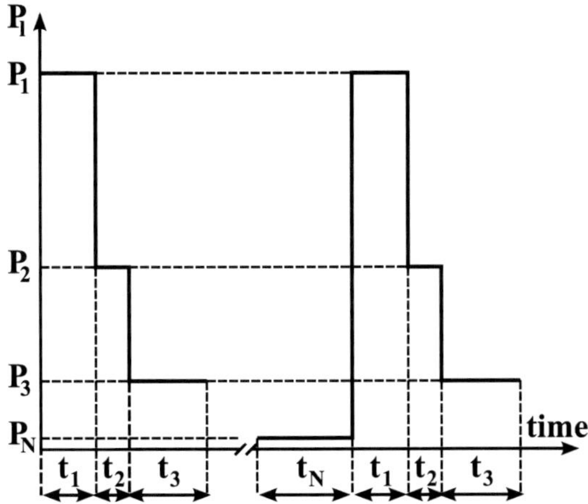

Figure 1.15 Power consumption of the load as a function of time.[13]

1.9.2 *Conditions for Energy Neural Operation with* N *Power Consumption Modes*

Figure 1.15 shows a load with N different power consumption modes. P_N is the lowest power consumption mode, whereas P_1 is the highest power consumption mode. The rest of the power consumption modes between these two values increase consecutively their power consumption from P_N to P_1. When this assumption is accepted, for the general case of N different consumption modes, it is calculated that

$$\tau = \sum_{i=1}^{N} x_i \tau, \tag{1.37}$$

where τ is period of the power consumed by the load and x_i is the percentage of τ where the power consumed by the load is P_i.

$$\rho_l = \sum_{i=1}^{N} P_i x_i \tag{1.38}$$

The expressions for σ_3 and σ_4 are

$$\sigma_3 = \tau \left((P_1 - \rho_l) x_1 + \sum_{i=2}^{N-1} \langle P_i - \rho_l \rangle x_i \right) = \tau \sum_{i=1}^{N} \langle P_i - \rho_l \rangle x_i$$

$$\tag{1.39}$$

$$\sigma_4 = \tau \left((\rho_l - P_N) x_N \sum_{i=2}^{N-1} \langle \rho_l - P_i \rangle x_i \right) = \tau \sum_{i=1}^{N} \langle \rho_l - P_i \rangle x_i \quad (1.40)$$

The addition of σ_3 and σ_4 gives the following as a result:

$$\sigma_3 + \sigma_4 = \tau \left(\sum_{i=1}^{N} \langle \rho_l - P_i \rangle x_i + \sum_{i=1}^{N} \langle P_i - \rho_l \rangle x_i \right)$$

$$= \tau \left((P_1 - \rho_l) x_1 \sum_{i=2}^{N-1} \langle \rho_l - P_i \rangle x_i \right.$$

$$\left. + \sum_{i=2}^{N-1} \langle P_i - \rho_l \rangle x_i + (\rho_l - P_N) x_N \right) \quad (1.41)$$

$(\rho_l - P_N) x_N \tau$ can be expressed as a function of the rest of the power consumption modes:

$$(\rho_l - P_N) x_N \tau = \rho_c T \left(1 - \sum_{i=1}^{N-1} x_i \right) - P_N x_N T$$

$$= \left(\sum_{i=1}^{N-1} (P_i - \rho_c) x_i \right) \tau \quad (1.42)$$

Therefore, substituting the above expression in Eq. 1.41, the following is obtained:

$$\sigma_3 + \sigma_4 = 2\tau \sum_{i=1}^{N-1} \langle \rho_l - P_i \rangle x_i \quad (1.43)$$

Therefore, the conditions to be fulfilled for energy neutral operation by an electronic load with N consumption modes are summarized here employing Eqs. (1.30) and (1.36).

$$\sigma_2 + T \sum_{i=1}^{N-1} \langle P_i - \rho_l \rangle x_i \leq B_0 \quad (1.44)$$

$$\sigma_1 + \sigma_2 + 2T \sum_{i=1}^{N-1} \langle P_i - \rho_l \rangle x_i \leq B \quad (1.45)$$

1.10 Conclusion

The different parts that compose an energy harvesting system are presented and general information about each of them is given. This information is extended in the following chapters of the book. Moreover, it is explained which blocks are optional and why.

The concept of continuous and discontinuous load operation has been introduced. In an energy harvesting system, the key approach is to energy and not to power since the input energy sources are not present all the time and will probably not deliver the required power to let the load operate in active mode all the time. Thus, it is desirable to achieve the energy neutral operation in an energy harvesting system since it assures that the energy requirements of the load can be achieved.

References

1. S. Roundy, P. Wright, and K. Pister. Micro-electrostatic vibration-to-electricity converters. In *Proceedings of ASME International Mechanical Engineering Congress and Exposition IMECE2002,* vol. 220, pp. 17–22 (November 2002).
2. M. Stordeur and I. Stark. Low power thermoelctric generator: self-sufficient energy supply for micro systems. In *Proceedings of the 16th International Conference on Thermo-electrics,* pp. 575–577, (1997).
3. N. Shenck and J. Paradiso, Energy scavenging with shoe-mounted piezoelectrics, *Micro, IEEE.* 21(3), 30–42, (2001).
4. S. Roundy. *Energy Scavenging for Wireless Sensor Nodes with a Focus on Vibration to Electricity Conversion.* PhD thesis, university of California, (2003).
5. M. Raju. Energy harvesting, ULP meets energy harvesting: A game-changing combination for design engineers. Technical report, Texas Instruments, (2008).
6. A. Reinders. Options for photovoltaic solar energy systems in portable products. In *proceedings of TCME 2002, Fourth International symposium,* (22–26 April, 2002).
7. C. Williams and R. Yates. Analysis of a micro-electric generator for microsystems. In *Proceedings of the 8th International Conference on Solid-State Sensors and Actuatros, and Eu-rosensor IX,* (1995).

8. T. von Bren, G. Troester, and P. Lukowicz. Kinetic energy powered computing. In *Proceedings of the Seventh IEEE International Symposium on Wearable Computers (ISWC'03)*, (2003).

9. T. von Bren, P. Mitcheson, T. Green, E. Yeatman, A. Holmes, and G. Troster, Optimization of inertial micropower generators for human walking motion, *Sensors Journal, IEEE.* 6(1), 28–38, (2006). ISSN 1530-437X.

10. T. Starner, Human-powered Wearable Computing, IBM *Systems Journal.* 35(3&4), (1996).

11. F. Moll and A. Rubio. An approach to the analysis of wearable body-powered systems. In *Mixed Signal Design Workshop* (June 2000).

12. L. Mateu, C. Villavieja, and F. Moll, Physics-based time-domain model of a magnetic induction microgenerator, *Magnetics, IEEE Transactions on.* 43(3), 992–1001 (March, 2007).

13. L. Mateu. *Energy Harvesting from Human Passive Power.* PhD thesis, Universitat Politècnica de Catalunya (June, 2009).

14. *Advances in Human-Powered Energy Systems in Consumer Products* (18–21 May, 2004). International Design Conference—Design 2004.

15. T. Starner and J. Paradiso. Human generated power for mobile electronics. In ed. C. Piguet, *Low-Power Electronics,* number 45. CRC Press, (2005).

16. T. Starner and Y. Maguire, Heat dissipation in wearable computers aided by thermal coupling with the user, *Mobile Networks and Applications.* 4(1), 3–13, (1999).

17. V. Leonov and R. Vullers. Thermoelectric generators on living beings. In *Proceedings of the 5th European Conference on Thermoelectrics* (September 2007).

18. V. Leonov, T. Torfs, P. Fiorini, and C. Van Hoof, Thermoelectric converters of human warmth for self-powered wireless sensor nodes, *Sensors Journal, IEEE.* 7(5), 650–657, (2007).

19. P. D. Mitcheson, T. C. Green, E. M. Yeatman, and A. S. Holmes, Architectures for vibration-driven micropower generators, *Journal. of Microelectromechanical Systems.* 13(3) (June 2004).

20. L. Mateu and F. Moll, Optimum piezoelectric bending beam structures for energy harvesting using shoe inserts, *Jouranl of Intelligent Material Systems and Structures.* 16(10), 835–845, (2005).

21. S. Roundy, P. K. Wright, and J. M. Rabaey, *Energy Scavenging for Wireless Sensor Networks with Special Focus on Vibrations* (Kluwer Academic Publishers, 2004).

22. H. Hu, H. Xue, and Y. Hu, A spiral-shaped harvester with an improved harvesting element and an adaptive storage circuit, *IEEE Transactions on Ultrasonics, Ferroelectrics and Frequency Control.* 54(6), 1177–1187 (June 2007).

23. C. Williams, C. Shearwood, M. Harradine, P. Mellor, T. Birch, and R. Yates, Development of an electromagnetic micro-generator, *Circuits, Devices and Systems, IEE Proceedings.* 148(6), 337–342 (December 2001).

24. W. Li, Z. Wen, P. Wong, G. Chan, and P. Leong. A micromachined vibration-induced power generator for low power sensors of robotic systems. In *Proceedings of Eight International Symposium on Robotics with Applications,* pp. 16–21 (June 2000).

25. N. N. H. Ching, G. M. H. Chan, W. J. Li, H. Y. Wong, and P. H. W. Leong. PCB integrated micro generator for wireless systems. In *Intl. Symp. on Smart Structures and Microsystems* (19–21 October 2000).

26. R. Amirtharajah and A. Chandrakasan. Self-powered low power signal processing. In *Proceedings of the Symposium on VLSI Circuits Digest of Technical Papers,* (1997).

27. S. Yuen, J. Lee, W. Li, and P. Leong, An AA-sized vibration-based microgenerator for wireless sensors, *IEEE Pervasive Computing.* 6(1), 64–72 January–March 2007).

28. S. Meninger, J. Mur-Miranda, R. Amirtharajah, A. P. Chandrasakan, and J. H. Lang, Vibration to electric energy conversion, *IEEE Trans, on VLSI.* 9(1) (February, 2001).

29. T. Sterken, K. Baert, R. Puers, and S. Borghs. Power extraction from ambient vibration. In *Proceedings of the Workshop on Semiconductor Sensors,* pp. 680–683 (November 2002).

30. M. Miyazaki, H. Tanaka, T. N. G. Ono, N. Ohkubo, T. Kawahara, and K. Yano. Electric-energy generation using variable-capacitive resonator for power-free LSI: efficiency analysis and fundamental experiment. In *Proceedings of the ISLPED 03,* pp. 193–198 (25–27 August 2003).

31. D. Jia and J. Liu, Human power-based energy harvesting strategies for mobile electronic devices, *Frontiers of Energy and Power Engineering in China.* 3(1), 27–46, (2009).

32. S. Angrist, *Direct Energy Conversion.* (Allyn & Bacon, 1982).

33. I. Stark and M. Stordeur. new micro thermoelectric devices based on bismuth telluride-type thin solid films. In *Proceedings of the 18th International Conference on Thermoelectron-ics,* pp. 465–472, (1999).

34. J. Stevens. Optimized thermal design of small <5t thermoelectric generators. In *Proceedings of the 34th Intersociety Energy Conversion Engineering Conference,* (1999).

35. S. I. Inc. Seiko Instruments Inc. http://www.sii.co.jp/info/eg/thermi-cjriain.html.

36. M. Kishi, H. Nemoto, T. Hamao, M. Yamamoto, S. Sudou, M. Mandai, and S. Ya-mamoto, Micro thermoelectric modules and their application to wristwatchesas an energy source, *Thermoelectrics, 1999. Eighteenth International Conference on.* pp. 301–307, (1999).

37. I. Stark. Thermal energy harvesting with thermo life. In *Proceedings of the International Workshop on Wearable and Implantable Body Sensor Networks (BSN'06)*, (2006).

38. S. Meninger, J. Mur-Miranda, R. Amirtharajah, A. Chandrakasan, and J. Lang, Vibration-to-electric energy conversion, *Very Large Scale Integration (VLSI) Systems, IEEE Transactions on.* 9(1), 64–76, (2001). ISSN 1063–8210.

39. U. of Energy. Guidelines for measurement of standby power use. (2002).

40. E. Lefeuvre, A. Badel, C. Richard, L. Petit, and D. Guyomar. Optimization of piezoelectric electrical generators powered by random vibrations. In *Dans Symposium on Design, Test, Integration and Packaging (DTIP) of MEMS/MOEMS.* Citeseer, (2006).

41. Y. Tan, J. Lee, and S. Panda. Maximize piezoelectric energy harvesting using synchronous charge extraction technique for powering autonomous wireless transmitter. In *IEEE International Conference on Sustainable Energy Technologies, 2008. ICSET2008,* pp. 1123-1128, (2008).

42. G. Ivensky, S. Bronstein, and S. Ben-Yaakov, A comparison of piezoelectric transformer AC/DC converters with current doubler and voltage doubler rectifiers, *IEEE Transactions on Power Electronics.* 19(6), 1446-1453, (2004).

43. C. Murray. Energy harvesting gets real (March, 2009). URL http: //www. designnews. com/article/189768-Energy_Harvesting_Gets_ Real.php.

44. A. Valenzuela. Batteryless energy harvesting for embedded designs (August 2009). URL http://www.powermanagementdesignline. com/219100013;jsessionid=Q3MWONBlKBZ5BQElGHRSKHWATMY32 JVN?printableArticle=true.

45. L. Mateu and F. Moll. Review of energy harvesting techniques for microelectronics. In *Proceedings of SPIE Microtechnologies for the New Millenium,* pp. 359–373 (May 2005).

46. Application report: Wireless sensor monitor using the ez430-rf2500 (September 2008). URL http://focus.ti.com/lit/an/slaa378b/slaa378b. pdf

47. R. Pallas-Areny and J. Webster, *Sensors and Signal Conditioning.* (John Wiley & Sons, 2001), 2nd edition.

48. E. Yeatman. Advances in power sources for wireless sensor nodes. In *International. Workshop on Wearable and Implantable Body Sensor Networks, London, UK,* (2004).

49. T. Torfs, V. Leonov, and R. Vullers, Pulse oximeter fully powered by human body heat, *Sensors & Transducers Journal.* 80(6), 1230–1238, (2007).

50. T. Torfs, V. Leonov, C. Van Hoof, and B. Gyselinckx. Body-heat powered autonomous pulse oximeter. In *Sensors, 2006. 5th IEEE Conference on,* pp. 427–430 (October, 2007).

51. D. Yates and A. Holmes, Micro power radio module, *DC FET Project ORESTEIA Deliverable* ND3. 2, (2003).

52. A. Kansal, D. Potter, and M. B. Srivastava. Performance aware tasking for environmentally powered sensor networks. In *SIGMETRICS '04/Performance '04: Proceedings of the joint international conference on Measurement and modeling of computer systems,* pp. 223–234, New York, NY, USA, (2004). ACM Press. ISBN 1-58113-873-3. doi: http://doi.acm.org/10.1145/1005686.1005714.

53. A. Kansal, J. Hsu, S. Zahedi, and M. B. Srivastava. Power management in energy harvesting sensor networks. Technical Report TR-UCLA-NESL-200603-02, Networked and Embedded Systems Laboratory, UCLA, (2006). URL http://nesl. ee. ucla. edu/fw/kansal/kansal/tecs.pdf.

54. Micro circuit labs (July, 2010). URL http://www.micro circuit labs. com/SDL-1.htm.

Chapter 2

Input Energy

Loreto Mateu,[a] William Kaal,[b] Monika Freunek Müller,[c] Birger Zimmermann,[d] and Uli Würfel[d]

[a] *Fraunhofer Institute for Integrated Circuits IIS, Nordostpark 93, 90411 Nuremberg, Germany*

[b] *Fraunhofer Institute for Structural Durability and System Reliability LBF, Bartningstr. 47, 64289 Darmstadt, Germany*

[c] *ZHAW Zürcher Hochschule für Angewandte Wissenschaften, Technoparkstrasse 2, 8401 Winterthur, Switzerland*

[d] *Fraunhofer Institute for Solar Energy Systems ISE, Heidenhofstr. 2, 79110 Freiburg, Germany*

loreto.mateu@iis.fraunhofer.de, william.kaal@lbf.fraunhofer.de, uli.wuerfel@ise.fraunhofer.de

Electrical energy can be harvested from multiple sources (mechanical, solar, thermal, et cetera) employing several transducers. The input energy of the transducer is characterized through various parameters. The measurement of the characterization parameters allows later on to reproduce the input energy of the transducer in the laboratory. In this way, it is possible to reproduce the ambient conditions to which the transducer will be exposed.

Handbook of Energy Harvesting Power Supplies and Applications
Edited by Peter Spies, Loreto Mateu, and Markus Pollak
Copyright © 2015 Pan Stanford Publishing Pte. Ltd.
ISBN 978-981-4241-86-1 (Hardcover), 978-981-4303-06-4 (eBook)
www.panstanford.com

2.1 Mechanical Energy

Different kinds of mechanical energy are present in the environment. Wherever mass is moving there is kinetic energy, e.g., in vibrating structures or in flowing fluids. Therefore, there is a high potential for energy harvesting applications. For achieving significant amounts of energy, the excitation must be oscillating; singular events are not taken into account. The oscillations may be mainly periodic, such as the vibrations of an engine, or random such as in most natural systems.

Traditionally, fluid flow has been the most exploited mechanical energy source. Wind and water mills converted ambient mechanical energy to useful forms of mechanical energy for many centuries. Later on, fluid flow has also been used to convert mechanical energy into electrical energy with electromechanical generators in power plants. However, for typical modern energy harvesting applications (as dealt with in this book) fluid flow has only recently been looked at. For example, some research has been done to design a piezoelectric windmill or a piezoelectric generator submerged in flowing water.[1]

Nevertheless, the vibration of structures is used to generate electrical power in most cases of energy harvesting applications, since mechanical vibrations are present almost everywhere, due to the fact that all technical systems are subject to vibrations in some way. Moving systems are excited to mechanical vibrations by the motion itself due to unevenness of the road or track surface. Stationary machines also vibrate due to rotating parts and their inherent unbalances and out-of-roundness or due to deficient mass balancing of oscillating parts. However, buildings too, show considerably high vibrations due to wind or other forms of excitation. Especially, large structures such as bridges or towers are susceptible for vibrations. Apart from technical systems, natural systems are also proposed as source for mechanical energy. Human body motion can be exploited, and researches have even thought of using the motion of animals to drive self-powered systems.

2.1.1 *Characterization of Parameters*

2.1.1.1 Characterization of vibrations

In mechanical vibrations, the state variable $q(t)$ changes with time: It can be, for example, a displacement, an angle, a force, or a pressure. A harmonic motion is the simplest form of periodic motion. Here the motion can be described by oscillatory functions such as the sine and the cosine function. The displacement, for example, can be written in the form of

$$q(t) = |A| \cos{(\omega t + \phi)}, \tag{2.1}$$

where $|A|$ is the amplitude of the motion in units of length, ω the circular frequency in radians per second and ϕ is an arbitrary phase angle in radians.

To characterize the mechanical energy of a vibrating source, its frequency range is important. In some cases, there are broadband vibrations such as in systems with rotating or oscillating parts that typically show one dominant frequency and higher order frequencies. In technical applications, there is of course a higher frequency range available than in natural systems. Vibrations excited by humans hardly exceed 2 Hz, whereas typical mechanical vibrations of machines are much higher. Vibrations between 20 Hz to 20 kHz are audible for the human ear, and therefore sound and vibrations are closely related in this area. Table 2.1 shows some exemplary vibration sources, their dominant frequencies, and peak acceleration.

Table 2.1 Some exemplary vibration sources[2,3]

Vibration source	Peak acceleration (m/s^2)	Frequency (Hz)
Car engine compartment	12	200
Base of 3-axis machine tool	10	70
Car instrument panel	3	13
Person tapping his heel	3	1
Wooden deck with people walking	1.3	385
Window next to busy road	0.7	100
Washing machine	0.5	109
AC vents in office building	0.2–1.5	60
Refrigerator	0.1	240

If alternating forces or motions are applied to a mechanical system, the system will vibrate with the frequency of the excitation (*forced vibration*). However, systems that are allowed to vibrate freely vibrate in their *natural frequencies*, also called *eigenfrequencies*. A bridge excited, for example, by wind and traffic will dominantly vibrate in its first few natural frequencies. The natural frequency of a system is dependent only on the stiffness of the structure and the mass that participates with the structure (including self-weight) and is independent from the load function. By means of modal analysis, the natural frequencies of a given system can be analyzed. The lowest natural frequency is called the *fundamental frequency*. It is useful to know the natural frequencies of a structure as it allows to tune an energy harvesting system exactly to one frequency. The device will then work in resonance, yielding the maximum energy.

In some technical applications, vibrations are necessary and even useful, for example, in vibratory feeder technology. Here an exploitation of the mechanical energy is only reasonable, if at all, to a certain extent. However, in most cases vibrations in technical systems are unwanted and can even cause damage. The most serious effect of vibration is that high alternating stresses produce fatigue failure in structural parts. Less serious effects include increased wear of parts and the propagation of vibration through foundations and buildings to locations where the vibration is intolerable either for human comfort or for the successful operation of sensitive measuring equipment. In these cases, harvesting mechanical energy and transforming it to electrical energy will either not interfere with the vibrating structure or even improve its performance by reducing the unwanted vibrations.

To characterize the system properties, an experimental modal analysis can be performed by applying a test force to the structure. An exciter system or an impact hammer serves for this purpose. Force transducers will measure the input force. One or more motion transducers measure the output vibration. The measured time domain signals are then transformed into the frequency domain, and the modal analysis calculates the eigenfrequencies, eigenvectors and damping parameters of the system. The knowledge of these modal

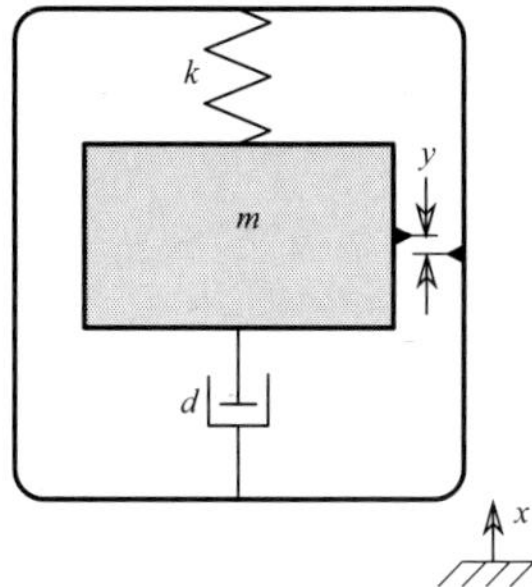

Figure 2.1 Kinetic linear energy harvesting model.

quantities allows a description of the dynamic behavior and is the basis of further numerical investigations.

2.1.1.2 Kinetic energy harvesting model

The maximum power provided to the energy harvesting device depends on the characteristics of the ambient vibration (frequency and amplitude of base acceleration) and the size of the device. A kinetic energy harvesting model is used for describing this relation analytically. The most commonly used model for this purpose is based on a second order spring and mass system with a linear damper as shown in Fig. 2.1.[3] A mass m is suspended in a rigid frame by a spring k and a damper d. The frame is exposed to a motion $x(t)$, resulting in a relative motion $y(t)$ of mass and frame. The damper dissipates mechanical energy and therefore represents the power conversion. Since in a mechanical damper the damping force is proportional to the velocity it describes better an electromechanical generator. However, the basic conclusions from this model are generally valid for every transducer mechanism.

The maximum power occurs at the natural frequency, that is at $\omega_n = \sqrt{k/m}$. The damping coefficient d of the model leads to a damping ratio $\zeta = d/2m\omega_n$, which can be subdivided into the electrical induced damping ζ_e and the parasitic mechanical damping ζ_p. The maximum power is generated at the resonance frequency and can in terms of excitation amplitude X be written as

$$P_{\max} = \frac{\zeta_e}{4(\zeta_p + \zeta_e)^2} m\omega_n^3 X^2 . \tag{2.2}$$

The output power at resonance is proportional to the cube of the natural frequency, showing that when the amplitude of base excitation is fixed, the highest possible frequency should be exploited. Knowing that the excitation displacement and acceleration are directly coupled over $A = \omega_n^2 X$, Eq. 2.2 can be written as

$$P_{\max} = \frac{\zeta_e}{4(\zeta_p + \zeta_e)^2} m \frac{1}{\omega_n} A^2 . \tag{2.3}$$

Here, the power is inversely proportional to the natural frequency. Thus, if the acceleration is constant, the transducer should be designed to resonate at the lowest fundamental frequency. Since the generated power is proportional to the mass in any case, the converter should be as large as possible. Equations 2.2 and 2.3 also show that the power generated by a vibrating system is maximized when the electrical damping ratio is equal to the mechanical damping ratio. This model also shows that a highly damped system extracts energy over a wide bandwidth of frequencies. A less damped system would extract more power but in a smaller frequency range. For a more detailed study of the behavior of this model, see.[3,4]

2.1.1.3 Finding optimal positions on the mechanical structure

For the installation of an energy harvesting device, it is necessary to know the optimal position on the mechanical structure. Therefore, the whole dynamic behavior of the structure must be analyzed and the transducer placed at the position where the motion is greatest. According to the mechanism of the transducer, the optimal position is where maximum acceleration or maximum strain occurs. Considering a simple cantilever like the one shown in Fig. 2.2a will clarify the approach.

At its first eigenfrequency, the cantilever beam will vibrate in the first eigenform $\Phi_1(x)$, shown in Fig. 2.2b. In this case, the eigenform $\Phi(x)$ describes the vertical displacement of the beam at location x. The associated strain distribution is shown in Fig. 2.2e, which for this model is just the mirrored function. The eigenforms for the second and third eigenfrequencies $\Phi_2(x)$ and $\Phi_3(x)$ are shown in Figs. 2.2c+f and 2.2d+g, respectively. Each eigenform is characterized by the alternating succession of nodes

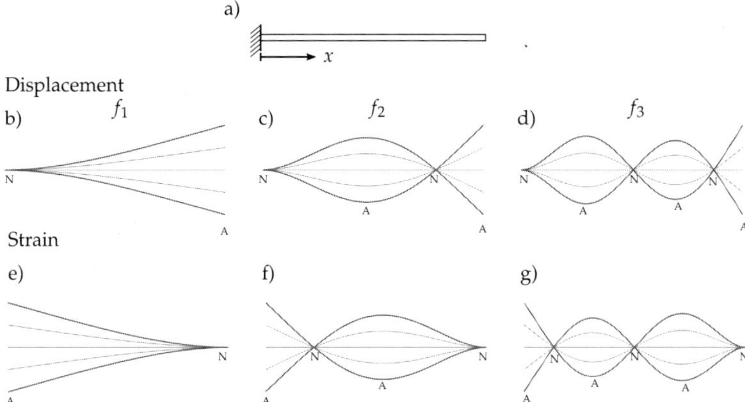

Figure 2.2 First three eigenforms of a cantilever beam, displacement, and strain.

(N) and antinodes (A). If only one frequency and therefore only one eigenform is considered, the transducer should be placed at the antinode with the largest amplitude for best results, positioning it at a node will obviously lead to no energy conversion at all.

However, if more than one frequency is of interest, the task of finding the optimal position becomes more complex. There are different mathematical methods that can be applied. Equations 2.4 and 2.5 show two possible approaches to find the optimal position for the general case of three-dimensional eigenforms $\Phi_i(x, y, z)$ of a vibrating volume V. All n relevant eigenforms can be multiplied (Eq. 2.4), the maximum of the overall product will show the best position. The disadvantage of this method is that positions that form nodes in only one eigenmode are underestimated. To avoid this problem, the sum of all eigenmodes can be calculated, using a weighting factor for each eigenmode to take into consideration different energy harvesting potentials of each mode (Eq. 2.5). The related frequency, the average amplitude, and Eqs. 2.2 and 2.3 are possible starting points for deriving the weighting factors.

$$\hat{\Phi}_P(x, y, z) = \prod_{i=1}^{n} \left| \frac{1}{V} \int_0^V \frac{\Phi_i(x, y, z)}{|\Phi_i|_{\max}} dV \right| \quad 0 < \hat{\Phi}_P < 1 \qquad (2.4)$$

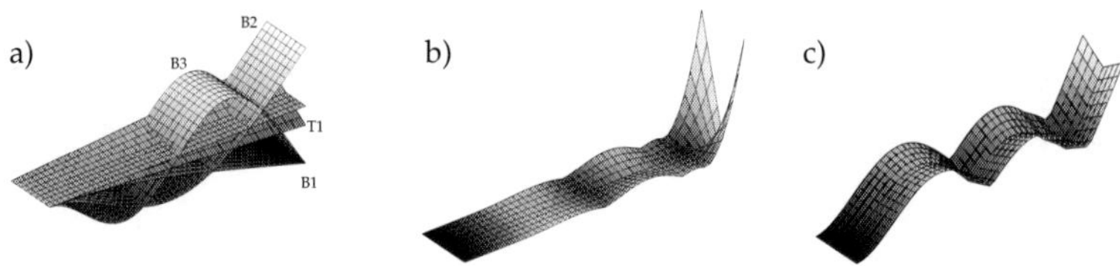

Figure 2.3 Eigenforms of a cantilever beam, product, and sum.

$$\hat{\Phi}_S(x, y, z) = \frac{1}{V}\frac{1}{\sum\limits_{i=1}^{n} G_i} \sum_{i=1}^{n} \left| G_i \int_0^V \frac{\Phi_i(x, y, z)}{|\Phi_i|_{\max}} dV \right| \qquad 0 < \hat{\Phi}_S < 1$$

$$(2.5)$$

To illustrate the outcome of these equations, the cantilever beam discussed before is extended to a two-dimensional model. The optimal position of an energy harvesting transducer is searched for, assuming that the harvesting device exploits the acceleration of the structure. The first four eigenmodes are taken into consideration, and it is further supposed that the generator does not affect the vibration. Figure 2.3a shows the first eigenmodes of the cantilever beam, three bending modes (B1-B3) and one torsional mode (T1). The implementation of Eq. 2.4 leads to the result shown in Fig. 2.3b, while Eq. 2.5 with weighting factors proportional to the respective eigenfrequency will result in Fig. 2.3c. In both cases, the best positions for energy harvesting are the top ends of the beam, as intuitively plausible. However, Eq. 2.5 shows that other positions on the beam with large strain at higher frequencies might also be considered as potential harvesting locations.

2.1.2 *Measurement Setup*

In order to characterize the ambient mechanical energy, different mechanical parameters must be measured. Various sensors have to be taken into consideration according to the quantity to be measured and the measuring frequency range. Figure 2.4 shows different types of sensors to measure displacement, velocity, and

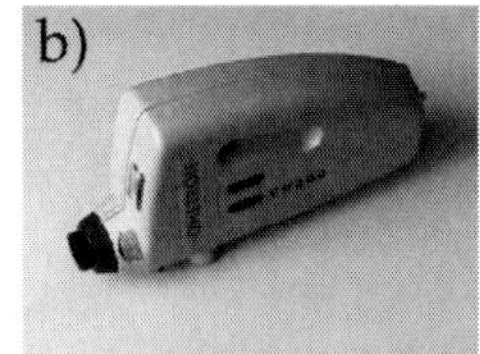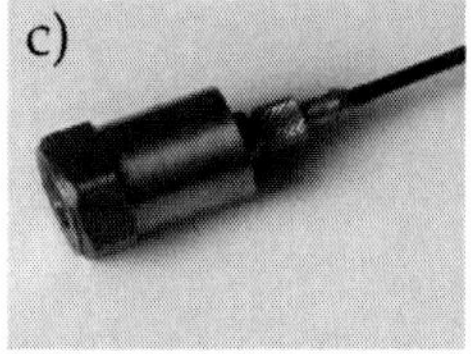

Figure 2.4 (a) laser triangulator, (b) laser vibrometer, (c) piezo accelerometer.

acceleration. The proper selection and knowledge of equipment and measurement instruments is important for yielding optimal results.

2.1.2.1 Accelerometers

For characterizing vibrations, accelerometers are the most employed sensors. They offer the advantage that they can measure the absolute acceleration and then can be applied directly to the vibrating structure. Displacement and velocity sensors must be applied to an external fixed point since they measure the relative motion. In many practical applications this can hardly be done, e.g., when measuring the oscillations on top of a large building.

Piezoelectric accelerometers are mostly used for measuring vibrations. Their high dynamics, immunity to wear, and small dimensions make them outperform other accelerometers in most cases. There are, however, other techniques for measuring vibrations that are sometimes used. Piezoresistive or capacitive accelerometers or strain gauge accelerometers can measure static acceleration, electrodynamic accelerometers are very large and can be used to measure very low frequencies.

Accelerometers are applied to a structure and normally measure the acceleration perpendicular to the surface. There are, however, special accelerometers that measure the acceleration in all three degrees of freedom in space. Usually, it is assumed that the mass of the accelerometers is small compared to the structure and that it does not affect the vibration.

Accelerometers feature a seismic mass which experiences a dynamic force, $F = ma$, due to its inertia when they are accelerated. This force is transformed into a voltage signal by means of some

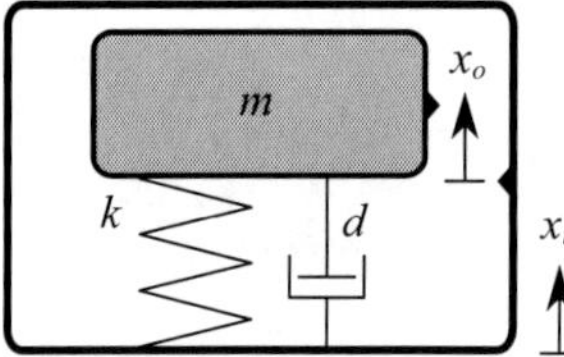

m = seismic mass [kg]
k = stiffness [N/m]
d = damping coefficient [Ns/m]
x_i = input signal: displacement of sensor [m]
x_o = output signal: relative displacement of mass to frame [m]

Figure 2.5 Basic principle of an accelerometer.

appropriate physical principles, e.g., the piezoelectric effect. The basic design of an accelerometer is shown in Fig. 2.5. The seismic mass m is suspended by an elastic element with stiffness k and provided with a damping d. The frame is exposed to the acceleration of the ground x_i, yielding the differential equation for the system:

$$\ddot{x}_o + 2\xi\omega_0\dot{x}_o + \omega_0^2 x_o = -\ddot{x}_i \tag{2.6}$$

with

$$\omega_0 = 2\pi f_0 = \sqrt{\frac{k}{m}}, \qquad \text{resonance frequency} \tag{2.7}$$

$$\xi = \frac{d}{2\omega_0 m}, \qquad \text{damping ratio} \tag{2.8}$$

From this differential equation, some basic principles for its application can be derived. In order to use this sensor to measure acceleration, the output x_o must be proportional to the acceleration $\ddot{x}_i$. This will be the case if $\omega_0^2 x_o$ is the dominating term on the left hand side of the differential equation. For high resonance frequencies, which can be realized by small masses and stiff springs, this condition holds true. Therefore, accelerometers have very high resonance frequencies ($f_0 = 10 \ldots 100$ kHz). However, the measurement range must be far below this frequency ($f_M \ll f_0$). The equation for the sensor can then be simplified to

$$x_o = -\frac{1}{\omega_0^2}\ddot{x}_i . \tag{2.9}$$

Nevertheless, in some cases other types of sensors must be used to characterize vibrations. If the structure is difficult to access or the surface too hot to apply a sensor directly, laser vibrometers are used. Here a laser beam is directed to the surface of interest, and the vibrometer measures the target velocity component along the direction of the laser beam. If the motion is too slow to induce appreciable accelerations, a laser triangulator can be used to measure displacements.

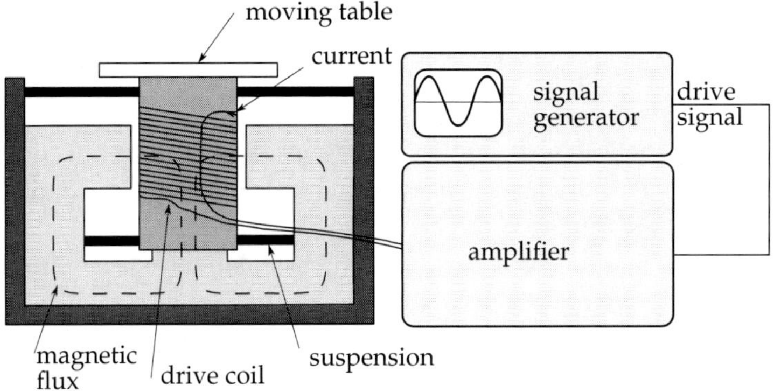

Figure 2.6 Basic setup of an electrodynamic shaker.

2.1.3 *Experimental Setup*

To reproduce the vibration conditions measured at the location of interest for the energy harvesting device, some kind of actuator must be used which can be driven with an electric input signal. In this way, realistic scenarios can be emulated in the laboratory.

There are several types of vibrational exciters (also known as shakers), for example, hydraulic, pneumatic, piezoelectric, and electrodynamic exciters or also rotating unbalance vibration exciters. For vibration analysis, electrodynamic exciters are in most cases the best option due to their high dynamic, broad frequency range and flexibility. Thus, this section will focus on them. For further literature about other exciters, see reference.[5]

2.1.3.1 Electrodynamic vibrators

In electrodynamic vibrators, a driving coil assembly, connected to a moving table (or a rod), is positioned in a magnetic field (see Fig. 2.6). A signal generator provides the drive signal, which is then amplified and fed to the driving coil. The electric current through the coil generates a mechanic force, leading to the excitation of the test structure.

The magnetic field is set up either by an electromagnet for large forces or by a permanent magnet for small forces. Varying the current i flowing through the coil, the output force will change proportionally. The electrical input signal can have any waveform since the forces generated by an electrodynamic vibrator can be of very different nature: periodic, transient, random, or impacts. For best results, a closed loop system should be set up. In this case, a sensor (typically an accelerometer) yields a mechanical output signal that is subsequently fed into a controller unit and compared to the desired input signal.

Vibrators are a convenient means for conducting vibration tests with sine wave or random signals. In the low frequency range the upper restriction is the maximum displacement, whereas for high frequencies the maximum force (or for a given mass the maximum acceleration respectively) is the limiting factor. In between, the maximum velocity can be critical with some vibrators. A typical graph with the limiting curves for displacement, velocity, and acceleration for an electrodynamic exciter is shown in Fig. 2.7. These limiting curves are the basis for choosing a proper exciter for a given task.

To excite the test system, electrodynamic exciters (Fig. 2.8a) can be used both in direct and inertial drive modes.[5] At direct drive, the coil is attached to the clamped test object and the vibrator body is

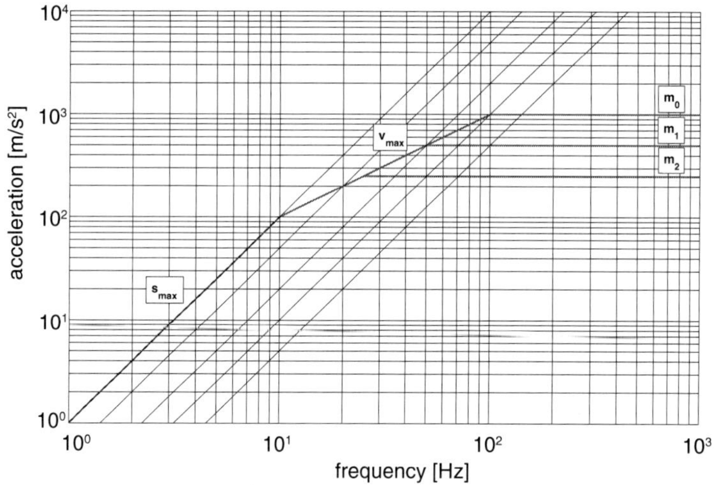

Figure 2.7 Limiting curves of an electrodynamic vibrator.

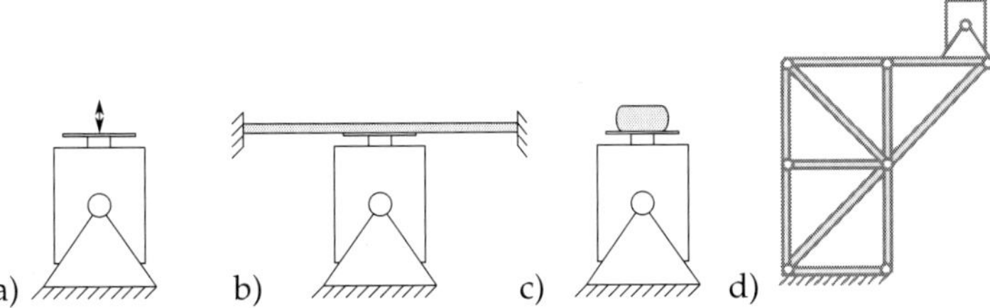

a) electrodynamic shaker
b) shaker on rigid base with clamped test object
c) shaker on rigid base with free test object
d) shaker with inertial mass on huge structure

Figure 2.8 Different shaker configurations.

mounted on a rigid base (Fig. 2.8b). Thus, a force is directly applied to the structure. Inertial drive can be performed in two ways: Either the shaker is mounted on a rigid base and the test object mounted on the moving table (Fig. 2.8c), or the exciter is applied to the test structure, exciting the structure with inertial forces (Fig. 2.8d). Generally, the arrangement in Fig. 2.8c is only used when a large vibrator is used with a comparatively small test component, e.g., to simulate a base excitation in the laboratory. The arrangement in Fig. 2.8d though is used to excite a large structure with a small exciter and is therefore used in experimental modal analysis and mobility measurements.

2.2 Light

Light is electromagnetic radiation. It consists of photons γ, which carry a quantum of energy:

$$E_\gamma = h\nu = \hbar\omega = \frac{hc_0/n}{\lambda} \tag{2.10}$$

where $h = 6.626 \times 10^{-34}$ Js is Planck's constant, $c_0 = 2.998 \times 10^8$ m/s is the vacuum speed of light, n the refractive index of the medium and ν the frequency of the light. When referring to an angular frequency $\omega = 2\pi\nu$, h is replaced by the reduced Planck constant $\hbar = h/2\pi$.

The photon current density $\phi_\gamma(\nu)$ per energy intervall and unit area is the spectral photon flux $\Phi_\gamma(\nu)$ per area A. The spectral energy current density or spectral irradiance is obtained, if the spectral photon current density is multiplied by the energy of the photons:

$$I_{e,\lambda}(\lambda) = \frac{hc_0}{\lambda}\phi_\gamma(\lambda). \tag{2.11}$$

The spectral photon and energy current density often appear foreshortened as spectrum of the light source. The integral over this spectral distribution yields an irradiance I_e.[a]

$$I_e = \int_0^\infty I_{e,\lambda}(\lambda)d\lambda = \int_0^\infty \frac{hc_0}{\lambda}\phi_\gamma(\lambda)d\lambda \tag{2.12}$$

From the irradiance I_e the radiant power P_{rad} incident on a surface with area A oriented normal to the direction of the incident light can be calculated to be

$$P_{rad} = I_e A. \tag{2.13}$$

2.2.1 *Spectra of Common Light Sources*

One type of light emitters are radiators that can be sufficiently described as Planckian radiators. Examples are the sun or halogen lamps. These radiators have a broad spectral distribution. The sun as radiation source can be well described as a black body with a surface temperature of approx. 5800 K and it covers a solid angle of $\Omega_S = 6.8 \times 10^{-5}$ when viewed from the earth. A black body is defined by having an absorptance of $a(\hbar\omega) = 1$ for all photon energies $\hbar\omega$. The spectrum of a black body can be calculated from Planck's law of radiation, which gives the spectral energy current density per photon energy intervall $d\hbar\omega$:

$$I_{e,\omega}(\hbar\omega) = \frac{(\hbar\omega)^3 \, d\Omega}{4\pi^3\hbar^3 (c_0/n)^2} \frac{1}{\exp[\hbar\omega/k_B T] - 1} \tag{2.14}$$

$d\Omega$ is the solid angle covered by the source as seen from the receiving area. For example, the solid angle of the sun seen from the

---

[a] In literature, I_e often is denoted as E_e. This has been avoided in this text in order to minimize confusion with the notation E for energy.

surface of the earth is $\Omega_S = 6.8 \times 10^{-5}$.

$$P_{\mathrm{rad}} = A \int_0^\infty I_{\mathrm{e},\omega}(\hbar\omega)\,\mathrm{d}\hbar\omega$$

$$= A \int_0^\infty \frac{(\hbar\omega)^3\,\mathrm{d}\Omega}{4\pi^3\hbar^3\,(c_0/n)^2} \frac{1}{\exp\left[\hbar\omega/k_B T\right] - 1}\,\mathrm{d}\hbar\omega. \quad (2.15)$$

On the way through the atmosphere of the earth, the solar radiation is partially absorbed. In the infrared region of the solar spectrum this absorption is almost entirely caused by gases such as water vapor (H_2O), carbon dioxide (CO_2), nitrous oxide (N_2O), methane (CH_4) and fluorinated hydrocarbons as well as by dust. In the ultraviolet region of the spectrum, it is altered by absorption of ozone and oxygen. The longer the path of the light through the atmosphere the more is absorbed. The so-called air mass coefficient AM is the path length l through the atmosphere divided by its thickness l_0. For solar radiation with an angle of incidence α to the surface normal, it is

$$\mathrm{AM} = l/l_0 = 1/\cos\alpha. \quad (2.16)$$

The spectrum of the sunlight just before entering the atmosphere of the earth is referred to as AM0. Thus, AM1 designates the one on the surface of the earth for normal incidence. The spectrum AM1.5 is the standard reference spectrum for evaluation of solar cells and corresponds to an angle of incidence of 48.19° relative to the surface normal.[6] Figure 2.9 shows the AM0 and the AM1.5 spectrum together with the spectrum of a black body with $T = 5800\,\mathrm{K}$. Another group of light emitters, mainly found in indoor applications, are narrow-band emitters with monochromatic laser light as an extreme case. More common are fluorescent tubes or different types of light-emitting diodes (LED), which have discrete line spectra. The fluorescence bulb in the inset of Fig. 2.9 is one example.

Solar light on earth reaches maximum intensities in the range of $1000\,\mathrm{Wm^{-2}}$. In office buildings, basic irradiance levels typically range below $10\,\mathrm{Wm^{-2}}$ from both natural and artificial light.[7] In micro energy harvesting applications, the main use of light energy is to convert it to electric energy. The most common converters are based on photoelectric effects in semiconducting materials.

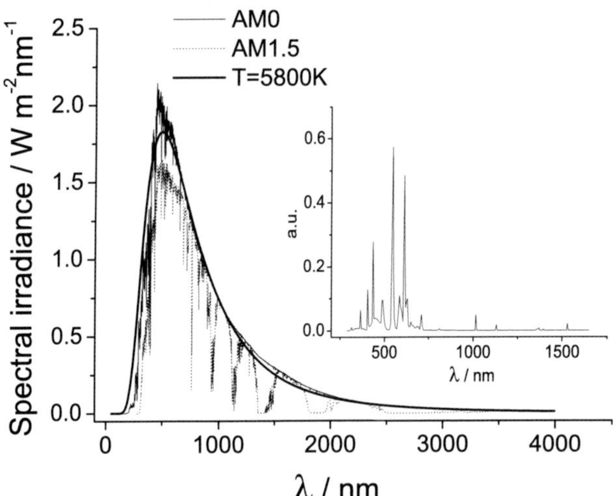

Figure 2.9 Extraterrestrial spectrum AM0 (thin line), terrestrial standard spectrum AM1.5 (dotted line) and the spectrum of a black body with $T = 5800$ K (heavy line). The inset shows a fluorescent tube (triband lamp).

As an alternative, the temperature difference between a surface heated up by an irradiance and its environment can be applied to a thermoelectric generator (TEG).

2.2.2 *Measurement Techniques*

Common measurement instruments for light are either based on thermal effects, such as pyranometers and pyrheliometers, or on photoelectric effects in semiconducting materials, such as silicon irradiance sensors. Due to different spectral distributions and a possible difference in intensity of three orders of magnitude, indoor light can require other measurement instruments than outdoor light.

2.2.2.1 Pyranometers

Based on thermoelectric arrays, these instruments measure diffuse and direct irradiation with a hemispherical field of view. Pyranometers have a spectral response wave band from 310 to 2800 nm,

Figure 2.10 Pyranometer for measurements of global irradiance. Source: IMTEK 2010.

and are standard instruments in outside measurements of solar radiation (Fig. 2.10). The unamplified output signal per Wm^{-2} is typically a few μV.

Provided with a shadow band, which excludes direct irradiance, pyranometers can be used to measure diffuse radiation only.

2.2.2.2 Pyrheliometers

In pyrheliometers, solar irradiance also heats a thermal sensing element. Having a narrow field of view due to a small opening angle, they measure direct irradiance. As a consequence, for varying angles of incident light as for sunlight measurements, these instruments need to track the light source.

2.2.2.3 Sunshine recorders

Sunshine recorders measure duration of a minimum level of sunshine. A common principle is concentrating the light via a glass sphere to burn marks on a measurement card. The minimum irradiance for a mark ranges from about $70\,Wm^{-2}$ to $280\,Wm^{-2}$.

Figure 2.11 Silicon-based irradiance sensor for outside measurements of solar irradiance. Source: IMT SOLAR-Ingenieurbüro Mencke und Tegtmeyer, 2014.

2.2.2.4 Silicon irradiance sensors

While at low light intensities, such as $0.1\,\mathrm{Wm^{-2}}$, more effects need to be considered, for many applications the short circuit current of a silicon solar cell is sufficiently described to be proportional to the incident irradiance. Therefore, silicon-based irradiance sensors range amongst the most common and affordable measurement instruments for irradiance. Depending on the manufacturer, they are available as handheld and mounted products and can also include electronic compensation of temperature effects.

2.2.2.5 Luxmeter

Light in buildings is optimized for the spectral sensitivity function of the human eye, $V(\lambda)$. Taking the product of the spectral irradiance $I_{e,\lambda}(\lambda)$, the human sensitivity function $V(\lambda)$ and the photometric equivalent $K_{\mathrm{m}} = 683\,\mathrm{lm/W}$ and integrating it over the corresponding wavelength range gives an illuminance $I_{\mathrm{V}}(\lambda)$ in the photometric unit Lux. $I_{\mathrm{V}}(\lambda)$ is measured with luxmeters. Their measurement principles and their use for photovoltaic application are discussed in Chapter 7. The deviating results on the displays of

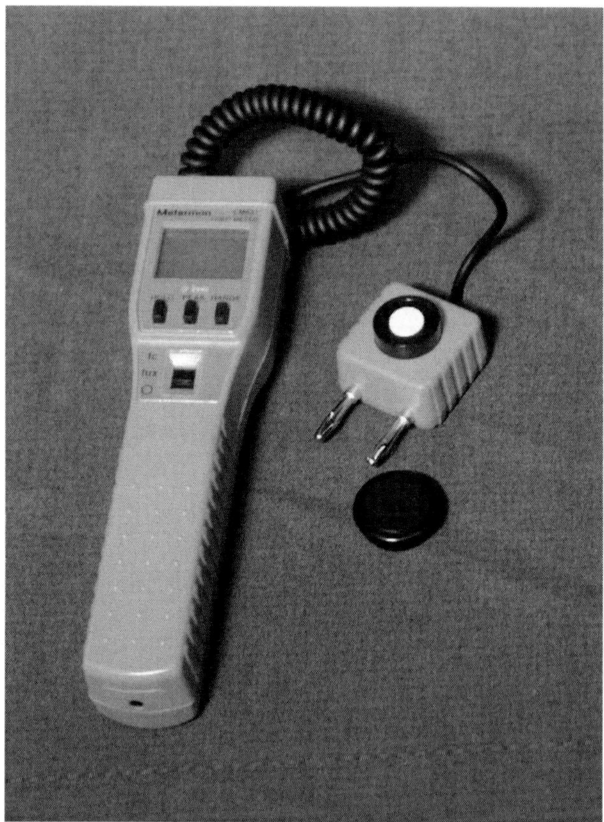

Figure 2.12 Exemplary luxmeter calibrated to a standard illuminant A (incandescent bulb). Source: IMTEK, 2010.

the luxmeters from Fig. 2.12 are due to the spectral mismatch to the light source.

2.2.2.6 Spectroradiometers

Spectral irradiance is measured with spectroradiometers. These measurement instruments are based on fiber optics coupling the light into CMOS/CCD-detector arrays. As for all optical measurement instruments, its spectral sensitivity needs to be adjusted to the spectral range of the measured light source.

2.2.2.7 Numerical approach: ray-tracing programs

Due to the amount of variables influencing the available irradiance, such as geographical position, orientation, shading, or, for indoor applications, user behavior, the use of ray-tracing programs is a valid approach. It is well established in the planning of solar outdoor plants. Examples of such programs are *DAYSIM* or *PV*SOL*. For indoor applications, this approach is feasible, e.g., by combining the ray tracers *DAYSIM*, *Radiance*, and a user behavior model (see Chapter 7).

2.2.3 *Experimental Setup*

Outside conditions and terrestrial solar light are reproduced using solar simulators. For indoor conditions, no standard procedures exist.

2.2.3.1 Solar simulators

In solar simulators, a standardized spectral irradiance of terrestrial sunlight is approximated with artificial light, e.g., xenon arc lamps. The light is either continuous, such as in steady-state solar simula-

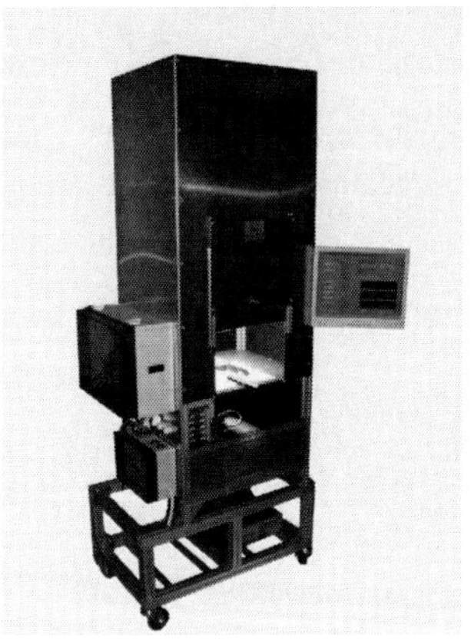

Figure 2.13 Small steady-state solar simulator for calibration measurements of modules with areas of a few cm^2. Source: IMT SOLAR-Ingenieurbüro Mencke und Tegtmeyer, 2014.

Figure 2.14 Steady-state solar simulator for calibration measurements of modules with an area of m². Source: IMT SOLAR-Ingenieurbüro Mencke und Tegtmeyer, 2010.

tors, or in intervals of milliseconds, such as in flash solar simulators, to minimize temperature effects during measurements. The size of solar simulators depends on the investigated module area, which ranges between m² and cm². Figure 2.13 shows a rollable steady-state simulator for measurements of small photovoltaic devices. Figures 2.14 and 2.15 depict a solar simulator for industrial solar module measurements.

A relevant standard concerning requirements of solar simulators is *IEC 60904-9*. The different classes of solar simulators *A*, *B*, and *C* describe the achieved accuracy of the spectral match to the standard solar spectrum, the spatial uniformity of the irradiance, and the temporal stability of the simulator.

2.3 Thermal Energy

Thermal energy is present in any location where there is a temperature gradient. Thermal energy is available, for example, in the industrial environment (machinery, pipelines, vehicles), in buildings and in the human body. A thermogenerator makes use of

Figure 2.15 Lighting construction for a steady-state solar simulator for calibration measurements of modules with an area of m². Source: IMT SOLAR-Ingenieurbüro Mencke und Tegtmeyer, 2010.

the temperature gradient between its hot and cold side to convert thermal energy into electrical energy using the Seebeck effect. For maintaining the thermal gradient, it is necessary to connect a heat source to one side of the thermogenerator (hot side) and a heat sink on the other side (cold side).

The maximum electrical output that can be delivered by a thermogenerator is:[8]

$$P_{\text{el}}^{\max} = \frac{A}{l} \left[\frac{1}{4} \frac{\alpha_{\text{m}}^2 \, (T_{\text{H}} - T_{\text{C}})^2}{\rho_{\text{m}}} \right] \tag{2.17}$$

where α_m is the Seebeck coefficient of the thermogenerator in volts per kelvin, T_H is the temperature at the hot side of the thermogenerator in kelvin, T_C is the temperature at the cold side of the thermogenerator in kelvin, ρ_m is the electrical resistivity in ohm. metre. A is the cross-sectional area in square metre and l is the length of one thermoelectric leg couple in metre.

The heat absorbed by the hot side of the thermogenerator is

$$q_H = \alpha_m T_H I - \frac{I^2 R_m}{2} + K_m \Delta T \qquad (2.18)$$

where $\alpha_m T_H I$ corresponds to the Seebeck power generation, $\frac{I^2 R_m}{2}$ corresponds to the Joule heating effect and the term $K_m \Delta T$ is due to thermal convection. ΔT is the thermal gradient between the hot and cold side ($\Delta T = T_H - T_C$), I is the current flowing through the thermogenerator. R_m is the electrical resistance and Km represents the thermal conductance of the thermogenerator.

The heat emitted by the cold side of the thermogenerator is

$$q_C = \alpha_m T_C I + \frac{I^2 R_m}{2} + K_m \Delta T \qquad (2.19)$$

2.3.1 *Characterization of Parameters*

It is necessary to measure the temperature at each side of the thermogenerator in order to characterize the performance of a TEG. To minimize the error due to local drops in temperature, the points of measurement should be as close as possible to the thermogenerator sides. The temperature is measured by a temperature sensor that can be a thermistor, a thermocouple, an resistance temperature detector (RTD), or similar depending on the accuracy, temperature operating range or stability required. Table 2.2 shows a comparison table of different properties for three different temperature sensors.

Thermistors have the advantage of a very high sensitivity to temperature changes, but the disadvantage of nonlinearity. Thermistors, because of their high sensitivity, are ideal for detecting small changes in temperature.[10] For extreme accuracy, the RTD is the best choice, but thermistors with 0.1°C accuracy are now available. Their advantage compared to RTDs are a faster response time and a greater output per °C[9].

Table 2.2 Temperature sensors properties[9]

Property	Thermocouple	RTD (Pt100)	Thermistor
Operating Range	$-200°$C to $2000°$C	$-250°$C to $850°$C	$-100°$C to $300°$C
Accuracy	Low ($1°$C)	Very high ($0.03°$C)	High ($1°$C)
Linearity	Medium	High	Low
Thermal Response	Fast	Slow	Medium
Cost	Low	High	Low to moderate
Noise Problems	High	Medium	Low
Long-term stability	Low	High	Medium
Cost of measuring instrument	Medium	High	Low

2.3.2 Measurement Setup

A thermal resistor or thermistor changes its electrical resistance with a variation in the temperature. There are thermistors with a positive temperature coefficient (PTC) and with a negative temperature coefficient (NTC) depending on whether the resistance increases or decreases with increasing temperature, respectively. The expression of the resistance of a NTC thermistor as a function of the temperature is[11]

$$R_T = R_{T0} \exp\left[\frac{B\,(T_0 - T)}{T\,T_0}\right] = R_{T0}\,f(T) \qquad (2.20)$$

where R_T is the resistance at temperature T expressed in Kelvin, R_{T0} is the resistance at ambient temperature and B is the B-constant of the thermistor, which is given by Eq. (2.22). The nominal resistance is the resistance of the thermistor at T_0, which is the reference temperature, usually $25°$C.

The temperature of the NTC thermistor is deduced from Eq. (2.20).

$$T = \left(\frac{1}{T_0} + \frac{\ln\frac{R_T}{R_{T0}}}{B}\right)^{-1} \qquad (2.21)$$

Rearranging Eq. (2.20), B-constant is expressed as

$$B = \ln\left(\frac{R_T}{R_{T0}}\right)\frac{T\,T_0}{T_0 - T} \qquad (2.22)$$

The temperature coefficient α of the resistance is defined as

$$\alpha = \frac{1}{R}\frac{dR}{dT} \qquad (2.23)$$

By using Eq. (2.20) in Eq. (2.23), α is obtained as a function of B-constant:

$$\alpha = -\frac{B}{T^2} \tag{2.24}$$

Eq. (2.20) is valid for the temperature range where a linear relation between $\ln(R)$ and $1/T$ can be considered . Another expression that relates temperature and resistance of an NTC thermistor is the Steinhart–Hart equation that employs a third order polynomial approximation:[11]

$$\frac{1}{T} = a + b\ln(R) + c\ln^2(R) + d\ln^3(R) \tag{2.25}$$

Rearranging the terms of Eq. (2.25), $\ln(R)$ is expressed as a function of temperature.

$$\ln(R) = a + \frac{b}{T} + \frac{c}{T^2} + \frac{d}{T^3} \tag{2.26}$$

Equation (2.25) can be simplified without significant loss of accuracy to:

$$\frac{1}{T} = a + b\ln(R) + d\ln^3(R) \tag{2.27}$$

The use of the previous equation in the temperature range of 0 to 70°C assures accuracy in a millidegree range.[12]

The parameters for selecting a NTC thermistor are the resistance R_{T0} at the reference temperature and the tolerance x of the B-constant and y of the resistance R_{T0}. The combination of the resistance and B-constant tolerances of the thermistor gives as a result the resistance range for a certain temperature. The following equations provide the resistance upper and lower limits as a function of temperature with a B-constant tolerance y and a resistance tolerance x.[11]

$$R_{\text{TLH}} = R_{T0}(1 + x)\exp[B(1 + y)(1/T_{\text{L}} - 1/T)] \tag{2.28}$$

$$R_{\text{THH}} = R_{T0}(1 + x)\exp[B(1 - y)(1/T_{\text{H}} - 1/T)] \tag{2.29}$$

$$R_{\text{TLL}} = R_{T0}(1 - x)\exp[B(1 - y)(1/T_{\text{L}} - 1/T)] \tag{2.30}$$

$$R_{\text{THL}} = R_{T0}(1 - x)\exp[B(1 + y)(1/T_{\text{H}} - 1/T)] \tag{2.31}$$

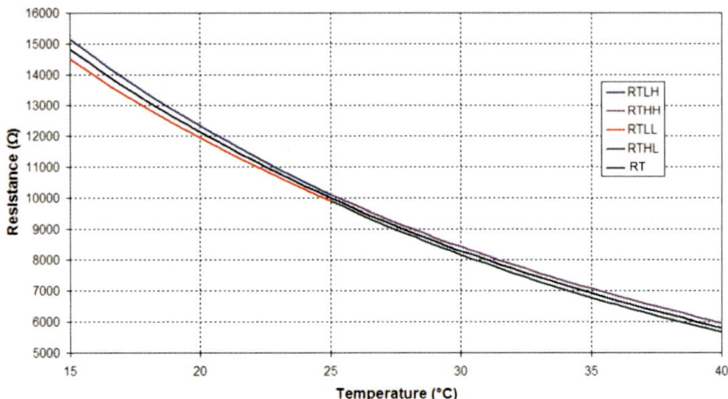

Figure 2.16 Effect of B-constant on a 10 kΩ NTC thermistor with a resistance tolerance of ±10% and a B-constant tolerance of ±5%.

where T_L is the temperature of the NTC thermistor lower than T_0 and T_H is the temperature of the NTC thermistor higher than T_0.

Figure 2.16 represents R_{TLH}, R_{THH}, R_{TLL}, and R_{THL} as a function of temperature for the case of a NTC thermistor with a nominal resistance of 10 kΩ at 25°C, a B-constant tolerance of ±5% and a resistance tolerance of ±10%. RT represents the value of the NTC thermistor without taking into account the B-constant and the resistance tolerances. Therefore, a calibration is necessary in most of the applications due to the tolerance of B-constant and the resistance. The value of B can be derived from a two-point calibration:[12]

$$B = \frac{\ln \frac{R_{T1}}{R_{T2}}}{\frac{1}{T_1} - \frac{1}{T_2}}, \tag{2.32}$$

where R_{T1} and R_{T2} are the resistances at the temperatures T_1 and T_2, respectively.

The Steinhart–Hart model needs a three-point calibration for calculating the constants a, b and d of the simplified version. However, Fraden[12] proposes a combined model of the logarithmic and Steinhart–Hart models where B-constant is a linear function of temperature and for a temperature range from 0 to 70°C, errors lower than 0.03°C are obtained with a two point calibration. Fraden[12] explains the methodology and equipment required for the calibration of NTC thermistors.

The NTC thermistors are going to be used for temperature measurement and also in the control of the experimental setup to reproduce the temperatures at both sides of a thermogenerator. To obtain a linear relation between resistance and temperature, a thermistor circuit is needed. The voltage divider is the simplest thermistor circuit. The output voltage of the voltage divider is

$$V_{\text{out}}(T) = V_{\text{in}}\left(\frac{R_1}{R_1 + R_T}\right) = V_{\text{in}}\left(\frac{1}{1 + R_T/R_1}\right) \qquad (2.33)$$

where V_{in} is the voltage of the input source, R_T is the resistance of the NTC thermistor and R_1 is the resistance of the fixed resistor.

The ratio between R_T and R_1 can be expressed as

$$\frac{R_T}{R_1} = \frac{R_{T0}}{R_1} f(T) = s f(T) \qquad (2.34)$$

Hence, the output voltage of the voltage divider as a function of s and $f(T)$ is

$$V_{\text{out}}(T) = V_{\text{in}}\left(\frac{1}{1 + sf(T)}\right) = V_{\text{in}} F(T) \qquad (2.35)$$

The value of $f(T)$ depends on the NTC thermistor employed while the value of s is chosen to reach linearity for the function F(T) in the temperature range under consideration.

A modification of the voltage divider circuit is shown in Fig. 2.17, where a second resistor R_s is connected in series to the NTC thermistor. Figures. 2.18 and 2.19 show the output voltage as a function of the temperature for a 10 kΩ NTC thermistor and an input voltage of 1.9 V.

Another thermistor circuit is the Wheatstone bridge (see Fig. 2.20), which is composed by four resistors and a voltage source V_{in}.[14,22] The output voltage of the circuit is

$$V_{\text{out}} = V_{\text{AB}} - V_{\text{AD}} = V_{\text{in}}\left(\frac{R_1}{R_1 + R_2} - \frac{R_4}{R_4 + R_3}\right)$$

$$= V_{\text{in}}\frac{R_1 R_3 - R_2 R_4}{(R_1 + R_2)(R_4 + R_3)} \qquad (2.36)$$

There are different configurations of the Wheatstone bridge depending on the location of the resistive sensor. The quarter bridge circuit contains only one resistive sensor and three resistors, the half bridge circuit has two resistive sensors and two resistors, and the

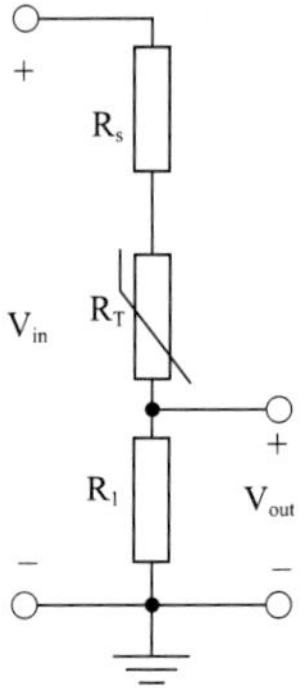

Figure 2.17 Voltage divider circuit containing a thermistor and two resistors.

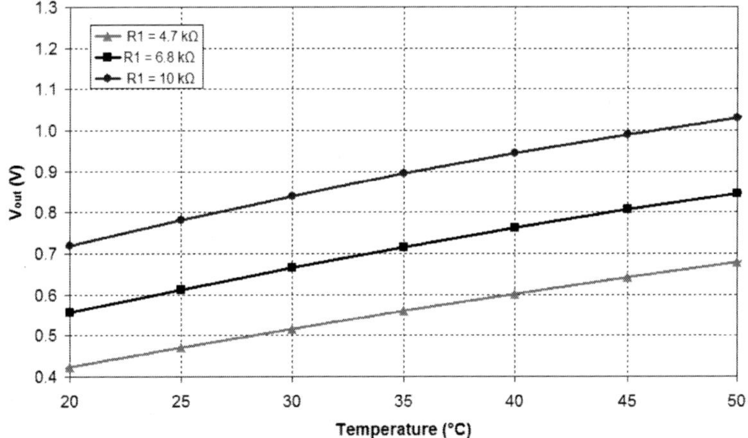

Figure 2.18 Output voltage as a function of the temperature for the resistive divider of Fig. 2.17 with $R_s = 0\ \Omega$.

full bridge circuit has four resistive sensors. From Eq. (2.36), it is deduced that the output voltage of the Wheatstone bridge is zero $(V_{out} = 0)$ when $\frac{R_1}{R_2} = \frac{R_4}{R_3} = \frac{1}{r}$.

Hence, the sensitivity of an initially balanced Wheatstone bridge is

$$S = \frac{V_{out}}{\Delta R_T} \tag{2.37}$$

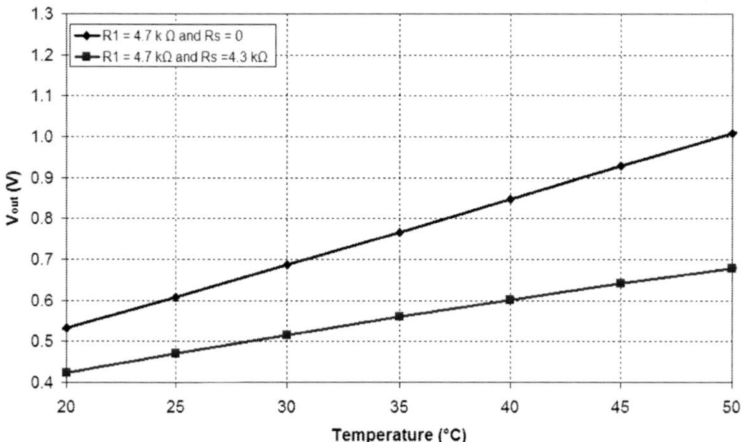

Figure 2.19 Output voltage as a function of the temperature for the resistive divider of Fig. 2.17.

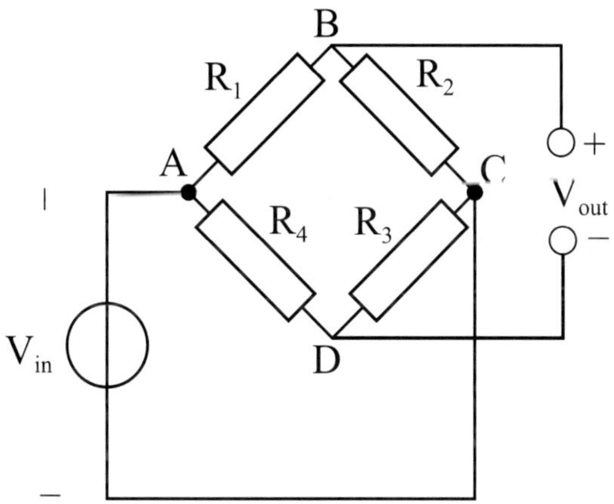

Figure 2.20 Wheatstone bridge circuit.

The value for ΔR_T, when R_T has the expression shown in Eq. (2.20) is

$$\Delta R_\mathrm{T} = R_{T0} \exp\left(B\left(1/T - 1/T_0\right)\right)\left(-B\Delta T/T^2\right) \qquad (2.38)$$

Combining the previous expression with Eq. (2.20), the following is obtained:

$$\frac{\Delta R_T}{R_T} = -B\frac{\Delta T}{T^2} \tag{2.39}$$

For a balanced Wheatstone bridge where the NTC thermistor is resistor R_3 and $R_1 = R_4$ and $R_2 = R_{T0}$, its sensitivity is given by

$$\frac{V_{\text{out}}}{\Delta R_T} = -\frac{R_1}{(R_1 + R_{T0})^2}V_{\text{in}} \tag{2.40}$$

Substituting Eq. (2.39) in Eq. (2.40), the following expression is derived:

$$\frac{V_{\text{out}}}{\Delta T} = \frac{B}{T^2}\left(R_1\frac{R_{T0}}{(R_1 + R_{T0})^2}\right)V_{\text{in}} \tag{2.41}$$

For a quarter bridge configuration where all the resistors have the same value, $R_1 = R_2 = R_4 = R_{T0}$, the expression for V_{out} is

$$V_{\text{out}} = -\frac{\Delta R_T}{4R_{T0}}V_{\text{in}} = B\frac{\Delta T}{4T^2}V_{\text{in}} \tag{2.42}$$

The output voltage range obtained with a Wheatstone bridge circuit is usually 10 mV to 100 mV and therefore, an amplification circuit is necessary.[15]

2.3.3 *Experimental Setup*

When a thermogenerator has to be evaluated under certain temperature conditions, it is necessary to fix the temperature of both sides. For doing this, two thermocoolers can be placed at each side of the thermogenerator under test. The temperature of the thermocoolers can be regulated with a proportional-integral-derivative (PID) controller. For achieving a correct regulation loop of the thermocoolers temperature, its equivalent circuit and transfer function are presented. Moreover, a PID controller circuit is introduced.

2.3.3.1 Thermoelectric cooler model

In thermoelectric coolers (TECs), the heat is absorbed by the cold side and is emitted by the hot side when electrical energy is applied to the electrical terminals of the coolers. Figure 2.21 shows the

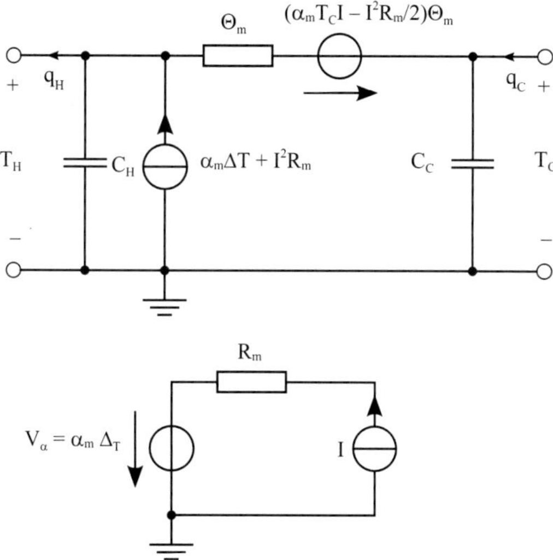

Figure 2.21   Thermocooler equivalent circuit.[16]

TEC equivalent circuit based on the model presented by Lineykin et al[16]. An equivalent model is presented by Chavez et al.[17]. Two different circuits can be distinguished in both models: the circuit that models the thermal behavior and the circuit that models the electrical behavior.

The circuit that models the electrical response has a temperature-dependent voltage source V_α and a resistor R_m. The value of the voltage source is proportional to the temperature gradient between the hot side (T_H) and the cold side (T_C) of the TEC and to the Seebeck coefficient α_m. R_m represents the equivalent electric resistance of the thermoelectric module (TEM). An external current source (I in Fig. 2.21) or voltage source must be connected to the electrical circuit in order to create a temperature gradient between both sides of the thermoelectric module.

The thermal part of the model, illustrated by Fig. 2.21, has two capacitors at each side of the TEM, C_H and C_C. C_H represents the thermal mass of the hot side whereas C_C is the thermal mass of the cold side of the Peltier element. The model is also composed by a voltage-controlled voltage source, a current-controlled current source and a thermal resistance Θ_m.

Table 2.3 Thermal parameters

Parameter	Definition	Unit
C_H	Thermal mass of the hot side of the Peltier module	[J/K]
C_C	Thermal mass of the cold side of the Peltier module	[J/K]
K_m	Thermal conductance	[W/K]
Θ_m	Thermal resistance	[K/W]
R_m	Electrical resistance	[Ω]
α_m	Seebeck coefficient	[V/K]

q_C is the heat absorbed by the cold side and q_H is the heat emitted at the hot side. The thermal part of the TEC model is described by Eqs. (2.43) and (2.44).

$$q_C = \alpha_m T_C I - \frac{I^2 R_m}{2} - K_m \Delta T \tag{2.43}$$

where $\alpha T_C I$ corresponds to the Seebeck power generation, $\frac{1}{2}I^2 R_m$ corresponds to the Joule heating effect, the term $K_m \Delta T$ (and $\frac{\Delta T}{\Theta_m}$) is due to thermal convection and $\Delta T = T_H - T_C$.

$$q_H = \alpha_m T_H I + \frac{I^2 R_m}{2} - K_m \Delta T \tag{2.44}$$

The electrical characteristic of the thermoelectric module can be modeled by a voltage source proportional to the temperature gradient between the hot and the cold side.

$$V_\alpha = \alpha_m (T_H - T_C) = \alpha \Delta T \tag{2.45}$$

When the TEM is employed as a TEC both sides of the TEM will be at ambient temperature at the beginning. The current or voltage delivered at the electrical terminals of the TEM will generate a temperature difference between the two sides of the TEM.

References

1. R. S. Anton and A. H. Sodano, A review of power harvesting using piezoelectric materials (2003–2006), *Smart Materials and Structures.* **16**(3), R1–R21, (2007). ISSN 0964–1726.

2. J. S. Roundy. *Energy Scavenging for Wireless Sensor Nodes with a Focus on Vibration to Electricity Conversion.* PhD thesis, University of California at Berkeley, Berkeley CA, USA, (2003).

3. A. K. Cook-Chennault, N. Thambi, and M. A. Sastry Powering mems portable devices: a review of non-regenerative and regenerative power supply systems with special emphasis on piezoelectric energy harvesting systems, *Smart Materials and Structures.* **17**(4), 043001, (2008). ISSN 0964–1726.

4. G. N. Stephen, On energy harvesting from ambient vibration, *Journal of Sound and Vibration.* **293**(1–2), 409–425, (2006).

5. G. Buzdugan, E. Mihaeailescu, and M. Rades, *Vibration measurement,* vol. 8, *Mechanics-Dynamical systems* (Nijhoff, Dordrecht, 1986). ISBN 9024731119.

6. National renewable energy laboratory: *Reference Solar Spectral Irradiance: Air Mass 1.5,* (10.12.2009). URL http://rredc.nrel.gov/solar/spectra/am1.5/

7. M. Müller, J. Wienold, W. Walker, and L. Reindl. Characterization of indoor photovoltaic devices and light. In *Photovoltaic Specialists Conference (PVSC), 2009 34th IEEE,* pp. 000738–000743 (7–12, 2009). doi: 10.1109/PVSC.2009.5411178.

8. M. Ryan and J. Fleurial, Where there is heat, there is a way: thermal to electric power conversion using thermoelectric microconverters, *The Electrochemical Society interface.* **11**(2), 30–33, (2002).

9. A. long, Improving the accuracy of temperature measurements, *Sensor Review.* **21**(3), 193–198, (2001).

10. Calibrating thermistor sensors. http://www.mstarlabs.com/sensors/thermistor-calibration.html.

11. NTC thermistors. http://www.gesensing.com/products/resources/whitepapers/ntcnotes.pdf.

12. J. Fraden. A two-point calibration of negative temperature coefficient thermistors, *Review of Scientific Instruments.* **71**, 1901–1905 (2000).

13. HFAN-08.2.1: PWM Temperature Controller for Thermoelectric Modules Keeps Components within 0.1°C. http://www.maxim-ic.com/app-notes/index.mvp/id/1757 (February 2003).

14. Wheatstone Bridges: Introduction. http://www.efunda.com/design-standards/sensors/methods/wheatstoneJbridge.cfm.

15. Signal Conditioning Wheatstone Resistive Bridge Sensors. http://focus.ti.com/lit/an/sloa034/sloa034.pdf (September, 1999).

16. S. Lineykin and S. Ben-Yaakov, SPICE compatible equivalent circuit of the energy conversion processes in thermoelectric modules, *Electrical and Electronics Engineers in Israel, 2004. Proceedings. 2004 23rd IEEE Convention of.* 346–349 (September, 2004).

17. J. A. Chavez, J. A. Ortega, J. Salazar, A. Turo, and M. J. Garcia, SPICE model of thermoelectric elements including thermal effects, *Instrumentation and Measurement Technology Conference, 2000. IMTC 2000. Proceedings of the 17th IEEE.* **2**, 1019–1023 (May, 2000).

18. A. Lima, G. Deep, L. de Almeida, H. Neff, and M. Fontana. A gain-scheduling pid-like controller for Peltier-based thermal hysteresis characterization platform. In *Proceedings of the 18th IEEE Instrumentation and Measurement Technology Conference, 2001. IMTC 2001.*, **2**, 919–924, (2001).

19. S. Lineykin and S. Ben-Yaakov, Analysis of thermoelectric coolers by a spice-compatible equivalent-circuit model, *IEEE Power Electronics Letters.* **3**(2), 63–66 June, 2005).

20. HFAN-08.2.0: Thermoelectric Cooler (TEC) Control, http://www.maxim-ic.com/app-notes/index.mvp/id/3318 (September, 2004).

21. D. Mitrani, J. Tome, J. Salazar, A. Turo, M. Garcia, and J. Chavez. Dynamic measurement system of thermoelectric module parameters. In *Thermoelectrics, 2003 Twenty-Second International Conference on-ICT,* 524–527, (2003).

22. F. Golnaraghi and B. Kuo, *Automatic Control Systems.* John Wiley & Sons, 2003).

23. MAX1968, MAX1969 Power Drivers for Peltier TEC Modules. http://www.maxim-ic.com/quick_view2.cfm/qv_pk/3377.

24. LTC1923—Higher Efficiency Thermoelectric Cooler Controller. http://www.linear.com/pc/productDetail.jsp?navId=HO,Cl,C1010,C1095,P1893.

25. Laird TECHNOLOGIES Temperature Controllers, http://lairdtech.thomasnet.com/viewitems/ tempera ture-controllers-2/ temperature-controllers?

26. Temperature Controller VE0016. http://www.premosys.com/en/electronics/ve0016-temperature-controller.htm.

27. TE TECHNOLOGY, INC. Temperature Controllers, http://www.tetech.com/Temperature-Controllers.html.

28. OVEN industries. Thermoelectric Module Controllers. http://www.ovenind.com/bv/Departments/Temperature-Controllers/Thermo-electric-Module-Controllers.aspx

Chapter 3

Piezoelectric Transducers

Bernhard Brunner,[a] Matthias Kurch,[b] and William Kaal[b]

[a] *Fraunhofer Institute for Silicate Research, Neunerplatz 2,
97082 Wuerzburg, Germany*
[b] *Fraunhofer Institute for Structural Durability and System Reliability,
Bartningstr. 47, 64289 Darmstadt, Germany*
bernhard.brunner@isc.fraunhofer.de, william.kaal@lbf.fraunhofer.de

This chapter provides a short history of piezoelectric materials
and describes their properties and the underlying phenomenon. It
goes on to suggest and discuss a mechatronic model to simulate
piezoelectric transducers, outlining some of the most important
properties required for energy harvesting applications.

3.1 History

The piezoelectric effect was first reported in 1880. The French
brothers Curie discovered that certain crystals such as quartz
become electrically polarized when subjected to mechanical strain.
In turn, applying an electric field to a piezoelectric material results
in a deformation of that material. Thus, two basic applications
became obvious: strain sensors based on the direct piezoelectric
effect and actuators based on the inverse piezoelectric effect. In
the following years, the piezoelectric effect was mathematically

Handbook of Energy Harvesting Power Supplies and Applications
Edited by Peter Spies, Loreto Mateu, and Markus Pollak
Copyright © 2015 Pan Stanford Publishing Pte. Ltd.
ISBN 978-981-4241-86-1 (Hardcover), 978-981-4303-06-4 (eBook)
www.panstanford.com

described by its constitutive equations. The first practical use of piezoelectrics took place during World War I. It was the invention of sonar devices, in which piezoelectric crystals vibrate in response to ultrasonic acoustic waves, that propelled the piezoelectric crystal out of the laboratory into the real world. From that period on, a wide variety of new applications were developed: microphones, accelerometers, actuators, ultrasonic motors, etc. In the 1950s, an exceptionally strong piezoelectric response was discovered in lead zirconate—leading to the development of zirconate-titanate (PZT) solid solution systems. Since then, modified PZT ceramics have become the dominant piezoelectric material for a variety of applications such as actuators in fuel injection systems or medical ultrasound devices.

Later on, also polymers such as polyvinylidene difluoride (PVDF) have been found to have piezoelectric properties as a result of the stretching of molecules when the material is formed into sheets. They have been used as transducers. Piezoelectric composite materials incorporating piezoelectric ceramics such as powders and fibers embedded in a passive polymer matrix have been developed in different configurations to meet specific requirements for instance fitting the transducer to the shape of concave or convex components. Today, among many others, most prominent applications of piezoceramics are in medicine as ultrasonic devices, in time measurement (quartz clock) and power engine technology for fuel injection. New applications in micro energy harvesting have been tested, but further engineering work is necessary to achieve technical and economic breakthroughs.

3.2 Material Processing

Single crystals such as quartz, $LiNbO_3$, $GaPO_4$ or Langasite ($La_3Ga_5SiO_{14}$) are less commonly used as piezoelectric devices than $Pb(Zr_xTi_{1-x})O_3$ (PZT) ceramics, but there are, nevertheless, certain applications in which they are useful, for instance those involving high frequencies or requiring resistance to high temperatures. Typically, the single crystals are grown by the Czochralski method or in the case of quartz, hydrothermally. Single crystals are not suitable for applications with large displacements at high frequencies because of crack formation.

The process for producing polycrystalline piezoceramic materials generally comprises two steps. After the preparation of the ceramic powder by the mixed oxide method, which involves firing a mixture of oxide powders (calcination) and then milling into fine powders, the ceramic is then sintered in the desired shape. In the preparation of $Pb(Zr_x Ti_{1-x})O_3$ the oxide powders PbO, ZrO_2, and TiO_2 are weighed out in appropriate proportions, mixed and then calcined at 800–900°C for 1–2 h. The milling step often leads to undesirable particle size distributions and a contamination of the powder by the milling media.

In the sintering process, the calcined powders are generally mixed with an appropriate binder and formed into the desired shape by pressing or extrusion or other casting methods. This pressed green body is subjected to a low temperature "burn-out process in order to volatilize the binder from the body. The final firing process at high temperatures (approx. 1200°C for 16 h for PZT) where measures have to be taken to prevent loss of PbO in which the ceramic attains its optimum density is called *sintering*. This process promotes diffusion of the constituent atoms on the particulate surfaces, which leads to crystal bonding. As a result, the ceramic body acquires the required mechanical strength while retaining its intended shape as it uniformly shrinks. More details in PZT ceramics preparation can be found in literature [1, 2].

More compositionally homogeneous ceramics may be produced by wet chemical methods such as the *alkoxide* or *sol-gel* methods. When metal alkoxides are mixed with alcohol in appropriate proportions and water is added, the hydrolytic reaction produces alcohol and a metal oxide or hydrate. This *sol-gel* method can produce very fine (particle size < 1 μm), high purity (99.98%) powders.

Polycrystalline piezoceramics have to be polarized in an electric field in order to align the electric diploes, thus enhancing the piezoelectric properties of the material. Doping the piezoceramic material with small amounts of impurities can significantly improve its properties. Currently, much research effort is being devoted to optimizing the properties by selecting appropriate doping formulas.

The most well-known difference is the distinction between the so-called hard and soft piezoceramic actuators. A material with a coercive field greater than 1 kV/mm is called a *hard piezoelectric* and a material with a coercive field between 0.1 and 1 kV/mm is called a *soft piezoelectric*. Soft PZT exhibits high induced strain and low hysteresis and is more prone to depolarization than hard PZT, which exhibits relatively low induced strains.

Tensile stresses can initiate crack formation which may cause fracture of the brittle ceramics. The maximum tensile stress of piezoelectric bulk material is 10 MPa; as a general rule it is recommended that tensile stress should be avoided. The maximum compressive stress is typically 10 times greater.

Large sheets of polyvinylidene difluoride containing trifluoroethylene (PVDF-TRFE) polymers can be manufactured and thermally formed into complex shapes. Crystallization from the melt produces the nonpolar α-phase, which can be converted into a stable polar β-phase by stretching on one or two axes. The resulting dipoles are then aligned by electrically poling the material above T_g. Due to their low piezoelectric strain coefficient and high g-constant (see Section 3.5), such materials are ideal for use in microphones or ultrasonic hydrophones.

For the electrical connection of the piezoelectric material, appropriate electrode materials and their related manufacturing processes have to be chosen. This is especially true for the manufacturing of complex piezoceramic actuators by cofiring. The most commonly used process employs silver-palladium, which is sputtered or printed as a polymer paste on the piezoelectric device.

The piezoelectric effect of any chosen material is limited by its Curie or phase transition temperature. For PZT, the Curie temperature varies between 250 and 400°C depending on its composition. New single crystal materials such as Langasite show a phase transition temperature up to 1400°C. Applications should be restricted to well below this temperature. Therefore, stacks of polycrystalline ceramic discs are used as actuators.

A simple classification of piezoelectric devices can be derived from the structure of the transducer. Those comprising only ceramic materials include simple disks or multilayer actuators that make use of the longitudinal strain (see Fig. 3.1).

Figure 3.1 1 PZT disc with silver electrodes on both sides (thickness 0.4 mm, diameter 30 mm).

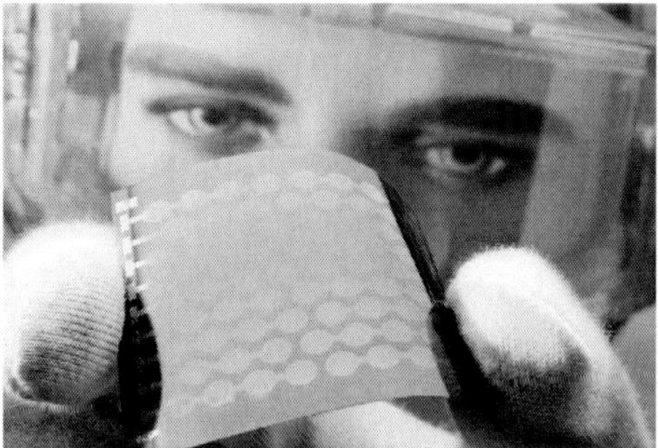

Figure 3.2 Flexible PZT thin film on a stainless steel substrate for an ultrasonic device.

More complex devices can be created by depositing or gluing the piezoceramic onto the surface of a flexible substrate to magnify the actuation displacement. Figure 3.2 shows a piezoelectric thin film (thickness 2 microns) on a stainless steel substrate (thickness 100 microns) which may be used to produce uni- or bimorph actuators or large area sensor/energy transducers. Piezoelectric composites, comprising a piezoelectric ceramic (as a powder, thin film, or fibers) and a polymer matrix are interesting devices for sensor and actuator applications. In active fiber composites (see Fig. 3.3), piezoelectric fibers are arranged in an epoxy resin. The composite structure is then sandwiched between interdigital electrodes (printed silver

Figure 3.3 Piezoelectric fiber composites in different shapes.

paste), which are coated with Kepton or epoxy resin for electrical insulation. Their high electromechanical coupling factors, good acoustic impedance, low thickness, and mechanical flexibility make these transducers useful for ultrasonic and in-plane stress sensing. They are also well suited for active vibration reduction applications, especially when incorporated in fiber-reinforced polymer devices.

3.2.1 *Physical Phenomena*

This section deals merely with the fundamental aspects of piezoelectricity: What is the physical phenomenon behind it and how can the effect be described? For a more detailed appraisal of piezoceramic transducers, please refer to the literature. Uchino and Giniewicz [3] provide a full and comprehensive explanation of the physical properties of piezoelectrics.

A necessary condition for a material to exhibit piezoelectricity is that its crystal structure lacks a center of symmetry. These materials possess intrinsic polarity as they are constituted of oriented dipoles. Subjecting piezoelectric materials to a mechanical stress or an electrical field changes the distance between the positive and negative dipoles. This leads to a net polarization and therefore an electrical charge or a macroscopic change of the dimensions of the transducer. Crystalline materials that show spontaneous polarization are called polar crystals. The magnitude of the polarization depends largely on the environment temperature. This is called the *pyro-electric effect.* Pyro-electric crystals whose polarization can be reversed by an electric field are called *ferroelectric crystals.* The term ferroelectric is used because of its analogy with ferromagnetism. Crystals that demonstrate the ferroelectric effect exhibit a strain linearly proportional to the applied electric field. When the strain is quadratically proportional to the applied electric field, the phenomenon is defined as *electrostriction.*

Most piezoelectric materials are crystalline or natural or synthetic monocrystals or polycrystalline materials such as ferroelectric ceramics. The most commonly used monocrystal is quartz. Quartz oscillators are used as frequency-stabilized clocks in computers, etc. The use of piezoelectric materials has significantly augmented since the discovery of piezoelectric ceramics. At the moment, piezoelectric ceramics are the most widely used piezoelectric materials as opposed to the traditional monocrystallines. The most well-known piezoceramic is lead zirconate titanate (PZT). Figure 3.4 shows the phase diagram of the lead zirconate titanate $Pb(Zr_x Ti_{1-x})O_3$ solid solution system with the temperature dependent morphotropic phase boundary between the tetragonal and rhombohedral phase.

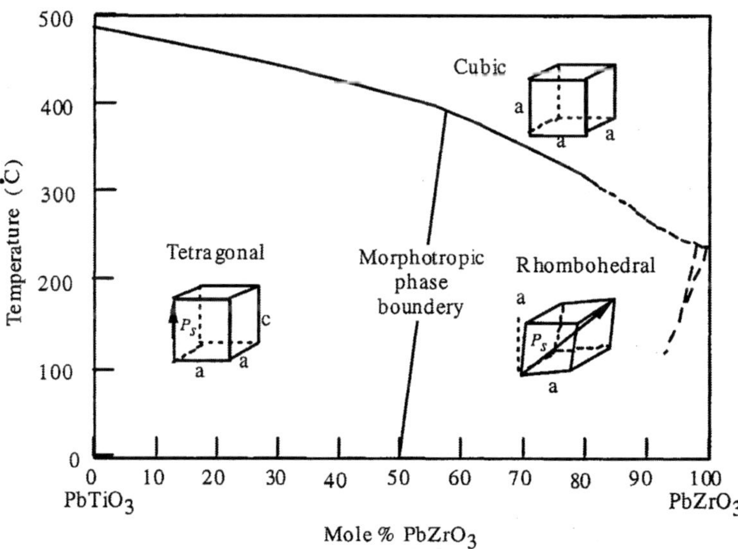

Figure 3.4 Phase diagram of the lead zirconate titanate $Pb(Zr_x Ti_{1-x})O_3$ system [1].

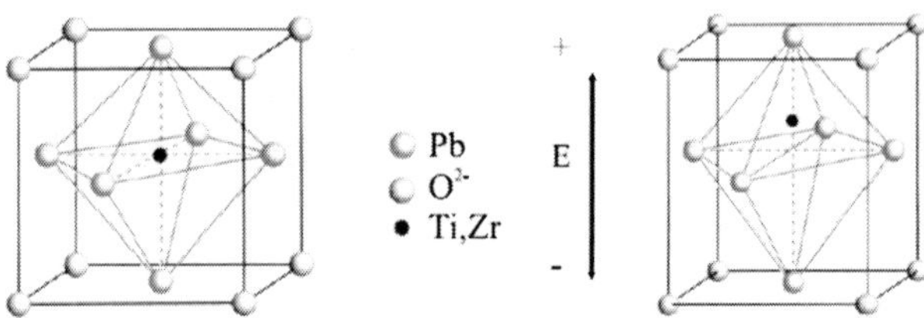

Figure 3.5 Crystal structure of PZT: above the Curie temperature T_C, the lattice is symmetric (a) and under T_C the structure is anisotropic (b)[4].

Other frequently used piezoceramic materials are lead metaniobate (PN) and lead nickel niobate (PNN). PZT has a relative maximum strain of 0.2% and is the most frequently used material for industrial applications. Piezoelectric ceramics can be seen as a large collection of randomly oriented small crystallites. After sintering, the ceramic is—from a macroscopic point of view— isotropic and does not show any piezoelectric effect due to the arbitrary orientation of the crystallites. To make the material macroscopically piezoelectric, the ceramic has to be polarized.

As mentioned before, the anisotropic structure of PZT crystals is the functional origin of the piezoelectric effect. Note that this anisotropy is only present below a characteristic temperature for the material in question, called the Curie temperature T_C. Below this temperature, PZT crystals have a permanent dipole moment.

The size of this dipole moment is related to the dimensional anisotropy of the PZT-perovskite crystal. Above the Curie temperature T_C the PZT crystal forms a symmetrical cube around a defined center (e.g., center-symmetric cubic), as can be seen in Fig. 3.5a. Cooling results in the formation of anistropic tetragonal crystals, as shown in Fig. 3.5b. In order to minimize the elastic energy, adjacent crystals orient themselves along to equal crystallographic axes. This creates the so-called Weiss domains, in which the material is aligned along parallel dipole moments. In non-polarized PZT materials, the Weiss domains are arbitrarily oriented, as illustrated in Fig. 3.6(①). The arbitrary orientation cancels out any net polarization. By applying a strong electric field, typically around > 3 kV/mm the electric dipoles align (②). When the electric field is removed, the dipoles tend to remain in their original orientation, resulting in

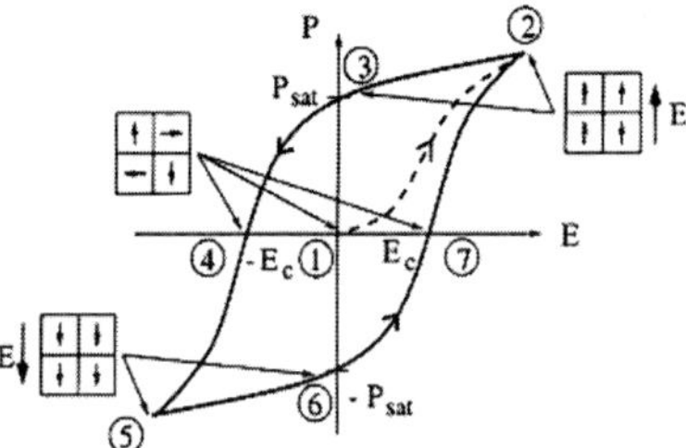

Figure 3.6 Hysteresis loop of PZT ceramic.

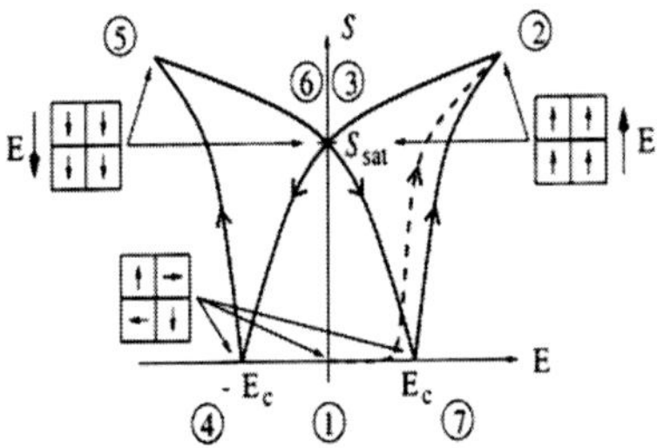

Figure 3.7 Strain–electrical field curve of PZT ceramic

remanent polarization (③). The application of a coercive electrical field in the opposite direction causes all dipoles to return to their original arbitrary oriented phase (④). The application of an electric field sufficiently strong to orient all dipoles in the opposite direction leads to saturation (⑤). The remanent polarization in the opposite direction (⑥) will be overcome when a positive electrical coercitive field is applied (⑦). When an electric field is applied to the polarized piezoceramic, the alignment of the electrical dipole moments of the different Weiss domains is proportionally increased or decreased as shown in the hysteresis curve Fig. 3.6 This results in a dimensional change of the PZT material as shown in Fig. 3.7, the so-called *butterfly curve.*

The properties of all piezoceramic materials are temperature-dependent. The piezoelectric constant d (see Section 3.5) changes about 0.2%/K. This effect can be ignored when the transducer operates close to room temperature. Another constraint concerns the Curie point. The operating temperature should be lower

Table 3.1 Properties of most popular piezoelectric materials

	ρ [g/cm^3]	d_{33} [pC/N]	g_{33} [mVm/N]	ε_{33} [ε_0]	k_t	Y [GPa]	T_C [$^\circ$C]
Quartz	2.65	2.3	57.8	5	0.09	77	537 (phase transition)
BaTiO$_3$	5.85	190	12.6	1700	0.38	105	120
PZT 4	7.9	289	26.1	1300	0.51	80	328
PVDF-TrFE	1.8	33	340	6	0.30	3	102
1-3 Piezo fiber composite	3	289	75	400	0.40	20	328

than half the Curie temperature. Note that at high frequency the piezoceramic actuators produce heat due to hysteresis behavior. Heat loss plays a subordinate role in energy harvesting applications because of the low generated voltages.

Table 3.1 summarizes the typical physical properties of the most popular piezoelectric materials such as PZT, BaTiO$_3$, PVDF, quartz, and piezo fiber composites. Here, ρ is the material's density, d_{33} is the effective piezoelectric strain coefficient, g_{33} is the effective piezoelectric voltage coefficient, ε_{33} is the effective dielectric constant, k_t is the electromechanical coupling factor in thickness mode, Y is Young's modulus of elasticity and T_C the Curie temperature. For explanation of these parameters see Section 3.5.

3.2.2 Mechatronic Model

From a mechatronical point of view, a piezoelectric material is a transducer, converting electrical energy to mechanical energy and vice versa. A piezoelectric transducer can therefore be represented by an element with an electrical and mechanical gate. This is shown in Fig. 3.8. The electrical gate is defined by two parameters: the electrical field strength E [V/m] and the dielectric displacement D [C/m^2] The mechanical gate is represented by the mechanical stress T [N/m^2] and the mechanical strain S [m/m]. The relation between the gate parameters is mathematically described by the constitutive equations. Equation (3.1) uses the mechanical stress T and the electrical field E as independent variables and is called

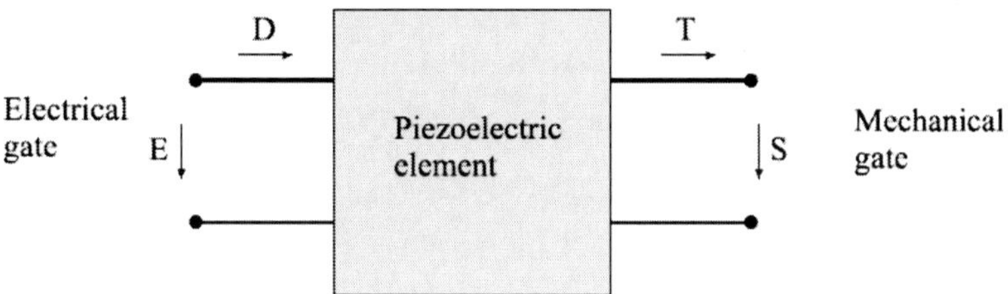

Figure 3.8 Representation of a piezoelectric element as a two-port element.

the d-formulation. This formulation is common for the reverse piezoelectric effect, when a piezoelement is used as an actuator.

$$\left.\begin{array}{l} S = s^E T + dE \\ D = dT + \epsilon^T E \end{array}\right\} - \text{d-formulation} \qquad (3.1)$$

Also commonly used is the g-formulation; see Eq. (3.2). Here the mechanical stress T and the dielectric displacement D are the independent variables. This formulation is common for piezoelectric sensors and energy harvesters.

$$\left.\begin{array}{l} E = -gT + \frac{D}{\varepsilon^T} \\ S = s^d T + gD \end{array}\right\} - \text{g-formulation} \qquad (3.2)$$

In these equations, the following material parameters are used:

d: The piezoelectric constant, expressed in m/V or C/N. This quantity is an important figure of merit for piezoelectric actuators because it represents their ability to convert an electrical voltage into a mechanical deformation.

s^E: The compliance measured with a constant electric field (the electrical voltage at the electrodes is constant), expressed in m/N.

ε^T: The dielectric permittivity measured when a constant mechanical load is applied, expressed in C/mV.

S^D: The compliance measured with a constant dielectric displacement (the electrical charge at the electrodes is kept constant), expressed in m/N.

g: The piezoelectric voltage coefficient, expressed in V m/N.

The latter quantity is a measure of the performance of piezoelectric sensors and energy transducers because it directly indicates their ability to convert the mechanical stress into an electrical voltage. As seen in Table 3.1, the piezoelectric polymer PVDF has the highest piezoelectric voltage constant. However, the low Curie temperature limits many technical applications.

The piezoelectric constant d has two related meanings. It represents the proportion of the strain S to the applied electrical field E (*converse piezoelectric effect*), but also the proportion of the dielectric displacement D to the applied mechanical tension T (*direct piezoelectric effect*). In other words, the piezoelectric constant d translates the reciprocity of the piezoelectric effect.

Since piezoelectric materials are anisotropic, their physical constants are tensor quantities. The mechanical and electrical gate variables are represented by vectors; see Eq. (3.3).

$$\mathbf{S} = \begin{bmatrix} S_1 \\ S_2 \\ S_3 \\ S_4 \\ S_5 \\ S_6 \end{bmatrix}, \ \mathbf{T} = \begin{bmatrix} T_1 \\ T_2 \\ T_3 \\ T_4 \\ T_5 \\ T_6 \end{bmatrix}, \ \mathbf{D} = \begin{bmatrix} D_1 \\ D_2 \\ D_3 \end{bmatrix}, \ \mathbf{E} = \begin{bmatrix} E_1 \\ E_2 \\ E_3 \end{bmatrix} \qquad (3.3)$$

For the vectors of the strain $\mathbf{S}$ and the stress $\mathbf{T}$, the indices 1,2,3 represent the x-, y-, and z-directions. The indices 4, 5, and 6 represent, respectively, the shear strain and shear stress. For example, S_4 indicates a shear strain in the y–z plane. The z-direction (3) is taken as the polarization direction. Figure 3.9 illustrates the index notation. The piezoelectric material parameters depend on the relative directions of the applied electrical field, the dielectric displacement, the mechanical stress and the elastic

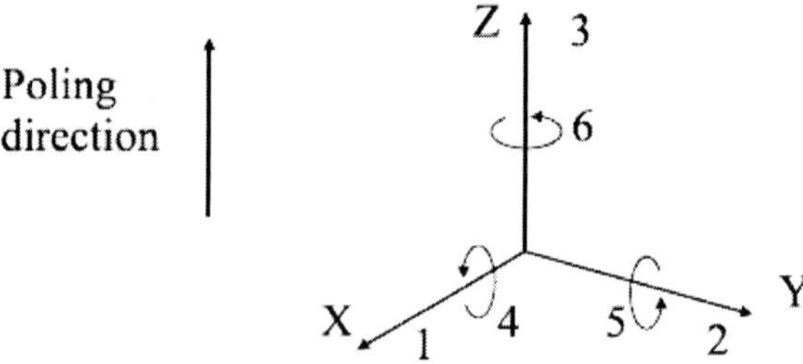

Figure 3.9 Used convention for the index notation.

strain. Due to the symmetry of the crystal structure, a reduced set of piezoelectrical, elastical, and dielectrical coefficients can be constructed. Specifically for polarized piezoelectrical ceramics, only six independent elastic constants, three piezoelectric constants, and two dielectric constants are sufficient to describe the behavior of the piezoceramic material. Equations (3.4) to (3.6) give the three matrices that are sufficient to fully describe a piezoelectric ceramic such as PZT.

$$
s^E = \begin{bmatrix} s_{11}^E & s_{12}^E & s_{13}^E & 0 & 0 & 0 \\ s_{12}^E & s_{13}^E & s_{13}^E & 0 & 0 & 0 \\ s_{13}^E & s_{13}^E & s_{33}^E & 0 & 0 & 0 \\ 0 & 0 & 0 & s_{44}^E & 0 & 0 \\ 0 & 0 & 0 & 0 & s_{66}^E & 0 \\ 0 & 0 & 0 & 0 & 0 & s_{66}^E \end{bmatrix}, \tag{3.4}
$$

$$
d = \begin{bmatrix} 0 & 0 & d_{31} \\ 0 & 0 & d_{31} \\ 0 & 0 & d_{33} \\ 0 & d_{15} & 0 \\ d_{15} & 0 & 0 \\ 0 & 0 & 0 \end{bmatrix}, \tag{3.5}
$$

$$
\varepsilon^T = \begin{bmatrix} \varepsilon_{11}^T & 0 & 0 \\ 0 & \varepsilon_{11}^T & 0 \\ 0 & 0 & \varepsilon_{33}^T \end{bmatrix} \tag{3.6}
$$

The constitutive equations result in static relations between physical variables that can be used to characterize the piezoelectric material. The variables of relevance to practical applications engineering are force, speed, electrical current, and voltage. When a voltage is applied, the piezoceramic element changes the dimensions such as an actuator. Similarly, the application of a mechanical force generates an electric charge. If the electrodes are not short-circuited, a voltage associated with the charge appears. Due to this effect, a piezoelectric element can be used as a sensor or an energy transducer.

All the piezoelectric parameters are either related to, or a function of, the direction of the mechanical or electrical excitation. Generally, if a piezoelectric transducer is assumed to expand in only one direction, the constitutive equations for a one-dimensional excitation and deformation of a piezoceramic element are less complex. Following the use of subscripts and superscripts in piezoelectricity, Eq. (3.2) defined for a piezoelectric sensor or energy harvester can be expressed more specifically as

$$S_3 = s_3^D T_3 + g_{33} D_3 \tag{3.7}$$

$$E_3 = -g_{33} T_3 + D_3/\varepsilon_{33}, \tag{3.8}$$

where D is electric density or flux density [C/m^2], E the Electric field [V/m], T the mechanical stress [N/m^2], S the mechanical strain [m/m], ε the dielectric permittivity of the material [F/m $=$ C^2/Nm^2], d the piezoelectric d-constant [m/V $=$ C/N], and s the mechanical compliance (inverse of Young's modulus) [m^2/N].

Starting from the above expressions, it can be demonstrated that the constitutive equations for the piezoelectric transducer are as follows:

$$Q_{piezo} = C_{piezo} V_{piezo} - d_{33} F_{piezo} \tag{3.9}$$

$$z_{piezo} = d_{33} V_{piezo} - \frac{1}{\kappa_{piezo}} F_{piezo}, \tag{3.10}$$

where κ_{piezo} is the stiffness of the piezoelement, F_{piezo} the acting force, z_{piezo} the displacement of the respective direction, Q_{piezo} the charge, and V_{piezo} the applied or generated voltage.

Considering the electrical energy produced by the electro-mechanical response of the piezoelectric material with respect to the mechanical energy supplied to the material, the so-called *electro-mechanical coupling factor k* is defined according to Eq. (3.11):

$$k^2 = \frac{[\text{Stored electrical energy}]}{[\text{Input mechanical energy}]} \tag{3.11}$$

The stored electrical energy per unit volume can be written as

$$\text{Stored electrical energy} = \frac{1}{2}\varepsilon_0 \varepsilon^T E^2, \tag{3.12}$$

where ε_0 is the dielectric constant in vacuum (8.85 $\times$ 10^{-12} C/Vm), ε^T is the relative dielectric permittivity of the piezoelectric transducer and E is the electrical field. The input mechanical energy per unit volume is given by Eq. (3.13), where S is the mechanical strain, s is the mechanical compliance, and d is the piezoelectric constant.

$$\text{Input mechanical energy} = \frac{1}{2}\frac{S^2}{s} = \frac{1}{2}\frac{(dE)^2}{s}. \tag{3.13}$$

Making these substitutions (Eq. (3.12) and Eq. (3.13)) into equation Eq. (3.11) gives

$$k^2 = \frac{d^2}{\varepsilon_0 \varepsilon^T s}. \tag{3.14}$$

Note that all the above equations are only valid for static events, e.g., when a constant force is acting on the piezoelectric transducer. In practice, most electrical energy can be obtained by applying a dynamic strain.

3.3 Power Conversion

The power conversion from ambient vibration energy to electrical energy is a crucial point for a successful design of energy harvesting devices. Hence, in this section, an electromechanical model is presented and the power conversion is analyzed.

In the last few years, several researchers from different disciplines have studied modeling of piczoelectric energy harvesters. Different methods for the modeling were proposed based on their various backgrounds. In general, a piezoelectric energy harvester is often modeled as a driven, damped vibrating system. This structure consists of a piezoelectric transducer coupled with the mechanical structure and connected to an energy storage system by an energy harvesting circuit. The model of an electromechanical subsystem can be obtained by means of finite element method, modal analysis, or analytical methods [5–7]. In either case, the analysis yields a system of N ordinary differential equations:

$$\mathbf{M}\ddot{\mathbf{q}} + \mathbf{B}\dot{\mathbf{q}} + \mathbf{K}\mathbf{q} = \mathbf{F}, \tag{3.15}$$

where

$$\mathbf{M} = \begin{pmatrix} M_{uu} & \mathbf{0} \\ \mathbf{0} & \mathbf{0} \end{pmatrix}, \quad B = \begin{pmatrix} B_{uu} & \mathbf{0} \\ \mathbf{0} & \mathbf{0} \end{pmatrix}, \quad \mathbf{K} = \begin{pmatrix} K_{uu} & K_{u\phi} \\ K_{\phi u} & K_{\phi\phi} \end{pmatrix}$$

are the global mass matrix, the global damping matrix, and the global stiffness matrix of the system, respectively. The sub-matrix K_{uu} is the stiffness matrix of the purely mechanical system and $K_{\phi\phi}$ holds the electrical term. The coupling between both domains is described by the sub-block $K_{u\phi} = K_{\phi u}^{T}$. The load vector is **F** and

$$\mathbf{q} = \begin{pmatrix} u \\ \phi \end{pmatrix} \tag{3.16}$$

is a vector of unknown degree of freedom. This vector is the concatenation of variables which describe mechanical and piezo-electric degrees of freedom, respectively. For example, these are the mechanical displacement u and the electric potential ϕ when the finite element formulation [5] is applied.

For further analysis, the matrix notation is not convenient. A better approach is to use the network analysis because it is well established for the development of efficient electrical power conversion and storage systems. In this context, the energy harvesting system has to be described in terms of a two-port (see Fig. 3.8) model. For this reason, the system of differential equations is reduced to a single-degree-of-freedom model. Due to the fact that an energy harvesting device is often tuned to a certain natural frequency such a device can be modeled as a single-degree-of-freedom mass-spring-damper system [8–10], without loss of generality. This equivalent system is schematically shown in Fig. 3.10. The governing equations of this oscillator are

$$m\ddot{u}(t) + b\dot{u}(t) + ku(t) + \Theta V(t) = F(t) \tag{3.17}$$

$$-\Theta\dot{u}(t) + C_{\mathrm{p}}\dot{V}(t) = -I(t), \tag{3.18}$$

where u is the displacement of the mass m and k is the total stiffness of the piezo transducer and any other stiffness connected

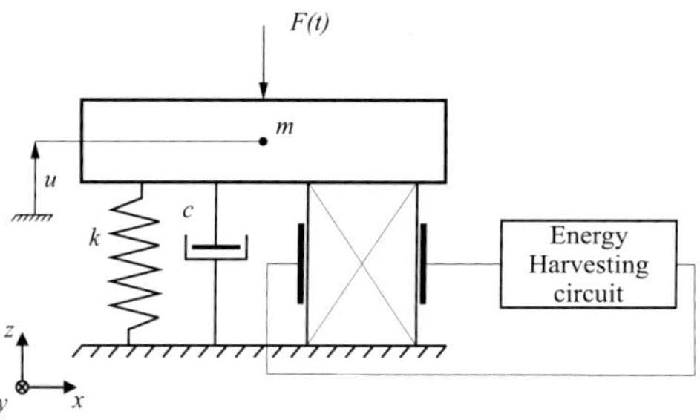

Figure 3.10 Schematic of the force-excited oscillator.

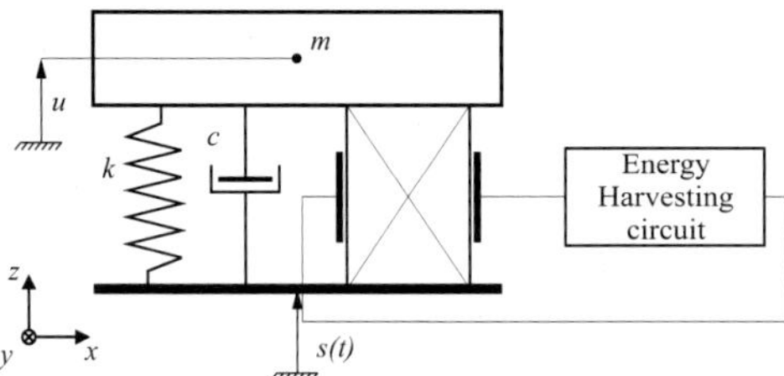

Figure 3.11   Schematic of the base-excited oscillator.

in parallel or series. To have a description that is more convenient for the subsequent electric analysis, the lower part of Eq. (3.17) was integrated with respect to the time. Now it can be expressed in terms of the electric current $I(t)$ and the voltage $V(t)$. The effective piezoelectric coefficient Θ and the capacitance C_p depend on the geometry of the transducer and the load direction. For stack-like transducers that are poled and loaded along their packing direction, $\Theta = \frac{d_{33}}{s_{33}^E}\frac{An}{l}$ and $C_\mathrm{p} = (s_{33}^3\varepsilon_{33}^T - d_{33}^2)\frac{An^2}{ls_{33}^E}$. Here A is the cross-sectional area, l is the length, and n is the number of layers of the stack.

Frequently, the energy harvesting system is applied on top of a vibration mechanism and in this configuration the modeling of a force excitation is not appropriate. So far, for this case a base excitation has to be taken into consideration instead. The schematic of the system is depicted in Fig. 3.11. The system-governing equations have to be modified as well. However, they can be transformed to an analog representation such as Eqs. (3.17) and (3.18):

$$m\ddot{q}(t) + b\dot{q}(t) + kq(t) + \Theta V(t) = -m\ddot{s}(t) \qquad (3.19)$$

$$-\Theta\dot{q}(t) + C_\mathrm{p}\dot{V}(t) = -I(t) \qquad (3.20)$$

Here $q(t) = u(t) - s(t)$ is the difference displacement of the mass m and the excitation of the basis. For a base excitation, the right hand side of Eq. (3.19) yields $-m\ddot{s}(t)$, which is a d'Alembert force induced by the acceleration of the ground.

The energy equation of the system is obtained by multiplying Eq. (3.17) by the velocity $\dot{u}$ and Eq. (3.18) by the voltage V. The integration of these two equations with respect to time yields a presentation of energy transformations in a system

$$\frac{1}{2}m\ddot{u}^2 + b\int \ddot{u}^2 dt + \frac{1}{2}ku^2 + \Theta \int V\dot{u}\,dt = \int F(t)\dot{u}\,dt \qquad (3.21)$$

$$\Theta \int V\dot{u}\,dt - \frac{1}{2}C_\mathrm{p}V^2 = \int V(t)I(t)dt, \qquad (3.22)$$

where $\int F(t)\dot{u}(t)dt$ is the provided ambient energy and the terms $\frac{1}{2}m\dot{u}^2$ and $\frac{1}{2}ku^2$ are the kinetic and the potential energy, respectively. The mechanical losses are represented by $b\int \dot{u}^2 dt$. The term $\int V(t)I(t)dt$ includes the energy absorbed by a connected electrical device, which is the sum of the energy stored in the piezoelectric transducer $\frac{1}{2}C_\mathrm{p}V^2$ and the transferred energy $\Theta \int V\dot{u}\,dt$. The equation of energy balance is obtained and terms of conservative energy are removed from the equations Eqs. (3.21) and (3.22). After combining both equations we see that the input energy is either converted into mechanical losses or powering the connected energy harvesting circuit.

$$\int F(t)\dot{u}\,dt = b\int \dot{u}^2 dt + \int V_p(t)I(t)dt \qquad (3.23)$$

Hence, it is a challenging task when designing an energy harvesting device to find a setup where the ratio of serviceable electrical energy to the mechanical losses is well balanced. In the following sections, analyses are presented to gain a better comprehension of the power conversion in piezoelectric transducers.

However, before the analysis is carried out, a few fundamentals about electrical and mechanical power have to be discussed. In general, power is the rate at which energy is converted. Hence, it is defined as the flow of energy passing a given point. For the system considered here, all electrical loads are linear and current and voltage are sinusoidal. Therefore, the power flow depends on the relation of current and voltage. When the load is purely resistive, then both quantities are in phase, which means that they reverse their polarity at the same time. Since the power is the product of voltage and current

$$S_e(t) = V(t)I(t) = \sqrt{P_e^2 + Q_e^2} \qquad (3.24)$$

the direction of the energy flow does not reverse. In this case, only real power P_e flows and energy is transferred to the load. Reactive loads such as ideal inductors and capacitors dissipate zero energy; they just store it. Hence, the peaks of voltage are reached at the times when the current crosses zero. In this case, the current and voltage are out of phase. However, this elements draw current and drop voltage, but the portion of stored energy returns to the source in each cycle. This rate of energy is known as reactive power Q_e. Transforming all quantities to the frequency domain gives the power equation with respect to frequency

$$\underline{S_e} = \underline{V}\,\underline{I}^* = P_e + jQ_e. \tag{3.25}$$

Mechanical systems have dissipating and conserving elements as well. Comparing the governing differential equation of both domains, one can find an analogy (see [11]) existing between electrical and mechanical systems. According to that, there exists a mechanical real power, P_e, which characterizes the energy transferred to elements such as damper, and a reactive power Q_e, which identifies the energy that is conserved as parts of the system

$$\underline{S_m} = \underline{F}\,\underline{v}^* = P_\mathrm{m} + jQ_\mathrm{m}. \tag{3.26}$$

However, for energy harvesting, only the real power is relevant. Consequently, this fraction is considered in the following sections.

3.4 Impedance of the Electric Network

From Eq. (3.22), we know that the provided power depends on the current $I(t)$ flowing into this circuit and its relation to the voltage $V(t)$. Since current and voltage depend on the design of the connected electrical network, we have to define an electric load. In a first analysis, a purely resistive and inductive electrical shunt (Fig. 3.12) is examined to demonstrate the power conversion. The generated voltage therefore depends on the load R and the inductance L. It is assumed that the system is driven by an external excitation force, which is sinusoidal with a frequency close to the natural frequency of the system with the loaded piezoelectric element. Under steady-state operation conditions, the solution to differential equation $u(t)$, $s(t)$, $F(t)$, and $V(t)$ has the following

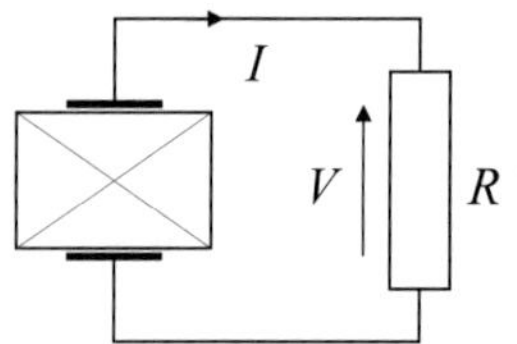

Figure 3.12 Standard AC device.

forms:

$$u(t) = \underline{u}e^{j\omega t}, \quad s(t) = \underline{s}e^{j\omega t}, \quad F(t) = \underline{F}e^{j\omega t}, \quad V(t) = \underline{V}e^{j\omega t} \tag{3.27}$$

Replacing the current with respect to the connected network

$$\underline{Z} = \frac{\underline{V}}{\underline{I}} = R + j\omega L \tag{3.28}$$

and substituting Eq. (3.27) into Eq. (3.18) yields the voltage of the piezoelectric element.

$$\underline{V} = \frac{\Theta \left(R + j\omega L\right) j\omega}{1 + C_{\mathrm{p}} \left(R + j\omega L\right) j\omega} \underline{u} \tag{3.29}$$

When Eq. (3.17) is written in the frequency domain as well, this gives an expression linking force, displacement, and voltage

$$\underline{F} = \left(-\omega^2 m + j\omega b + k\right)\underline{u} + \Theta\underline{V}. \tag{3.30}$$

To simplify this equation, we introduce the dimensionless frequency $\eta = \frac{\omega}{\omega_0}$ where $\omega_0^2 = \frac{k}{m}$ is the angular frequency of the undamped oscillator. The systems mechanical losses are represented by the damping ratio $D = \frac{b}{2m\omega_0}$.

$$\underline{u} = \frac{1}{1 + 2Dj\eta - \eta^2} \frac{\underline{F} - \Theta\underline{V}}{k} \tag{3.31}$$

Now the voltage is substituted using Eq. (3.29) and the alternative electromechanical coupling coefficient $c^2 = \frac{\Theta^2}{kC_{\mathrm{p}}}$ is introduced. This quantity is not to be confused with the electromechanical coupling coefficient k_{33} (see Eq. (3.11)). The two coupling coefficients are related through

$$c^2 = \frac{k_{33}^2}{1 - k_{33}^2} = \frac{d_{33}^2}{c_{33}^E \varepsilon_{33}^E}. \tag{3.32}$$

With the normalized resistance $r = C_\mathrm{p} R \omega_0$ and the inductance $l = C_\mathrm{p} L \omega_0^2$, it is possible to define the dimensionless impedance $\underline{z} = r + j\eta l$. Putting all this together yields the frequency response function

$$H_f(\eta, D, \underline{z}, c) = \frac{\underline{u}}{\underline{u}_0} = \frac{1}{1 + 2Dj\eta - \eta^2 + \frac{c^2 \underline{z} j \eta}{1 + \underline{z} j \eta}} \qquad (3.33)$$

of the force-excited energy harvesting system. Here, $\underline{u}_0 = \frac{F}{k}$ is the static deflection of the system. Note that this equation is similar in appearance to the mechanical system [12] aside from the term $\frac{c^2 \underline{z} j \eta}{1 + \underline{z} j \eta}$. The frequency response function of the base-excited harvesting device is derived analogously thereby the inertial force is set to $\underline{F} = m\omega^2 \underline{s}$ in the previous equation:

$$H_b(\eta, D, \underline{z}, c) = \frac{\underline{q}}{\underline{s}} = \frac{\eta^2}{1 + 2Dj\eta - \eta^2 + \frac{c^2 \underline{z} j \eta}{1 + \underline{z} j \eta}}. \qquad (3.34)$$

Applying the dimensionless impedance and the alternative electro-mechanical coupling coefficient to Eq. (3.29) gives

$$\underline{V} = \sqrt{\frac{k}{C_\mathrm{p}}} \frac{c \underline{z} j \eta}{1 + \underline{z} j \eta} \underline{u}. \qquad (3.35)$$

Considering Eq. (3.29), the harvested power is determined, which is a function of the displacement amplitude, when the voltage $\underline{V}$ is multiplied by its conjugate $\underline{V}^*$ and divided by the conjugate of the networks impedance $\underline{Z}^*$:

$$\underline{S}_e = \frac{\underline{V}\,\underline{V}^*}{\underline{Z}^*} \qquad (3.36)$$

3.4.1 *Weak Coupling*

Piezoelectric devices are frequently modeled as current sources. The capacitance C_p of their internal electrodes is considered in parallel with the load resistance R as depicted in Fig. 3.13 Assuming that the internal current source is independent of the impedance from the external load, then the term ΘV can be dropped from Eq. (3.17). This assumption is equivalent to the statement that the coupling is very weak or does not exist. Hence this analysis is generally known as uncoupled analysis. The governing equations change to

$$m\ddot{u}(t) + b\dot{u}(t) + ku(t) = F(t) \qquad (3.37)$$

$$-\Theta \dot{u}(t) + C_\mathrm{p} \dot{V}(t) = -I(t). \qquad (3.38)$$

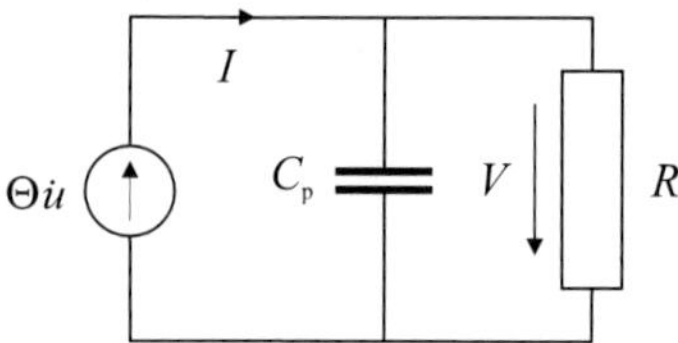

Figure 3.13 An equivalent circuit for the weak coupled model.

Now the displacement can be solved independently using a harmonic analysis, which provides the frequency response function of the mechanical mass-spring-damper oscillator:

$$H_f(\eta, D, \underline{z}) = \frac{\underline{u}}{u_0} = \frac{1}{1 + 2Dj\eta - \eta^2}. \tag{3.39}$$

Then $\Theta\dot{u}(t)$ is treated as a known current source. The provided voltage V and the average harvested power P_e are calculated by Eq. (3.29) and Eq. (3.36) and are as follows:

$$\underline{V} = \sqrt{\frac{k}{C_p}} \frac{crj\eta}{1 + rj\eta}\underline{u}, \tag{3.40}$$

$$P_e = \frac{c^2 r\eta^2}{\left(1 + r^2\eta^2\right)\left(\left(1 - \eta^2\right)^2 + (2D\eta)^2\right)} k\omega_0 |\underline{u_0}|^2. \tag{3.41}$$

For this model, the calculation of the optimal resistance for obtaining the maximal power is straightforward. Therefore, Eq. (3.41) is differentiated with respect to r, set to zero and solved. That gives the optimal resistance $R_{opt} = \frac{1}{C_p\omega}$. Since the amplitudes of the oscillator becomes maximal in the natural frequency, ω has to be set to be equal ω_0 to harvest the maximal power.

$$R_{opt} = \frac{1}{C_p\omega_0} \tag{3.42}$$

The surface of the power function is depicted in Fig. 3.14 Here the natural frequency is set to $f_0 = \omega_0/2/\pi = 40$ kHz. The capacitance of the piezo transducer, the mass, the force amplitude, and the damping and the effective piezoelectric coefficient are assumed as follows: $C_p = 1.5$ µF, $m = 10$ kg, $\underline{F} = 1$ N, $D = 2\%$ and $\Theta = 1$ N/V. It can be seen that in this plot the power has a line where the power becomes larger when the system

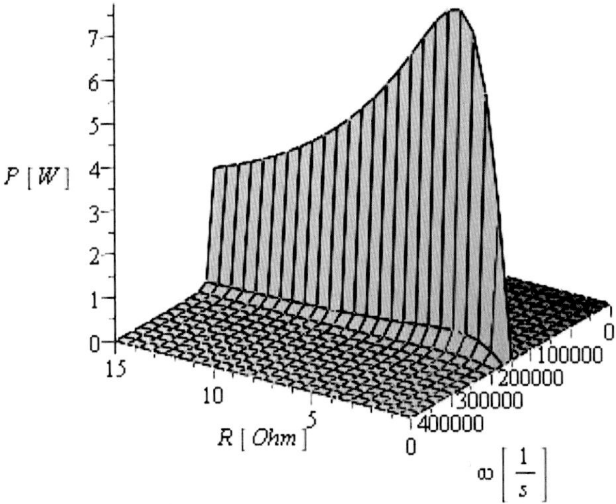

Figure 3.14 Power over angular frequency and resistance.

oscillates in its natural frequency. However, when the resistance of the shunt is set to $R = 2.65\ \Omega$ the obtained power becomes maximal. However, this is a basic model, which is only valid for weak coupling. However, for energy harvesting, it does not seem to be the optimal connection. Consequently, the influence of the dynamics terms effected by the piezo transducer's capacitance and other time depending dimensions should be included in the analysis.

3.4.2 *Optimal Resistance and Power*

Considering that the coupling has a relevant influence, then the piezo transducer's characteristics produce a shift of the natural frequency of the first mode, which depends on the connected network. Assuming that the attached circuit is a purely resistive load, there exist two limits of the magnitude of the resistance. First of all, the transducers electrodes may be shorted, then the shunted resistance approaches zero. On the other hand, the piezoelectrodes may be open and the resistance is approximately infinite. In both conditions, the system has an associated natural frequency. This is caused by the piezoelectric effect of the transducer. When the transducer is mechanically stressed, the piezoelectric material

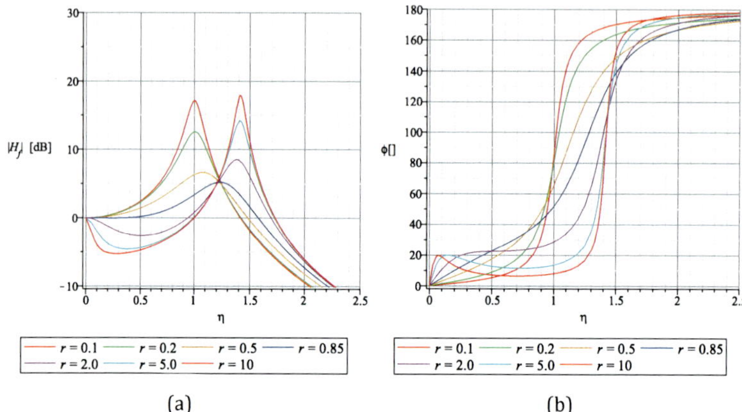

(a)   (b)

Figure 3.15 Bode plot of the base force oscillator.

produces an electrical charge. If this electric charge cannot be drained from the piezoelement, it produces an electrical response that counteracts to the resultant strain. As a result, the effective Young's Modulus with an open circuit piezoelectric is higher as the modulus of a shorted piezoelectric. This implies that the natural frequency of the first mode increases from the first to the second limit case and gives the two the frequency ratios

$$\eta_E = 1 \quad \text{and} \quad \eta_D = 1 + c^2, \tag{3.43}$$

where η_E is related to the shorted piezoelectric and η_D to the open one, respectively. This effect is depicted in Fig. 3.15 and Fig. 3.16 For the force- excited oscillator and the base excited, it can be seen that for both natural frequencies the displacement amplitude becomes maximal, when the systems are driven by a sinusoidal input force.

However, in the intermediate range, the resistance introduces damping. This effect becomes maximal when the amplitude is minimal.

One method to determine the optimal resistance for energy harvesting is to estimate the systems structural damping, which is small for a wide range of mechanical structures, and then to calculate the resistance where electric power becomes maximal. That is how the ratio of the mechanical input power to the electrical output power is optimized. Assuming the piezo is connected to a resistance, determining the electrical power is straightforward since no reactive

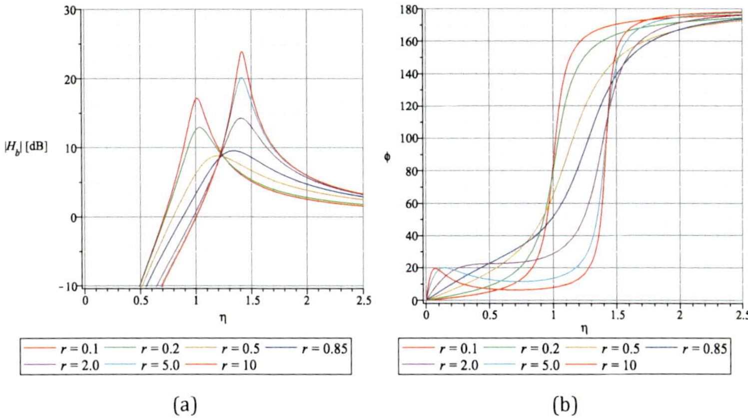

Figure 3.16 Bode plot of the base-excited oscillator.

power occurs. The equation for the real part of the power Eq. (3.36) looks as follows:

$$P_{\mathrm{e}} = \frac{|V|^2}{R} = \frac{c^2 r \eta^2}{1 + r^2 \eta^2} k\omega_0 |\underline{u}|^2. \tag{3.44}$$

The displacement is determined either by Eq. (3.33) or by Eq. (3.34) depending on which excitation is modeled. For the base excitation, the absolute displacement $\underline{u}$ has to be substituted with the relative displacement $\underline{q}$. The graphs of the electrical power with respect to the frequency are plotted in Fig. 3.17 It is quite evident that the harvestable power becomes maximal next to the natural frequency of the system because this power depends on the displacement. Furthermore, for a sinusoidal excitation the optimal resistance depends on the input frequency, because next to the frequency η_{D} high resistance derives more energy whereas lower resistances maximize the obtained power next to the frequency η_{E}. To determine the efficiency of the power conversion, the input power has to be considered next. For mechanical systems, this value is defined by $P_{\mathrm{m}}(t) = F(t)\frac{dx(t)}{dt}$. Substituting Eq. (3.27) into the previous equation leads to power representation in the frequency domain. Therefore, only the real part is considered since this is the fraction of the power that can be transformed into electrical power because the exciting force and the velocity of the system are in phase.

$$P_{\mathrm{m}} = \mathrm{Re}\,(j\eta H_{\mathrm{f}})\,\frac{\omega_0}{k}\underline{F}^2. \tag{3.45}$$

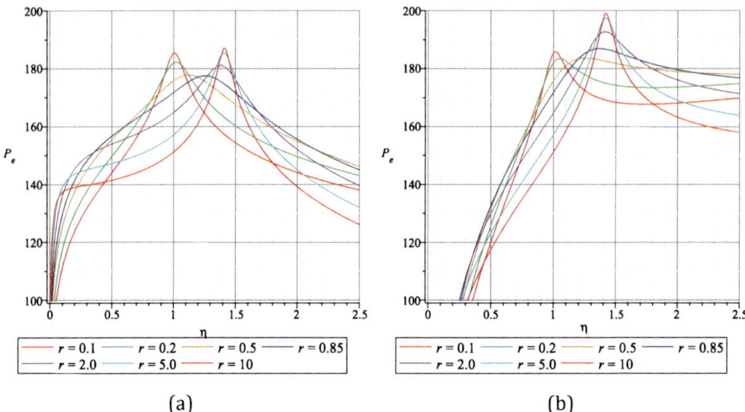

Figure 3.17 Electrical output power P_e (a) of the force-excited and (b) base-excited oscillator.

In this way, the previous equation describes the mechanical input power for the force-excited energy harvesting device. After applying a couple of simplification rules, the equation is changed to

$$P_m = \frac{\left(2D\eta + \frac{c^2 r\eta}{1+r^2\eta^2}\right)\eta}{\left(1 - \eta^2 + \frac{r^2\eta^2 c^2}{1+r^2\eta^2}\right)^2 + \left(2D\eta + \frac{c^2 r\eta}{1+r^2\eta^2}\right)^2} \frac{\omega_0}{k} \underline{F}^2. \qquad (3.46)$$

The mechanical power of the base-excited energy harvesting model can be treated in the same way, which gives

$$P_m = \mathrm{Re}\left(j\eta^3 H_b\right)\omega_0 k\underline{s}^2. \qquad (3.47)$$

To see how the power depends on the resistance and excitation frequency both power functions were plotted in Fig. 3.18 for different resistances. In each picture, the mechanical energy introduced into the system is obviously maximal when one of the natural frequencies is met. However, for low frequencies, no appreciable amount of energy supplied. To estimate the fraction of transformed power, the degree of efficiency has to be considered, which is the rate of electrical real power to mechanical real power

$$\eta_G = \frac{P_e}{P_m}. \qquad (3.48)$$

In Fig. 3.19, two sets of curves are shown, where the degree of efficiency is plotted with respect to the frequency. The plotted curves

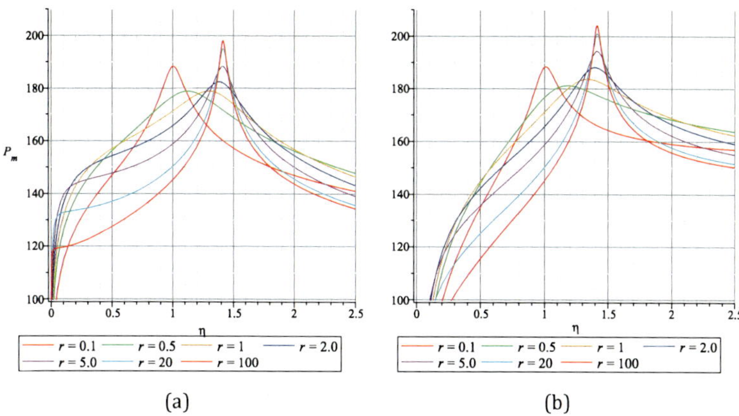

(a)   (b)

Figure 3.18 Mechanical input power P_m (a) of the force-excited and (b) base-excited oscillator.

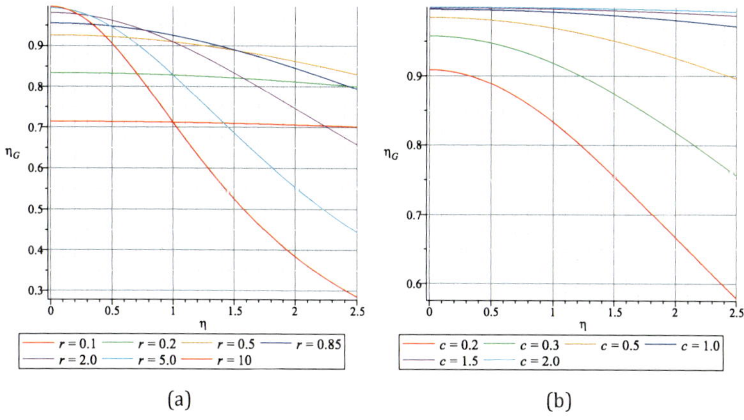

(a)   (b)

Figure 3.19 Degree of efficiency η_G of the base-excited oscillator (a) for a set of resistances r and (b) for a set of coupling coefficients c.

are obtained from an analysis of a base-excited device, but they are equivalent for the force excitation, because the dissimilar terms cancel out. The plot in Fig. 3.19(a) shows the degree of efficiency for a set of different values of the resistances. It can been seen that η_G decreases proportional to an increase in frequency, due to the mechanical damping losses. This effect should be kept in mind when

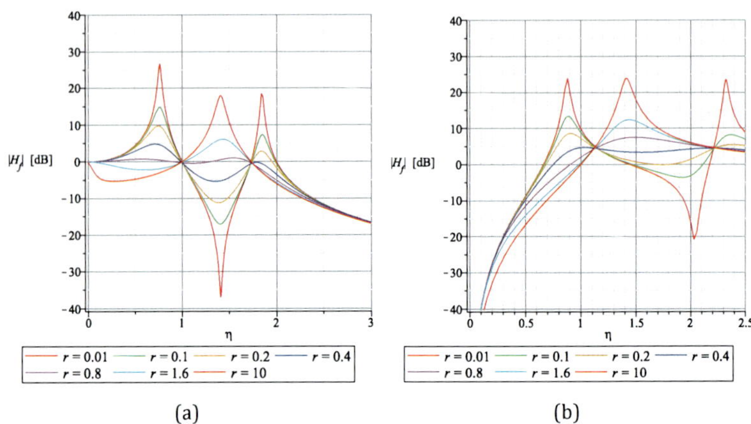

Figure 3.20   Frequency response function of the RL-shunt (a) for the force and (b) the base excitation.

designing energy harvesting systems. When loads are applied that are mainly resistive, the total resistance should be rather small. In Fig. 3.19(b) a set of curves shows the dependency on the coupling c. It is an obvious fact that for small values of c the rate of electrical to mechanical power is also small. As a result, the coefficient c should be maximized such that the piezoelectric coupling factor k is roughly 1.

To connect a purely resistive load to the piezo transducer is interesting for fundamental studies of the power conversion. However, current energy harvesting circuits have inductive and capacitive characteristics as well. Therefore, the analysis has to be extended due to the fact that these terms introduce new dynamic effects. When an inductance is connected in series to the resistance, these elements form in combination with the piezo transducer's capacitance a resonant circuit. This effect is presented in Fig. 3.20, where the frequency response functions of system excited by a force (Fig. 3.20(a)) and on the base (Fig. 3.20(b)) are depicted. The former natural frequency at $\eta = 1$ vanishes and two new frequencies occur, when r is relatively small. This dynamic effect is well known and often utilized for semi-active vibration absorbers. The value of the

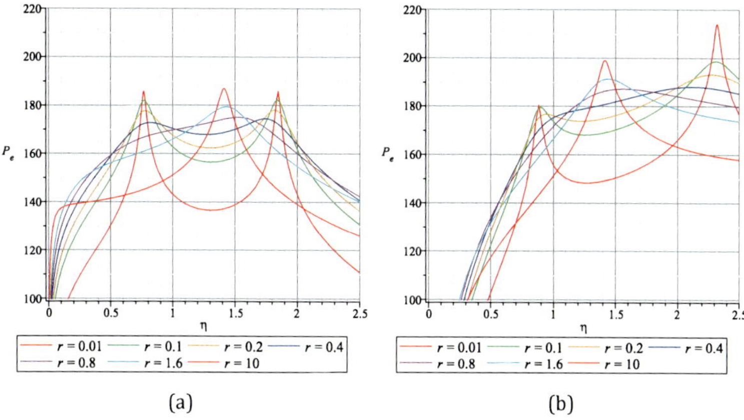

Figure 3.21 Electrical output power P_e (a) of the force-excited and (b) base-excited oscillator.

inductance is calculated by

$$\omega_e = \sqrt{\frac{1}{LC} - \left(\frac{R}{2L}\right)^2} \qquad (3.49)$$

so that the natural frequency of the network ω_e matches the natural frequency of the oscillator. For this network the generated power $\underline{S_e}$ of the shunt is determined from Eq. (3.36) which changes to

$$\underline{S_e} = \frac{c^2 \underline{z} \eta^2}{1 + \underline{z}^2 \eta^2} k\omega_0 |\underline{u}|^2. \qquad (3.50)$$

This equation considers the complex power of the system, but for energy harvesting only the in-phase part is useable. The characteristics of this real power P_e, which is the fraction converted in the resistor, is depicted in Fig. 3.21 The mechanical input power is calculated in the same way as before using Eq. (3.45) or rather Eq. (3.47). The run of the curves is shown in Eq. (3.22). Using these results, the degree of efficiency is computed. In contrast to the purely resistive load, the rate of output to input (see Fig. 3.23) does not decrease continuously with respect to the frequency. In Fig. 3.23(b) it can be observed again that for high coupling factors most of the energy is obtained. However, in both curves there is a single frequency where efficiency becomes maximal. This effect becomes more apparent when the

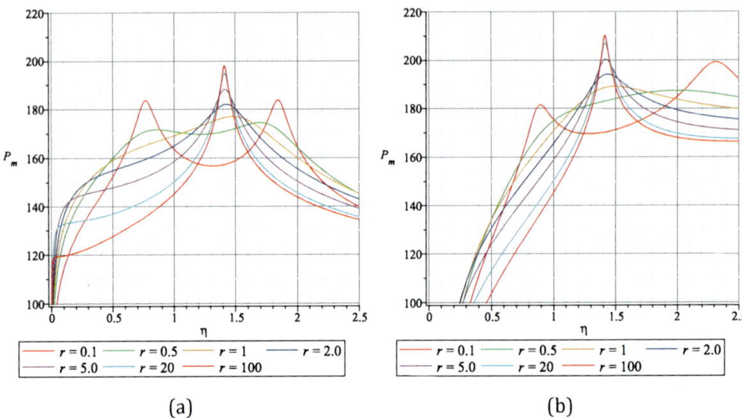

Figure 3.22 Mechanical input power (a) of the force-excited and (b) base-excited oscillator.

theory of tuned vibration absorbers is considered. As mentioned above, the electrical shunt is tuned to the natural frequency of the system. Hence, a new two-degree-of-freedom system results. This system has two natural frequencies, corresponding to its modes of vibration. When a relatively small resistance is chosen the oscillator becomes an undamped vibration absorber. See therefore the red line ($r = 0.01$) in Fig. 3.20(a). For such an absorber, both masses move in phase for the first natural frequency. In the second natural frequency the masses oscillate out of phase, which means that they move in opposite directions. Between these frequencies there exists a zero in the frequency response function. This is the tuned resonant frequency of the vibration absorber. When this device is applied to a structure, the original system stops moving and just the absorber is vibrating because all energy is transferred into it. For the energy harvester model, this implies that at this frequency the mechanical system stands still and most energy is absorbed by the energy harvesting circuit. When the resistance is increased this effect will become weaker until it is no longer apparent. If the resistance is increased until it is so high that only marginal current flow exists, then the system becomes once again a single degree of freedom oscillator. This effect can be seen in Fig. 3.20. The natural frequency converges to the structure with open circuit electrodes.

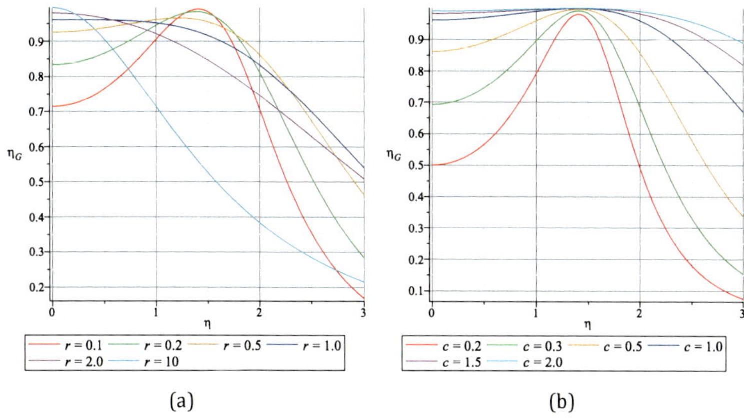

(a)   (b)

Figure 3.23   Degree of efficiency η_G of the force-excited oscillator (a) for a set of resistances r and (b) for a set of coupling coefficients c.

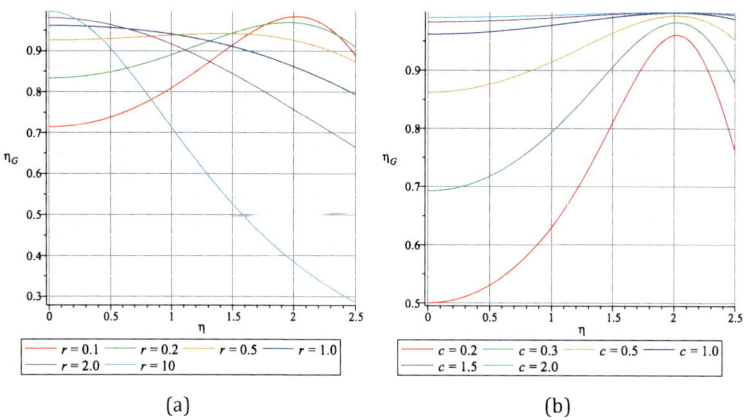

(a)   (b)

Figure 3.24   Degree of efficiency of the base-excited oscillator (a) for a set of resistances r and (b) for a set of coupling coefficients c.

3.5 Application of Several Identical Transducers

In the previous section, it was assumed that only one piezo transducer is applied to the vibrating structure. However, to improve the performance of energy harvesting systems with piezoelectric transducers, it might be required to apply more than one transducer on the structure. For practical reasons, the transducers cannot

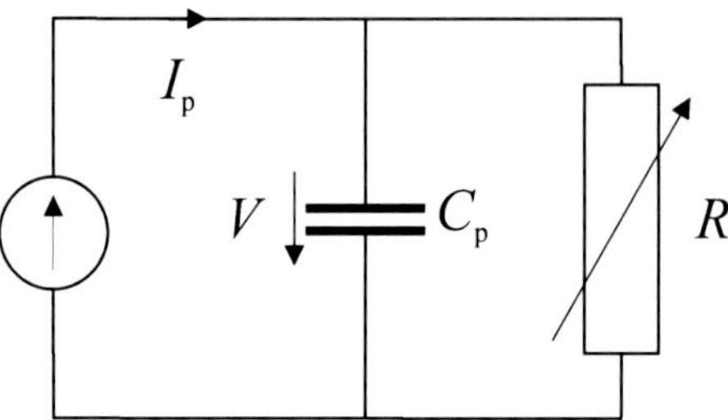

Figure 3.25   Equivalent circuit for a piezoelectric transducer.

be connected to different harvesting circuits but have to be interconnected in series or parallel.

Piezoelectric bending beams have proven to be a promising approach for microgenerators that convert vibration energy into electrical energy. They can be used especially for the designing of energy self-sufficient sensor nodes. Since these generators have a characteristic eigenfrequency at which they yield a maximum of electric power, they usually are tuned to the dominating frequency of their environment. In this section, the application of several identical generators on one vibrating structure to the amount of harvested energy is examined. This section will present analytical considerations of these combinations and prove the results by means of an experimental analysis.

3.5.1 *Analytical Consideration*

Piezoelectric transducers working as generators can as a first approximation (uncoupled analysis) be characterized by the equivalent circuit shown in Fig. 3.25. C_p is the capacitance of the piezoelectric transducer and I_p the electric current resulting from the mechanic excitation of the piezoceramics. For a piezo patch of width b and length l the electric current can under the assumption of plane stress be written as

$$I_\mathrm{P}(t) = \frac{d_{31}}{s_{11} + s_{12}} bl \frac{d\epsilon}{dt} . \tag{3.51}$$

Assuming a harmonic excitation of the transducer $\left(I_\mathrm{p}(t) = \hat{I}_\mathrm{p} \sin(\omega t)\right)$ the equivalent circuit yields for current and voltage in complex notation:

$$\underline{I}_R = \frac{1}{1 + j\omega RC_p}\hat{I}_p, \qquad \underline{V}_R = \frac{R}{1 + j\omega RC_p}\hat{I}_p. \tag{3.52}$$

Varying the connected resistance R leads to a change in the voltage amplitude; an optimal resistance value can be found at which the highest electrical power is gained:

$$P = \frac{\hat{I}_p^2 R}{1 + \left(R\omega C_p\right)^2}, \tag{3.53}$$

$$R_{opt} = \frac{1}{\omega C_p}, \qquad P_{max} = \frac{\hat{I}_p^2}{2\omega C_p}. \tag{3.54}$$

Considering a system with two identical piezoelectric transducers that are excited at the same frequency and in phase but with different amplitudes due to different local strain conditions raises the question of how effectively a combination of both generators presents itself with regard to the overall amount of harvested energy. For the electric combination two configurations are possible. The transducers can be connected in series or in parallel. It shall be examined which configuration is more promising for energy harvesting and how the performance of a combination of two transducers comes off compared to two separate systems.

3.5.2 *Series Connection of Two Generators*

Assuming that $C_1 = C_2 = C$ and that I_1 and I_2 are in phase, we obtain for a series configuration (Fig. 3.26)

$$\underline{I}_R = \frac{\left(\hat{I}_1 + \hat{I}_2\right)}{2 + Rj\omega C}, \qquad V_{max} = \frac{\left(\hat{I}_1 + \hat{I}_2\right)}{\omega C}, \tag{3.55}$$

$$R_{opt} = \frac{2}{\omega C}, \qquad P_{max} = \frac{1}{4}\frac{\left(\hat{I}_1 + \hat{I}_2\right)^2}{\omega C}. \tag{3.56}$$

The maximum voltage is the sum of the two maximum voltages of the individual generators and the optimal resistance value is twice the optimal resistance value of each single generator. The maximal power of the combination does not equate the sum of the maximum powers of the individual generators. Parallel

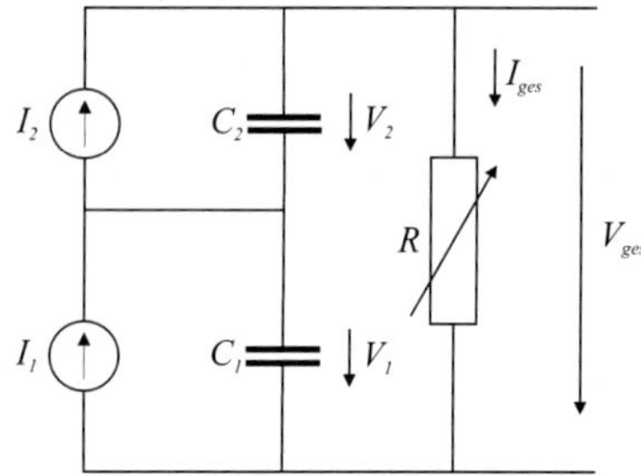

Figure 3.26 Equivalent circuit series connection.

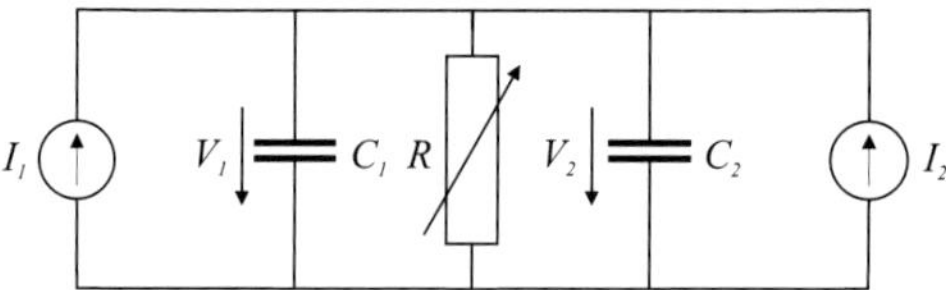

Figure 3.27 Equivalent circuit parallel connection

configuration (Fig. 3.27) results with the same assumptions in

$$\underline{I_R} = \frac{\left(\hat{I}_1 + \hat{I}_2\right)}{1 + 2Rj\omega C}, \quad V_{\max} = \frac{\left(\hat{I}_1 + \hat{I}_2\right)}{2\omega C}, \tag{3.57}$$

$$R_{\mathrm{opt}} = \frac{1}{2\omega C}, \quad P_{\max} = \frac{1}{4}\frac{\left(\hat{I}_1 + \hat{I}_2\right)^2}{\omega C}. \tag{3.58}$$

The maximal voltage is the arithmetic mean value of the individual maximal voltages. The optimal resistance value is half the size of the optimal resistance for the single generator and consequently four times smaller compared to the series connection. The power at the optimum is the same as with series connection.

3.5.3 *Discussion of the results*

Plotting the voltage against the resistance (Fig. 3.28(a)) demonstrates the influence of the configuration on the open circuit voltage, giving either the sum or the mean value of the individual maximal voltages. The power diagram (Fig. 3.28(b)) is symmetric if the resistance is plotted logarithmically. The power maxima with series

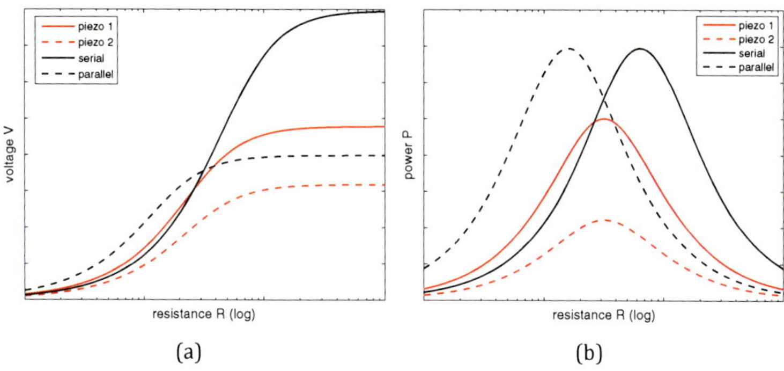

Figure 3.28   Voltage and power over resistance.

or parallel setup are arranged with factor two to the left and right of the power maxima of the single generator at the same power level. With respect to maximal energy output, series and parallel connection are therefore identical. The optimal resistance and hence the current and voltage values, however, are different, which will give one of the two alternatives priority in practical applications. It is noticeable though that the values of the maximal power do not simply add up. Denoting the ratio of the amplitudes of the two current sources with $\eta_I = \hat{I}_2/\hat{I}_1$, we can see the connection between the individual powers (red lines), the power with electric combination (black line) and the sum of individual powers (green line) as shown in Fig. 3.29 It can be seen that by combining two generators less power is generated than with two individual systems. Only for the convenient case of equal amplitudes ($\eta_I = 1$) the harvested overall power equals the sum of the individual powers. For other ratios the generator performance of the combination is always worse than that of two separate generators. For

$$\eta_I < \sqrt{2} - 1 \qquad \text{and for} \qquad \eta_I > \frac{1}{\left(\sqrt{2} - 1\right)}$$

respectively, the combination yields even less energy than one single generator. In this case, the addition of another piezoelectric transducer leads to a negative effect on the overall energy balance. If several transducers are to be combined to one single generator,

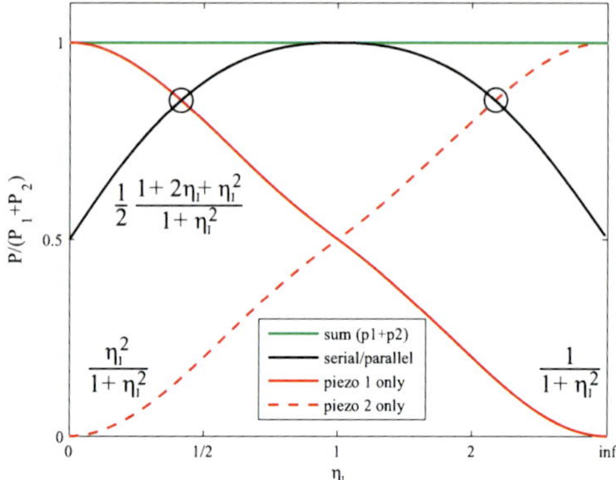

Figure 3.29 Relative power with two piezo generators.

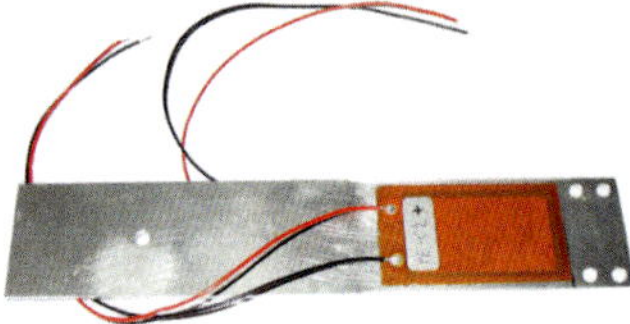

Figure 3.30 Aluminum beam with piezo patches on both sides.

it is therefore necessary to ensure that the differences in the strain amplitudes of the individual generators do not become too large.

3.5.4 *Experimental Verification*

In order to prove the analytically obtained results experimentally a series of measurements have been carried out. Piezoelectric patches have been bonded to an aluminum beam at different locations. It has been clamped at one end and harmonically excited at the other with frequencies far below the first bending eigenfrequency. Hence, an in-phase excitation of the patches has been ensured (or inversely phased in case of opposite application, which can by means of interchanging circuit points be transferred to an in-phase mode). The results have shown that the power sums up in the case of equal amplitudes (Fig. 3.31(a)). In the case of different amplitudes the

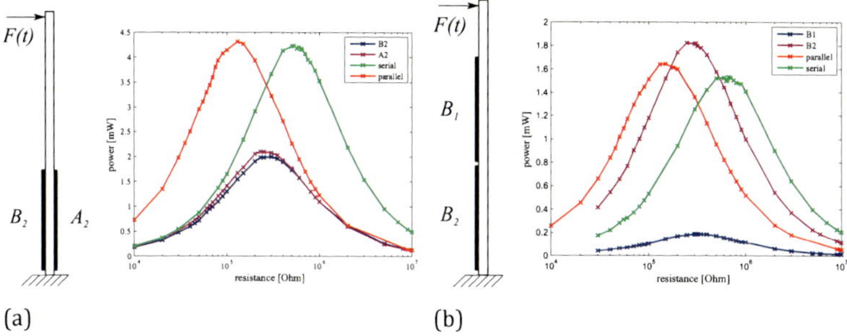

Figure 3.31 Generators with equal strain

power gained with the combination even drops under the power gained with the higher loaded patch only (Fig. 3.31(b)).

3.6 Conclusion

Piezoelectric transducers have been used for a lot of energy harvesting applications. Since they are solid elements that convert mechanical energy directly into electric energy and vice versa, they are applied to a vibrating structure to produce electrical energy.

In this chapter, the history, material properties, and physical phenomena were presented. In this context, it should be highlighted that piezo electric composites exist that are very bendable. These composites can be applied on double curved structures just as well as in flexible structures.

Two models for piezo transducers were introduced to demonstrate the power conversion from mechanical to electrical energy. Therefore, we used a linear mass-spring-damper system, which was loaded by a resistance, which is admittedly an academic example, but otherwise suitable to understand the fundamentals of piezo power conversion. In addition, a resonate circuit was applied to the model of the structure to show how the dynamics of the connected circuit influences the amount of harvestable power. Furthermore, it has been presented how the application of several identical piezoelectric generators applied on a vibrating structure influence the total amount of harvested energy.

When more complex networks for power management and storage are connected (see Chapter 7) to the transducer this analytical method will become inapplicable. The same thing will happen when the geometry of the mechanical structure is more sophisticated, which is common in many engineering applications. Then we recommend the use of numerical methods such as the finite element method, electronic circuit simulator, or other computer-aided methods to find the optimal operating parameters.

References

1. L. Kong, W. Zhu, and O. Tan, Preparation and characterization of $Pb(Zr_{0.52}Ti_{0.48})O_3$ ceramics from high energy ball milling powders, *Mater. Lett.* **42**, 232–239 (2000).

2. R. Ostertag, G. Rinn, G. Tunker, and H. Schmidt, Preparation and properties of sol-gel-derived PZT, *Br. Ceramic Proc.* **41**, 11–20 (1989).

3. K. Uchino and J. R. Giniweicz, *Micromechatronics* (Marcel Dekker, Inc., 2003).

4. *Catalogue from Physik Instrumente (PI)*. Physik Instrumente (PI) GmbH & Co. KG (2001). URL http://www.physikinstrumente.com.

5. H. Allik and T. Hughes, Finite element method for piezoelectric vibration, *Int. J. Numerical Methods Eng.* **2**(2), 151–157 (1970).

6. N. W. Hagood, W. H. Chung, and A. V. Flotow, Modelling of piezoelectric actuator dynamics for active structural control, *J. Intell. Mater. Syst. Struct.* **1**(3), 327–354 (1990). doi: 10.1177/1045389X9000100305.

7. H. A. Sodano, Estimation of electric charge output for piezoelectric energy harvesting, *Strain.* **2004**(40), 49–58 (2004).

8. G. Ottman, H. Hofmann, A. Bhatt, and G. Lesieutre, Adaptive piezoelectric energy harvesting circuit for wireless remote power supply, *IEEE Trans. Power Electron.* **17**(5), 669–676 (2002). ISSN 0885-8993.

9. C. D. Richards, M. J. Anderson, D. F. Bahr, and R. F. Richards, Efficiency of energy conversion for devices containing a piezoelectric component, *J. Micromech. Microeng.* **14**(5), 717–721 (2004). URL http://stacks.iop.org/0960-1317/14/717.

10. D. Guyomar, A. Badel, E. Lefeuvre, and C. Richard, Toward energy harvesting using active materials and conversion improvement by

nonlinear processing, *IEEE Trans. Ultrason. Ferroelectr. Freq. Control.* **52**(4), 584–595 (2005).

11. E. J. Skudrzyk, Vibrations of a system with a finite or an infinite number of resonances, *J. Acoust. Soc. Am.* **30**(12), 1140–1152 (dec, 1958). URL http://link.aip.org/link/?JAS/30/1140/1.

12. D. J. Ewins, *Modal Testing: Theory, Practice and Application* (Wiley, August 2001), 2 edition.

Chapter 4

Electromagnetic Transducers

Dirk Spreemann and Bernd Folkmer

Institute of Micro- & Information Technology, Hahn-Schickard Society e.V.,
Wilhelm-Schickard-Str. 10, 78052 Villingen-Schwenningen, Germany
bernd.folkmer@hsg-imit.de

This chapter covers the design of vibration energy harvesters based on the electromagnetic working principle. In the recent years a large number of such transducers have been developed by different scientific organizations. A conclusion of the state of the art leads to the analytical description and basic tools for the design of resonant vibration transducers with its mechanical and electromagnetic subsystems. The design and optimization of the overall transducer system is shown on a sample for automotive applications.

PART 1: STATE OF THE ART

4.1 Literature Review and "State of the Art" in Electromagnetic Vibration Transducers

In recent years, a multiplicity of electromagnetic-based vibration transducers have been developed by numerous research facilities. The transducers basically differ in size, electromagnetic coupling

Handbook of Energy Harvesting Power Supplies and Applications
Edited by Peter Spies, Loreto Mateu, and Markus Pollak
Copyright © 2015 Pan Stanford Publishing Pte. Ltd.
ISBN 978-981-4241-86-1 (Hardcover), 978-981-4303-06-4 (eBook)
www.panstanford.com

architecture, excitation condition and output performance. Actually, electromagnetic resonant vibration transducers are already commercially available [2, 3, 5]. These commercial vibration transducers are normally add-on solutions, typically greater than 50 cm^3 and they fulfil standard industry specifications such as ingress protection (IP code, IEC 60529), operation conditions (like the temperature range, shock limit and so on) or the requirements for electrical equipment in hazardous areas (ATEX/IECEx). Nevertheless, requests from the industry show that beyond add-on solutions, there is a great demand for application-specific solutions. This is because the required output power, the available vibration level and the overall mass and volume will be significantly different in application. Moreover, it is often necessary that the system must be integrated in an existing subassembly. Especially for applications where the volume of the transducer is a critical parameter, these facts show that the available output power of vibration transducers can only be used optimally for application-specific customized developments. The primary purpose of this introduction is to give a basic overview of the recent research achievements in electromagnetic vibration conversion. One suitable way to classify the transducers is to use the coupling architecture. The most common architectures are shown in Table 4.1, where the oscillation mass is usually (not always) provided by the magnet. Note that this table is not complete. Indeed a basic characteristic of electromagnetic vibration conversion is that there are various possibilities for the implementation of the coupling architecture.

In architecture A I, a magnet oscillates inside a coil. A silicon micromachined implementation of this architecture with discrete magnets has been presented in [11] and [58]. In MEMS-based systems, the springs are usually made of silicon or polymer such as SU-8. In spite of this, the most commonly PCB material FR4 has been used as spring material to assemble a transducer based on A I in [31]. In order to achieve a broadband behaviour a multi-modal resonating power transducer has been presented in [47]. A I has also been considered for a transducer as an implantable middle ear hearing aid [53]. A modification of A I with linear suspended magnetic spring elements as shown in Fig. 4.1 is used in [15, 26, 32]. Note that beside the nonlinear magnetic spring forces there is a

Table 4.1 Commonly used electromagnetic coupling architectures

<table>
<tr><td>(a)</td><td colspan="5">“Magnet in-line coil” architectures</td></tr>
<tr><td colspan="3">Without back iron</td><td colspan="2">With back iron</td></tr>
<tr><td>A I</td><td>A II</td><td>A III</td><td>A IV</td><td>A V</td></tr>
<tr>
<td>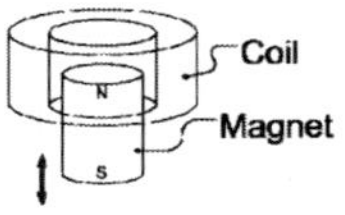</td>
<td>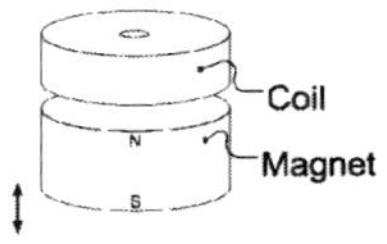</td>
<td>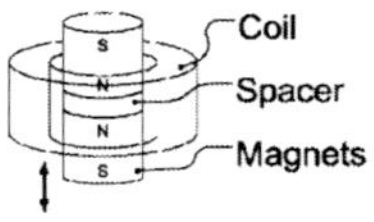</td>
<td>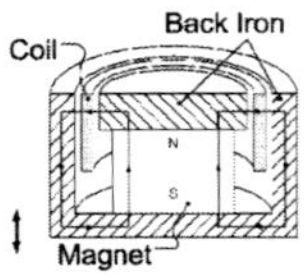</td>
<td>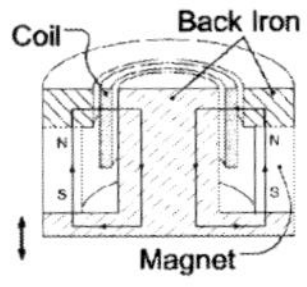</td>
</tr>
<tr><td>(b)</td><td colspan="4">“Magnet across coil” architectures</td></tr>
<tr><td colspan="2">Without back iron</td><td colspan="3">With back iron</td></tr>
<tr><td>A VI</td><td>A VII</td><td colspan="3">A VIII</td></tr>
<tr>
<td>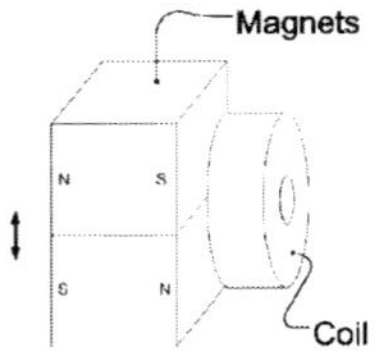</td>
<td>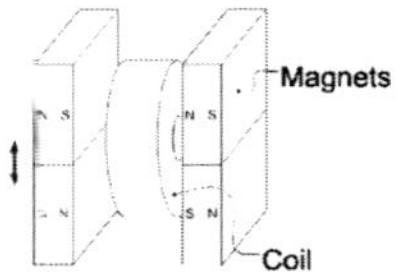</td>
<td colspan="3">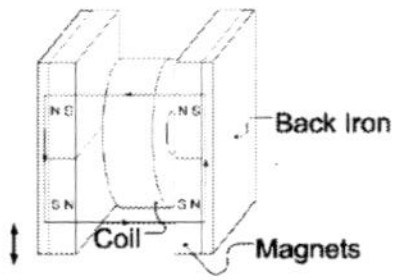</td>
</tr>
</table>

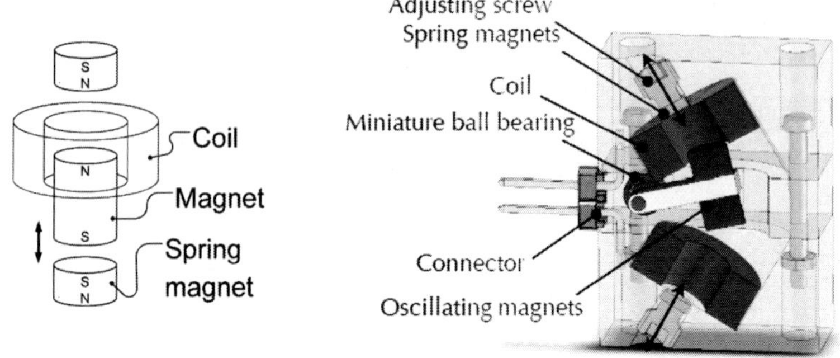

Figure 4.1 (a) Modification of architecture A I with linear suspended magnetic spring. (b) A rotary suspended version has been used to reduce the friction and to adjust the resonance frequency in [18].

torque which tries to rotate the magnet. This torque usually causes high friction loss between the magnet and the cylinder in which the magnet oscillates. To reduce the friction a rotary suspended version has been presented in [18]. Therein it has also been shown that the nonlinear spring force can moreover be used to manually adjust the resonance frequency in the range of 30–60 Hz.

In architecture A II, the magnet oscillates towards the coil without an immersion. This architecture is particularly used in microfabricated electromagnetic vibrational energy harvesters. A fully integrated fabrication of such a transducer has been presented in [45]. Therein micromagnet technology based on embedded NdFeB powder has been applied. Another transducer based on micromachined flexible polyimide film and planar micro-coils has been presented in [16]. Both magnetically functionalized SU-8 and discrete NdFeB magnets with a polyimide membrane on a glass substrate have been investigated. The same architecture with coils on both sides of the magnet has been implemented using silicon micromachining (nickel planar springs) and microelectroplating techniques in [49]. Fine-mechanical implementations of A II have been presented in [4] and [9]. In the latter publication, different vibration modes of a beam structure have been used in order to achieve a broadband behaviour.

Architecture A III is quite similar to A I. However, there are two magnets with opposite direction of the polarization. Due to the assembling complexity this architecture is rather suitable for fine-mechanical implementations. With respect to vibration conversion, A III has first been applied in [40]. With the special aim of supplying power to body-worn sensor nodes, A III has also been used in [59]. In spite of the most other publications, extensive optimization calculations have been applied to find the most efficient coil and magnet dimensions. Architectures A IV and A V are the first architectures with back-iron. They are typically used in moving-coil loudspeakers. Because of this, there exists a lot of design experience and a lot of theoretical work has already been done as well. Nevertheless, there are not too many groups which have applied these architectures so far [21, 34, 37].

The remaining architectures of the "magnet across coil" class are quite similar. In A VI, two opposite polarized rectangular magnets oscillate across a coil. In A VII, the magnets are arranged on both sides of the coil. Consequently, in A VIII, back iron is used to close the magnetic circuit. A fine-mechanical multi-pole version of A VI has been presented in [52]. Herein the use of multiple magnetic pole arrays has been investigated with the intent to enhance the voltage output. A silicon micromachined moving-coil version of A VII has been presented in [29]. In [56], the same group built up an A VIII transducer based on discrete components combined with micro fabricated components. A pure fine-mechanical transducer designed to convert human motions like arm swinging, horizontal foot movement or up-down centre of gravity movement during walking has been presented in [48]. In [44], a moving-coil version of A VIII has been used together with piecewise linear springs in order to realize a wideband vibration transducer.

4.2 Conclusions from the Literature

The previous section gave a short qualitative overview of the state of the art in electromagnetic vibration transducers. More detailed quantitative review focused on electromagnetic energy conversion is given in [17]. However, it must be said that the comparison of existing work is to some extent rather delicate. This is because it is sometimes not obvious whether effective or amplitude values are denoted for the output signal, the performance is referred to AC or DC values (before/after rectification), different convention holds (i.e. what exactly contains the denoted volume?) or important data is even omitted. Nevertheless, the overview based on the coupling architecture indicates the huge diversity in the field of electromagnetic vibration conversion. Consequently a very basic question arises from this: Which architecture performs the best?

For application-oriented developments, this is a complicate question because it results in a multi-objective optimization. For example, a vibration transducer should be cheap, batch-processed and should have a certain form factor (e.g. flat, cylindrical) and so on, but first and foremost the transducer must be able to deliver enough power to drive the application. Based on a construction volume of 1 cm^3 (which contains the magnet, coil and the back-iron parts at the resting position of the oscillating mass) and typical boundary conditions of mesoscale vibration transducers, the optimization and comparison of the architectures shown in Table 4.1 have been presented in [20] and [21]. A basic result is that the dimensions of the magnet, of the coil and of the back iron parts have a significant influence on the output performance. Moreover, it has been shown that each architecture can be designed on a voltage- or a power optimum (Fig. 4.2). For effective design, it is essential to be able to determine these optimal dimensions. This is especially important for applications with small construction volume or small vibration level where the generation of useful voltage levels (at least 1 V is essential for effective rectification [13]) evokes a challenge. Another fact from the literature review is that in the most of all cases the approach for the development is to first build up the vibration transducer and afterwards adjust the vibration for the characterization. However, in application-oriented developments, one needs to go the other way round. Here the development is strongly dependent on the given vibration characteristic. This is due to the fact that the most of all vibration energy harvesting devices are so far unfortunately restricted to narrowband operation due to their resonance nature. To overcome this drawback there is a great effort in developing active tunable vibration transducers [10, 25] or to develop conversion mechanisms with broadband behaviour [24]. For active tunable devices, the primary challenge is to drive the control and the active part with the available power.

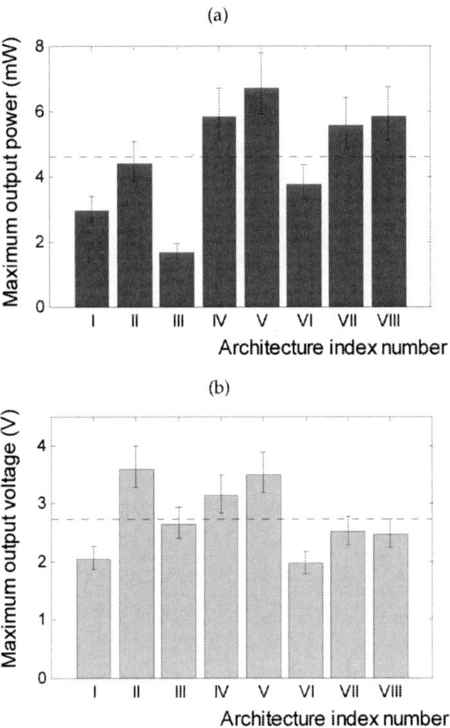

Figure 4.2 Comparison of the maximal output performance of the architectures in Table 4.1 [20, 21]: (a) optimal output power and (b) output voltage. The dashed curve indicates the mean value of all architectures.

Part 2 covers the analytical treatment of resonant electromagnetic vibration transducers. The mechanical subsystem and the electromagnetic subsystem will be investigated separately. Afterwards the subsystems are combined to an overall model of resonant electromagnetic vibration transducers. The results lead to first-order power estimation. The handling of machinery-induced vibration will be discussed as well. Part 3 discusses an optimal electromagnetic coupling design for a prototype vibration transducer based on A II. This development is based on measured stochastic vibration in a car engine compartment. Consequently Part 4 discusses the assembling and the experimental characterization of the prototype vibration transducer.

PART 2: ANALYTICAL DESCRIPTION—BASIC TOOLS FOR THE DESIGN OF RESONANT VIBRATION TRANSDUCERS

4.3 Introduction

The presented review of existing work on electromagnetic inertial vibration transducers in Section 4.1 shows that there has been much recent interest in the design of vibration energy harvesting devices. The basic analytical theory of the most of all presented devices is commonly known in the energy harvesting society. It is based on a well-understood linear second-order spring-mass-damper system with base excitation. Specific analysis for vibration transducers was first proposed by Williams and Yates [12]. Since then, the theory has been modified and described in various kinds even though the basic findings are more or less the same. These are namely the calculation of the maximum output power that can be extracted from a certain vibration and moreover the optimization of parameters such as the optimal load resistance in consideration of realistic constrains (e.g. the limitation of the inner displacement of the seismic mass [51]). However, it must be said that especially for the latter case, it is rather difficult unless impossible to use the outcome from the analytical modelling directly for the design process of application-oriented developments. This is because the theory is based on simplifying assumptions, which does often not correlate well with the "real world" (e.g. random vibration instead of harmonic excitation or complex load circuit instead of simple resistance). However, for all that, the analytical modelling is useful to understand the most important system parameters and furthermore offers a deeper insight into the overall system behaviour.

4.4 Mechanical Subsystem

4.4.1 *Linear Spring System*

The aim of a vibration transducer is the conversion of vibration energy into electrical energy. Usually it is assumed that the energy conversion as well as the mass of the transducer has no effect

(a)

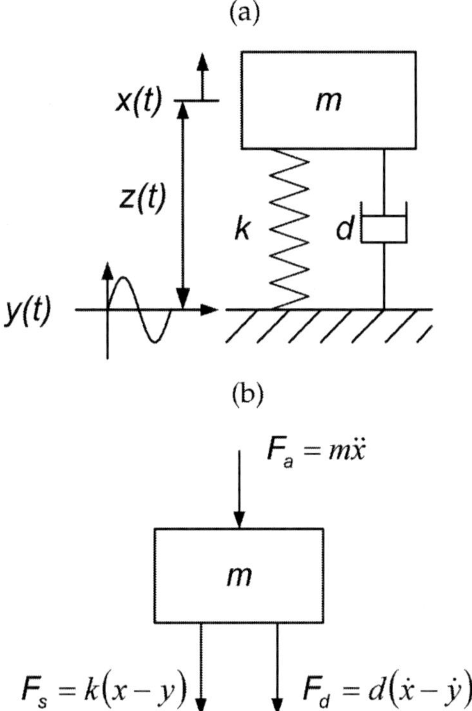

Figure 4.3 (a) Linear single degree-of-freedom (SDOF) spring-mass-damper model of a resonant vibration transducer with harmonic base excitation. (b) Dynamic free body diagram of the oscillating mass with the inertia force F_a and the forces associate with the spring F_s and the damping element F_d .

on the vibration source. This assumption is fulfilled as long as the mass of the transducer is much smaller than the mass of the vibration source. A commonly used linear single degree of freedom mechanical model of a vibration transducer is shown in Fig. 4.3. It consists of a seismic mass m attached to a spring with spring rate k and a damping element with damping factor d. The governing equation of motion for relative coordinates $z = x - y$ is

$$m\ddot{z} + d\dot{z} + kz = -m\ddot{y} = m\omega^2 Y \sin(\omega t), \qquad (4.1)$$

where Y is the amplitude of the excitation. The theory of this second-order differential equation is well known and discussed in a multiplicity of textbooks [27, 60]. The solution to Eq. (4.1) can easily be found in frequency domain with the Laplace transformation:

$$\left(ms^2 + ds + k\right) Z\left(s\right) = -ms^2 Y\left(s\right).$$ (4.2)

Rearranging together with the natural frequency of the undamped oscillation $\omega_n = \sqrt{k/m}$, the normalized damping factor $\zeta = d/2m\omega_n$ and substituting $s = j\omega$ the transfer function becomes

$$G_{\text{mech}}\left(\omega\right) = \frac{\omega^2}{\left(\omega_n^2 - \omega^2\right) + 2\zeta\omega_n\omega j}.$$ (4.3)

In general, this solution can be applied if the excitation is a harmonic function or if the excitation can be represent as a Fourier series of harmonic functions. The steady-state solution of Eq. (4.1) in time domain yields the amplitude and phase of steady-state motion:

$$Z = \frac{m\omega^2 Y}{\sqrt{\left(k - m\omega^2\right)^2 + (d\omega)^2}} = \frac{r^2 Y}{\sqrt{\left(1 - r^2\right)^2 + (2\zeta r)^2}},$$

$$\tan\varphi = \frac{d\omega}{k - m\omega^2} = \frac{2\zeta r}{1 - r^2}.$$ (4.4)

The curves are plotted in Fig. 4.4.

4.4.2 Nonlinear Spring System

For the design of vibration transducers, it is for various reasons also important to understand the behaviour of non-linear spring systems. The first reason is that non-linearity appears for any real spring if the deflection is only high enough. Beyond this it is in some cases possible to enhance the output power using non-linear spring systems. This potential has first been investigated in [23]. A more analytical treatment of non-linear spring systems for energy harvesting devices has been presented in [14]. This section gives a short introduction on the behaviour of non-linear vibration [33].

The considered non-linearity appears whenever the restoring force exerted by the spring is no more proportional to the spring rate k but can be expressed by

$$f\left(x\right) = kx\left(1 + \mu x^2\right),$$ (4.5)

where μ is the nonlinearity parameter, which is positive for hardening and negative for softening spring, respectively (Fig. 4.5).

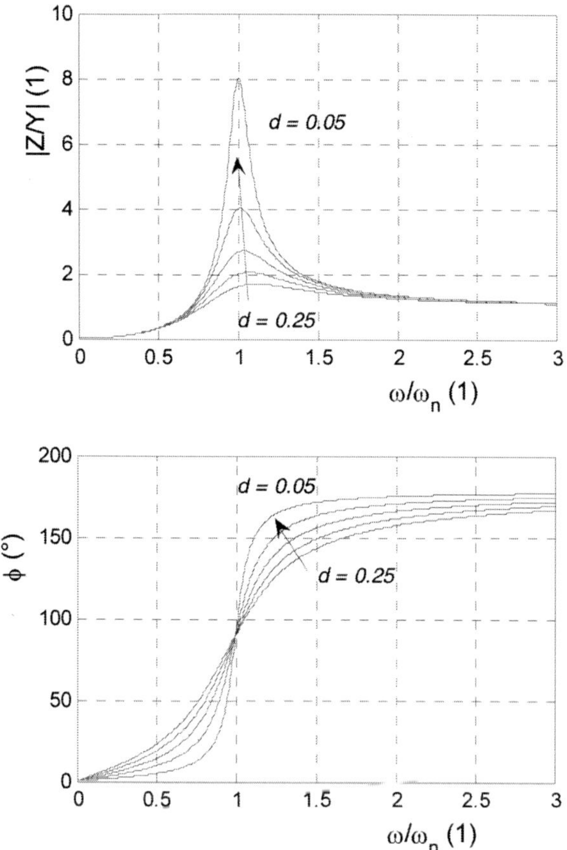

Figure 4.4 Amplitude and phase of steady-state motion in relative coordinates.

Using a forcing function with amplitude f_0 as excitation together with Eq. (4.5) the corresponding equation of motion can be written as

$$\ddot{x} + 2\beta\dot{x} + \omega_n^2 x + \omega_n^2 \mu x^3 = \frac{f_0}{m}\sin(\omega t + \Phi), \qquad (4.6)$$

where $\beta = d/2m$ and ω_n indicates the natural frequency of the undamped motion. This nonlinear differential equation can be solved using the Ritz-averaging method. However, the theory to solve this equation is beyond the scope of this chapter. The most important result at this point is the displacement amplitude which

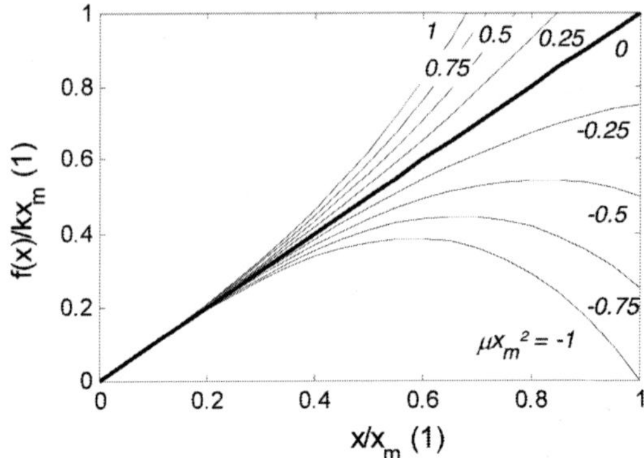

Figure 4.5 Nondimensional plot of the non-linear cubic restoring force (Eq. (4.5)). Softening behaviour results for $\mu x_\mathrm{m}^2 < 0$ and hardening behaviour for $\mu x_\mathrm{m}^2 > 0$. A pure linear spring is obtained for $\mu x_\mathrm{m}^2 = 0$.

can be written as:

$$\left[\left(k - m\omega^2\right) x_\mathrm{m} + \frac{3}{4} k\mu x_\mathrm{m}^3 \right]^2 + x_\mathrm{m}^2 d^2 \omega^2 = f_0^2. \tag{4.7}$$

Typical plots of the frequency response for nonlinear hardening ($\mu > 0$) and nonlinear softening ($\mu < 0$) systems are given in Figs. 4.6 and 4.7, respectively. For increasing frequency the amplitude function passes through A, B, C and D and for decreasing frequency through D, E, F and A. This effect is also known as the jump

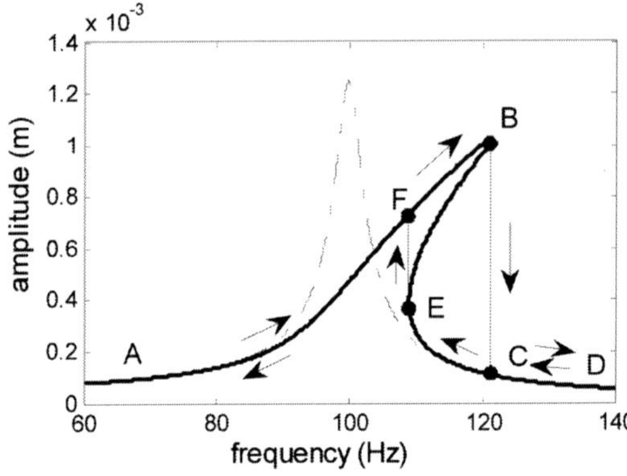

Figure 4.6 Frequency response for nonlinear hardening system. The dash dotted curve indicates the response of a linear system.

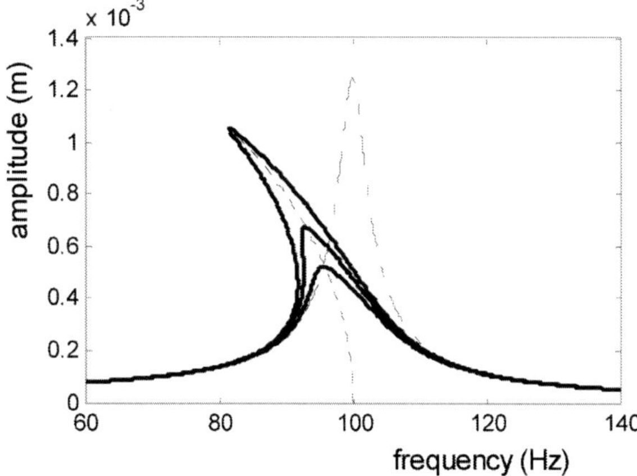

Figure 4.7 Frequency response of typical nonlinear system for different damping factors. The dotted line is the so-called backbone or skeleton curve.

phenomenon. Hence, in spite of linear systems, the state of nonlinear systems is dependent from the history. Note that even though the amplitude response of a nonlinear system suggests a higher bandwidth it is very delicate to claim a general advantage in this case. The reason for this is that it might be difficult to guarantee that the system operates between the branch F and B and not between E and C in application!

4.5 Electromagnetic Subsystem

4.5.1 *Basics on Electromagnetic Induction*

The output power performance of electromagnetic vibration transducers is strongly dependent on the design of the electromagnetic coupling. Factors such size, material properties or geometric configuration of magnet, coil and magnetic circuit parts take, therefore, a vital key role in the design process. However, conclusions found in literature are often based on simplifying assumptions. This is because the analytical calculation of magnetic field is rather complicate for ironless systems and even impossible for systems with back iron. Nevertheless, it is necessary to understand the basic

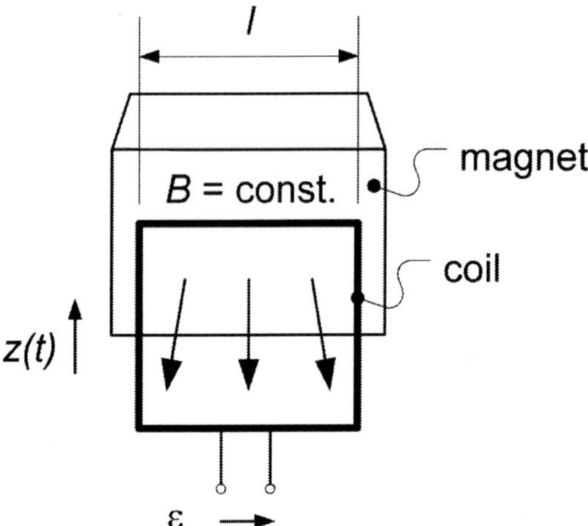

Figure 4.8 Popular model for linearized electromagnetic transducer analysis. Induction coil arranged in constant magnetic field with rectangular cross section. Beyond the magnet boundary, the magnetic field is assumed to be zero.

theory of the magnetic transduction mechanism based on Faraday's law of induction. With the magnetic flux

$$\varphi = \iint\limits_{A} B \, dA, \tag{4.8}$$

the induced voltage (the so-called electromotive force EMF) is given by

$$\varepsilon = -\frac{d\varphi}{dt} = -\left(\frac{dA}{dt} B + \frac{dB}{dt} A \right), \tag{4.9}$$

where A indicates the area enclosed by the wire loop and B is the magnetic flux density. This equation shows that for magnetic induction it does not matter whether the magnetic field is changing during constant area or the area is changing during constant magnetic field. Actually this characteristic offers a wide range of implementations of the electromagnetic coupling. A basic arrangement is shown in Fig. 4.8 [51]. It can be referred to the first term in brackets of Eq. (4.9). This arrangement is often used for analytical analysis due to the simplicity in calculation in spite of the arrangements where the EMF is produced through a diverging magnetic field according to the second term in brackets of Eq. (4.9). Such arrangement will be studied in detail in Section 4.8. The change of overlapping area for a coil with N windings follows $N \cdot dA/dt = Nl \cdot dz/dt = Nl\dot{z}$. Therewith the EMF voltage becomes

$$\varepsilon = -NBl\dot{z}. \tag{4.10}$$

Using the chain rule, Eqs. (4.9) and (4.10) can be extended to give

$$\varepsilon = -\frac{d\varphi}{dz} \cdot \frac{dz}{dt} = -NBl \cdot \dot{z} = k_t \cdot \dot{z}, \tag{4.11}$$

where k_t is the so-called transduction factor.

4.5.2 *Electrical Network Representation*

In application, the electrical load will be a complex analog and digital electronic load circuit. In analytical analysis of vibration transducers the ohmic loss of the load circuit can be represented by a resistor R_{load} in series to the coil resistance R_{coil} and coil inductance L_{coil} as shown in Fig. 4.9. The governing equation is given by

$$L_{coil}\dot{i}(t) + (R_{coil} + R_{load})\,i(t) = -NBl\dot{z}(t). \tag{4.12}$$

In frequency domain with the voltage at the load resistance this differential equation becomes

$$G_{emag}(s) = \frac{V(s)}{Z(s)} = \frac{-NBl\,R_{load}s}{L_{coil}s + R_{coil} + R_{load}}. \tag{4.13}$$

Inductances and resistances of air cored coils can easily be calculated if the copper fill factor of the coil winding process is known. The copper fill factor is thereby defined as the ratio of the

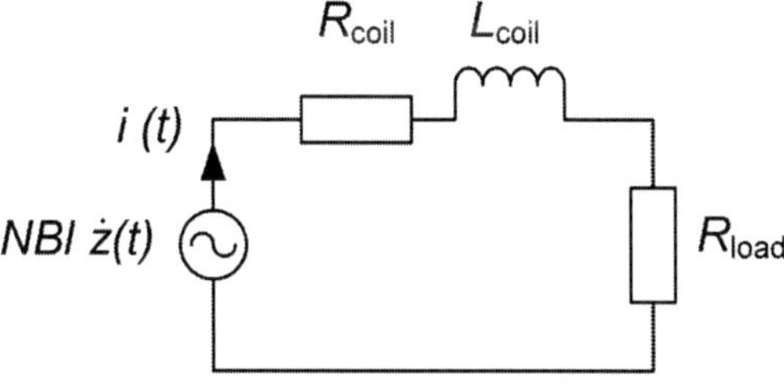

Figure 4.9 Circuit diagram representation of the electromagnetic subsystem for analytical analyses.

overall wire area (without isolation) A_l to the cross section of the winding area A_w [41]:

$$k_{co} = \frac{A_l}{A_w} = \frac{\pi \cdot d_{co}^2 \cdot N}{4 \cdot A_w}.$$ (4.14)

Note that there is no possibility for the theoretical calculation of the copper fill factor for wire diameters smaller than 200 μm because the wire layers are unsystematic and random winding occurs. However, in miniaturized vibration the wire diameters are typically smaller than 200 μm. In this case, k_{co} needs to be determined experimentally. For given copper fill factor, the number of windings in longitudinal N_{long} and lateral N_{lat} direction can be calculated using

$$\lfloor N_{long} \rfloor = \frac{2 \cdot h_{coil}}{d_{co}\sqrt{\pi/k_{co}}}, \lfloor N_{lat} \rfloor = \frac{2 \cdot (R_o - R_i)}{d_{co}\sqrt{\pi/k_{co}}}.$$ (4.15)

With the resistance per meter R' (enamelled copper wires are based on IEC 60317), the resistance is simply given by

$$R_{coil} = N \cdot 2\pi \cdot \frac{(R_o - R_i)}{2} \cdot R',$$ (4.16)

where the number of windings N is the product of N_{long} and N_{lat}. The inductance can be determined according to the Wheeler approximations [51]:

$$L_{coil} = \frac{3.15 \times 10^{-5} \cdot R_m^2 \cdot N^2}{6R_m + 9h_{coil} + 10(R_o - R_i)},$$ (4.17)

where R_m denotes the middle radius given by $(R_o + R_i)/2$. For low-frequency applications, further simplification of the electrical network can be done. In this case, the reactance of the coil is much greater than the inductance. Figure 4.10b gives the ratio of the resistance (real part) to the reactance (imaginary part) of the inductance for a frequency of 100 Hz and different winding areas. The result shows that even for unusual large winding areas the reactance is well below 10% of the resistance. For simplification matters, it is therefore appropriate to neglect the reactance. In this case, impedance matching condition and therefore maximum power transfer to the load can be obtained using the EDAM (electrical domain analogue matching) [46]:

$$R_{load,opt} = R_{coil} + \frac{k_t^2}{d_m},$$ (4.18)

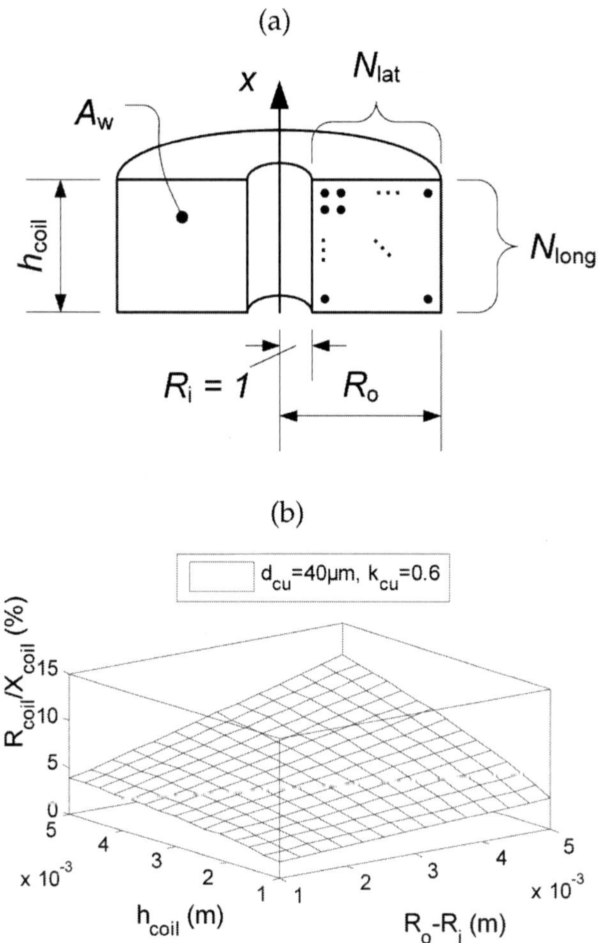

Figure 4.10 (a) Example of cylindrical air cored coil with inner radius of 1 mm (b) Ratio of resistance R_{coil} to reactance X_{coil} in percent for different winding areas A_w of the example coil at 100 Hz. For typical winding areas the resistance dominates the impedance.

where d_m denotes the unwanted mechanical damping effects. Note that beyond the commonly known resistance matching in electrical domain this approach includes also the electrical analogue of the mechanical damping.

4.6 Overall System

4.6.1 *General Behaviour*

The previous sections discuss the mechanical and electromagnetic subsystem of the vibration transducer. These subsystems can now be combined to an overall system model. The mechanical domain (input force and relative velocity of the mass) and the electromagnetic domain (EMF and induced current) are related via the transduction factor k_t. For closed circuit condition the EMF voltage will cause a current to flow. This current in turn creates a magnetic field which opposes the cause according to Lenz's law. The feedback electromechanical force (together with Eq. (4.11) and Ohm's law) is given by

$$F = k_t i = \frac{k_t^2}{R_{\text{coil}} + R_{\text{load}}} \dot{z}. \qquad (4.19)$$

Hence, the dissipative feedback electromechanical force can be represented by a velocity proportional viscous damping element with the damping coefficient:

$$d_e = \frac{k_t^2}{R_{\text{coil}} + R_{\text{load}}} = \frac{(NBl)^2}{R_{\text{coil}} + R_{\text{load}}}. \qquad (4.20)$$

Now Eqs. (4.2) and (4.13) can be combined to an overall input force to output voltage transfer function for the electrodynamic transducer:

$$G_{\text{overall}}(s) = \frac{V(s)}{F(s)}$$

$$= \frac{NBl\,R_{\text{load}}s}{\left(ms^2 + d_m s + k\right)\left(Ls + R_{\text{load}} + R_{\text{coil}}\right) + \frac{(NBl)^2}{R_{\text{load}}}s}$$

$$(4.21)$$

The displacement of the vibration $Y(s)$ has therein been replaced by the forcing function $F(s) = -mY(s)s^2$. If the inductance can be neglected, Eq. (4.21) reduces to

$$G_{\text{overall}}(s) = \frac{NBl\,R_{\text{load}}s}{\left(ms^2 + \left(d_m + \frac{(NBl)^2}{R_{\text{load}}}\right)s + k\right)\left(R_{\text{load}} + R_{\text{coil}}\right)}. \qquad (4.22)$$

For both systems and the subsystems, the magnitude and phase response of a 100 Hz system are shown in Fig. 4.11. The dashed

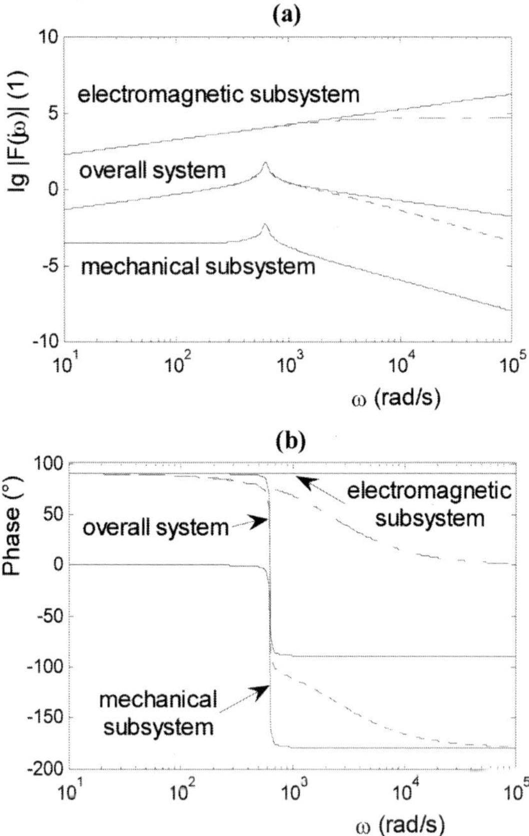

Figure 4.11 (a) Magnitude and (b) phase response for the subsystems and the overall transducer system. The dashed curves indicate the influence of the inductance.

curves indicate the influence of the inductance. In the resonance frequency range there is nearly no influence of the inductance observable. A corresponding block diagram for simulation in Matlab/Simulink® is shown in Fig. 4.12.

4.6.2 First-Order Power Estimation

Inspection of Eq. (4.19) shows that the electromagnetic transduction mechanism can be represented by a dissipative velocity propor-

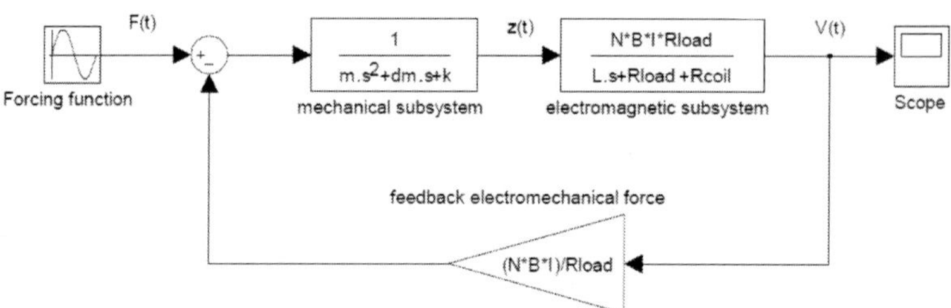

Figure 4.12 Block diagram of the overall transducer model for simulation in Matlab/Simulink®.

tional damping element. Hence, the maximum power that can be extracted from the vibration is related to the power dissipated in that damper. Together with $F(t) = d_e \dot{z}$, the instantaneous dissipated power becomes

$$P(t) = \frac{1}{2} d_e \dot{z}^2. \qquad (4.23)$$

The velocity is simply obtained by the first derivative of the steady-state amplitude in Eq. (4.4). Therewith the electrical generated power becomes

$$P = \frac{m\zeta_e \left(\frac{\omega}{\omega_n}\right)^3 \omega^3 Y^2}{\left(1 - \left(\frac{\omega}{\omega_n}\right)^2\right)^2 + \left(2\left(\zeta_e + \zeta_m\right)\frac{\omega}{\omega_n}\right)^2}. \qquad (4.24)$$

This basic equation has first been published by Williams and Yates in [12]. For operation at resonance ($\omega = \omega_n$) (4.24) can be reduced to:

$$P = \frac{m\zeta_e \omega_n^3 Y^2}{4\left(\zeta_e + \zeta_m\right)^2} = \frac{m^2 d_e \omega_n^4 Y^2}{2\left(d_e + d_m\right)^2}. \qquad (4.25)$$

Recall that the base excitation is a pure harmonic input the output power may also be expressed as a function of the input acceleration amplitude $\left|\ddot{Y}\right| = Y\omega^2$:

$$P = \frac{m\zeta_e \ddot{Y}^2}{4\omega_n\left(\zeta_e + \zeta_m\right)^2} = \frac{m^2 d_e \ddot{Y}^2}{2\left(d_e + d_m\right)^2} \qquad (4.26)$$

Hence, beyond the parameters of the vibration which cannot be influenced, the output power depends on the oscillating mass

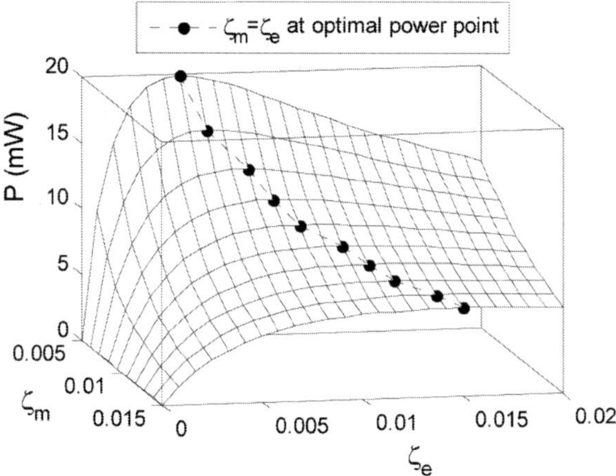

Figure 4.13 Output power as a function of the electromechanical and mechanical damping.

and the damping coefficients. The optimal ratio for the damping factors (obtained by setting $dP/d\zeta_e = 0$) is given by $\zeta_e = \zeta_m$, which is also apparent from the plot of Eq. (4.25) in Fig. 4.13 [55]. However, for high oscillation mass, operation at high vibration amplitudes or small mechanical damping factor the optimal value of the electromagnetic damping could violate the maximum allowable inner displacement. In this case, the electromagnetic damping needs to be increased and the maximum power will be reduced. Detailed analysis on this is given in [51]. For constrained condition, the minimum allowable damping factor is given by the rearrangement of the steady-state amplitude in Eq. (4.4):

$$\zeta_{e,\min} = \frac{1}{2r}\sqrt{r^4\left(\frac{Y}{Z_1}\right)^2 - \left(1-r^2\right)^2} - \zeta_m, \tag{4.27}$$

where Z_1 denotes the maximum relative displacement. This expression can now be substituted into Eq. (4.24) to obtain the displacement constrained output power. Consequently, the maximum power for constrained condition is again obtained at resonance with $\omega = \omega_n$ and $r = 1$:

$$P_{cs,\max} = \frac{1}{2}m\omega_n^3 Z_1 \cdot (Y - 2Z_1\zeta_m) \tag{4.28}$$

4.7 Characterization and Handling of Machinery-Induced Vibration

So far, the vibration source is assumed to be a pure harmonic function. Because this kind of excitation is deterministic the response of the system is also deterministic. Beyond this, there is no difficulty in expressing the response in closed form for any deterministic excitation function as long as the system is linear. However, in the real world, the vibration source might be a non-deterministic random function. In this case it is not possible to express the response in closed form. Nevertheless, it is important to assess most energetic resonance frequency for the transducer and to be able to make first-order power estimation. This can easily be done in frequency-domain rather than time-domain analyses.

The base for the development of vibration transducers is in the most of all cases an ensemble of measured acceleration time histories of the vibration source. In the following, it is assumed that any single sample function is "representative" and can be used as the basis for the development of the vibration transducer (ergodic process). Three different example vibration spectra from completely different applications are shown in Fig. 4.14. Measurement of Fig. 4.14a was taken on an air-filter housing of a car engine during country driving route. In the frequency content, the order-related phenomenon of the car engine is apparent [19]. The acceleration profile in Fig. 4.14b was measured close to the chisel of a tunnel boring machine and Fig. 4.14c as an example for noisy harmonic vibration source was measured on a mould for the fabrication of concrete products. For a multiplicity of technical vibration sources the highest acceleration amplitudes are predominant below 200 Hz. According to Eq. (4.26), the maximum output power is directly proportional to the input acceleration amplitude. The most energetic vibration frequencies respectively the preferred resonance frequency of the vibration transducer can immediately take out from the acceleration spectra. If there are two or more frequency ranges with equal acceleration level as in Fig. 4.14a, it is

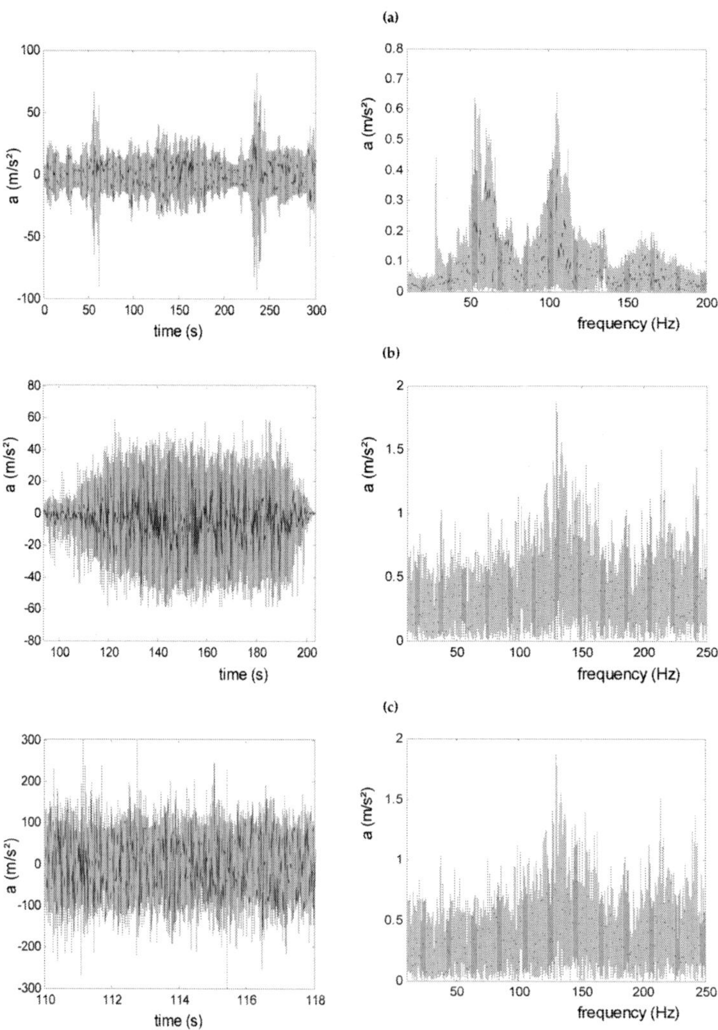

Figure 4.14 Acceleration (left) and frequency content (right) of different example vibration sources. (a) air-filter housing of a car engine during country driving route, (b) tunnel boring machine close to the chisel and (c) mould for fabrication of concrete products.

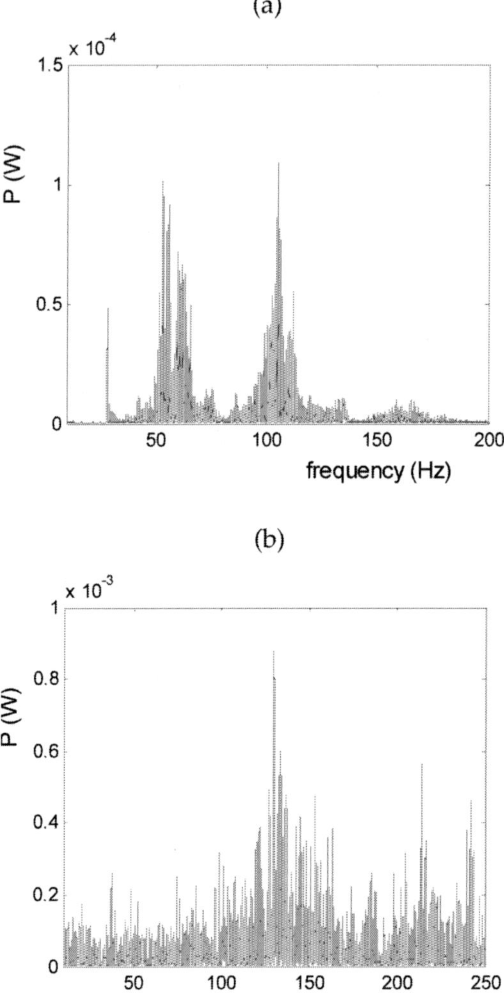

Figure 4.15 First-order power estimation for unconstrained oscillation for the vibration of the air-filter housing (a) and the vibration of the tunnel borer machine (b).

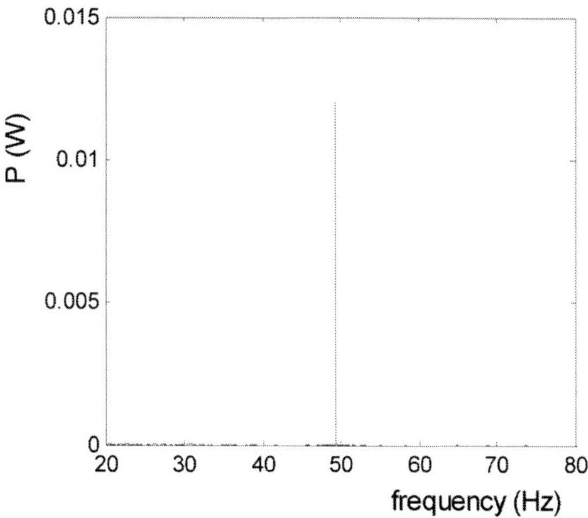

Figure 4.16 Due to the high acceleration amplitude of the mould vibration, it is appropriate to apply constrained condition for the first-order power estimation.

advisable to choose the frequency range with higher bandwidth in order to reduce the necessity that the resonance frequency matches the vibration frequency. Now the question is: How much power can be extracted?

A first-order power estimation can be obtained by substituting the acceleration amplitude spectra into Eq. (4.26). Actually this is a simplification which assumes that the vibration transducer acts like a narrow band-pass filter. For the vibration of the air-filter housing and the tunnel borer machine this first-order power estimation is shown in Fig. 4.15. For the mould vibration it is assumed that the high acceleration amplitude violates the inner displacement constrained. In this case it is obvious to use Eq. (4.28) for the power estimation (Fig. 4.16). Note that the point of interest here is to find the maximum feasible output power. However, in application it will in the most of all cases not be possible to design such an electromagnetic damping element. An alternative to protect the vibration transducer from too high inner displacement is to operate beside the resonance frequency which moreover reduces the necessity that the resonance frequency matches the vibration frequency. Nevertheless, the maximum output power will drop.

4.8 Conclusions from Analytical Analyses

In this section, a basic summarization of analytical electromagnetic vibration transducers analysis has been presented. The results were used to understand the behaviour of the mechanical- and electromagnetic subsystem and to identify the most important parameters for power extraction of the overall system. Based on results from literature first-order power estimation has been carried out for both harmonic and machinery-induced random excitation. As often stated in literature, it has been verified that the maximum possible output power can be achieved if the mechanical damping equals the electromagnetic damping coefficient. However, it must be said that even though this conclusion is valid from the theoretical point of view it has only marginal relevance for the design process of vibration transducers. This is due to the fact that the underlying analytical theory does not consider any volume-dependent effects so far. However, in the majority of cases, the construction volume is limited for application-oriented developments. According to the basic Eq. (4.25), the output power is squarely dependent on the oscillating mass. Beyond the density the oscillating mass can primarily be increased by its volume, which means less space for the implementation of the electromagnetic damping. The electromagnetic damping can in turn primarily be increased by the number of windings which also needs space no more available for increasing the mass. In other words, for limited construction volume conditions the mass and electromagnetic damping are no more independent which evokes a trade off. The major task is to find the dimensions which yield the optimal output power. This central issue can be explained using a model as shown in Fig. 4.17, where a magnet moves relative to a thin coil with rectangular cross section. The construction volume defined by a, b and l should be fixed. That means an increasing of the oscillating mass by increasing α' results in a decreasing of α'' and therewith less windings. Now the question to be answered is: Which ratio of α'/α'' yields the maximum output power?

The number of windings of the thin coil N is simply given by the multiplication of α'', with the number of turns per unit length

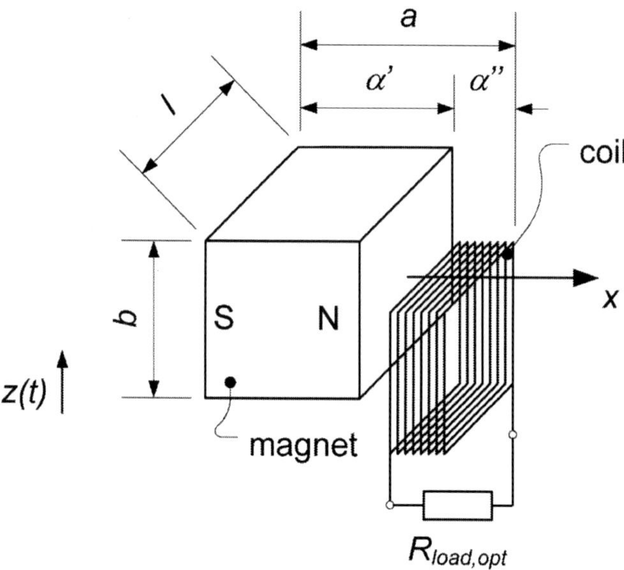

Figure 4.17 Model of resonant electromagnetic vibration transducer adapted to investigate volume-dependent effects.

η. With the lateral length l of the rectangular coil together with the resistance per unit length R' the total resistance is then given by $R_{\text{coil}} = 4l R' N$. As a first approximation the decrease of magnetic field can be taken into account using the magnetic field component in the x-direction on the centre axis of the magnet which is given by

$$B_x\left(x\right) = \frac{B_r}{\pi}\left(\begin{array}{c} a \tan\left(\frac{b \cdot l}{2x\sqrt{4x^2+b^2+l^2}}\right) \\ -a \tan\left(\frac{b \cdot l}{2(\alpha'+x)\sqrt{4(\alpha'+x)^2+b^2+l^2}}\right) \end{array} \right), \tag{4.29}$$

where B_r indicates the residual flux density of the magnet. The transduction factor can then be written as

$$k_t\left(x\right) = Nl \cdot \frac{1}{\alpha''}\int_0^{\alpha''} B_x\left(x\right) dx. \tag{4.30}$$

A plot of this function is shown in Fig. 4.18. Note that the transduction factor has a maximum in the vicinity of $\alpha'/\alpha'' = 1$ for this calculation example. Now the optimal load resistance can be

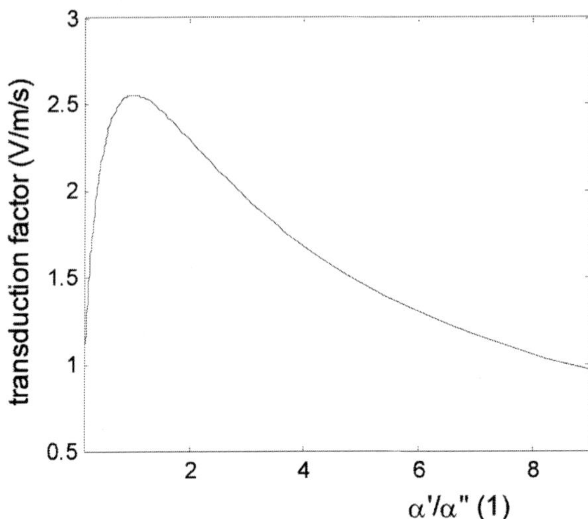

Figure 4.18 Magnetic flux gradient of the volume constrained model with respect to the ratio of α' and α''.

calculated according to Eq. (4.18). Results for different mechanical damping factors are shown in Fig. 4.19. According to Faraday's law of induction (4.11) in consideration of the displacement amplitude (4.4) together with the electromagnetic damping (4.20), the output voltage at resonance can be obtained by applying Ohm's law to the voltage divider:

$$V_{R_{\text{load,opt}}} = k_{\text{t}}(x)\, \frac{m\omega^2 Y}{\left(\dfrac{k_{\text{t}}(x)^2}{R_{\text{coil}}+R_{\text{load,opt}}}\right) + d_{\text{m}}} \cdot \frac{R_{\text{load,opt}}}{R_{\text{coil}}+R_{\text{load,opt}}}. \qquad (4.31)$$

Figure 4.20 is a plot of the output voltage for various mechanical damping factors. Again it is apparent that there exists a ratio of α'/α'' where the output voltage gets maximal. With respect to the transduction factor, this maximum is shifted to higher values of α'/α'', at higher oscillating velocities! In a final step, the resulting output power can be calculated using Joule's law:

$$P_{R_{\text{load,opt}}} = \frac{V_{R_{\text{load,opt}}}^2}{R_{\text{load,opt}}}. \qquad (4.32)$$

Consequently, the results in Fig. 4.21 show that there is also an optimal ratio of α'/α'' where the output power gets maximal.

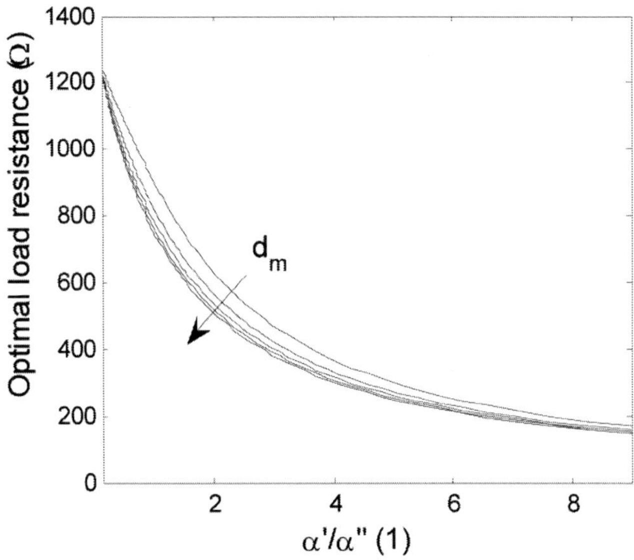

Figure 4.19   The optimal load resistance calculated with EDAM is dependent on the mechanical damping factor.

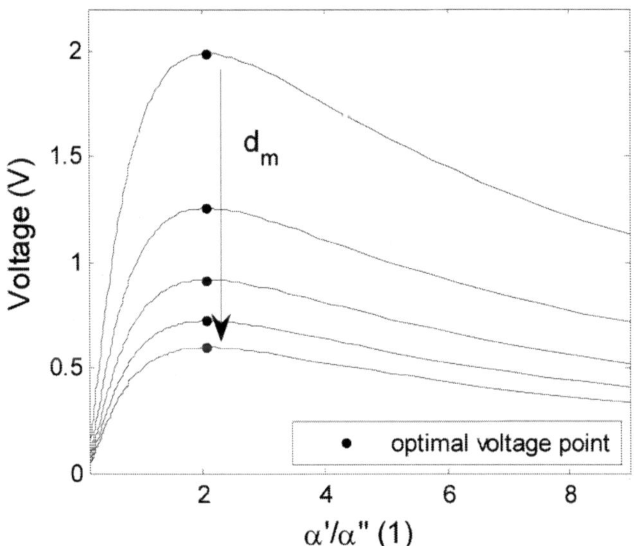

Figure 4.20   Maximum output voltage of the vibration transducer results for certain ratio of α' and α''.

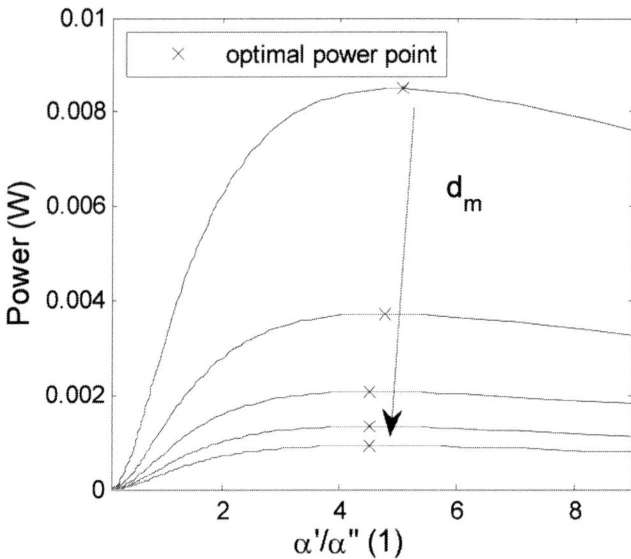

Figure 4.21 Output power as a function of the geometrical parameters in the constrained construction volume.

With respect to the output voltage this optimal ratio is once again shifted to higher values of α'/α'' at to smaller resistances! The damping ratios at optimal voltage and optimal power points are given in Fig. 4.22. In comparison to the optimal damping ratio of $d_e/d_m = 1$ stated from the basic analytical theory the optimal damping ratios of this construction volume constrained calculation example are well below and even lower than the maximal obtainable damping ratios. In spite of this, the optimal voltage points correspond to the maximum damping ratios. As mentioned, the reason for this inconsistence is that whenever the construction volume is constrained the oscillating mass and the electromagnetic coupling are not independent. In this case the task is to find the geometrical parameters which maximize the output voltage or output power respectively. In analytical form such geometrical parameter optimization can only be performed for rather simplified assumptions and only for a few coupling architectures. This is because the efficient design of electromagnetic devices requires

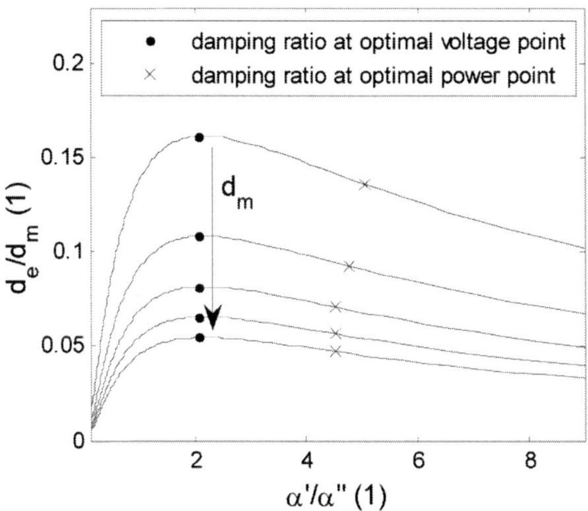

Figure 4.22 The optimum output voltage points match the maximal possible damping ratios in the constrained construction volume. In spite of this, optimum output power occurs beyond the maximum possible damping ratios.

numerical methods especially if ferromagnetic flux-conducting parts are used.

All the results presented in this chapter lead to the following conclusions for the usage of the analytical theory in the design process of electromagnetic vibration transducers:

- The outcome from the basic analytical theory is important to understand the behaviour of the subsystems, to identify most important parameters of the overall system and to investigate those dependencies.
- The output power equations are essential to make first-order power estimation under consideration of basic practical constrains like the inner displacement limit (especially for applications with random excitation).
- In literature it is often stated that the output power gets maximal if the electromagnetic damping factor equals the mechanical damping factor. Even though this is correct from the theoretical point of view it is an inapplicable design guide line for constrained construction volume condition.

- Whenever the construction volume is limited, the task is to find the optimal geometrical parameters of magnet and coil rather than optimizing damping ratios.
- It is important to know that there exists a separate optimum for voltage and power optimized design.
- Realizing an optimized design requires the calculation of magnetic field distribution. Analytical solutions based on Maxwell's equations are very difficult and sometimes impossible. Therefore it is necessary to involve numerical methods in the design process.
- Once the vibration transducer is complete and frequency response measurements have been performed the analytical theory is in the most of all cases proper for accurate simulation of the measurement results as realized in many publications.

PART 3: APPLICATION-ORIENTED DESIGN OF AN ELECTROMAGNETIC VIBRATION TRANSDUCER

4.9 Introduction

The literature review in the introduction section shows that a lot of different electromagnetic coupling architectures have been applied by numerous research facilities. However, in the multiplicity of publications, the selection criterion of the coupling architecture is often omitted. Beyond this, the geometrical parameters of the magnet, the coil and if existent the back iron parts are in the most of all cases based on rough simplifying analytical assumptions, experience if not intuition. This chapter describes the optimized dimensioning and design of a fictive application-oriented electromagnetic vibration transducer. Note that the underlying theory of the optimization approach is mainly based on the analytical analysis of Section 4.6. However, an essential part of the optimization is the accurate calculation of the transduction factor. For this purpose the spatial magnetic field calculation of cylindrical permanent magnets is used.

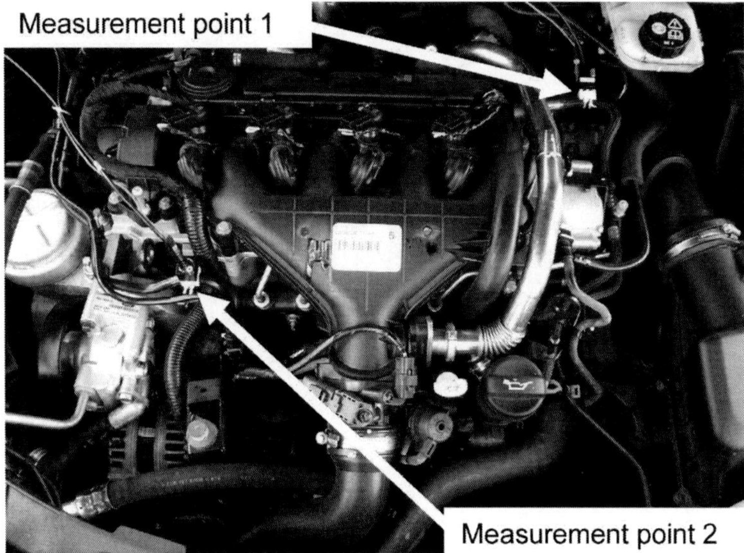

Figure 4.23 Vibration measurements in the engine compartment of a four-cylinder in-line diesel engine.

4.10 Available Vibration: The Basis for Development

The compartment of a car engine (see Fig. 4.23) is the intended example operational environment for the development of the resonant vibration transducer described in this chapter. In this environment the vibration characteristic is correlated to the load condition of the engine. Hence, the vibration will change over time in an unforeseeable manner. Nevertheless, it is possible to define energetic vibration frequencies using the order domain analysis which correlates the vibration frequency with the revolution speed. The investigated four-cylinder in-line diesel engine has eigenfrequencies of n-th order, which can be calculated using

$$f = \frac{\mathrm{rpm}}{60} \cdot n, \qquad (4.33)$$

where rpm denotes revolutions per minute. Statistical evaluation of rpm measurements during typical load profiles (e.g. city, country and highway driving route) show the most frequent rpm of about 1800 1/min at the given type of engine [19]. In a final step, this result

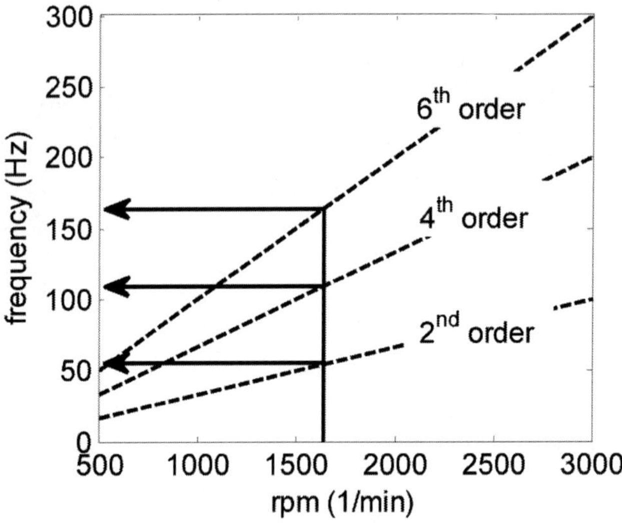

Figure 4.24 Expected predominant vibration frequencies at the considered four-cylinder in-line engine based on the order domain analyses (Campbell diagram). Frequency marked for rpm $= 1650$ min^{-1}.

can be used to determine the most energetic vibration frequencies (respectively the desired resonance frequency of the harvester). Figure 4.24 shows the plot of the eigenfrequencies of n-th order as a function of the number of revolutions (so-called Campbell diagram). Most energetic resonance frequencies are 60, 120 and 180 Hz for the conversion of the second, fourth and sixth order, respectively. Note that these frequencies are also visible in the acceleration profile in Fig. 4.14a.

4.10.1 *Coupling Architecture and Boundary Condition*

The architectures presented in Section 4.1 are compared with respect to their output voltage and output power performance. Consequently, the architectures with the highest output performance should be preferred in application whenever possible. However, there might be additional or different parameters as a benchmark like the packaging, the costs and so on. For the considered example, vibration transducer architecture A II has been chosen

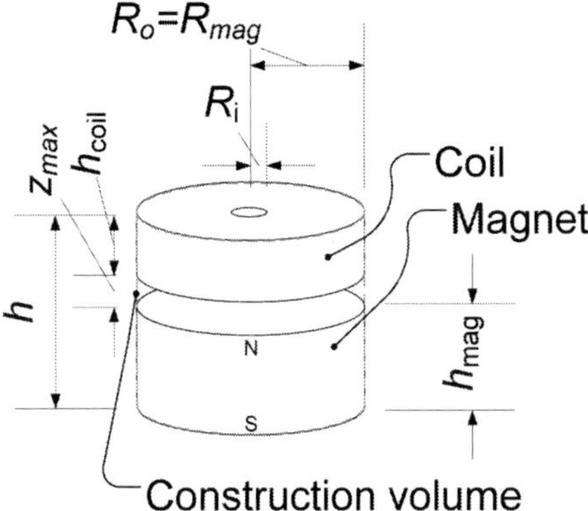

Figure 4.25 Coupling architecture used for the design of the vibration transducer.

because of the simple assembly and the high voltage generation capability. Generally the output voltage takes a vital key role in the development of small vibration transducers (<1 cm^3) especially if the excitation amplitudes are also quite small. Nevertheless, characteristic drawback of the architecture is the moderate power generation capability and the fact that the magnetic circuit is not closed. Due to this nonlinear magnetic forces may disturb the oscillation and eddy current loss takes place.

The underlying boundary conditions, based on typical mesoscale vibration transducers, are given in Table 4.2. The construction volume (see Fig. 4.25) is defined as the cylinder that encompasses the coil and the magnet at the resting position. The radius R_o is specified to be 7.5 mm and the height $h = 14.1$ mm (2.5 cm^3). The volume of spring and housing is ignored because they can be implemented in several kinds. Another fixed geometrical parameter is the gap between the static and the oscillating parts. This minimum gap size is necessary to adjust form- and position tolerances. The excitation is considered as a pure harmonic vibration with 2 m/s^2 at 60 Hz (typical for the considered operation environment). The most important material parameters of the permanent magnet are the residual flux density and the density of rare earth iron boron magnet (NdFeB). This type of rare earth magnet is often favoured in electromagnetic vibration transducers due to the high

Table 4.2 Fixed parameters of the overall boundary conditions used for the optimization

Symbol	Description	Value	Unit
	Geometry		
V_{constr}	Construction volume	2.5	cm^3
R_o	Outer radius	7.5	mm
h	Height	14.1	mm
Z_{max}	Maximum inner displacement	1.5	mm
	Magnet		
B_r	Residual flux density	1.1	T
ρ_{mag}	Density of magnet	7.6	g/cm^3
	Coil		
k_{co}	Copper fill factor	0.6	1
d_{co}	Wire diameter	40	μm
R'	Resistance per unit length	13.6	Ω/m
	Other		
Y_{acc}	Excitation amplitude	2	m/s^2
F	Excitation frequency	60	Hz
d_m	Mechanical damping	0.1	N/m/s

residual magnetic flux density. A comparison of different permanent magnet materials together with their maximum energy product development potential is shown in Fig. 4.26 [6]. Dependent on the material grade, NdFeB magnets can operate up to 220°C as shown in Fig. 4.27. The coil should be made of 40 μm diameter enamelled copper wire with typical copper filling factor of 0.6.

4.11 Optimization Procedure

4.11.1 *Calculation of Magnetic Flux Gradient*

In the considered architecture, the coil moves parallel to the diverging field of the magnet. Therewith EMF occurs via the change of magnetic field through the constant area of the coil turns. Usually the calculation of this kind of electromagnetic coupling is rather extensive because it implies numerical methods. Here the magnetic field is calculated and afterwards integrated over all windings of the coil. Another possibility is to use the mutual inductance [54].

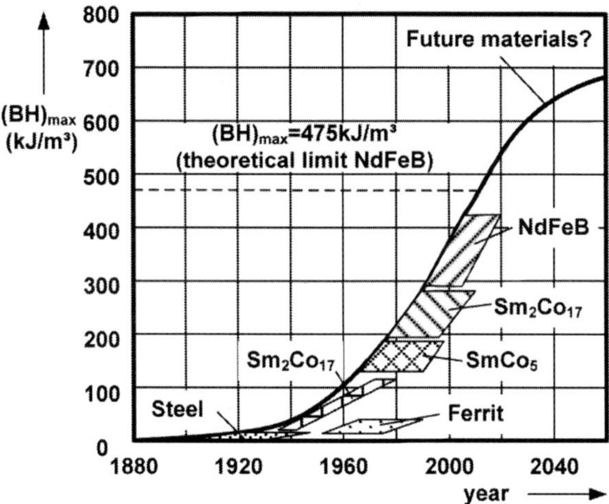

Figure 4.26   Development potential of $(BH)_{max}$ for permanent magnets [6].

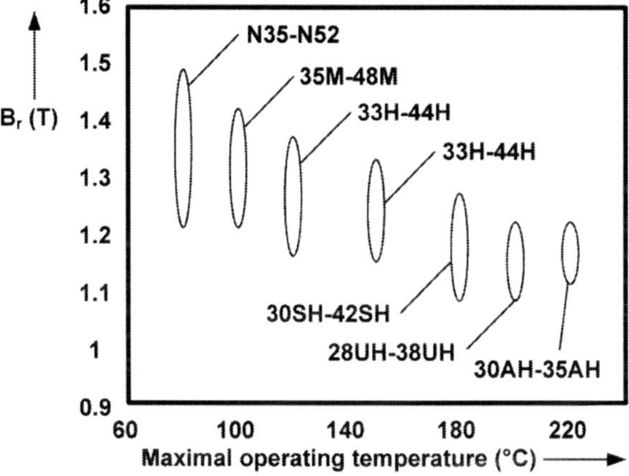

Figure 4.27   Maximal operating temperature and residual magnetic flux density for Neodymium magnet material grades.

For the calculation of permanent magnetic field distribution, there are basically two models that can be used. These are namely the scalar potential model and the vector potential model. The scalar potential model assumes fictitious magnetic charges. The vector potential model based on the molecular current viewpoint is used

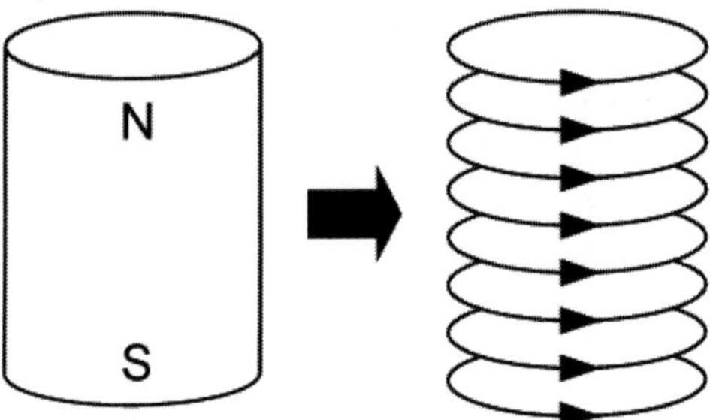

Figure 4.28 In the vector potential model, the magnet is equivalent to a current sheet.

in this chapter. Therein the magnetic flux density distribution of cylindrical permanent magnets is equivalent to that of a current sheet respectively a single-layered cylindrical coil. The underlying semi-analytical approach of static magnetic field calculation for cylindrical permanent magnets used in this chapter is given in [39]. Because the surface normal of the turns is in the same direction as the vector of the magnetization, only the field component parallel to the surface normal of the windings will cause a magnetic flux change in the coil turns. Hence, it is the only component of interest. According to the basic Eq. (4.8), the magnetic flux is obtained by integrating the axial component of the magnetic field over the area enclosed by each coil turn. Subsequently the magnetic flux of the coil is obtained by summarizing the magnetic flux of each coil turn. To reduce the computation complexity the winding area of the coil is divided into a number of cells as shown in Fig. 4.29. The magnetic flux for the coil turn in the centre of the cell is computed. This magnetic flux is representative for all the turns this cell. Therewith the computation time can be reduced significantly whereas the result is still accurate. The magnetic flux gradient is than given by the derivative of the magnetic flux function.

4.11.2 *General Calculation Method*

For optimization purpose, it is necessary to be able to calculate the output power and output voltage independent on the geometrical parameters of each architecture. Based on the analytical tools which

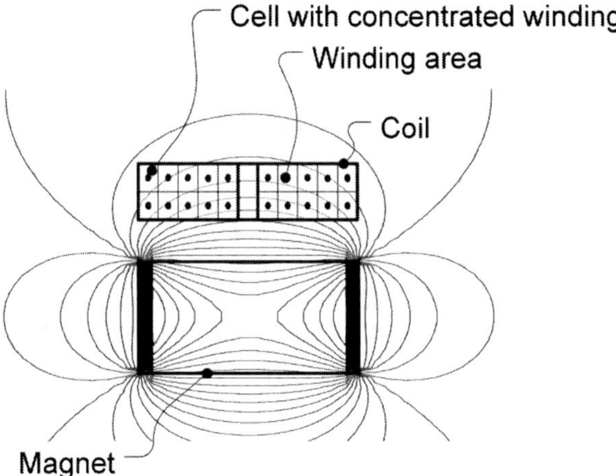

Figure 4.29 Cross section of the coupling architecture. To reduce the computation time the winding area is divided into cells and the number of windings in a cell is assumed to be concentrated in the centre.

have been introduced in Section 4.6, the steps of the analyses to perform this task are shown in the flowchart of Fig. 4.30. With respect to the overall boundary condition, the first step is to define a set of possible solution candidates. For the considered architecture, this yields a two-dimensional search space where the geometrical parameters R_i and h_{coil} need to be optimized. Note that the dimensions of the magnet follow from them. For every solution candidate the coil calculation can be done using Eqs. (4.15) and (4.16). Depending on the resulting number of windings and the internal resistance, the magnetic field and magnetic flux in the coil at the resting position of the magnet is determined as described in the previous section. The result of the coil resistance together with the magnetic flux gradient (which equals the transduction factor k_t) is necessary to determine the load resistance which yields the maximum output power according to EDAM (4.18). In this regard the transduction factor is assumed to be constant which results in a linearization of the model. Now the electromagnetic damping coefficient (4.20), the resultant relative amplitude of motion as well as the EMF (4.11) are calculated. In a final step, the resulting output power and output voltage is calculated according to Eqs. (4.31) and

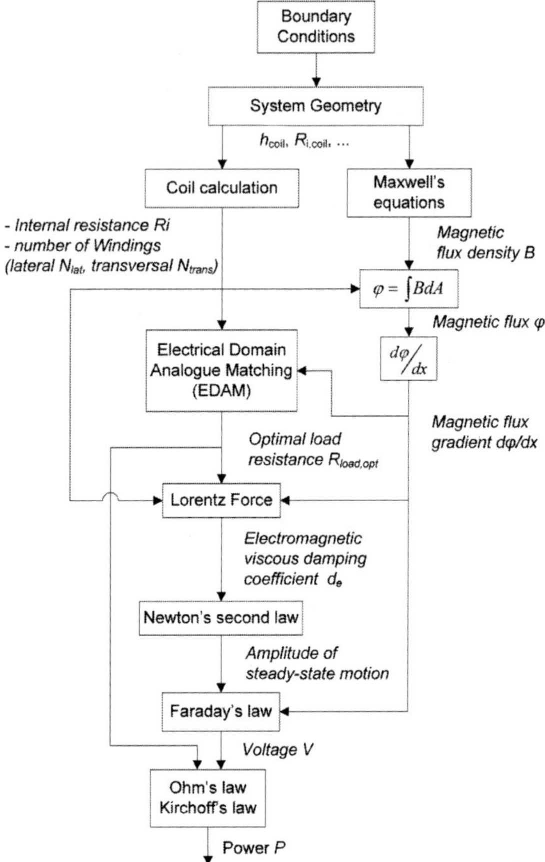

Figure 4.30 Flow chart of the calculation procedure used in the optimization approach. For each set of geometry parameters, the output power and output voltage is calculated.

(4.32). Note that in this respect, the output voltage is related to the load resistance which yields maximum output power.

4.11.3 *Optimization Results*

The resulting output parameters for different dimensions of the coil radius and height (two-dimensional search space) are represented in contour plots in Fig. 4.31. Starting with the inner displacement

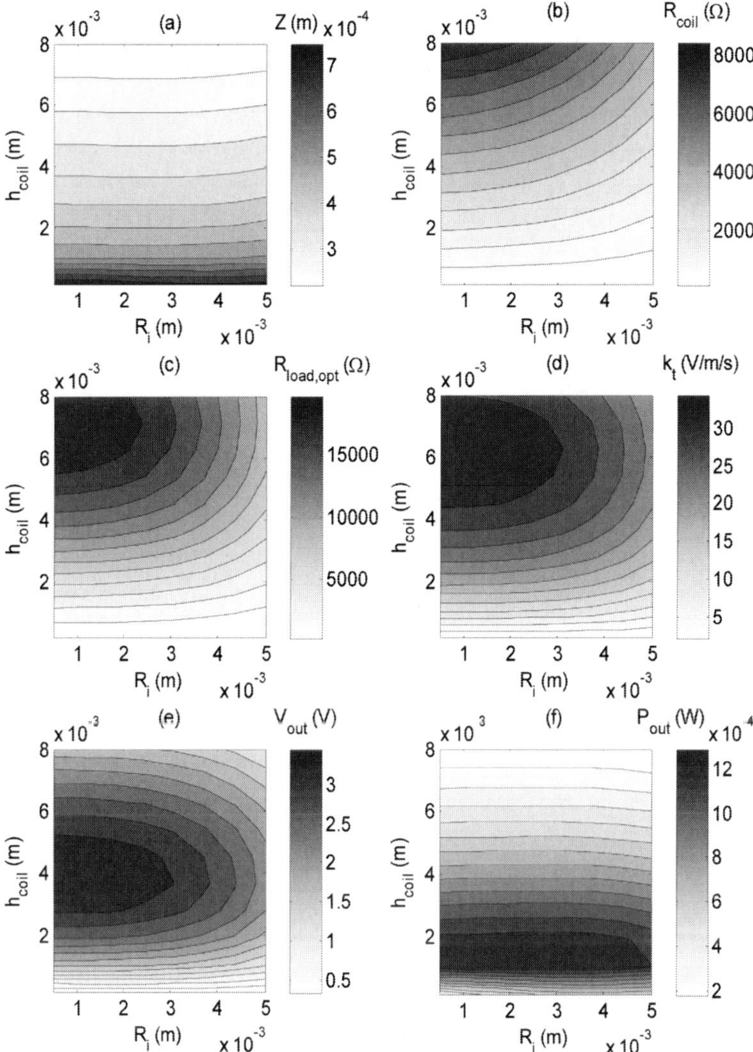

Figure 4.31 Optimization results in a construction volume of 1 cm^3. The figure shows the resulting (a) inner displacement amplitude, (b) coil resistance, (c) optimal load resistance for different dimensions of the coil. There are definitely different optimal dimensions for maximising (d) the magnetic flux gradient, (e) the output voltage and (f) the output power.

(Fig. 4.31a) it is evident that the higher the oscillating mass the higher the inner displacement will be. Because the oscillating mass only depends on the coil height but not on the inner radius, the isolines of the relative inner displacement amplitudes are almost horizontal isolines. However, for large inner radius the number of windings, the electromagnetic coupling and hence also the electromagnetic damping decreases. For this reason, the inner displacement amplitudes slightly increase. In spite of the inner displacement, the highest internal resistances result for large winding area (h_{coil} large and R_i small). This applies also for the optimal load resistance (Fig. 4.31b,c). Note that consistent with the EDAM, the absolute value of the optimal load resistance is greater than the internal resistance of the coil due to the additional term of the mechanical damping electrical analogue. Even more exciting is the result of the transduction factor (Fig. 4.31d) where an optimum is inside the defined design domain. However, the goal at this point is not the optimization of the transduction factor but the optimization of the output voltage and the output power (Fig. 4.31e,f). With respect to the transduction factor the optimum of the output voltage is shifted to smaller coil height or in other words to higher oscillation amplitudes! This is plausible because the EMF depends on both the transduction factor and the oscillating velocity. The optimum for the output power is consequently further shifted to smaller resistances with respect to the output voltage optimum. These results show that there are separate design optima for the transduction factor, the output voltage and the output power. For the given boundary conditions the highest possible output power is 1.38 mW at 1.88 V for a coil with 3.07 mm inner radius and 1.70 mm height. The output voltage is maximized for a coil with 0.50 mm inner radius and 3.67 mm height. With these dimensions, 3.73 V can be obtained at a power level of 0.96 mW. The remaining parameters can be picked up from the contour plots. Both the output power and the output voltage optimized dimensioning are rather sensitive to the coil height than to the inner radius of the coil. Note that the resulting inner displacement amplitude for voltage optimized design (400 μm) is smaller than for power optimized design (560 μm). This effect is due to the volume constraint and has already been observed

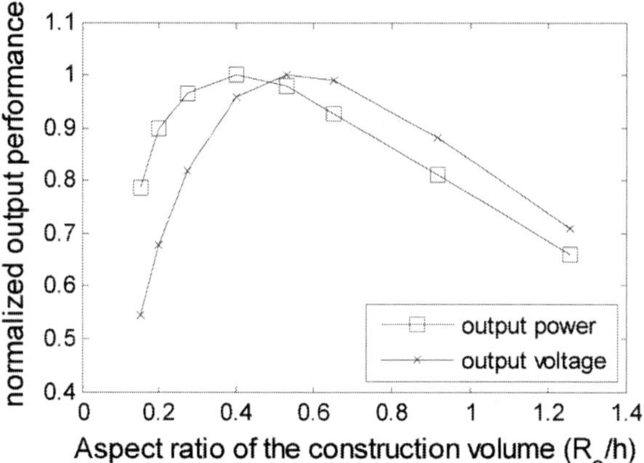

Figure 4.32 There are different optimal aspect ratios for the construction volume where the output power and output voltage are maximal.

for the simplified model where the damping at the voltage optimum is greater than for the power optimum (Fig. 4.22).

Altogether the results show that for construction volume constrained condition there are separate optimal dimensions for the output power and the output voltage. To maximize the output performance, it is of great importance to be able to find these optimal dimensions. Otherwise the available excitation will not be used efficiently. For example, a coil with 3 mm height (instead of 1.07 mm) already results in a drop of the maximal output power of about 40%! In this optimization, the dimensions of the construction volume yield an aspect ratio of 0.53. This is a good compromise between the optimal aspect ratio for power and voltage optimized design. The maximum output power and output voltage for other aspect ratios is shown in Fig. 4.32.

4.12 Resonator Design

For the design of the resonator, it is first of all essential to know the inertial mass and the desired resonance frequency. According to the optimization results (the considered vibration transducer should operate at the power optimum), the optimal magnet height

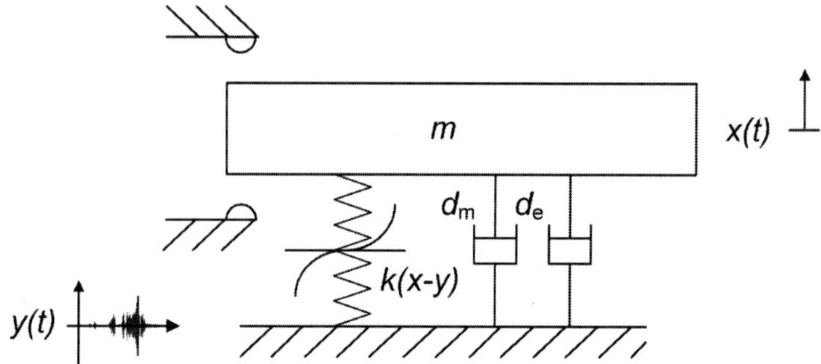

Figure 4.33 Overall model of the vibration transducer used for transient simulations.

is 10.90 mm. This value is rounded down to 10 mm in order to use standard magnets. Hence, the total mass of the magnet is 13.43 g. Note that it is appropriate to round down rather than round up in order to shift the dimensions into the direction of the voltage optimum. The desired resonance frequency has already been defined in the boundary conditions. Beyond this it is worthwhile to investigate the effect of nonlinear spring behaviour as defined in Eq. (4.5) in order to increase the output power (especially for stochastic excitation). This can be done using transient simulations. Basic features of the simulation model used here (shown in Fig. 4.33) are that the measured stochastic acceleration data can directly be used as excitation, the nonlinearity of the spring can be adjusted and the oscillation range is limited by mechanical stoppers (as it is for any application-oriented device!). In the simulation, the latter can be included using partially elastic collision [8]. For different linear and nonlinear (hardening and softening) springs, the RMS value of the power dissipated in the viscous electromagnetic damping element has been recorded. Typical simulation results for different inner displacement limits are shown in Fig. 4.34. Note that the most energetic resonance frequencies are in agreement with the order domain and the frequency domain analyses. For inner displacement limit of (a) 1 mm and (b) 1.5 mm (as specified in the boundary conditions), the conversion of the second order is more efficient than the conversion of the fourth order and nonlinear springs are not advantageous. Nevertheless, if higher inner displacement

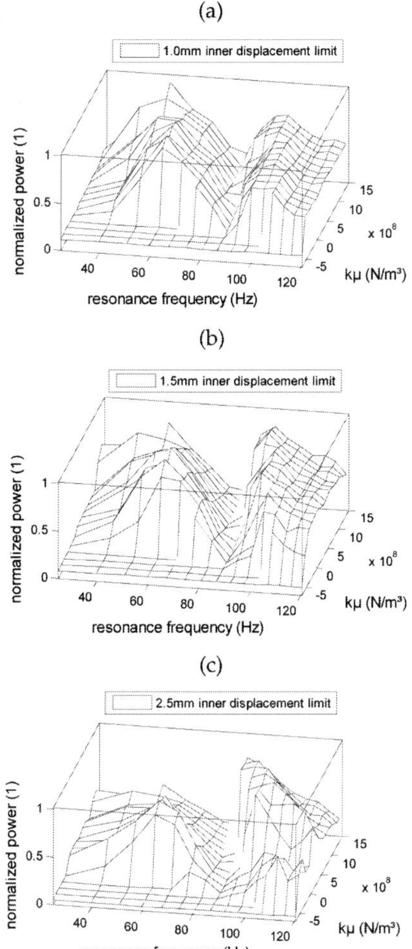

Figure 4.34 Normalized output power dependent on the resonance frequency and the nonlinearity of the spring for different inner displacement limits. The second and fourth order is clearly apparent. At 1.5 mm inner displacement the fourth order conversion become more efficient and the highest output power is obtained with a nonlinear hardening spring.

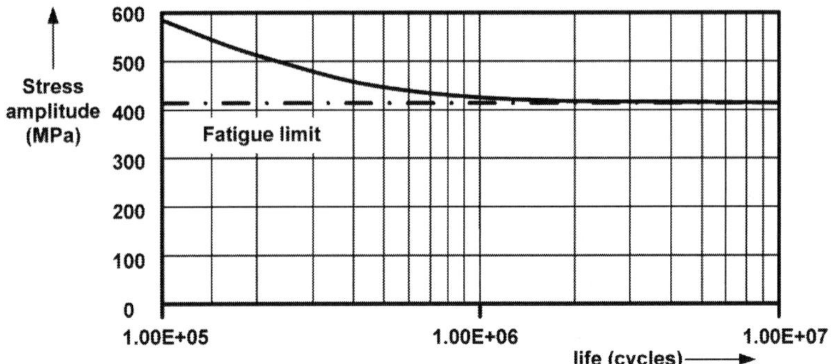

Figure 4.35 Wöhler curve of typical spring material $CuSn_6$ (R550) [7].

limit could be applied the conversion of the fourth order would be more efficient and nonlinear hardening springs are capable of generating the highest output power (c). However, the inner displacement cannot be increased arbitrary without violating the fatigue limitation caused by the spring material. The spring material used here is $CuSn_6$. The corresponding Wöhler curve (also known as the S–N curve) is shown in Fig. 4.35. According to this curve, the mechanical stress level has to be well below 410 MPa in order to enable unlimited lifetime of the spring element. Note that due to the considerable amount of cycles in vibration transducers this is a rather critical design criterion. A FEA-based modal analysis of the resonator system (containing the spring element, the magnet and a support part which provides the distance between the magnet and the spring [see Fig. 4.36]) is shown in Fig. 4.37. The width

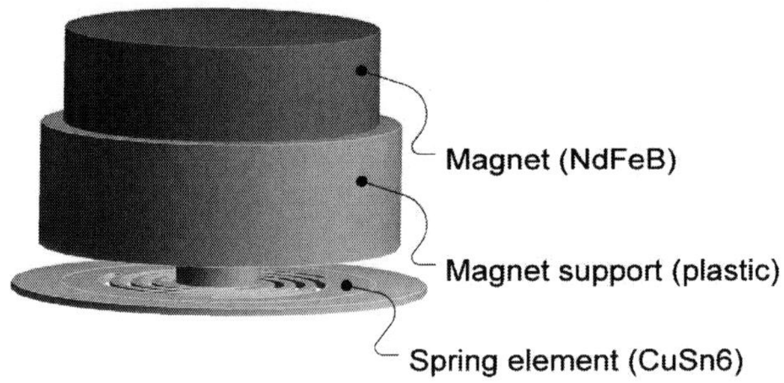

Figure 4.36 Considered resonator assembly

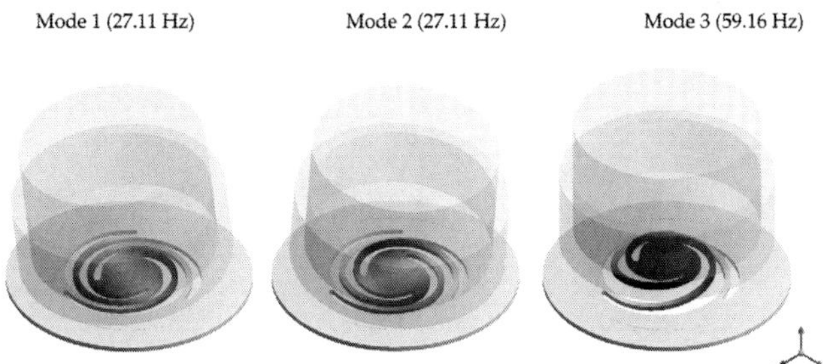

Figure 4.37 First three modes of the resonator system for second order conversion. The linear oscillation of the third mode is used for energy conversion.

and the length of the spring beams define the resonance frequency. However, in the considered application it is not so important to match the resonance frequency exactly. Tolerances of 60± 5 Hz are still acceptable. The third mode (pure linear oscillation) is used for the energy conversion whereas the first two modes of the resonator are unwanted. The stress in the spring for mode 3 is shown in Fig. 4.38. Due to the high oscillation mass, the spring constant is also rather high. Thus the expected stress level at about 1 mm inner displacement already causes fatigue problems. Hence, the inner displacement needs to be reduced in order to operate well below the fatigue limit. In the strict sense this fact changes the boundary conditions and the optimization calculations should be repeated. This underlines that the development of an optimal vibration transducer is not straight forward but an iterative process.

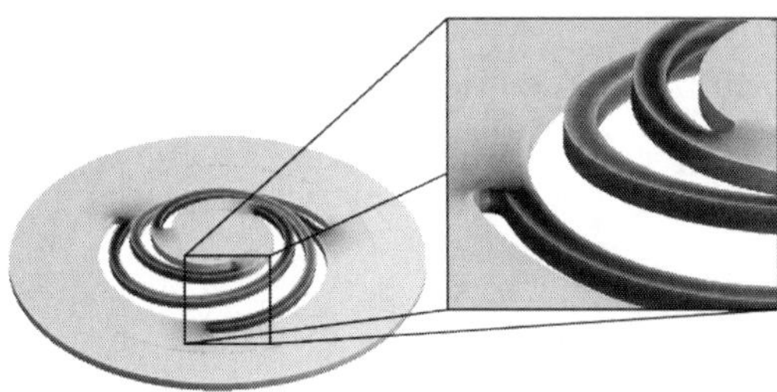

Figure 4.38 Equivalent (von-Mises) stress in the spring element. The highest stress level appears where the beam merges into the outer ring.

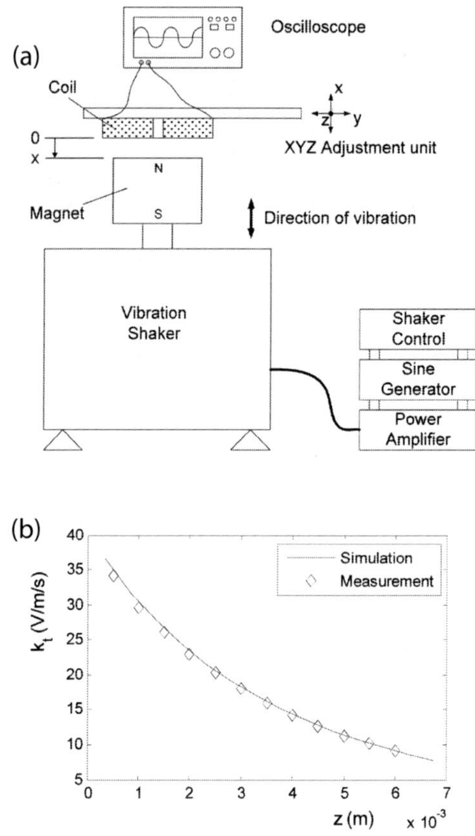

Figure 4.39   (a) Measurement set-up to measure the transduction factor at different distances between magnet and coil. (b) Measured and simulated transduction factor of the prototype vibration transducer.

PART 4: PROTOTYPE PERFORMANCE

4.13 Transduction Factor

The transduction factor can simply be measured by attaching the magnet (without resonator) to a shaker unit with accelerometer feedback, whereas the coil is mounted on a fixed XYZ adjustment unit as shown in Fig. 4.39a. The coil of the prototype vibration

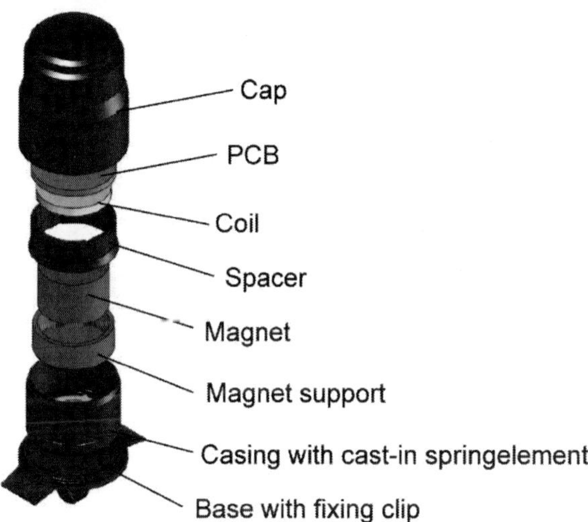

Figure 4.40 Photograph and exploded view of the prototype vibration transducer assembled with casting components.

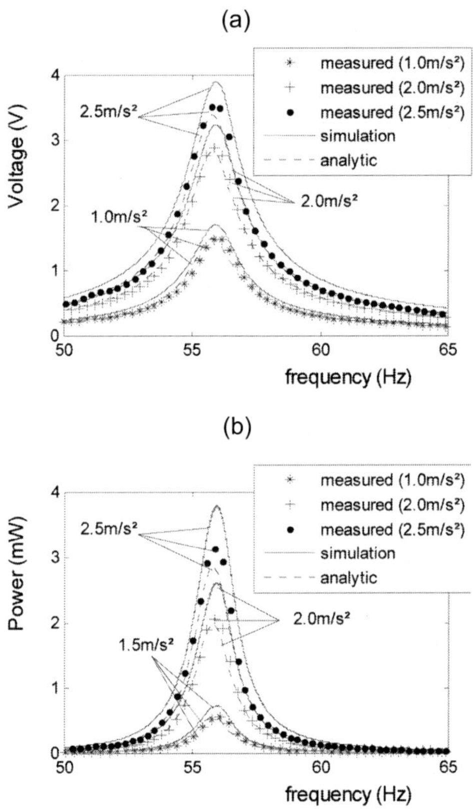

Figure 4.41 Measured, simulated and analytically calculated frequency response of the prototype vibration transducer for different excitation amplitudes and a load resistance of 4 kΩ.

transducer has an outer radius of 8 mm, an inner radius of 1.5 mm and a height of 2 mm. For a wire diameter of 40 μm, this results in a total resistance of 1925 Ω and an inductance of 789 mH. Afterwards, the shaker is forced to vibrate at a predefined controlled oscillation velocity (defined by the amplitude and the frequency of the harmonic vibration). The transduction factor is consequently given by the rearrangement of (4.11) where the emf can, for example, be measured with an oscilloscope. Note that due to the nonlinearity of the transduction factor the amplitudes of the shaker vibration should be small (<0.1 mm) in order to obtain accurate

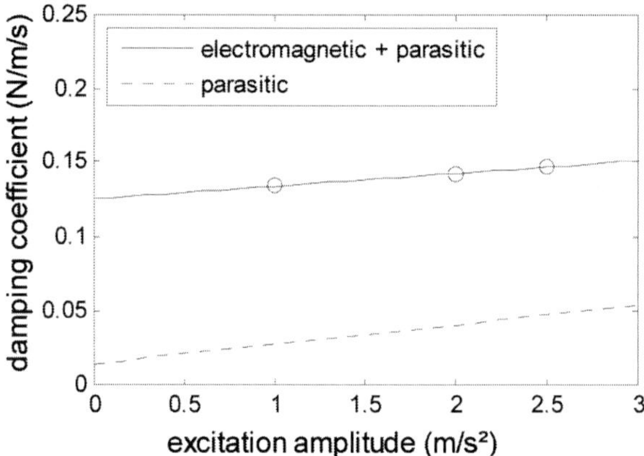

Figure 4.42 Measured overall damping coefficient (electromagnetic plus parasitic) at different excitation amplitudes. The pure parasitic damping (dashed curve) has been deduced using the analytical expression for the electromagnetic damping (4.19).

results. A comparison of measured and simulated transduction factor values for the prototype vibration transducer is shown in Fig. 4.39b.

4.14 Frequency Response Characterisation

With the assembled prototype vibration transducer (shown in Fig. 4.40) on the shaker unit, frequency response measurements have been performed for different excitation amplitudes. The Q-factor of the measured response curves can easily be extracted using the resonance frequency and the $(-3\,\text{dB})$ bandwidth. Consequently the damping coefficients are given by:

$$d_{\text{tot}} = d_{\text{m}} + d_{\text{e}} = \frac{m \cdot 2\pi \cdot f_0}{Q_{\text{tot}}}. \tag{4.34}$$

Using Eq. (4.20) and the total damping coefficient d_{tot} the parasitic damping can be separated from the electromagnetic damping. Note that due to the nonlinearity in the transduction factor the electromagnetic damping is nonlinear as well. In the first glance, this effect has been neglected here. Voltage and power response measurements with 1.0, 2.0 and 2.5 m/s^2 excitation amplitude and a load resistance of 4 kΩ are shown in Fig. 4.41. Even at 1 m/s^2 the voltage level is obviously higher than 1 V The corresponding parasitic damping coefficient (Fig. 4.42) has been extracted from the measurements and afterwards applied to the analytical Eq. (4.24) and the transient simulation model (already used to determine most efficient spring characteristic in 3.5).

References

1. A. Timotin and M. Marinescu, Die optimale Projektierung eines magnetischen Kreises mit Dauermagnet für Lautsprecher, *Archiv für Elektrotechnik*, 54, 229–239, 1971.

2. Available in the internet: http://www.ferrosi.com/, state February 2010.

3. Available in the internet: http://www.kcftech.com/, state February 2010.

4. Available in the internet: http://www.lumedynetechnologies.com/, state February 2010.

5. Available in the internet: http://www.perpetuum.com/, state February 2010.

6. Available in the internet: http://www.vacuumschmelze.de, state February 2010.

7. Available in the internet: http://www.wieland.de/, state February 2010.

8. Available in the internet: www.ba-horb.de/~ga/PDF_Files/SimuMech Ansch.pdf, state February 2010.

9. B. Yang et al., Electromagnetic energy harvesting from vibrations of multiple freuquencies, *J. Micromech. Microeng.*, 19(3), March 2009, 035001.

10. C. Peters, D. Maurath, W. Schock, F. Mezger, and Y. Manoli: A closed-loop wide-range tunable mechanical resonator for energy harvesting systems, *J. Micromech. Microeng.*, 19(9), 2009, 094004.

11. C. Serre, A. Perez-Rodriguez, N. Fondevilla, E. Martincic, S. Martinez et al., Design and implementation of mechanical resonators for optimized inertial electromagnetic microgenerators, *Microsyst. Technol.*, 14(4–5), 653–658, April 2008.

12. C. B. Williams, R. B. Yates, Analysis of a micro-electric generator for microsystems, in Transducers '95, Eurosensors IX, The 8th International

Conference on Solid-State Sensors and Actuators, and Eurosensors IX, Stockholm, Schweden, pp. 369–372, June 1995.

13. C. Peters, D. Spreemann, M. Ortmanns, and Y. Manoli, A CMOS integrated voltage and power efficient AC/DC converter for energy harvesting applications, *J. Micromech. Microeng.*, 18(10), September 2008, 104005.

14. C. R. Mc Innes, D. G. Gorman, and M. P. Cartmell, Enhanced vibrational energy harvesting using non-linear stochastic resonance, *J. Sound Vibration*, 318(4–5), 655–662, December 2008.

15. C. R. Saha, T. O'Donnel, N. Wang, and P. Mc Closkey, Electromagnetic generator for harvesting energy from human motion, *Sens. Actuators A*, 147, 248–253, 2008.

16. D. Hoffmann, C. Kallenbach, M. Dobmaier, B. Folkmer, and Y. Manoli, Flexible polyimide film technology for vibration energy harvesting, in *Proceedings of PowerMEMS 2009*, pp. 455–458, Washington DC, USA, December 2009.

17. D. P. Arnold, Review of microscale magnetic power generation, *Trans. magn.*, 43(11), 3940–3951, November 2007.

18. D. Spreemann et al., Tunable Transducer for low frequency vibrational energy scavenging, *20th Eurosensors* Conference, Göteborg, 2006.

19. D. Spreemann, A. Willmann, B. Folkmer, and Y. Manoli, Characterization and in situ test of vibration transducers for energy-harvesting in automobile applications, in *Proceedings of PowerMEMS 2008*, pp. 261–264, Sendai, Japan, November 2008.

20. D. Spreemann, B. Folkmer, and Y. Manoli, Comparative study of electromagnetic coupling architectures for vibration energy harvesting devices, in *Proceedings of PowerMEMS 2008*, pp. 261–264, Sendai, Japan, November 2008.

21. D. Spreemann, B. Folkmer, and Y. Manoli, Optimization and comparison of back iron based coupling architectures for electromagnetic vibration transducers using evolution strategy, in *Proceedings of PowerMEMS 2009*, pp. 372–375, Washington DC, USA, December 2009.

22. D. Spreemann, D. Hoffmann, B. Folkmer, and Y. Manoli, Numerical optimization approach for resonant electromagnetic vibration transducer designed for random vibration, *J. Micromech. Microeng.*, 18(10), October 2008.

23. D. Spreemann, D. Hoffmann, E. Hymon, B. Folkmer, and Y. Manoli, Über die Verwendung nichtlinearer Federn für miniaturisierte Vibrationswandler, *Mikrosystemtechnik Kongress*, 2007, Oct. 15–17, Dresden, 2007 (in German).

24. D. Spreemann, Y. Manoli, B. Folkmer, D. Mintenbeck, Non-resonant vibration conversion, *J. Micromech. Microeng.*, 16(9), 679–685, August 2006.

25. D. Zhu, S. Roberts, J. Tudor, and S. Beeby, Closed loop frequency tuning of a vibration-based micro-generator, in *Proceedings of PowerMEMS 2008*, pp. 229–232, Sendai, Japan, November 2008.

26. D. J. Domme, Experimental and analytical characterization of a transducer for energy harvesting through electromagnetic induction, Virginia State University, Master thesis, April 2008.

27. Daniel J. Inman, *Engineering Vibration*: Second Edition, Prentice Hall, 2000, USA, ISBN 0-13-726142-X.

28. E. Bouendeu, A. Greiner, P. J. Smith, and J. G. Korvink, An efficient low cost electromagnetic vibration harvester, in *Proceedings of PowerMEMS 2009*, pp. 372–375, Washington DC, USA, December 2009.

29. E. Koukharenko et al., Microelectromechanical systems vibration powered electromagnetic generator for wireless sensor applications, *Microsyst. Technol*, 12, 1071–1077, 2006.

30. E. I. Rivin, *Passive Vibration Isolation*, ASME Press, New York, USA, 2003, ISBN 0-79-180187-X.

31. G. Hatipoglu, H. Ürey, FR4-based electromagnetic energy harvester for wireless sensor node, *Smart Mater. Struct.*, 19(1), 015022, 2010.

32. G. Naumann, Energiewandlersystem für den Betrieb von autarken Sensoren in Fahrzeugen, TU Dresden, Phd thesis, December 2003 (in German).

33. Giancarlo Genta, *Vibration of Structures and Machines: Practical Aspects*, Springer, 3rd edition, 1998, ISBN 978-0-387-98506-0.

34. H. Toepfer et al., Electromechanical design and performance of a power supply for energy-autonomous electronic control units, 51st IWK—Internationales Wissenschaftliches Kolloquium, TU Illmenau, September 2006.

35. H. A. Wheeler, Simple inductance formulas for radio coils, *Proc. IRE*, 16(10), 1398–1400, October 1928.

36. I. Rechenberg, *Evolutionsstrategie—Optimierung technischer Systeme nach Prinzipien der biologischen Evolution*, Frommann Holzboog, Stuttgart, 1973.

37. J. K. Ward and S. Behrens, Adaptive learning algorithms for vibration energy harvesting, *Smart Mater. Struct.*, 17(3), 035025, June 2008.

38. J. T. Tanabe, *Iron Dominated Electromagnets: Design, Fabrication, Assembly and Measurement*, World Scientific, Singapore, 2005, ISBN 981-256327-X.

39. K. Foelsch, Magnetfeld und Induktivität einer zylindrischen Spule, *Electrical Eng. (Archiv für Elektrotechnik)*, 30(3), March 1936 (in German).

40. K. Takahara, S. Ohsaki, H. Kawaguchi, and Y. Itoh, Development of linear power generator: conversion of vibration energy of a vehicle to electric power, *J. Asian Electric Vehicles*, 2(2), 639–643 December 2004.

41. E. Kallenbach, R. Eick, P. Quendt, T. Ströhla, K. Feindt, M. Kallenbach, and O. Radler, *Elektromagnete: Grundlagen, Berechnung, Entwurf und Anwendung, Teubner*, Germany, 2003, ISBN 3-519-16163-X.

42. M. Bousonville, Optimierung von Lautsprechermagnetsystemen mit dem Finite-Elemente-Verfahren, diploma thesis, Department of Engineering Sciences, Fachbereich Informationstechnologie und Elektrotechnik, Wiesbaden.

43. M. Rossi, Acoustics and Electroacoustics, Artech House, 1988, ISBN-10 0890062552.

44. M. S. M. Soliman, E. M. Abdel-Rahman, E. F. El-Saadany, and R. R. Mansour, A wideband vibration-based energy harvester, *J. Micromech. Microeng.*, 18(11), 115021 doi:10.1088/0960-1317/18/11/115021, November 2008.

45. N. Wang and D. P. Arnold, Fully batch-fabricated MEMS magnetic vibrational energy harvesters, in *Proceedings of PowerMEMS 2009*, pp. 348–351, Washington DC, USA, December 2009.

46. N. G. Stephen, On energy harvesting from ambient vibration, *J. Sound Vibration*, 293, 409–425, December 2005.

47. N. N. H. Ching, H. Y. Wong, W. J. Li, P. H. W. Leong, and Z. Wen, A laser-micromachined multi-modal resonating power transducer for wireless sensing system, *Sens. Actuators A,* 97–98, 685–690, 2002.

48. P. Niu, P. Chapman, L. DiBerardino, and E. Hsiao-Wecksler, Design and Optimization of a Biomechanical Energy Harvesting Device, Power Electronics Specialists Conference PESC, pp. 4062–4069, June 2008.

49. P. Wang, X. Dai, X. Zhao, and G. Ding, A micro electromagnetic low level vibration energy harvester based on MEMS technology, *Microsyst. Technol.*, 15(6), 941–951, June 2009.

50. P. D. Mitcheson, Analysis and Optimisation of Energy-Harvesting Micro-Generator Systems, PhD thesis, University of London, 2005.

51. P. D. Mitcheson, T. C. Green, E. M. Yeatman, and A. S. Holmes, Architectures for vibration-driven micropower generators, *IEEE/ASME J. Microelectromech. Syst.*, 13(3), 429–440, June 2004.

52. S. Cheng, D. P. Arnold, A study of a multi-pole magnetic generator for low-frequency vibrational energy harvesting, *J. Micromech. Microeng.*, 20(2), 025015 doi:10.1088/0960-1317/20/2/025015, February 2010.

53. S. Park and K. C. Lee, Design and analysis of a microelectromagnetic vibration transducer used as an implantable middle ear hearing aid, *J. Micromech. Microeng.*, 12(5), 505 doi:10.1088/0960-1317/12/5/301, September 2002.

54. S. I. Babic and C. Akyel, New analytic-numerical solutions for the mutual inductance of two coaxial circular coils with rectangular cross section in air, *Trans. Magn.*, 42(6), 1661–1669, June 2006.

55. S. J. Roundy, Energy Scavenging for Wireless Sensor Nodes with a Focus on Vibration to Electricity Conversion, PhD thesis, University of California, Berkley, 2003.

56. S. P. Beeby et al., A micro electromagnetic generator for vibration energy harvesting, *J. Micromech. Microeng.*, 17(7), 1257–1265, 2007.

57. T. Bäck, *Evolutionary Algorithms in Theory and Practice*, Oxford University Press, UK, 1996, ISBN 0-19-509971-0.

58. T. Lai, C. Huang, and C. Tsou, Design and fabrication of acoustic wave actuated microgenerator for portable electronic devices, Symposium on Design, Test, Integration and Packaging of MEMS/MOEMS, pp. 28–33, April 2008.

59. T. von Büren and G. Tröster, Design and optimization of a linear vibration-driven electromagnetic micro-power generator, *Sens. Actuators A*, 135, 765–775, 2007.

60. William T. Thomson, *Theory of vibrations with applications*, Prentice Hall, USA, 1998, ISBN 0-13-651068-X.

61. X. Cao, W. J. Chiang, Y. C. King, and Y. K. Lee, Electromagnetic energy harvesting circuit with feedforward and feedback DC-DC PWM boost converter for vibration power generator system, *Trans. Power Electron.*, 22(2), 679–685, March 2007.

62. X. Gou, Y. Yang, and X. Zheng, Analytic expression of magnetic field distribution of rectangular permanent magnets, *Appl. Math. Mechan.*, 25(3), 297–306, March 2004.

Chapter 5

Electrostatic Transducers

Daniel Hoffmann and Bernd Folkmer

*Institute of Micro- & Information Technology, Hahn-Schickard Society e.V.,
Wilhelm-Schickard-Str. 10, 78052 Villingen-Schwenningen, Germany*
bernd.folkmer@hsg-imit.de

5.1 Physical Principle

5.1.1 Introduction

The choice of a specific transducer mechanism (electromagnetic,
piezoelectric, or electrostatic) for harvesting electrical energy from
kinetic motion, in particular mechanical vibrations, is heavily dependent on both the operating conditions (amplitude and frequency
spectrum of the excitation source) and available space given by the
application environment.

The selection process may be further rationalized by considering
the scaling behavior of electromagnetic and electrostatic forces
[1]. The electromagnetic coupling coefficient scales at a different
rate than the electrostatic coupling coefficient [2]. Therefore, when
the transducer system decreases by a factor of 100 in size, the
electrostatic coupling coefficient decreases by factor of 100, whereas
the electromagnetic coupling coefficient decreases by a factor of

Handbook of Energy Harvesting Power Supplies and Applications
Edited by Peter Spies, Loreto Mateu, and Markus Pollak
Copyright © 2015 Pan Stanford Publishing Pte. Ltd.
ISBN 978-981-4241-86-1 (Hardcover), 978-981-4303-06-4 (eBook)
www.panstanford.com

1000 [2]. From this it follows, that the electrostatic conversion mechanism is more efficient for transducer devices with a size of a typical MEMS device (< 100 mm^3). On the contrary, an electrostatic transducer device may be outperformed by an electromagnetic transducer for larger devices sizes (>1 cm^3).

Moreover, based on the fact that the technology for manufacturing electrostatic transducers such as capacitive-based sensors and actuators (e.g., accelerometers, gyroscopes, comb drives) is well established, it is beneficial to utilize the same technology (standard MEMS technology) for manufacturing electrostatic energy harvesting devices. This will allow the production of devices in large numbers at low cost. Moreover, an easy integration with electronics is also provided since standard packages as used for inertial sensors are employed for packaging. Therefore, electrostatic transducer devices can be handled and manipulated as any other electronic component providing a high level of integration.

In conclusion, if miniaturization of a kinetic energy transducer is an important issue due to little space available and generator devices are required at a large production scale, electrostatic MEMS transducers are the preferred choice. However, it must be considered that the output power of any inertial energy transducer is proportional to the oscillating mass. Consequently, miniaturized energy transducers having a very small proof mass will provide only low power levels in the range of μW in the first place. Therefore, the application of electrostatic MEMS transducers requires a consumer load, which is based on ultra-low-power devices.

The idea of electrostatic energy conversion goes back to 1976 when O. P. Breaux filed a patent about a rotary non-resonant conversion system [3]. In 1995, Williams et al. first published the concept of the electrostatic energy conversion method, together with the piezoelectric and electromagnetic conversion method, to be used as a transduction technology in the field of inertial energy harvesting [4]. Since then a number of research groups have been working on the area of electrostatic energy conversion systems for kinetic energy harvesting applications [5–12].

In order to convert kinetic vibration energy into usable electrical energy by means of energy harvesting, two functional mechanisms are required:

- a transduction mechanism to convert mechanical energy into electrical energy
- a mechanical mechanism to couple ambient vibrations to transduction mechanism

The transduction mechanism enables the transformation of kinetic energy into the electrical domain. In this respect an electrical current or voltage is generated which can be used to power electronic components. For the transduction mechanism to work effectively, a relative motion between two components must be available. Therefore, a mechanical mechanism is necessary in order to produce that relative motion required by the transduction mechanism. A very widespread principle to realize inertial vibration generators is based on a mechanical resonator: A proof mass is supported by a suspension, which is attached to a frame. The frame acts as a fixed reference and is attached to the vibration source. The inertia of the proof mass results in a relative displacement when the frame experiences acceleration. Once the proof mass is displaced from its equilibrium position, the restoring force of the suspension starts to act on the proof mass and oscillations occur. As a consequence, a relative motion between proof mass and frame is generated. The oscillations of the proof mass can be damped by a suitable transducer mechanism and thus kinetic energy is converted into electrical energy.

5.1.2 *Energy Conversion Mechanism*

The energy conversion mechanism of an electrostatic transducer is based on the physical coupling of the electrical and the mechanical domain by an electrostatic force. The electrostatic force is induced between opposite charges stored on two opposing electrodes. The amount of charge Q that accumulates on the electrodes is a function of the potential difference V between the electrodes and the capacitance C according to the relation $Q = C \cdot V$. The energy stored in the electric field that establishes between the electrodes follows from the equation $E = \frac{1}{2} \cdot C \cdot V^2$ or $E = \frac{1}{2 \cdot C} \cdot Q^2$, respectively. In order to convert mechanical energy into electrical energy by means of the electrostatic transduction mechanism, a variation of capacitance over time must occur [4, 13]. The physical principle of the energy conversion cycle then depends on how the variable capacitor is connected within the electrical circuitry.

There are two main schemes that may be applied for incorporation of a variable capacitor into an electrical circuit to form an electrostatic transducer system: switched systems and continuous systems [14]. Switched transducers are further classified according to their mode of operation, which utilizes either a charge-constrained or a voltage-constrained conversion cycle. The two operation modes of a switched transducer system may be considered as special cases of a continuous system where a variable capacitor is continuously connected to the electrical circuitry. In this respect, an electrostatic transducer operated in a charge-constrained mode by a switching system is equivalent to a continuous transducer operated with infinitely high load impedance (open circuit). In analogy, the voltage-constrained transducer is comparable with a continuous transducer operated with a load impedance of zero (short circuit) [14]. However, no work can be done when either the generated current or the generated voltage is zero and thus a switching circuit is required to make them operate.

5.1.3 *Switched Operation Scheme*

In a switch-operated transducer system the condition of the variable capacitor is changed in a discontinuous manner during the conversion cycle through the operation of switches. As previously described, in a switched scheme, the transducer may be operated either in a constant charge mode or in a constant voltage mode [13]. Figure 5.1a shows the ideal energy conversion cycle of a charge-constrained cycle in a Q–V diagram. When the variable capacitor is at its maximum capacitance, a switching operation (SO1) is carried out to charge the capacitor to a low voltage V_{low} in the first part of the cycle. The energy for charging the variable capacitor must be provided by a reservoir. The amount of charge Q_{max} stored in the capacitor is equal to the product $C_{\text{max}}V_{\text{low}}$. The capacitor is then disconnected (SO2) from the reservoir so that no charge may flow in or out. Since the capacitor is now in a charge-constrained condition, a reduction of its capacitance from C_{max} to C_{min} will

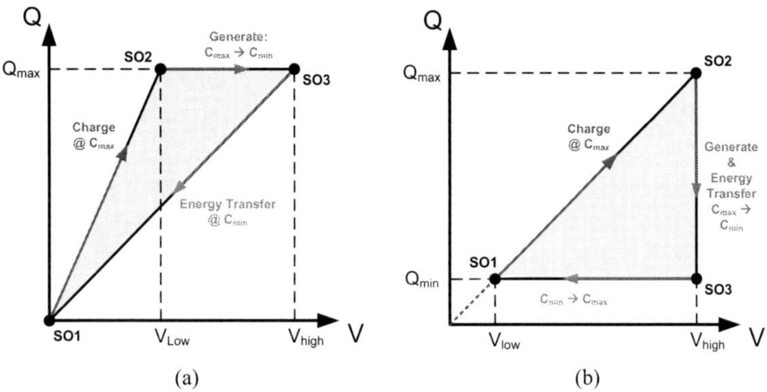

Figure 5.1 Operation mode of electrostatic transducers within a switched scheme: (a) Charge-constrained conversion cycle, (b) voltage-constrained conversion cycle.

result in a voltage increase from V_{low} to V_{high}. As a consequence, the energy content in the capacitor will increase by the factor C_{max}/C_{min} [15]. This energy gain comes from the force required to change the capacitance. The generated energy is equal to the shaded area enclosed by the Q–V diagram. In the third part of the conversion cycle the capacitance is discharged by connecting it (SO3) to a reservoir. Finally, the capacitance is increased from C_{min} to C_{max} and becomes then ready for the cycle to restart.

An alternative energy conversion cycle is shown in Fig. 5.1b using a voltage-constrained operation mode. In this cycle, a variable capacitor is initially charged at its maximum capacitance to a high voltage V_{high} by a voltage source. Again, the charge at this point will be $Q_{max} = C_{max}V_{high}$. However, in this case the capacitor is now switched (SO2) to a reservoir at a constant voltage V_{high}. When the capacitance is then decreased from C_{max} to C_{min}, charges are forced to move from the capacitor into the reservoir since the voltage is maintained at an equal level V_{high} [15]. This is the generation part of the conversion cycle. The capacitor is then disconnected (SO3) from the reservoir and the capacitance is increased at a constant voltage. Again, the area enclosed by the Q–V diagram represents the energy generated.

Comparing the charge-constrained cycle and the voltage-constrained cycle, the total conversion of energy per cycle will depend on both the values of C_{max}, C_{min}, V_{low} and V_{high}, which are limited by practical constraints, as well as the circuit implementation of the conversion cycle [15, 16].

The electronic circuitry for realizing a switched transducer scheme may be rather different depending on the type of the energy conversion cycle while the implementation of the voltage-constrained cycle requires a more complex circuit than the charge-constrained conversion cycle. Figure 5.2 presents a possible circuit implementation for each operation mode as proposed by Mur-Miranda [15]. In each case, the circuit is based on a charge pump,

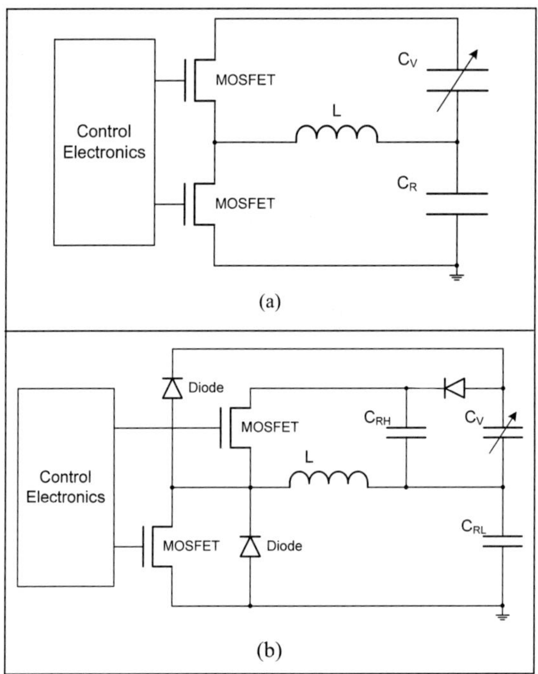

Figure 5.2 Circuit implementations of electrostatic transducers operated in the switched scheme [15]: (a) Charge-constrained cycle, (b) voltage-constrained cycle. The switches are implemented by MOSFET devices. The capacitor C_V denotes the variable capacitor, whereas C_R denotes a reservoir capacitor (C_{RL}: low voltage reservoir C_{RH}: high voltage reservoir).

which transfers harvested energy from the variable capacitor into a reservoir. The switches are realized by MOSFET devices, which are operated by some control electronics. The electronic circuit implementing the voltage-constrained cycle must perform three different tasks: charging the variable capacitor to the voltage V_{high}, keeping the voltage constant while reducing the capacity from C_{max} to C_{min}, and transferring energy from the high-voltage reservoir to the low-voltage reservoir. The implementation of a voltage-constrained cycle is more complex and requires a reservoir at a higher voltage in addition to the low-voltage reservoir (Fig. 5.2b). A charge-constrained cycle is easily implemented with a charge-pump (Fig. 5.2a). The associated circuit must realize the following functions: charging the capacitor to a voltage level V_{low}, keeping the charge in the capacitor constant while the capacity reduces from C_{max} to C_{min}, and transferring charge from the variable capacitor to a reservoir. The phase of constant charge is realized by simply disconnecting the variable capacitor. The disconnection of the variable capacitor may be realized by either electronic or mechanical switches. A mechanical switch is easily implemented when the variable capacitor itself is used. In this respect, the movable electrodes of the variable capacitor are only in contact with the circuitry when the displacement extremes are reached [16–18]. A number of different other circuit implementations are proposed in the literature [15, 19–21].

5.1.4 *Continuous Operation Scheme*

In contrast to switched systems, continuous systems do not require controlled switches to realize the operation of the transducer. In a continuous system, the variable capacitor is continuously connected to the circuitry, which includes the load and the polarization voltage for biasing the variable capacitor. A change in capacitance will always result in a charge transfer through the load resistor, which causes power being dissipated. As previously discussed, the operation of a variable capacitor in constant voltage or constant charge mode within a continuous system is not effective for conversion of mechanical to electrical energy. As a result, a continuous transducer system must be operated in between these

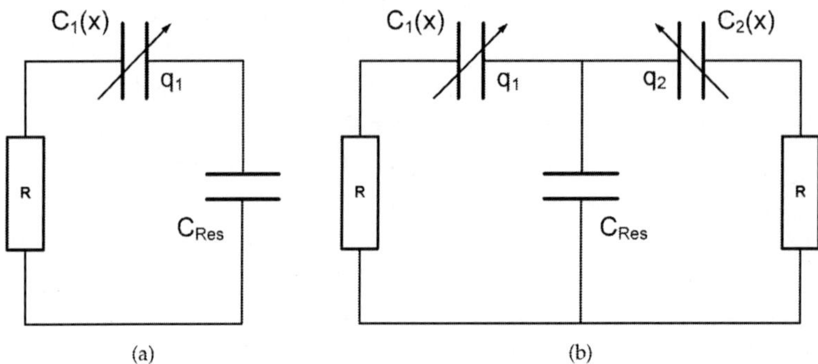

Figure 5.3 Circuit implementations of electrostatic transducers operated in the continuous scheme: (a) Single variable capacitor, (b) two complementary variable capacitors.

cases, which is the case when using the optimum load resistance [14]. This type of electrostatic transducer operation produces a velocity damped characteristic, in which the damping force is proportional to the relative velocity between the proof mass and the frame.

An advantage of a continuous scheme is that the transducer system can be implemented without the use of switches. The utilization of switches requires some extra circuitry to control them and precious energy is consumed by that control circuitry.

There are two fundamental schemes to realize a continuous system. In one scheme, a single variable capacitor is used in series with a voltage source (providing the required bias voltage) and a load resistor (Fig. 5.3a). In this case, charge flows through the voltage source. An alternative method (Fig. 5.3b) implements two complementary capacitors whose capacitances vary in an opposite manner [22–25]. This method offers some advantages in contrast to a single variable capacitor and was proposed by Sterken et al. [22]. One of the advantages is that the transduction is quite insensitive to parasitic capacitances. Also, in theory, no charge flows via the voltage source.

The circuit diagrams shown in Fig. 5.3 represent very basic implementations of the continuous transducer scheme, which are generally used for characterization purposes. The energy converted by the transducer is measured through the power dissipated in the load resistor. However, in order to store the harvested energy for use at a later time, some additional circuit components are required.

5.2 Implementation

5.2.1 *General Design Considerations*

The simplest design of a capacitor is a parallel plate capacitor. The capacity of a parallel plate capacitor is determined by the overlapping area A and the distance g of the plates as well as by the dielectric material property ε_r between the plates according to Eq. (5.1):

$$C = \frac{\varepsilon_0 \cdot \varepsilon_r \cdot A}{g},\tag{5.1}$$

where ε_0 is the vacuum permittivity. A variation of these parameters results in a change of capacity of the parallel plate capacitor. Since it may not be straightforward to vary the relative permittivity ε_r of the material by kinetic motion, the area A and the gap g are commonly used parameters to realize a variable area-overlap or a variable gap-closing capacitor [13, 14].

In general, there are two different methodologies for realization of an electrostatic transducer, either as a macro device [19, 26] or as a microdevice. The decisive characteristic that differentiates the two types is the minimum feature size of the function structure. In this respect, a device is considered as a macro device, when its manufacturing does not require any technologies, which are classified as microfabrication technologies. Therefore, a macro device may be manufactured by precision engineering and other fine mechanic processes only, whereas microdevices require the employment of micromachining techniques (microsystem technologies) for their fabrication. As previously discussed, the effectiveness of the electrostatic energy conversion reduces with increasing device size in contrast to the electromagnetic conversion. For that reason, electrostatic macro devices appear not worthwhile for further consideration and therefore the main focus in this chapter is on microdevices based on MEMS wafer technology. In

Table 5.1 Overview of design principles for variable capacitors realized by bulk micromachining

Motion direction	# Electrode layers	Capacitance variation
Out-of-plane motion	1 layer	Area overlap
	2 layers	Gap-closing
In-plane motion	1 layer	Area overlap, Gap-closing
	2 layers	Area overlap

this respect, the functional structures are manufactured utilizing surface micromachining and bulk micromachining technologies [27]. Preferred materials include silicon, glass, and polymers.

In general, gap-closing and area-overlap variable capacitors may be classified by the direction of motion of the movable electrodes with respect to the substrate surface. This motion may be either in-plane or out-of-plane. In both cases, it is possible to have movable and stationary electrodes either together in a single electrode layer or in two separate electrode layers. In the case of a single electrode layer, movable and stationary electrodes must be electrically isolated from each other. In general, there are five different design principles feasible as shown in Table 5.1, whereas principles incorporating one electrode layer require a different set of manufacturing technologies than principles incorporating two electrode layers. Which principle is to be chosen depends on the availability of the associated microfabrication technologies and processes. The fabrication of moveable and stationary electrodes in one device layer is based on the silicon-on-insulator (SOI) technology, which is widely accessible in semiconductor and MEMS foundries. Therefore, a transducer device utilizing one device layer for integration of movable and stationary electrodes is chosen for further considerations in this chapter. In the following, electrode geometries for realization of a variable capacitance are discussed.

5.2.2 *Electrode Geometry*

Well-known electrode geometries for electrostatic transducers (also named capacitive transducers) include area-sensitive and gap-sensitive structures, which utilize a specific number of interdigitated

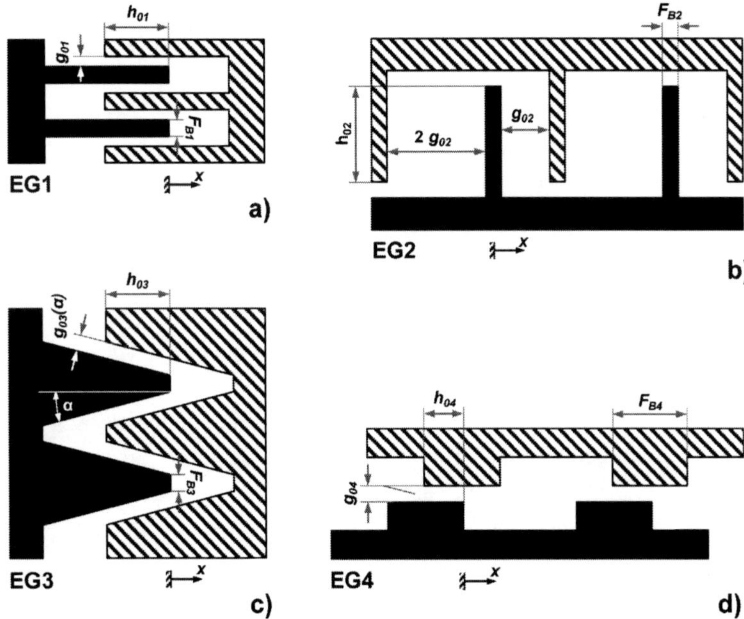

Figure 5.4 Top view of different electrode geometries, where the solid electrode is movable while the shaded electrode is stationary: (a) Interdigitated area-overlap, (b) interdigitated gap-closing, (c) interdigitated triangular-shaped (area-sensitive and gap-sensitive), (d) area-overlap. The positive motion direction x is indicated by an arrowhead below each geometry.

electrode elements (Fig. 5.4a,b) [22, 28]. Area-sensitive structures are usually employed to make an actuator (comb drive) while gap-sensitive structures are used to sense mechanical motion (e.g., accelerometer, gyroscope).

These common electrode geometries may be adapted for electrostatic energy harvesting devices, which require a variable capacitance. Both, area-sensitivity and gap-sensitivity may also be combined by incorporating a modified electrode geometry with angled sidewalls as shown in Fig. 5.4c [29]. A further geometry describes an area-sensitive electrode structure, where finger elements are not interdigitated (Fig. 5.4d). This geometry, when properly designed, has the capability to produce multiple repetitions of capacitance variations (minima, maxima) within one mechanical oscillating period. This characteristic may have advantages due to the multiple occurrence of the conversion cycle in the time of a single mechanical oscillation cycle [25].

Since there are different electrode geometries available for realization of a variable capacitor, the question that arises is which geometry provides the best performance. The performance will depend on the operation scheme (switched scheme versus continuous scheme), and in case of a switched scheme, it will also depend on the type of the conversion cycle (charge-constrained versus voltage-constrained). The operation conditions (excitation frequency and acceleration amplitude) will furthermore influence the choice of the electrode geometry. In principle, both the total change of capacitance as well as the change of capacitance per unit displacement are important system parameters that will influence the effectiveness of the electrostatic energy conversion. The capacitance characteristic for each of the four geometries is described in more detail below.

The schematic cross sections as depicted in Fig. 5.4 show each movable electrode in its position at rest ($x = 0$). The motion of the movable electrode therefore occurs in positive and negative x-direction whereas the position x is restricted to the displacement amplitude x_{max}. The parameters g_{01} (Fig. 5.4a) and g_{04} (Fig. 5.4d), respectively, describe the fixed distance between the surfaces of the movable and the stationary electrode. This parameter is limited by the capability of the photolithography process and cannot be smaller than the minimum feature size that the manufacturing facility is able to produce. The parameter g_{02} (Fig. 5.4b), which describes the initial gap between the gap-closing electrodes is a function of the displacement amplitude:

$$g_{02} = x_{max} + s_{02},\qquad(5.2)$$

where s_{02} accounts for the minimum gap that occurs when the electrode displacement of geometry EG2 reaches the displacement amplitude x_{max}. In theory, the parameter s_{02} may be chosen very small (in the range of nm), since it is not depended on the minimum feature size of the fabrication facility. However, in order to avoid a short circuit during operation the value for this parameter should be chosen not to small (e.g., >500 nm). The previous consideration for

s_{02} applies also for the parameter s_{03}, which defines the minimum gap between the electrodes for geometry EG3. The initial gap g_{03} of EG3 is a function of the displacement amplitude and the angle α as described by Eq. (5.3):

$$g_{03} = \left(\frac{s_{03}}{\sin(\alpha)} + x_{\max} \right) \cdot \tan(\alpha) \cdot \cos(\alpha), \qquad (5.3)$$

where the angle α specifies the angle of the sidewalls of the triangular shaped electrode geometry EG3. The capacitance $C(x)$ as a function of displacement x is described by Eqs. (5.4) to (5.7) for each of the four different electrode geometries:

$$C_{\mathrm{EG1}}(x) = 2 \cdot \varepsilon \cdot H_{\mathrm{F}} \cdot \frac{(x_{\max} + x)}{g_{01}} \qquad (5.4)$$

$$C_{\mathrm{EG2}}(x) = \frac{\varepsilon \cdot h_{02} \cdot H_{\mathrm{F}} \cdot 3 \cdot (x_{\max} + s_{02})}{(x_{\max} + s_{02} - x) \cdot (2 \cdot (x_{\max} + s_{02}) + x)} \qquad (5.5)$$

$$C_{\mathrm{EG3}}(x) = 2 \cdot \varepsilon \cdot H_{\mathrm{F}} \cdot \left[\frac{x_{\max} + x}{(s_{03} + (x_{\max} - x) \cdot \sin(\alpha)) \cdot \cos(\alpha)} + \tan(\alpha) \right]$$
$$(5.6)$$

$$C_{\mathrm{EG4}}(x) = \frac{\varepsilon \cdot H_{\mathrm{F}} \cdot \left(x + \frac{F_{B4}}{2} \right)}{g_{04}}, \ F_{B4} = x_{\max} \qquad (5.7)$$

where $\varepsilon = \varepsilon_0 \varepsilon_{\mathrm{r}}$ is the permittivity of the material between the electrodes, H_{F} is the height of the electrode structure and h_{02} is the fixed electrode overlap for EG2. The initial overlap h_{01} (Fig. 5.4a) is equal to the displacement amplitude $x_{\max}$. It must be noted that the height H_{F} is a technology parameter, which is equal to the thickness of the device layer. Therefore, the value of H_{F} is dependent on the material (e.g., SOI substrates) used. In case of customized SOI substrates, the device layer thickness depends on the fabrication process and manufacturing facilities.

In order to allow an impartial comparison of the four different electrode geometries, a consistent set of parameter values is used. In this respect, a thickness H_{F} of 50 µm and a displacement amplitude $x_{\max}$ of 20 µm is chosen for all geometries. A minimum feature size of 2.5 µm was chosen to define the gaps g_{01} and g_{04}. The parameters s_{02} and s_{03} have a value of 590 nm. The angle α is 11.31°. Figure 5.5a shows the change of capacitance over the displacement

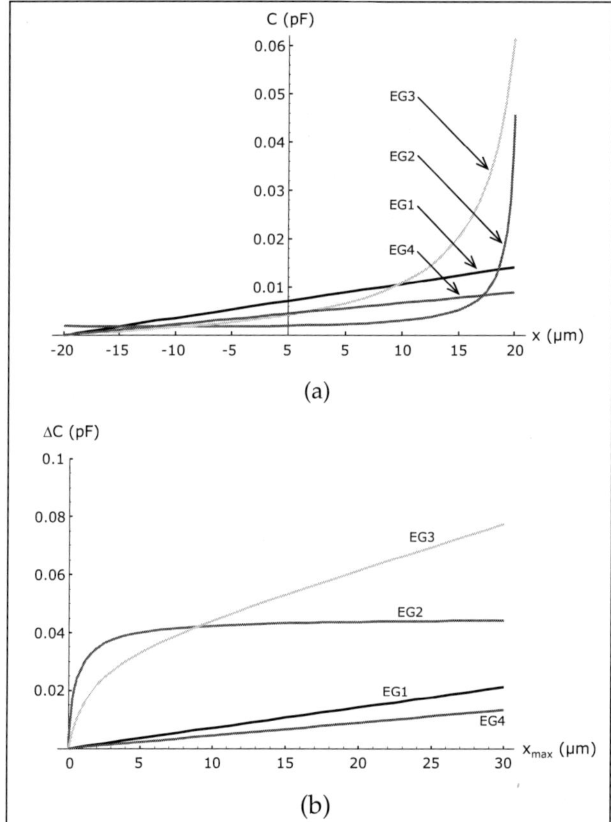

Figure 5.5 Capacitance characteristics of different electrode geometries: (a) Capacitance as a function of displacement with a maximum displacement of 20 μm, (b) change of capacitance as a function of the displacement amplitude x_{max}.

x for a single electrode element. Electrode geometries EG1 and EG4 show a linear change of capacitance while geometries EG2 and EG3 show a nonlinear characteristic with the largest capacitance C_{max}. The change of capacitance $\Delta C = C_{\mathrm{max}} - C_{\mathrm{min}}$ (where $C_{\mathrm{max}} = C(x_{\mathrm{max}})$ and $C_{\mathrm{min}} = C(-x_{\mathrm{max}})$) is shown in Fig. 5.5b as a function of the displacement amplitude x_{max}. The change of capacitance ΔC increases with the displacement amplitude x_{max} except for geometry EG2 ($x_{\mathrm{max}} > 5$ μm). This is because the electrode overlap area $h_{02} \cdot H_F$ and the minimum gap s_{02} are not dependent on the displacement amplitude. Again, the capacitance characteristics shown in Fig. 5.5 are for a single electrode element only. For a transducer design of a specific size the number of electrode elements that can be accommodated by that design depends on the electrode geometry and may also be a function of the displacement amplitude. Therefore, a comparison between different electrode geometries based on a single electrode element is not sufficient.

The number of electrode elements that can be arranged in a specific transducer area is given for each of the four different electrode geometries by Eqs. (5.8)[25] to (5.11)[25]:

$$N_{EG1} = \frac{L_t - W_{F1}}{2 \cdot (W_{F1} + g_{01})} \tag{5.8}$$

$$N_{EG2} = \frac{L_t - W_{F2}}{2 \cdot W_{F2} + 3 \cdot (x_{max} + s_{02})} \tag{5.9}$$

$$N_{EG3} = \frac{L_t - 4 \cdot W_{F3}}{2 \cdot \left(\frac{s_{03}}{\sin(\alpha)} + 2 \cdot x_{max}\right) \cdot \tan(\alpha) + 2 \cdot W_{F3}} \tag{5.10}$$

$$N_{EG4} = \frac{L_g}{2 \cdot W_{F4}} \tag{5.11}$$

Here, L_t describes the total usable length, which is available for placement of a specific number of electrode elements. The total usable length is a design parameter and depends on the layout of the transducer. The parameters W_{F1} to W_{F3} describe the width of the electrode element for EG1, EG2 and EG3 whereas $W_{F4} = x_{max}$ defines the electrode width for EG4. The total capacitance $C_{EGi\_t}(x)$ results from the product of $C_{EGi}(x)$ and N_{EGi} according to Eq. (5.12):

$$C_{EGi\_t}(x) = N_{EGi} \cdot C_{EGi}(x) \tag{5.12}$$

In order to capture the influence of the displacement amplitude on both the number of electrode elements N_{EGi} and the total change of capacitance ΔC_{EGi_t} $(C_{EGi\_t}(x_{max}) - C_{EGi\_t}(-x_{max}))$ a specific transducer layout with a finite value of the usable length must be considered. A possible layout is shown in Fig. 5.6, which incorporates two complementary variable capacitors $C_1(x)$ and $C_2(x)$. The total usable length is represented by the dotted areas (red

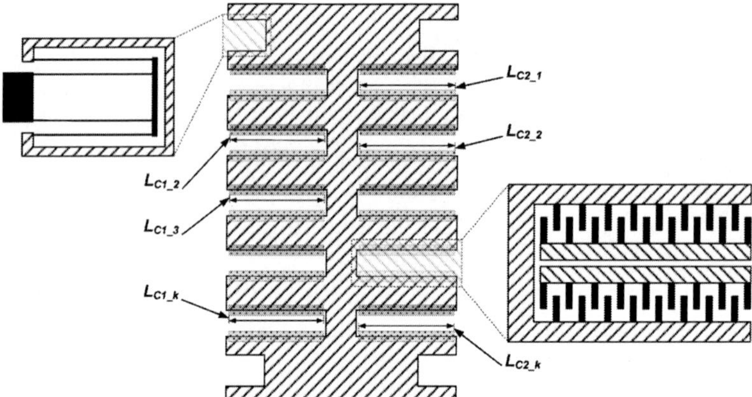

Figure 5.6 Transducer layout: A rectangular shape contains several cut-outs, which provide space for incorporation of the suspensions and the electrode elements. The shaded area in the center defines the proof mass. The suspensions are placed at the corners of the proof mass as indicated on the upper left. The electrode elements are placed within the oblong cut-outs as indicated on the lower right. Altogether, there is space for ten electrode units. Each electrode unit provides space (more precisely two usable length segments L_{C1_k} and L_{C2_k}) for placement of a specific number of electrode elements. The total usable length L_t results from the summation of the segments L_{C1_k} and L_{C2_k}, respectively.

for C_1 and green for C_2). According to Fig. 5.6, the total length where active transducer structures may be placed is the summation of the segments L_{Ci_k} for each capacitor C_1 and C_2:

$$L_t = \sum_k L_{Ci\_k} \tag{5.13}$$

To allow appropriate comparison between the different electrode designs, the minimum gaps s_{02} and s_{03} are chosen to be equal (590 nm) at the corresponding inner displacement amplitude x_{max} and $-x_{max}$, respectively. Also, the gaps g_{01} and g_{04} are chosen to be as small as possible (2.5 µm) with respect to the technological limitations of the microfabrication facilities.

Figure 5.7a shows the number of electrode elements N_{EGi} that can be accommodated by the transducer design (as presented in Fig. 5.6) with a total usable length L_t of 13.6 mm. It is evident that the number significantly dependents on the displacement amplitude except for geometry EG1. For a displacement amplitude $x_{max} =$

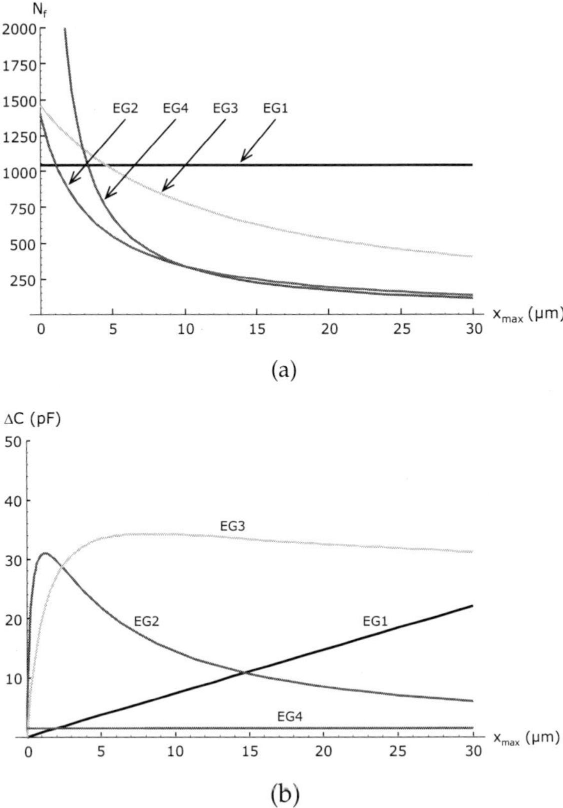

Figure 5.7 Characteristic of a specific transducer design: (a) Number of electrode elements as a function of displacement amplitude, (b) total change of capacitance as a function of displacement amplitude.

20 µm, the number of electrodes are as follows: $N_{EG1} = 1038$, $N_{EG2} = 190$, $N_{EG3} = 520$ and $N_{EG4} = 164$. The electrode overlap h_{02} of geometry EG2 is 60 µm. When graphing the total change of capacitance as a function of displacement amplitude (Fig. 5.7b) a completely different characteristic occurs in contrast to Fig. 5.5b.

In general, if x_{max} is not a definite constraint then it can be used as a variable design parameter to maximize the total change of capacitance. Electrode geometry EG3 is suitable for a wide range of inner displacement amplitudes larger than 5 µm since ΔC_{EGi_t} declines only very slowly. However, when designing electrostatic

Table 5.2 Parameters used for calculation of the capacitance characteristic (Fig. 5.8) with respect to a specific transducer layout (Fig. 5.6)

	L_t (mm)	N_{EGi}	h_{0i} (µm)	W_{Fi} (µm)	H_F (µm)	L_{Fi} (µm)	s_{0i} (nm)	g_{0i} (µm)
EG1	13.6	1038	20	4	50	44	—	2.5
EG2	13.6	190	60	4	50	64	590	20.59
EG3	13.6	520	20	4	50	43	590	4.51
EG4	13.6	164	20	20	50	—	—	2.5

transducers with $x_{max} < 4$ µm, electrode design EG2 is more applicable. The total capacitance change of geometry EG1 increases linear with the displacement amplitude. From the capacity point of view geometry EG1 and EG3 should be preferably used.

In Fig. 5.8 the capacitances $C_{EGi-t}(x)$ are shown for a specific number of electrode elements according to a transducer with a total usable length L_t of 13.6 mm and a displacement amplitude x_{max} of 20 µm. The specific parameters are summarized in Table 5.2. In comparison to Fig. 5.5a, electrode geometry EG3 still achieves the largest capacitance at $x = x_{max}$, however, geometry EG2 shows a much lower value.

Another interesting point, which becomes evident from Fig. 5.8 is, that the change of capacitance per unit displacement changes considerably for geometry EG2 and EG3. Looking at the geometry EG2, most of the capacitance change takes place for displacements in the region between 15 and 20 µm. Consequently, if the proof mass oscillates with smaller displacements than the displacement amplitude (e.g., 15 µm) almost no capacitance change occurs. From this point of view electrode geometry EG1 is the most preferable geometry since the change of capacitance per unit displacement is constant over the entire displacement range.

In conclusion, when comparing different electrode geometries, it is not sufficient to focus on a single electrode element only, but the complete transducer design must be considered. Still, from Fig. 5.8, it cannot be concluded that the highest effectiveness can be achieved by using geometry EG3 as an electrostatic transducer structure. Therefore, it is inevitable to perform simulations capturing the system behavior of the transducer device in order to optimize the transducer parameters.

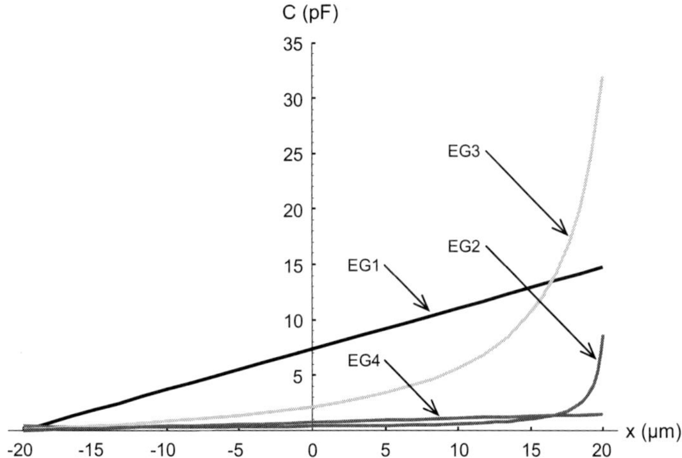

Figure 5.8 Total capacitance as a function of displacement for a specific transducer layout. The displacement amplitude is 20 μm. The geometry parameters of the different electrodes are given in Table 5.2.

5.3 Analytical and Numerical Models

5.3.1 *Analytical Description*

In order to understand and predict the dynamic behavior of an electrostatic transducer an adequate model description must be derived. The model that is presented here is based on the analytical description of the transducer system including two energy domains, a mechanical domain and an electrical domain, which are electromechanically coupled by an electrostatic field. In the following, a configuration with two complementary variable capacitors (Fig. 5.9) is considered whereby electrode geometry EG1 is used to realize an area-overlap capacitance variation. However, the following procedure can also be used to model a configuration with one variable capacitor or to investigate the influence of different electrode geometries on the performance of the transducer device. For a one-capacitor-model the set of two differential equations represented by Eq. (5.14) reduces to a single differential equation. For consideration of different electrode geometries (as shown in Fig. 5.4) one only needs to replace the expressions given

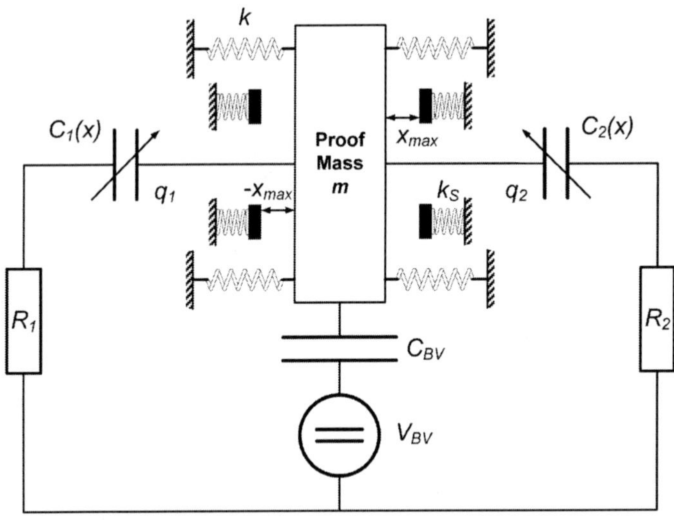

Figure 5.9 Electromechanical model of an electrostatic transducer device with two complementary variable capacitors including mechanical stoppers, a bias voltage and two load resistors.

in Eq. (5.15) by the respective capacitance function (e.g., Eqs. (5.4) to (5.7)) under consideration of Eqs. (5.8)[25] to (5.11).

According to Fig. 5.9, a set of two nonlinear differential equations can be established by means of Kirchhoff's second law applied to the meshes of the circuit. Thus, the state of the charge q_i on the two variable capacitors can be written as

$$R_1 \frac{dq_1}{dt} + \frac{q_1}{C_1(x)} + \frac{q_1 + q_2}{C_{BV}} - V_{BV} = 0$$
$$R_2 \frac{dq_2}{dt} + \frac{q_2}{C_2(x)} + \frac{q_1 + q_2}{C_{BV}} - V_{BV} = 0 \qquad (5.14)$$

where R_i is the load resistance, q_i is the charge on the variable capacitors, C_{BV} and V_{BV} are the capacitance and the voltage of the bias voltage source, respectively, and C_i is the capacitance of the two variable capacitors.

The capacitances C_i of the variable capacitors with an area-overlap characteristic (EG1) are given by

$$C_1(x) = 2 \cdot N_F \cdot \varepsilon \cdot H_F \cdot \frac{x_{max} + x}{g_{01}}$$
$$C_2(x) = 2 \cdot N_F \cdot \varepsilon \cdot H_F \cdot \frac{x_{max} - x}{g_{01}}, \qquad (5.15)$$

where N_F is the number of fingers of the comb electrodes, ε is the permittivity of the medium between the fixed and movable electrodes, H_F is the height of the finger equal to the thickness of the device layer, g_{01} is the gap between the fingers and x_{max} is the initial overlap of the fingers, which is equal to the maximum displacement amplitude.

The electrostatic forces that provide the feedback in the electromechanical system are calculated as follows:

$$F_{ES1} = \frac{1}{2} \cdot V_{C1}^2 \cdot \frac{dC_1(x)}{dx}$$

$$F_{ES2} = \frac{1}{2} \cdot V_{C2}^2 \cdot \frac{dC_2(x)}{dx}$$

(5.16)

where V_{Ci} is the voltage over the variable capacitors. It is evident from Eq. (5.16) that the electrostatic force is proportional to the bias voltage squared and to the rate of change of the capacitance. As previously described, the present electromechanical transducer system is a continuous system, where the voltage over the variable capacitors is not constrained and thus does change with time. Consequently, electrostatic forces are not constant over time unlike it is the case for constant voltage area-overlapping capacitors. The voltage for each capacitor is described as

$$V_{C1} = \frac{q_1}{C_1(x)}$$

$$V_{C2} = \frac{q_2}{C_2(x)}$$

(5.17)

where q_i is the charge on the variable capacitor C_i. Both, the charge stored in the variable capacitors and the capacitance value change with time.

We also included elastic stoppers in our model, because they have a significant impact on the dynamic behavior and the performance of the device. Since mechanical stoppers are necessary to be implemented in a real device, they must also be considered in the model. The elastic stoppers are modeled as described by T_{vedt} [24], where the stoppers are represented by springs that come into effect when the displacement of the proof mass is larger than the predefined displacement limit x_{max}:

$$F_S = \begin{cases} 0 & , \ -x_{max} \leq x \leq x_{max} \\ -k_S(x + x_{max}) & , \ x < -x_{max} \\ -k_S(x - x_{max}) & , \ x > x_{max} \end{cases}$$

(5.18)

In this expression, k_S is the spring stiffness of the stoppers, which should be chosen much larger than k to account for the rigid characteristic of the stoppers. In Eq. (5.18), it is assumed that there is no damping involved when the stoppers come into effect, hence, the stoppers act purely elastic. This ideal behavior of the stoppers is justified in this case since the effect of the stoppers themselves (limitation of the displacement amplitude) on the device behavior is the main focus of this investigation.

Mechanical damping, e.g., due to internal friction within the material (suspension beams) and viscous gas flow in the cavity, must be considered when modeling resonant electromechanical systems. The mechanical damping coefficient b can be described as a function of the quality factor Q of the resonator:

$$b = \frac{1}{Q} \cdot \omega_0 \cdot m, \tag{5.19}$$

where ω_0 is the mechanical angular Eigen frequency and m is the mass of the proof mass (including the movable electrode elements). The Q factor used in the simulation was estimated from frequency response measurements of fabricated prototype devices in accordance with Eq. 5.7:

$$Q = \frac{f_R}{\Delta f}, \tag{5.20}$$

where f_R is the resonance frequency and Δf is the bandwidth of the frequency response curve. Finally, the motion of the proof mass m (Fig. 5.9) is described by Newton's second law:

$$m\ddot{x} = -b\dot{x} - kx - F_S + F_{ES1} + F_{ES2} - ma, \tag{5.21}$$

where k is the spring stiffness and ma is the excitation force due to the acceleration of the device. The motion of the proof mass as described by Eq. (5.21) is inherently nonlinear since electrostatic forces (Eq. (5.16)) depend on the voltage squared (Eq. (5.17)) which varies over time. Also, the discontinuity of the stopper force (Eq. (5.18)) contributes to the nonlinear behavior of Eq. (5.21).

5.4 Numerical Model

The analytical description of the transducer model is implemented as a signal-flow model in Matlab/Simulink. This model considers a

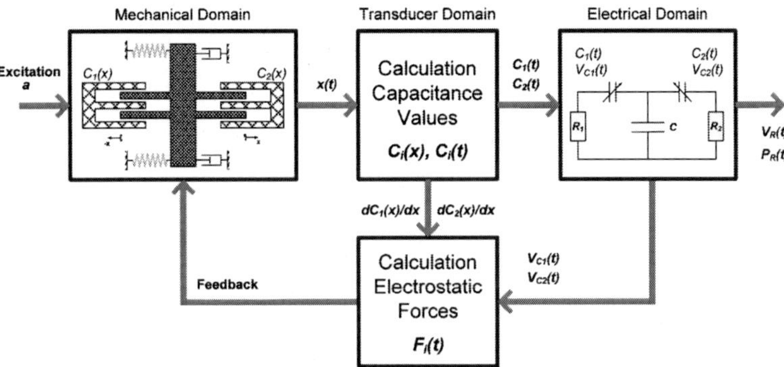

Figure 5.10 Schematic of the signal-flow model representing an electrostatic transducer.

nonlinear regime, where electrostatic forces couple back from the electrical domain to the mechanical domain, and therefore must be solved numerically. In addition, the effect of mechanical stoppers limiting the displacement of the proof mass is also taken into account.

A schematic view of the signal-flow diagram of the electrostatic transducer model is shown in Fig. 5.10. The mechanical domain, having one degree of freedom, is supplied with an acceleration signal a, which can be harmonic or random. The displacement x of the proof mass is then used to calculate the instantaneous capacitances C_i of the two variable capacitors. In the electrical domain, having two degrees of freedom, the charge q_i on the variable capacitors is determined on the basis of the instantaneous capacitance values C_i by solving Eq. (5.14). The voltages V_{Ci} over the variable capacitors can now be calculated using Eq. (5.17). The instantaneous voltage V_{Ri} over the load resistor follows from Kirchhoff's voltage law (all variables are time-dependent):

$$V_{Ri} = V_{BV} - V_{Ci} - V_{BC},\qquad(5.22)$$

where V_{BC} is the voltage over the bias capacitor. The peak value and the root-mean-square (rms) value of the generated power are calculated using the following equation:

$$P_{\text{peak,rms}} = \frac{V^2_{\text{peak,rms}}}{R_i}\qquad(5.23)$$

By using the space derivatives of C_i together with the voltages V_{Ci} the electrostatic forces coupling back into the mechanical domain are determined. The total simulation time is chosen in such a manner that the system is well in the steady state region when the simulation finished. The output parameters of the system model are voltage, current and power with respect to the load resistor.

The fully parameterized model described here can be used to investigate the behavior of electrostatic transducer devices for different device parameters and operation conditions. In addition, it provides the feasibility to carry out device optimization, which delivers optimal device parameters for a specific set of operation conditions.

5.5 Power Output and Device Behavior

5.5.1 *Device Design*

In this chapter the output parameters as well as the device behavior is discussed with respect to a specific transducer design. The design is based on the layout as shown in Fig. 5.6 where the proof mass is shaped in a particular way (fish bone structure) in order to obtain a large value for the total usable length L_t [2]. A summary of relevant parameter specifications is given in Table 5.3. The capacitive electrodes comprise interdigitated comb structures with a constant gap varying area-overlap characteristic (EG1). Fixed and movable comb fingers are designed with a gap g_0 of 2.5 µm and an initial overlap L_0 of 20 µm. An active device layer (H_F) of 50 µm thickness is chosen. Therefore, a maximum variation of capacitance ΔC of 13.3 pF (analytically calculated) for each capacitor is achieved utilizing 936 electrode elements. The displacement x of the proof mass m is limited to 20 µm (x_{max}) by mechanical stoppers. The suspension beams have a width of 4 µm and a length of 310 µm. Each suspension unit is designed with two folded beams, thus the total spring constant k of the resonator is 72.6 kg/s^2 (analytically calculated). With a total effective mass m of 642 µg the resonance frequency yields to 1692 Hz. The effective mass includes the proof mass as well as the mass of the beams and the trusses.

Table 5.3 Parameter definitions of a specific transducer design

Parameter	Symbol	Value
Mechanical parameters		
Mass of proof mass	m	642 µg
Total spring stiffness	k	72.6 N/m
Total stiffness of mechanical stoppers	k_S	10000 N/m
Displacement limit	x_{max}	20 µm
Suspension beam width	W_B	4 µm
Suspension beam length	L_B	310 µm
Number of folded beams	N_{fb}	8
Capacitor Parameters		
Number of electrode elements	N_F	936
Initial Overlap	L_0	20 µm
Gap between fingers	g_{01}	2.5 µm
Finger length	L_F	30 µm
Finger height (device layer thickness)	H_F	50 µm
Total change of capacitance	ΔC	13.3 pF
Circuit parameters		
Load resistor	R_1, R_2	560 kOhm
Bias capacitor	C_{BV}	1 µF

5.5.2 *Device Behavior*

In general, the simulation shows, that the generated voltage (Fig. 5.11a) increases linearly with the acceleration amplitude a until a certain excitation level is reached where the output voltage starts to level off. The output power (Fig. 5.11b) behaves in the same manner but increases initially with the square of a. The excitation level a at which the voltage and hence the generated power stop to increase depends on the value of the bias voltage V_{BV}. This behavior is due to the fact that the proof mass starts to impact at the mechanical stoppers if a certain excitation level is exceeded. A further increase of excitation does not cause a further increase of displacement or capacitance change, respectively. Instead, the stoppers cause the output power to decline continuously at a very low rate. From Fig. 5.11, it becomes evident that the maximum amount of energy that can be generated is strongly dependent on the strength of the electrical damping force, which is dependent on the bias voltage

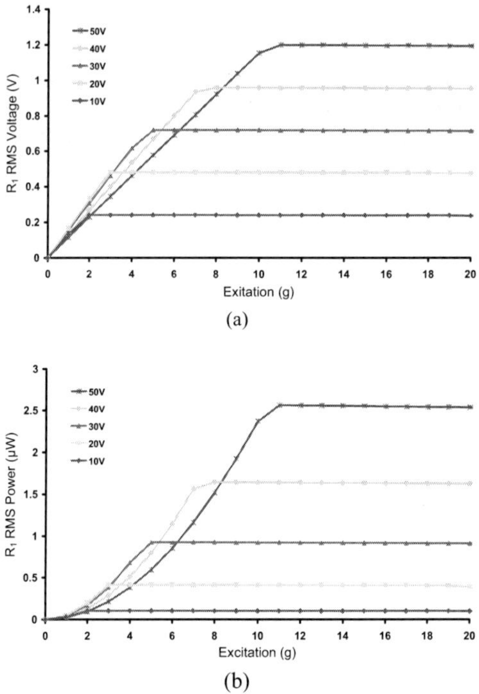

Figure 5.11 Transducer parameters for different bias voltages: (a) Voltage as a function of excitation, (b) power as a function of excitation.

(Eq. (5.16)). Therefore, the amount of mechanical energy required to operate the transducer device at its maximum inner displacement amplitude, must increase with higher bias voltages. In conclusion, the output power will always increase with increasing bias voltage if the excitation level is raised accordingly. However, the value of the bias voltage is limited by practical constraints (e.g., breakdown voltage of variable capacitor; maximum possible voltage of the bias voltage source limited by electret or breakdown voltage of bias capacitor).

The influence of the bias voltage on the output parameters (voltage and power) is shown in Fig. 5.12. For a specific excitation level (constant acceleration) an optimal bias voltage can be found at which the generated voltage (and power), reaches a maximum. At this point of maximum power, the proof mass oscillates with

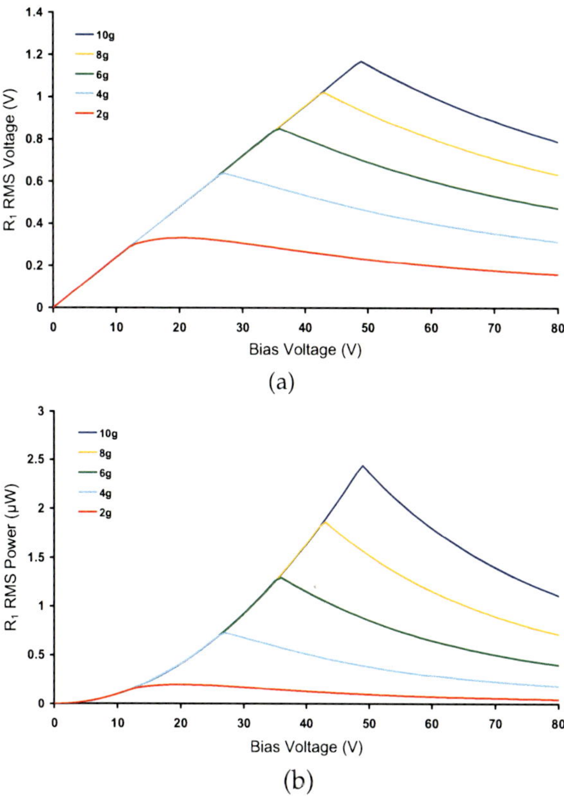

Figure 5.12 Influence of bias voltage on the output parameters at constant load resistance ($R_1 = R_2 = 560\,\text{kOhm}$): (a) Output voltage, (b) output power.

the maximum possible displacement, without the occurrence of an impact at the mechanical stoppers. Below the optimal bias voltage the generator operates in impact mode, i.e., the motion of the proof mass is impeded by the mechanical stoppers and thus impacts occur. For bias voltages higher than the optimal value, the displacement amplitude of the proof mass starts to decline since the influence of the electrostatic damping force becomes increasingly stronger. Therefore, an electrostatic transducer may be adjusted to the operation conditions (e.g., excitation level) by choosing the optimal value of the bias voltage. However, the optimal bias voltage also depends on the value of the load resistor. In practice, the load is usually represented by a rather complex circuit (in comparison to a simple load resistor) and its impedance cannot be changed once the circuit is manufactured. Therefore, it is more useful to adjust the polarization voltage for different application scenarios while using the same harvester circuit. Yet, it is necessary to identify the influence of the load resistance on the output performance in order to maximize the output parameters voltage and power for a given application scenario.

Figure 5.13 shows the influence of the load resistance on the output voltage and output power for specific excitation conditions

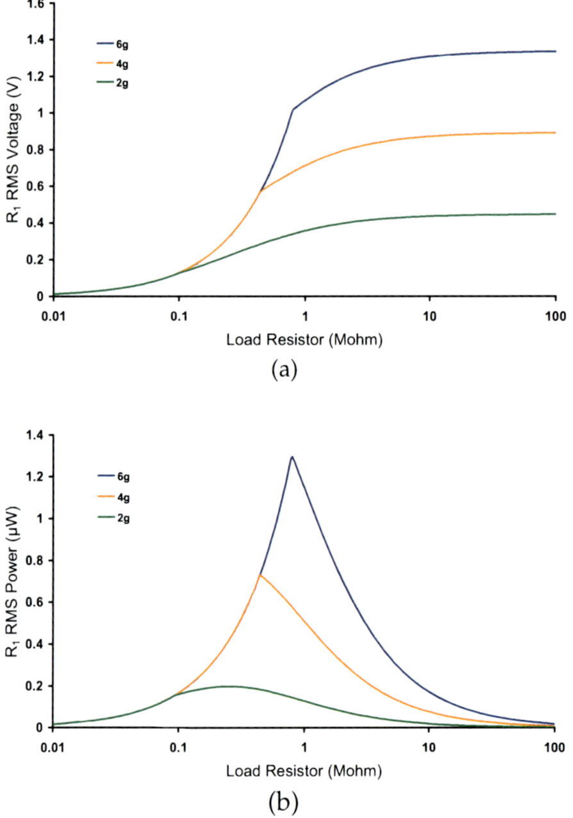

Figure 5.13 Influence of load resistance R_1 and R_2 on the output parameters at constant bias voltage ($V_{\mathrm{BV}} = 30$ V): (a) Output voltage, (b) output power.

and a bias voltage of 30 V. The voltage increases at a progressive rate until an optimal load resistance is reached (Fig. 5.13a). At this point, the power gains a maximum (Fig. 5.13b). With further increasing load resistance, the voltage starts to increase at a degressive rate whereas the power declines significantly. From Fig. 5.13b, it is also evident that the optimal load resistance is dependent on the excitation level. For higher excitation accelerations, the optimum resistance shifts towards higher values. In conclusion, optimizing the load resistance to a specific set of operation conditions enables to further increase the output power. For instance, at an acceleration amplitude of 6 g and a bias voltage of 30 V, the output power is 1.3 µW when using an optimized load resistance of 800 kOhm (Fig. 5.13b). At the same operation conditions (6 g, 30 V) but a load resistance of 560 kOhm, an output power of 0.92 µW is obtained (Fig. 5.12b). However, when optimizing the bias voltage (to a value of 36 V) for a load resistance of 560 kOhm and an acceleration amplitude of 6 g, the maximum power results to 1.3 µW (Fig. 5.12b). Consequently, for a fixed acceleration amplitude either the bias voltage or the load resistance can be optimized to maximize the output power. This provides a flexible way of optimization with respect to practical constraints.

The dynamic behavior of the transducer is characterized by means of a frequency sweep with a fixed harmonic acceleration amplitude a. Figure 5.14 shows the frequency response of the transducer device for different excitation levels and a fixed bias voltage of 30 V. The dynamic behavior may differ significantly depending on the excitation acceleration. For excitations of 5 g and lower a typical resonance behavior occurs when performing a frequency sweep upward (Fig. 5.14). However, if the excitation exceeds a certain level ($>$5 g), the transducer follows continually the excitation frequency within a specific frequency band. In that region, the amplitude of the generated voltage and generated power does not drop down beyond the resonance frequency but increases slightly at a progressive rate. This behavior is also present for bias voltages lower or higher than 30 V and is due to the fact the amplitude of the proof mass reaches the displacement limit at or before the resonance frequency. As a consequence, the transducer starts to operate in impact mode, i.e., persistent collisions occur between the proof mass and the mechanical stoppers. Assuming a pure elastic characteristic of the mechanical stoppers, the dynamic behavior of the generator is comparable with that of a piecewise-linear oscillator [30]. Piecewise-linear oscillators exhibit a broader bandwidth characteristic for harmonic up-sweep excitations. The increase in bandwidth may enhance the performance of a vibration transducer in case of harmonic excitations where the excitation frequency varies over time while the excitation amplitude remains constant, although the increased bandwidth is not attained if the transducer is initially excited at frequencies above the resonance frequency.

5.6 Device Fabrication and Characterization

5.6.1 *Fabrication*

There are quite a number of different ways to manufacture an electrostatic transducer device. The manufacturing process will also depend on the choice of materials to be used. In the following, a microfabrication process is described, where the device is realized in silicon utilizing customized SOI substrates. The process flow is adapted from a manufacturing process used to fabricate acceleration sensors as well as gyroscope sensors (SCRESOI-50 process) [2, 31]. This process uses a single active layer of 50 μm thickness. First, a substrate wafer is dry-etched to create a cavity of 50 μm depth for free movement of the proof mass including the movable electrodes attached to the proof mass (Fig. 5.15b). Then a 2000 nm thermal oxide layer is produced to provide isolation between the substrate and the device layer. Subsequently, a highly p-doped device wafer is bonded onto the substrate wafer and thinned to the required thickness by chemical mechanical polishing (Fig. 5.15c). This produces a customized SOI material, which incorporates a buried cavity. In order to realize capacitive structures in the device layer, fixed and movable electrodes require being electrically isolated from each other.

This is achieved using a trench-refill technology as described in [31, 32], where isolation trenches are created as shown in Fig.

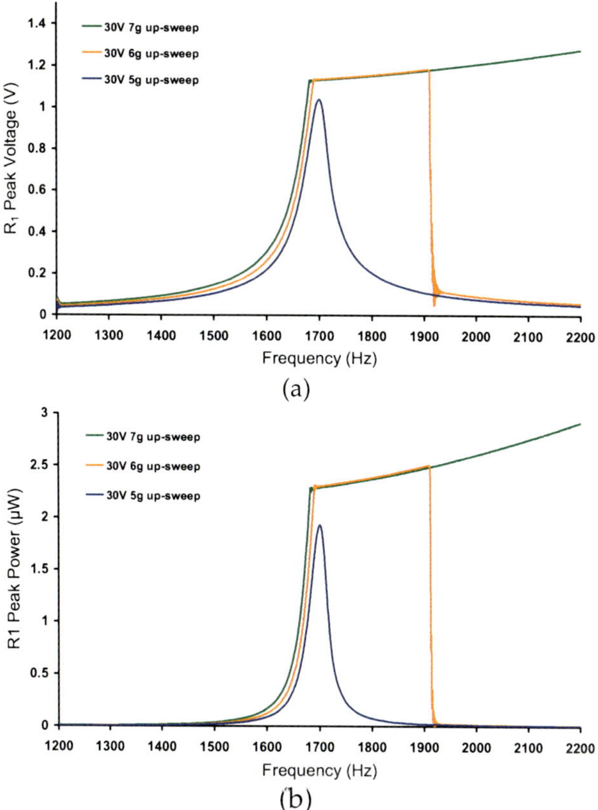

Figure 5.14 Dynamic behavior of the transducer under the influence of mechanical stoppers during a frequency up-sweep: (a) Output voltage, (b) output power.

5.15d. The trench-refill technology also allows the fabrication of track crossings, which may be required from a particular transducer design. Conductor tracks for contacting the fixed and movable electrodes are formed by wet-etching of a 500 nm-thick aluminum layer (Fig. 5.15e). Prior to the deposition of the aluminum layer a thermal oxide of 200 nm thickness is generated and structured to prevent short-circuits between the metal tracks. As a last step, the device layer is dry-etched to create the proof mass, suspensions, and comb electrodes (Fig. 5.15f). For encapsulation of the device wafer, a

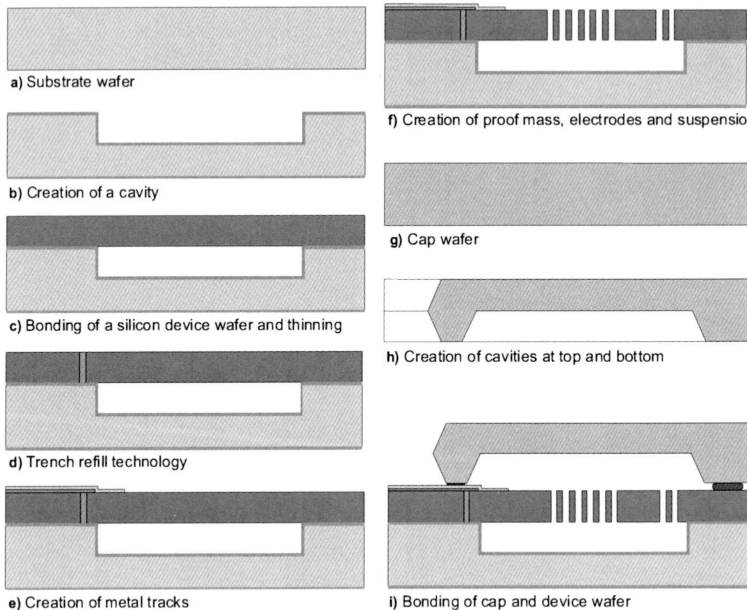

Figure 5.15 Fabrication process flow.

cap wafer is wet-etched to form cavities at the front side (for access to the bond pads) and the back side (to allow free movement of the proof mass) of the wafer (Fig. 5.15h). The device wafer and the cap wafer are then bonded together using a glass frit bonding technology [31]. This generates a hermetic seal, which protects the transducer structures from environmental influences and allows operating the generator in a defined vacuum (Fig. 5.15i). Finally, the wafer is diced to separate the devices, which are then ready for packaging.

The encapsulated transducer device chips are packaged into ceramic chip carriers using a conductive epoxy adhesive (Fig. 5.16a). This allows connecting the substrate to the ground in order to avoid any unwanted charging of the substrate. The chip includes five bond pads for connecting the transducer to the circuitry. Conventional wire-bonding technology is used to connect the bond pads to the CLCC package. For protection of the wire bonds and the device chip, a ceramic lid may be attached onto the package. At this stage, the packaged transducer devices are ready for integration on PCB

(a)

(b)

Figure 5.16 Electrostatic microgenerator: (a) Transducer chips packaged in CLCC packages, (b) packaged transducer integrated on PCB test board

level together with the electronic circuitry. Figure 5.16b shows a packaged device integrated on a PCB test board. The circuitry realized on the test board complies with Fig. 5.9. Therefore, the test board contains two load resistors R_1 and R_2 (560 kΩ each) and a pair of multilayer ceramic capacitors generating a total capacity of 1 µF for biasing the device.

5.6.2 *Characterization*

Figure 5.17 shows a microscopic close-up view of a fabricated electrostatic transducer device including one of the four mechanical suspension units and one of the ten electrode units. Each electrode unit comprises two comb electrodes, which consist of a specific number of interdigitating electrode elements. In order to form the two variable capacitors C_1 and C_2, corresponding comb electrodes

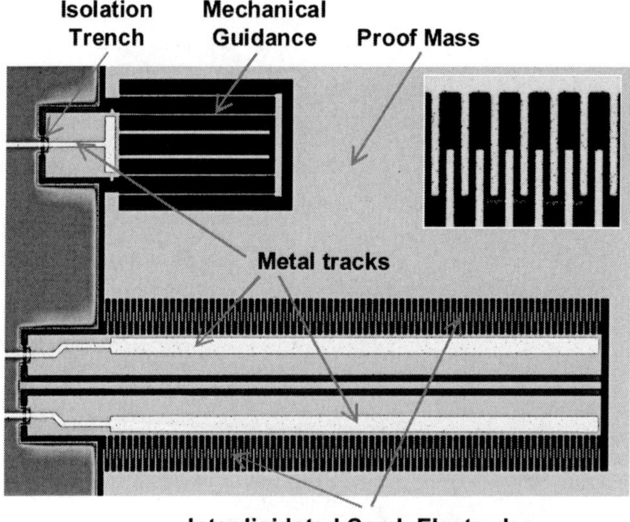

Figure 5.17 Microscopic close-up view showing a detail of the transducer structure including one of the four mechanical suspensions and 1 of the 10 electrode units (see Fig. 5.6). The inset in the upper right shows a detailed view of the interdigitated electrode structures, which implement an area-overlap characteristic (EG1).

are interconnected on chip via conductor paths. For contacting the movable comb electrodes attached to the proof mass and the fixed comb electrodes, metal tracks must cross the isolation trenches in order to make contact with the corresponding material. The inset in Fig. 5.17 provides a more detailed view of the interdigitated finger electrodes.

The maximum possible bias voltage that may be used for characterization is restricted to the maximum allowable voltage of the ceramic capacitors, which is 50 V. The maximum output power applying the maximum possible bias voltage was achieved at an excitation level of approximately 13 g. Excitation at the resonance frequency (harmonic excitation) generated a voltage of 1 V_{rms} over the load resistor. Since the induced current of 1.8 μA_{rms} is driven through the two load resistors R_1 and R_2, a total maximum power of 3.5 μW_{rms} was generated [2].

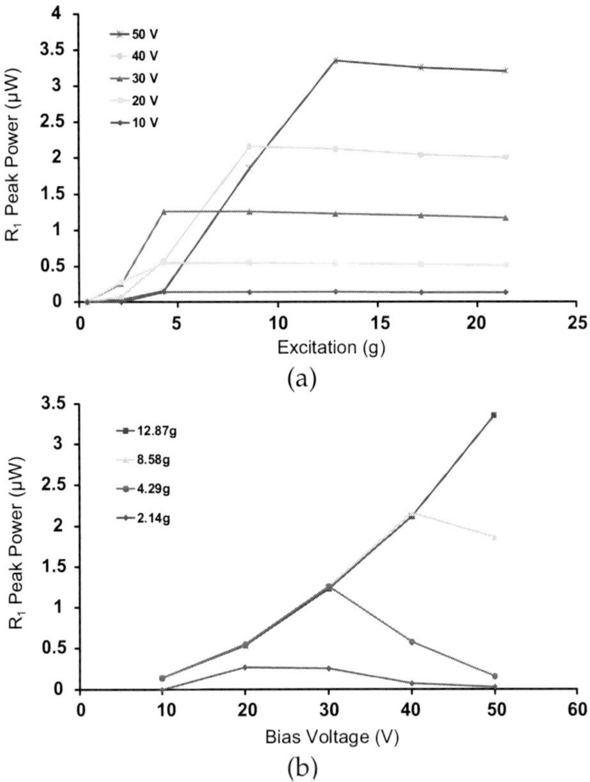

Figure 5.18 Experimental data: (a) Output power as a function of excitation, (b) output power as function of bias voltage.

In Fig. 5.17, the output power of a fabricated transducer device is shown. The predictions of the device behavior by the transducer model (Fig. 5.11b, and 5.12b) are in good agreement with the experimental data. According to Fig. 5.18a, the output power levels off with a further decreasing characteristic as predicted by the numerical simulations. The presence of an optimal bias voltage is also verified by the experimental data (Fig. 5.18b).

Figure 5.19 shows the frequency response of a transducer device for different bias voltages. The dynamic behavior resulting from frequency up-sweeps (sweeps from lower to higher frequencies) comply with the predictions from the numerical simulations of

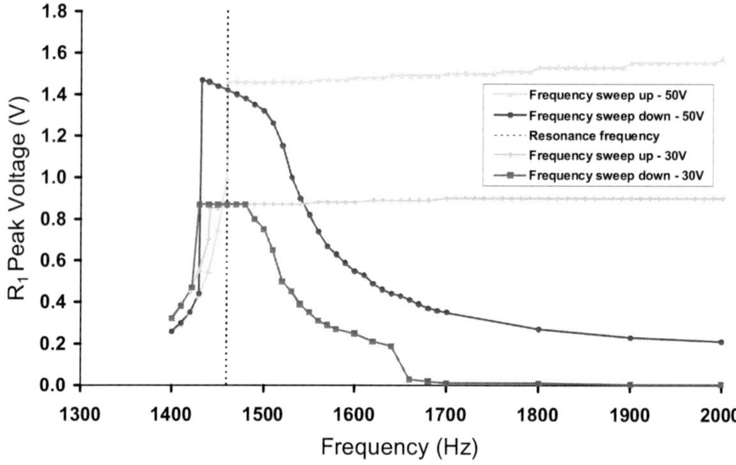

Figure 5.19 Experimental data: Dynamic device behavior during frequency sweeps at a fixed excitation amplitude y of 1.5 μm. The acceleration amplitude a is equal to $y\omega^2$.

the transducer model. Here, the output power keeps increasing for a very large frequency band beyond the resonance frequency. Again, this is due to the fact that the device operates in the impact mode. Figure 5.19 shows also a down-sweep characteristic. When sweeping the frequency from higher values to lower values across the resonance frequency a much lower frequency band occurs in comparison to a frequency up-sweep. This confirms the statement that the broadband characteristic of an impact-operated transducer will only occur for a frequency up-sweep whereby the initial start frequency must be below the resonance frequency.

5.7 Optimization Considerations

As stated earlier, the amount of energy that can be extracted is heavily dependent on the balance between the strength of the electrical damping force and the excitation force. Therefore, an optimal value of the electrostatic damping force must be designed with respect to the excitation conditions in order to achieve a maximum of power generation. The electrostatic damping force

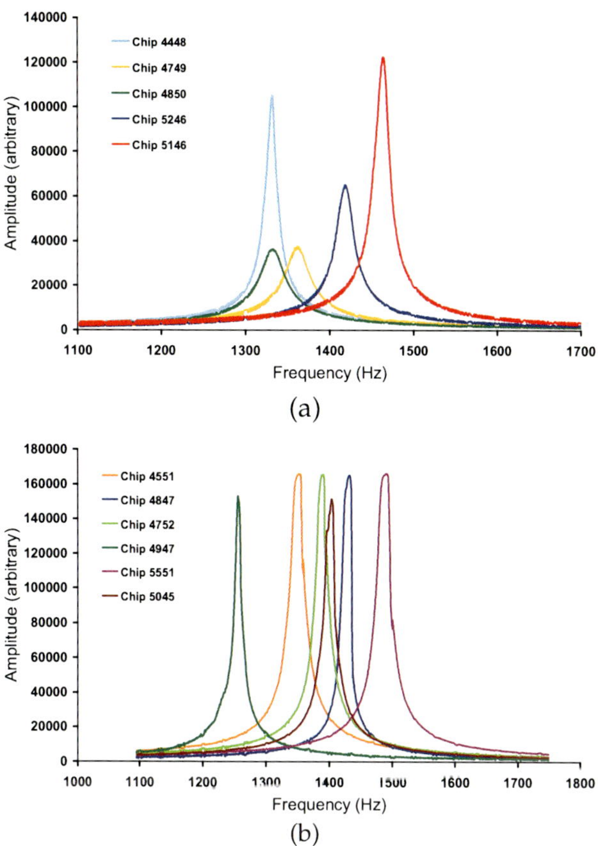

Figure 5.20 Variance of resonance frequency: (a) Batch A, (b) batch B.

is a function of several parameters including the geometry of the transducer (number of electrode elements, minimum gap between electrode elements, inner displacement amplitude, etc.) and the electrical operation conditions (bias voltage, load resistance). These parameters must be optimized for a specific operation frequency and acceleration amplitude.

The output power is directly proportional to the mass of the vibrating proof mass. However, a larger proof mass requires a larger chip size, which is a critical parameter with respect to the costs per chip. The ultimate advantage of batch-processing is the production of a large number of chips at the same time within a single process cycle. Therefore, the smaller the chip size, the more chips can be placed on a wafer layout and the lower the cost per chip. Consequently, the parameter chip size requires optimization with respect to the two conflicting target functions (large proof mass against cost per chip). In conclusion, the main benefit of electrostatic transducers, which is supposed to be the very small device size (which is feasible because of the availability of microfabrication technologies), creates a drawback at the same time. Due to the small proof mass of the transducer, only very low power levels in the range of μW are achievable. Therefore, the systems to be powered by electrostatic MEMS transducers must be ultra low-power systems.

There may also optimization of the fabrication process required to reduce technology imperfections. A trench widening effect, which may be a result of not perfectly tuned process parameters, can produce a gap sizing of up to 400 nm. This means that the gap between structures is larger than designed on the mask layout. The trench widening effect leads also to a beam sizing of the suspension beams and thus the beam width is smaller than the design value. Moreover, if the beam sizing is not uniform across the wafer a large variance of the resonance frequency may occur (Fig. 5.20). In order to increase the device yield for application-specific operation conditions, the beam sizing must be reduced by further improvement of the fabrication process [2].

References

1. Trimmer W. S. N. (1989) Microrobots and micromechanical systems, *Sens. Actuators*, 19, 267–287.

2. Hoffmann D., Folkmer B., and Manoli Y. (2009) Fabrication, characterization and modelling of electrostatic micro-generators, *J. Micromech. Microeng.*, 19, 094001. DOI:10.1088/0960-1317/19/9/094001.

3. Breaux O. P. (1976) Electrostatic energy conversion system United States Patent US04127804.

4. Williams C. B. and Yates R. B. (1995) Analysis of a micro-electric generator for microsystems, *Proc. Transducers/Eurosensors 1995* (Stockholm, Sweden, June 1995), 1, 369–372.

5. Tashiro R., Kabei N., Katayama K., Ishizuka Y., Tsuboi F., and Tsuchiya K. (2000) Development of an electrostatic generator that harnesses the motion of a living body (Use of a Resonant Phenomenon), *JSME Intl. J. Series C*, 43, 916–922.

6. Mitcheson P. D., Yeatman E. M., Rao G. K., Holmes A. S., and Green T. C. (2008) Energy harvesting from human and machine motion for wireless electronic devices, *Proc. IEEE*, 96, 1457–1486.

7. Tsutsumino T., Suzuki Y., Kasagi N., and Sakane Y. (2006) Seismic power generator using high-performance polymer electrets, *Proc. MEMS 2006* (Istanbul, Turkey, Jan. 2006) 98–101.

8. Tsutsumino T., Suzuki Y., Kasagi N., Kashiwagi K., and Morizawa Y. (2006) Micro seismic electret generator for energy harvesting, *Technical Digest PowerMEMS 2006* (Berkeley, USA, Nov. 2006), 133–136.

9. Sterken T., Altena G., Fiorini P., and Puers R. (2007) Characterisation of an electrostatic vibration harvester, *DTIP of MEMS & MOEMS* (Stresa, Italy, April 2007).

10. Sterken T., Baert K., Puers R., and Borghs S. (2002) Power extraction from ambient vibration, *Proc. SeSens* (Workshop on Semiconductor Sensors, Veldhoven, Netherlands, Nov. 2002), 680–683.

11. Miao P., Mitcheson P. D., Holmes A. S., Yeatman E. M., Green T. C., and Stark B. H. (2005) MEMS Inertial power generators for biomedical applications, *DTIP of MEMS & MOEMS* (Montreux, Switzerland, June 2005).

12. Naruse Y., Matsubara N., Mabuchi K., Izumi M., and Honma K. (2008) Electrostatic micro power generator from low frequency vibration such as human motion Technical, *Digest PowerMEMS 2008* (Sendai, Japan, Nov. 2008), 19–22.

13. Meninger S., Mur-Miranda J. O., Amirtharajah R., Chandrakasan A. P., and Lang J. H. (2001) Vibration-to-electric energy conversion, *IEEE Trans. VLSI Syst.*, 9, 64–76.

14. Mitcheson P. D., Sterken T., He C., Kiziroglou M., Yeatman E. M., and Puers R. (2008) Electrostatic microgenerators, *Meas. Control,* 41, 114–119.

15. Mur-Miranda J. O. (2004) Electrostatic Vibration-to-Electric Energy Conversion PhD Thesis, MIT.

16. Mitcheson P. D., Miao P., Stark B. H., Yeatman E. M., Holmes A. S., and Green T. C. (2004) MEMS electrostatic micropower generator for low frequency operation, *Sens. Actuators A*, 115, 523–529.

17. Noworolski J. M. and Sanders S. R. (1998) Microresonant devices for power conversion, *SPIE*, 3514, 260–265.

18. Miao P., Holmes A. S., Yeatman E. M., and Green T. C. (2003) Micro-machined variable capacitors for power generation, *Proc. Electrostatics'03*, Edinburgh, UK.

19. Yen B. C. and Lang J. H. (2006) A variable-capacitance vibration-to-electric energy harvester, *IEEE Trans. Circ. Syst.*, 53, 288–295.

20. Basset P., Galayko D., Paracha A. M., Marty F., Dudka A., and Bourouina T. (2009) A batch-fabricated and electret-free silicon electrostatic vibration energy harvester, *J. Micromech. Microeng.*, 19, 115025 doi:10.1088/0960-1317/19/11/115025.

21. Galayko D., Basset P., and Paracha A. M. (2008) Optimization and AMS modeling for design of an electrostatic vibration energy harvester's conditioning circuit with an auto-adaptive process to the external vibration changes, *DTIP of MEMS & MOEMS* 2008.

22. Sterken T., Baert K., Puers R., and Borghs S. (2002) Power extraction from ambient vibration, *Proc. SeSens 2002*, Veldhoven, the Netherlands, pp. 680–683.

23. Sterken T., Baert K., Puers R., Borghs G., and Mertens R. (2003) A new power MEMS component with variable capacitance, *Proc. Pan Pacific Microelectronics Symposium 2003*, Edina, USA, pp 27–34.

24. Tvedt L. G. W., Blystad L. C. J., Halvorsen E. (2008) Simulation of an electrostatic energy harvester at large amplitude narrow and wide band vibrations, *DTIP of MEMS & MOEMS*.

25. Mahmoud M. A., El-Saadany E. F., and Mansour R. R. (2006) Planar electret based electrostatic micro-generator, *Proc. PowerMEMS 2006* (Berkeley, USA, Nov. 2006).

26. Bartsch U. (2006) Electrostatic Energy Harvesting using Ambient Vibration Diploma Thesis, IMTEK, University of Freiburg, Germany.

27. Madou M. J. (1998) *Fundamentals of Microfabrication*, CRC-Press.

28. Beeby S. P., Tudor M. J., and White N. M. (2006) Energy harvesting vibration sources for microsystem applications, *Meas. Sci. Technol.*, 17, R174–R195.

29. Hoffmann D., Folkmer B., and Manoli Y. (2007) Design considerations of electrostatic electrode elements for in-plane micro-generators, *Technical Digest PowerMEMS 2007* (Freiburg, Germany, Nov. 2007), 133–136.

30. Soliman M. S. M., Abdel-Rahman E. M., El-Saadany E. F., and Mansour R. R. (2008) A wideband vibration-based energy harvester, *J. Micromech. Microeng.*, 18, 115021.

31. Knechtel R. (2005) Halbleiterwaferbondverbindungen mittels strukturierter Glaszwischen-schichten zur Verkapselung oberflächenmikromechanischer Sensoren auf Waferebene Dissertation, Technical University of Chemnitz, Germany.

32. Gormley C., Yallup K., and Nevin W. A. (999) State of the art deep silicon anisotropic etching on SOI bonded substrates for dielectric isolation and MEMS applications, *Proc. Tech. Dig. 5th Int. Symp. Semiconductor Wafer Bonding* (Honolulu, Oct. 1999), 350–361.

Chapter 6

Thermoelectric Generators

Robert Hahn[a] and Jan D. König[b]

[a] *Fraunhofer Institute Reliability and Microintegration,*
Gustav-Meyer-Allee 25, 13355 Berlin, Germany
[b] *Fraunhofer IPM, Heidenhofstrasse 8, 79110 Freiburg, Germany*
Robert.Hahn@izm.fraunhofer.de, Jan.Koenig@ipm.fraunhofer.de

6.1 Physical Principles

The discovery of thermoelectricity was done a long ago. In 1821, Thomas J. Seebeck observed that the needle of a compass was deflected in the vicinity of two metallic conductors connected to one another when different temperatures prevailed at the joints. The degree of deflection here was proportional to the temperature difference. The reason for the movement of the compass needle was an electrical field that had apparently been created by the difference in temperature between the conductors. The effect observed by Seebeck also functions in the opposite direction and was first described by Jean C. A. Peltier in 1834: If electricity is applied to the two connected conductors, a temperature gradient occurs at the contact points. Heat energy is transported from one connection point to the other, leading to a cooling effect.

Handbook of Energy Harvesting Power Supplies and Applications
Edited by Peter Spies, Loreto Mateu, and Markus Pollak
Copyright © 2015 Pan Stanford Publishing Pte. Ltd.
ISBN 978-981-4241-86-1 (Hardcover), 978-981-4303-06-4 (eBook)
www.panstanford.com

6.1.1 *The Seebeck Effect*

The Seebeck effect is the phenomenon underlying the conversion of heat energy into electrical power. Its physical significance can be appreciated by considering the effect of imposing a temperature gradient along a finite conductor. Without temperature gradient, the carriers in the conductor have a distribution according to the thermal equilibrium Fermi–Dirac distribution. An inhomogeneous carrier distribution occurs in the presence of a temperature gradient, because carriers at the hot end will have a greater kinetic energy and tend to diffuse to the cold end. Due to the carrier diffusion, an electric field opposite to the carrier movement is building up.

The junctions of a circuit are formed from two dissimilar conductors A and B (Fig. 6.2), which are connected electrically in series but thermally parallel and are maintained at different temperatures T_1 and T_2 and $T_1 > T_2$. An open circuit potential difference V without current flow is developed and is given by

$$V = \alpha \left(T_1 - T_2 \right), \tag{6.1}$$

and $\alpha = V/(T_1 - T_2)$ defines the differential Seebeck coefficient α_{ab} between the elements a and b. For small temperature differences, the relationship is linear.

Although by convention α is the symbol for the Seebeck coefficient, S is also sometimes used and the Seebeck coefficient referred to as the thermal emf or thermopower. The sign of α is positive if the emf causes a current to flow in a clockwise direction around the circuit and is measured in V/K or more often $\mu V/K$.

For small temperature differences, the relationship is linear and defines the relative Seebeck coefficient S_{AB} for the junction. $S_{AB} = S_A - S_B$ is the resulting Seebeck coefficient of the complete circuit by the thermovoltages of material A and material B. The sign of the Seebeck coefficient is determined by the direction of the current flow and is considered to be positive, if the conventional current tends to flow from A to B at the hot junction. Hence, the sign is determined by the carrier type of the investigated material, e.g. a p- or n-type semiconductor. Thus, the sign of the Seebeck coefficient as well as its magnitude depends upon the choice of materials.

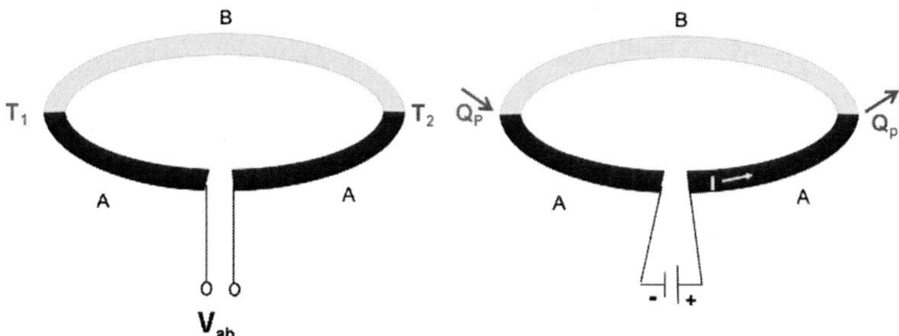

Figure 6.1 Schematic set-up for the measurement of thermoelectric effects: (a) Seebeck coefficient (b) Peltier coefficient.

6.1.2 *Peltier Effect*

The Peltier effect is the phenomenon used in thermoelectric refrigeration. The Peltier effect arises from different potential energies of the charge carriers in the materials on either side of a junction. Energy must be interchanged with the surroundings in order to maintain conservation of energy and charge, when a current flows through the junction (Fig. 6.1b). A current I flows from material A to material B and then to material A. At junction 1, the heat Q per unit time is absorbed from the surroundings according to

$$\frac{dQ}{dt} = \Pi_{AB} I \qquad (6.2)$$

$\Pi_{AB} = \Pi_A - \Pi_B$ is the relative Peltier coefficient of the junction given by the absolute Peltier coefficients of the two materials.[a] By convention, Π_{AB} is taken as positive, when current flows from A to B and Q is absorbed from the surroundings. The absorbed heat Q transported in relation to the electric current is given by $\Pi_{AB} I$. At least the same amount of heat is emitted to the surroundings at junction 2.

6.1.3 *Thomson Effect*

The Thomson effect relates to the rate of reversible heat generation, which results from the passage of a current along a portion of a

---

[a] In analogy to thermopower measurement, the Peltier coefficient of a material is determined by measuring Π_{AB} of a circuit, where the Peltier coefficient of one material is known.

single conductor over which a temperature difference ΔT is applied. Due to the temperature difference, heat per unit time is absorbed according to

$$\frac{dQ}{dt} = \beta I \Delta T,\tag{6.3}$$

where β is the Thomson coefficient. The origin of the effect is essentially the same as the Peltier effect. Here the temperature gradient along the conductor is responsible for differences in the potential energy of the charge carriers. The Thomson effect is not of primary importance in thermoelectric devices, but should not be neglected in detailed calculations.

6.1.4 Kelvin Relation

The three thermoelectric coefficients are related to each other by the Kelvin relationships, whereas Seebeck and Peltier coefficient are properties of a junction (relative coefficients) and the Thomson coefficient of a single material (absolute coefficient):

$$S_{AB} = \frac{\Pi_{AB}}{T};\ \frac{dS_{AB}}{dT} = \frac{\beta_A - \beta_B}{T}\tag{6.4}$$

The validity of these relationships has been demonstrated for a number of thermocouple materials. The second relationship enables a definition to be derived for the Seebeck coefficient of a single material by

$$S = \int_0^T \frac{\beta}{T}\,dT\tag{6.5}$$

The Seebeck coefficient is negative for n-type semiconductors and positive for p-type semiconductors.

6.2 Conversion Efficiency and Figure of Merit

6.2.1 Thermoelectric Generation Efficiency

A thermoelectric converter is a heat engine and like all heat engines, it obeys the laws of thermodynamics. If we first consider

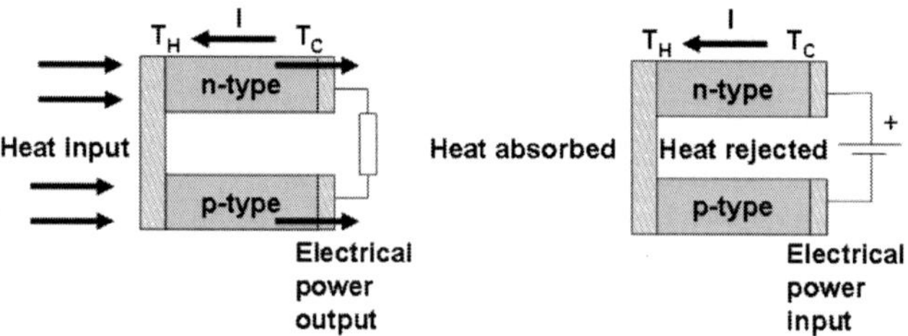

Figure 6.2 Thermoelectric generator (left); thermoelectric refrigerator (right).

the converter operating as an ideal generator in which there are no heat losses, the efficiency is defined as the ratio of the electrical power delivered to the load to the heat absorbed at the hot junction. Expressions for the important parameters in thermoelectric generation can readily be derived by considering the simplest generator consisting of a single thermocouple with thermo-elements fabricated from n- and p-type semiconductors as shown in Fig. 6.2. The thermocouples are built up of two branches: one n-type, one p-type material with the length L_n and L_p and constant cross sections A_n and A_p. The two branches are attached to metallic conductors of negligible electrical resistance.

The branches are connected electrically in series and thermally parallel. It is important to note that heat is transferred only by conduction along the branches of the thermocouple.

The thermocouple can be used in two ways. On one hand, it can be used as a generator to produce an electric current due to the Seebeck voltage caused by the temperature difference between heat source and drain. The generated electric power can be used by a resistive load R_L. On the other hand, the resistive load can be replaced by a current source. Here the thermocouple acts as a heat pump. Heat is pumped from heat source at temperature T_1 to heat sink at temperature T_2 by means of the Peltier effect.

The efficiency of a thermoelectric generator for the conversion of thermal into electric energy is defined by

$$\phi = \frac{\text{energy available at } R_L}{\text{heat energy absorbed at hot junction}}. \tag{6.6}$$

The energy taken from the resistive load R_L is calculated by the current I and R_L:

$$W = R_\mathrm{L} I^2 \qquad (6.7)$$

The heat transported through the branches q_p and q_n from heat source to heat drain is related to two effects. On one hand, heat flows through the branches because of thermal conduction related to the thermal conductivity λ_n and λ_p of the materials of the two branches. On the other hand, heat is transported together with a current flow due to the Peltier effect, when it has to be dissipated or absorbed as the current passes into another conductor. The amount of heat transported from heat source to drain is

$$q_\mathrm{p} = \Pi_\mathrm{p} I - \lambda_\mathrm{p} A_\mathrm{p} \frac{dT}{dx} \overset{\text{Eq.6.4}}{=} S_\mathrm{p} I T - \lambda_\mathrm{p} A_\mathrm{p} \frac{dT}{dx}$$

$$q_\mathrm{n} = \Pi_\mathrm{n} I - \lambda_\mathrm{n} A_\mathrm{n} \frac{dT}{dx} \overset{\text{Eq.6.4}}{=} \underbrace{S_\mathrm{n} I T}_{\text{Peltier}} - \underbrace{\lambda_\mathrm{n} A_\mathrm{n} \frac{dT}{dx}}_{\text{heat conduct}}, \qquad (6.8)$$

where Π is the absolute Peltier coefficient and S the absolute Seebeck coefficient. The Peltier coefficients are replaced by S*T according to the Kelvin relations 6.4. It should be noted that S_p is positive and S_n is negative, so that the Peltier heat flow is opposed by the heat conduction.

Additionally heat is generated by the electric current flowing through the branches according to Joule heating. The rate of heat generation per unit length is $I^2/\sigma A$, where σ is the electrical conductivity. The heat generation implies a non-uniform gradient. The Seebeck coefficient is assumed to be independent of temperature, so that the Thomson effect is absent:

$$-\lambda_\mathrm{p} A_\mathrm{p} \frac{d^2 T}{dx^2} = \frac{I^2}{A_\mathrm{p} \sigma_\mathrm{p}}; \quad -\lambda_\mathrm{n} A_\mathrm{n} \frac{d^2 T}{dx^2} = \frac{I^2}{A_\mathrm{n} \sigma_\mathrm{n}} \qquad (6.9)$$

This equation can be integrated once with the boundary conditions $T = T_1$ at $x = 0$ (heat source) and $T = T_2$ at $x = L_\mathrm{p}$ or $x = L_\mathrm{n}$ (drain). Combined with equation before the rate of heat flow at $x = 0$ (heat source, $T = T_1$),

$$q_\mathrm{p}\,(x = 0) = S_\mathrm{p} I T_1 - \frac{\lambda_\mathrm{p} A_\mathrm{p}\,(T_2 - T_1)}{L_\mathrm{p}} - \frac{I^2 L_\mathrm{p}}{2 A_\mathrm{p} \sigma_\mathrm{p}}$$

is $\qquad (6.10)$

$$q_\mathrm{n}\,(x = 0) = S_\mathrm{n} I T_1 - \frac{\lambda_\mathrm{n} A_\mathrm{n}\,(T_2 - T_1)}{L_\mathrm{n}} - \frac{I^2 L_\mathrm{n}}{2 A_\mathrm{n} \sigma_\mathrm{n}}$$

At least the complete amount of heat per unit time W_Q absorbed at $x = 0$ is

$$W_Q = q_p\,(x = 0) + q_n\,(x = 0) = (S_p - S_n)\,I\,T_1 - G\,(T_1 - T_2) - \frac{I^2 R}{2}$$

$$(6.11)$$

R is the total electric resistance of the two branches electrically in series and G the thermal conductance of two branches thermally parallel:

$$R = R_p + R_n = \frac{L_p}{\sigma_p A_p} + \frac{L_n}{\sigma_n A_n} \tag{6.12}$$

$$G = G_p + G_n = \frac{\lambda_p A_p}{L_p} + \frac{\lambda_n A_n}{L_n} \tag{6.13}$$

The generated current is calculated by the resulting Seebeck voltage and the sum of resistive load R_L and R:

$$I = \frac{u_p - u_n}{R + R_L} = \frac{(S_p - S_n)\,(T_1 - T_2)}{R + R_L} \tag{6.14}$$

The efficiency ϕ is expressed with the assumption of constant electrical conductivity, thermal conductivity and thermopower in one branch and negligible contact resistance:

$$\phi = \frac{I^2 R_L}{(S_p - S_n)\,I\,T_1 + G\,(T_1 - T_2) - \frac{1}{2}I^2 R} \tag{6.15}$$

In thermoelectric materials, the temperature dependence of σ, G, and S in both generation and refrigeration has to be taken into account. However, the simple expression obtained for the efficiency can still be employed with an acceptable degree of accuracy, if approximate average values are adopted for these parameters over the temperature range of interest.

6.2.2 *Thermoelectric Figure of Merit*

According to Eq. 6.15, the efficiency is a function of the ratio of resistive load and generator resistance. At maximum power, output it is given by

$$\phi_p = \frac{T_1 - T_2}{T_1}\,\frac{1}{\frac{3}{2} + \frac{T_2}{2T_1} + \frac{4}{Z_c T_1}} = \eta_{\text{Carnot}}\,\frac{1}{\frac{3}{2} + \frac{T_2}{2T_1} + \frac{4}{Z_c T_1}}. \tag{6.16}$$

Here, η_{Carnot} is the Carnot efficiency, $\eta_{Carnot} = (T_1 - T_2)/T_1$, which is the maximum efficiency obtainable for a temperature difference $T_1 - T_2$. Maximum power output is obtained if $R = R_L$. Additionally, the efficiency of the thermoelectric circuit depends on the figure of merit of the couple

$$Z_c = \frac{(S_p - S_n)^2}{RG}.\tag{6.17}$$

This shows that in addition to the Carnot efficiency, ϕ_p depends on thermopower, resistance and thermal conductance of the branches. The figure of merit is a measure how far the Carnot efficiency can be approximated by a thermoelectric circuit.

If the geometry of the two branches is matched to minimize heat absorption, Z_c is determined by

$$Z_c = \frac{(S_p - S_n)^2}{\left[\left(\frac{\lambda_p}{\sigma_p}\right)^{1/2} + \left(\frac{\lambda_n}{\sigma_n}\right)^{1/2}\right]^2}.\tag{6.18}$$

If one further assumes that the two arms of the junction have similar material constants, the figure of merit for a material

$$Z = \frac{\sigma S^2}{\lambda}\tag{6.19}$$

is employed. Here, σS^2 is referred to as the electrical power factor.

Often it is more convenient to define the dimensionless factor ZT, (Z multiplied by the absolute temperature) as the figure of merit

$$ZT = \frac{\sigma S^2}{\lambda}T\tag{6.20}$$

In Fig. 6.3, the efficiency is plotted against temperature difference for different values of Z assuming the cold junction temperature T_2 to be 300 K. It is apparent that for optimum power-generating efficiency, the factor ZT should be maximized. This means that a high value of Z should be combined with a great difference between T_1 and T_2. As a ballpark figure, a thermocouple fabricated from thermo-element materials with an average figure of merit of 3×10^{-3}/K would have an efficiency of ca. 20% when operated over a temperature difference of 500 K.

It should be noted that Z alone determines the efficiency only in case of the optimal electric current. Although very instructive, such

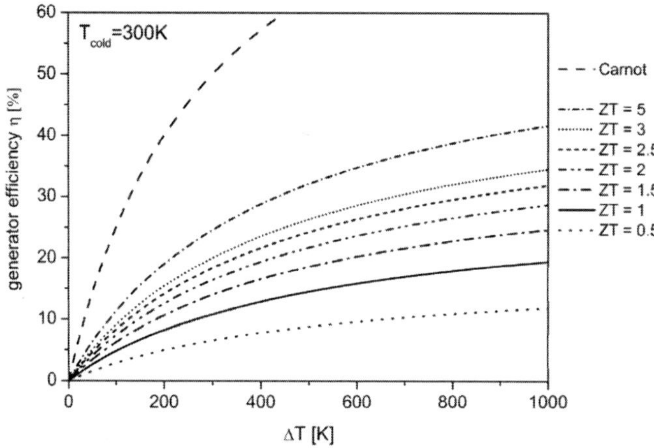

Figure 6.3 Conversion efficiency for different Z-values. The efficiency rises with an increasing figure of merit Z and increasing difference between hot and cold side. (The cold side temperature is 300 K.)

simplifications ignore the effect of thermoelectric compatibility in real thermoelectric devices. The thermoelectric compatibility factor is the reduced electric current, which is necessary to achieve the highest efficiency determined by Z. Because the compatibility factor changes with temperature while the electric current is constrained, the efficiency of a real device will be less than that calculated from z. The effect of thermoelectric compatibility is most important for segmented thermoelectric generators, but it also affects the exact calculation of performance for all thermoelectric devices. To calculate the exact performance of a thermoelectric generator analytically, on method is to use a reduced variables approach that will separate the so called intensive properties and variables (such as temperature gradient, Seebeck coefficient, current density, heat flux density) from the so called extensive ones (e.g., voltage, temperature difference, power output, area, length, resistance, load resistance). This approach allows a definition of a local, intensive efficiency in addition to the traditional system efficiency as well as the derivation of the compatibility factor [1]. Practical calculation guidelines are given in Sections 6.4 and 6.6.

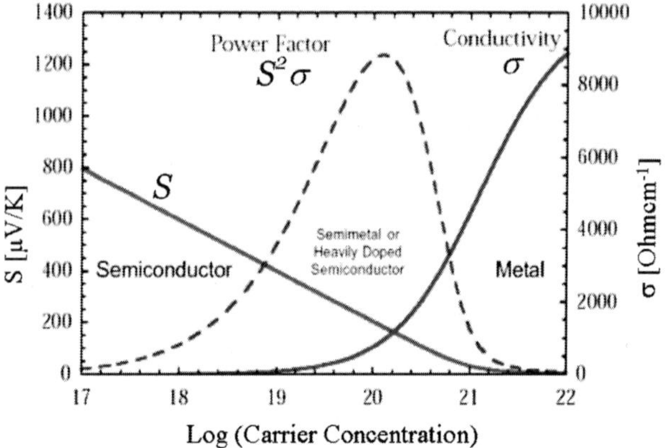

Figure 6.4 Schematic dependence of electrical conductivity, Seebeck coefficient and power factor on concentration of free carriers [1].

6.3 Thermoelectric Materials

6.3.1 *Theoretical Material Aspects*

All three parameters which occur in the figure of merit (6.19) are functions of the carrier concentration. The electrical conductivity increases with increase in carrier concentration as shown in Fig. 6.4, while the Seebeck coefficient decreases. Thus, the electrical power factor maximizing at a carrier concentration of around $10^{19}/cm^3$.

The electronic contribution to the thermal conductivity λ_e, which in thermoelectric materials is generally around 1/3 of the total thermal conductivity, also increases with carrier concentration. In metals the ratio of the electronic contribution to the thermal conductivity and the electrical conductivity is proportional to the temperature which is stated in the Wiedemann–Franz law.

Evidently the figure of merit optimizes at carrier concentrations which corresponds to heavily doped semiconductor materials or semimetals. Consequently, semiconductors are the materials most researched for thermoelectric applications.

Thermoelectric phenomena are exhibited in almost all conducting materials (except for superconductors below T_c). As stated above, because the figure of merit varies with temperature in most cases the dimensionless figure of merit ZT is used as measure of

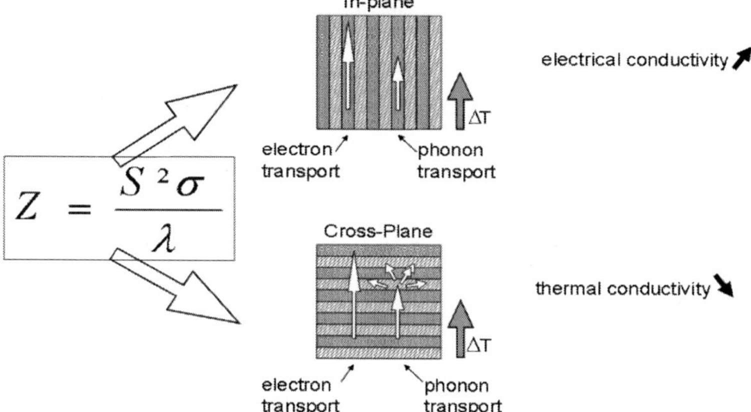

Figure 6.5 Designing materials with nanoscaled multilayer structures. Decreased thermal conductivity and high electrical conductivity in case of cross-plane charge transport.

performance. Only those materials which possess a $ZT > 0.5$ are usually regarded as thermoelectric materials.

According to Eq. 6.19 a good figure of merit could be achieved for a material were σ is high and λ is low but they are proportional in nature. Thus, modern high ZT materials manage to trick nature to a certain degree: They have a fabricated atomic configuration, in which the inner structure of the material restricts the mobility of the phonons and hence its thermal conductivity, (phonon blocking) while not obstructing or even promoting that of the electrons (electron transmitting). Thus, efforts have focused on improving the figure of merit by reducing the lattice thermal conductivity as illustrated in Fig. 6.5.

6.3.2 Materials Research

Two research avenues are currently being pursued. One is a search for so-called "phonon glass-electronic crystals" in which it is proposed that crystal structures containing weakly bound atoms or molecules that "rattle" within an atomic cage should conduct heat like a glass but conduct electricity like a crystal. Candidate materials receiving considerable attention are the filled skutterudites and the clathrates. During the past decade, material scientists have been optimistic in their belief that low-dimensional structures such as quantum wells (materials which are so thin as to be essentially of two dimensions (2D), quantum wires (extremely small cross section and considered to be of one dimension (1D), and referred to as nanowires) quantum dots which are quantum confined in all directions and superlattices (a multiple-layered structure of quantum wells) will provide a route for achieving significantly improved thermoelectric figures-of-merit. The expectation is that the reduced dimensions of these structures will result in an increase in phonon interface scattering and a consequent reduction in lattice thermal conductivity.

Although low-dimensional structures would find immediate application in microelectronics, at present the technology is expensive and applying it to bulk devices is problematic. In some respects, nanowires appear to be a more attractive proposition for thermoelectric applications than quantum well superlattices because the geometry of the current flow is more favourable and the fabrication process more compatible with integrated technology than molecular beam epitaxy (MBE).

Attempts are also being made to improve the competitiveness of thermoelectric material in directions other than the figure of merit. Efforts have focused for example on increasing the electrical power factor, decreasing cost, and developing environment-friendly materials. As examples, when the fuel cost is low or essentially free, as in waste heat recovery, then the cost per watt is mainly determined by the power per unit area and the operating period. The rare-earth compound $YbAl_3$, although possessing a relatively low figure of merit, has a power factor almost three times that of bismuth telluride, while MgSn has almost the same performance but costs less than a quarter of the price.

The ZT value had been stagnating for decades at below 1. Now, thanks to new classes of materials, laboratory values of up to 3.5 have been achieved. Values of around 1.5 to 2 are regarded as the threshold of profitability for using TEG cost-effectively in larger applications.

The variety of TE materials extends from monocrystalline or polycrystalline solids via semiconductors and metalloids, ceramic oxides to thin-layer super lattices.

Nanotechnologically produced materials are regarded as especially promising. They are manufactured on the basis of already familiar thermoelectric materials by, for example, embedding nanoparticles in a macroscopic matrix.

Nanowires made of metalloids, such as bismuth, in which the charge carriers can only move in one direction along the axis of the wire, are another focus of this research. In the laboratory ZT values of up to 3 are reported for wires with diameters of less than 15. Nanoscaled multilayer structures, often called superlattices, have the advantage that they can be directly transformed into the conventional vertical construction elements. Such structures exhibit physical effects that raise the ZT value. Heat and charge carriers flow within the layers or perpendicular to them. In the case of cross-plane charge transport, i.e. transport perpendicular to the layers, the numerous boundary surfaces scatter the heat-conducting phonons and thus considerably reduce thermal conductivity. Transport of the electrical charge carriers takes place largely without interference. These concepts can be transferred also to bulk materials. Another effective form of nanostructuring in bulk materials is not based on a target arrangement of layers or particles from outside, but on the spinodal decomposition of thermoelectric materials on the nanoscale. Altogether a considerable improvement results in the ZT value as illustrated in Fig. 6.6.

Preferential use is made of V–VI components, for which the highest nanoscale ZT values have been published to date, or IV–VI materials (PbTe based) as well as oriented, structurally ordered V–VI/IV–VI composites.

Even without nanostructuring, higher efficiency of materials can be achieved. Promising candidates are complex chalcogenides, clathrates, Zintl phases, half-Heusler compounds, ceramic oxides, or the space-capable high-temperature group of materials derived from $CoSb_3$, known as skutterudites, after the Norwegian town of Skutterud where they occur. In the latter case, scattering centres for phonons occur due to heavy atoms in the free spaces of the cubic crystal lattice. Figure 6.6 gives an overview of the figure of merit

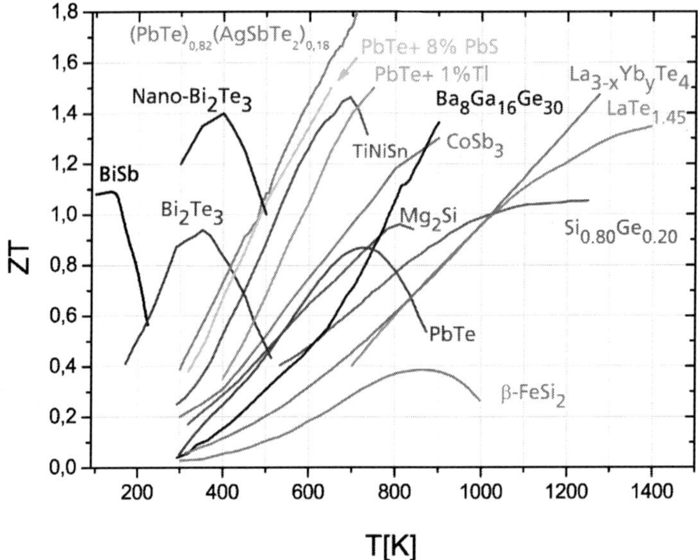

Figure 6.6 Overview of the thermoelectric figure of merit *ZT* of several new developed materials [20, 21].

as a function of temperature for several novel, bulk thermoelectric materials.

6.3.3 *Technical Relevant Materials*

The temperature-dependent figures of merit of the state-of-the-art thermoelectric materials are shown in Fig. 6.7. The most important thermoelectric materials are Bi_2Te_3-based solid solutions for room-temperature applications, PbTe-based solid solutions for the mid-temperature range between 600 and 800 K and $Si_{1-x}Ge_x$ for temperatures above 1000 K.

The suitability of a material system for thermoelectric applications depends not only on the figure of merit. So the materials' costs can be an important factor. Other important parameters are mechanical stability, thermal diffusion stability, contact stability (no inter-diffusion at contacts) or oxidation resistance in air. Material systems such as n-type BiSb alloys, p-type TAGS (tellurium–antimony–germanium–silver) and $FeSi_2$ have good thermoelectric

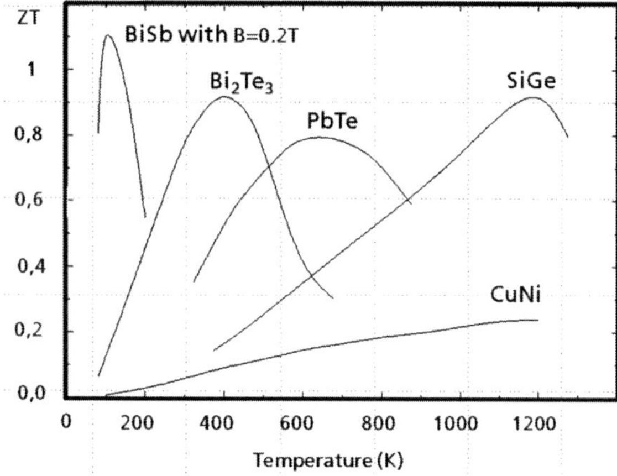

Figure 6.7   Overview of technical relevant thermoelectric materials [22].

properties but are not widely used because of various practical difficulties such as high sublimation rates, poor mechanical strength or absence of homologous n-type or p-type material. $ZT_{max} \approx 1$ is the actual limit over the whole temperature range between 100 and 1500 K, although significant improvements of thermoelectric properties in different material systems have been achieved in laboratory and promises higher conversion efficiencies in future.

6.4 Thermoelectric Module Construction

In practice, a large number of thermocouples are connected electrically in series and thermally in parallel by sandwiching them between two high thermal conducting and electrical isolating plates to form a module. This is because the voltage output from semiconductor thermocouples remains relatively low, in the order of hundreds of microvolts per degree. The module is the building-block of a thermoelectric conversion system. A typical example of a module and a schematic are shown in Figs. 6.8 and 6.9, respectively.

The performance of the module is not only a function of the thermoelectric material but it is influenced to a great extend on the electrical contact resistance between the thermocouples

Figure 6.8 Thermoelectric module, source: Fraunhofer IPM.

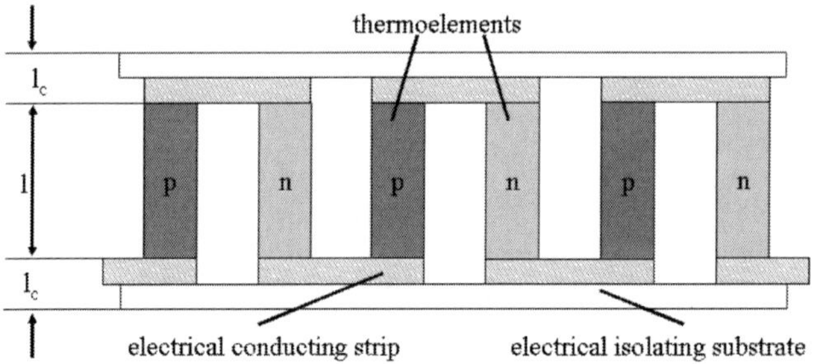

Figure 6.9 Principal configuration of a thermoelectric module.

and their electrical interconnection and on the thermal resistance of the interconnections and the module plates. The electrical contact resistance will increase the internal impedance while the thermal resistances reduce the temperature difference over the thermocouples. Both result in a lower output voltage. Based on the module configuration of Fig. 6.9, a simplified model has been developed which takes into account the thermal and electrical contact resistances [2]. It can be shown that when the module operates with a matched load, the output voltage V and current I

are given by

$$V = \frac{N\alpha\,(T_\mathrm{h} - T_\mathrm{c})}{1 + 2r\frac{l_\mathrm{c}}{l}} \tag{6.21}$$

$$I = \frac{A\alpha\,(T_\mathrm{h} - T_\mathrm{c})}{2\rho\,(n + l)\left(1 + 2r\frac{l_\mathrm{c}}{l}\right)}, \tag{6.22}$$

where N is the number of thermocouples in a module, α the Seebeck coefficient, and ρ the electrical resistivity, T_h and T_c are temperatures at the hot- and cold-sides of the module respectively, A and l are the cross-sectional area and thermoelement length l_c is the thickness of the contact layer,

$$n = 2\frac{\rho_\mathrm{c}}{\rho} \tag{6.23}$$

is the electrical contact parameter and

$$r = \frac{\lambda}{\lambda_\mathrm{c}} \tag{6.24}$$

is the thermal contact parameter where ρ_c is the electrical contact resistivity, λ_c the thermal contact conductivity, and λ the thermal conductivity of thermoelement materials. These can be estimated using a method described in [3]. For commercially available modules, appropriate values are $n = 0.1$ mm and $r = 0.2$. Figure 6.10 shows the current-per-unit-area of a thermoelement, I/A; and the voltage per thermocouple, V/N; as a function of thermoelement length l for different temperature differences. The voltage increases with an increase in thermoelement length, while the current exhibits a maximum at a shorter length.

The power output P and conversion efficiency φ of a thermoelectric module, when operated with a matched load, can be expressed as [4].

$$P = \frac{\alpha^2}{2\rho}\,\frac{AN\,(T_\mathrm{h} - T_\mathrm{c})^2}{(n + l)\left(1 + 2r\frac{l_\mathrm{c}}{l}\right)^2} \tag{6.25}$$

$$\phi = \frac{\left(\dfrac{T_\mathrm{h} - T_\mathrm{c}}{T_\mathrm{h}}\right)}{\left(1 + 2r\frac{l_\mathrm{c}}{l}\right)^2\left[2 - \frac{1}{2}\left(\dfrac{T_\mathrm{h} - T_\mathrm{c}}{T_\mathrm{h}}\right) + \left(\dfrac{4}{Z\,T_\mathrm{h}}\right)\left(\dfrac{l + n}{l + 2rl_\mathrm{c}}\right)\right]} \tag{6.26}$$

Figure 6.11 shows the power density $p = P/AN$ and conversion efficiency as function of the thermoelement length for different

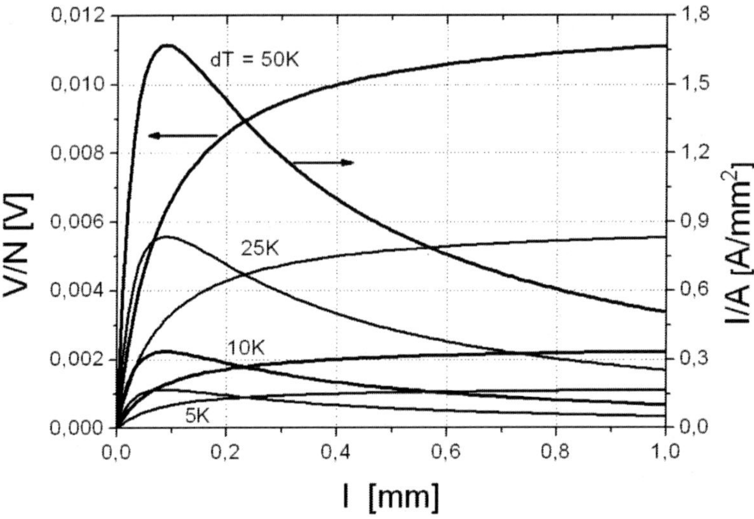

Figure 6.10 Voltage per thermocpouple and module current density as function of thermoelement length for different temperature differences according to equations (6.21) and (6.22). ($\rho = 10^{-5}$ Ωm, $\alpha = 240 \times 10^{-6}$ V/K, $n = 0.1$ mm, $r = 0.2$, $l_c = 0.2$ mm.

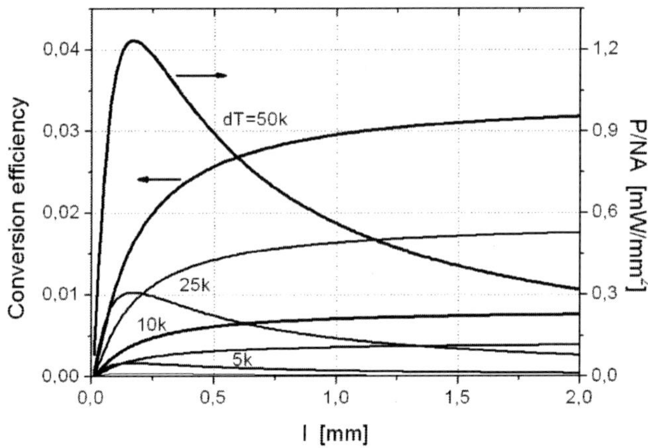

Figure 6.11 Power densisty and conversion efficiency vs. thermoelement length for different temperature differences. Material and design parameters as used for Fig. 6.10.

temperatures according to Eqs. (6.25) and (6.26). It can be seen that in order to obtain high conversion efficiency, the module should be designed with long thermoelements. However, if a large power-per-unit-area is required, the thermoelement length should be optimized at a relatively shorter length. For energy-harvesting application, in most cases the efficiency is not of concern but only the output power which is required to run the energy autarkic system. From the equations above, the required number of the thermocouples, N, and the cross-sectional area can be determined for a given specification and thermoelement material. The optimal thermoelectric length is independent of temperature. However, the determination of thermoelement length involves a rather complicated optimization procedure driven by its economic viability which is strongly dependent on the manufacturing process. For a better optimization of a specific TEG integration, the thermal impedance of the integration environment has to be considered as well as shown in Section 6.6.

Thus, appropriate selection of contact materials and formation of electrical and thermal junctions are important factors in the design and fabrication of thermoelectric modules. Rewriting Eq. 6.25,

$$P = F N \Delta T^2 \left(\frac{\alpha^2}{2\rho} \right) \left(\frac{A}{l} \right) \tag{6.27}$$

where

$$F = \frac{1}{\left(1 + \frac{n}{l}\right)\left(1 + 2r\frac{l_c}{l}\right)^2} \tag{6.28}$$

Here F is referred to as the fabrication quality factor. In ideal case, F approaches unity when contact properties n and r approach zero. However, in practice, contact resistances always exist, resulting in $F < 1$. Evidently, a key objective in module fabrication is to develop suitable fabrication technologies and procedures that minimize contact resistances. It can also be seen that once the contact properties n and r are given, F will be affected by thermoelement length and contact layer thickness. It can also be seen from Eq. 6.27 that if contact parameters are neglected, the ideal power of a TEG is the same for all thermolegs with the same area to length ratio. This is also true for the electrical and thermal resistance of the TEG.

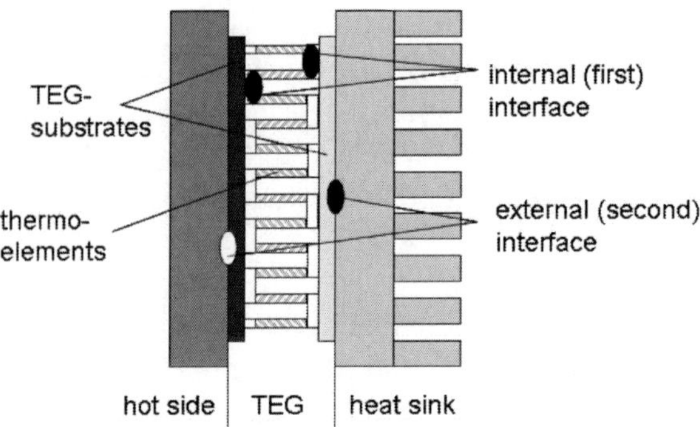

Figure 6.12 Integration of thermoelectric generators. Internal and external thermal interfaces.

This analysis is of special importance for microreactors. The length of the thermoelements of microreactors which are deposited according to the arrangement of Fig. 6.9 are likely to be very short (a few to tens of micrometres). From Eq. 6.28, it can be seen that very small contact parameters n and r are essential to obtain an adequate F factor. Clearly, successful realization of a vertically structured microconverter depends on substantial improvement in the electrical and thermal contact properties.

A typical integration of a thermoelectric generator into an application is shown in Fig. 6.12. Low-thermal-resistivity interfaces between the TEG-substrates and the hot side and heat sink at the cold side (external interfaces) are of importance as well. Since each surface has specific roughness, only point contacts will act for the transport of the heat flux. To reduce the thermal resistance several interface materials have been developed. They fill the cavities between the surfaces and can reduce the thermal interface resistance substantially as shown in Fig. 6.13.

The external interface areas are much higher in comparison to the interface of an individual thermoelement. Thus, the differences in the coefficient of thermal expansion (CTE) of the TEG-substrates and the hot and cold side materials have to be considered. In case of large CTE difference, ductile interface materials have to be used which can compensate the expansion mismatch. An overview of the

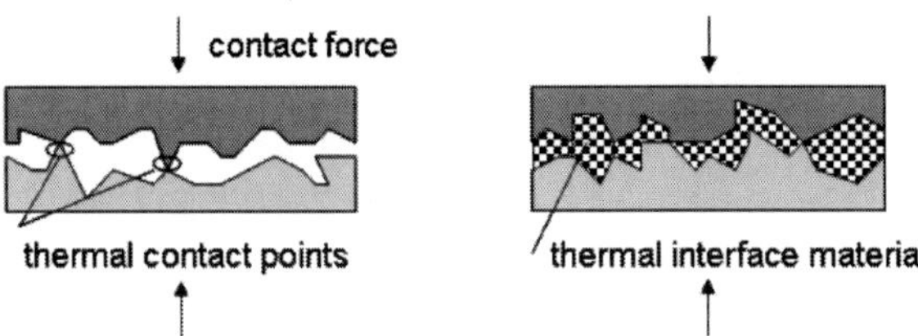

Figure 6.13 Improvement of the heat transfer with help of thermal interface material.

thermal interface resistance R_{jx} of practical interface materials is given in Fig. 6.14.

In recent years, a large variety of interface materials have been developed which have high thermal conductivity and develop a low thermal contact resistance. While a stable, self-sufficient junction can be achieved with help of solders or thermally conducting adhesives, in all other cases like thermal greases or phase change materials (PCM), a definite force has to be applied to the stack of TEG and cooling and heating bodies. It has to be ensured that the fixing elements which are required to apply the contact force do not act as thermal bypass (like K_{air2} shown in Fig. 6.25). In most cases, the lowest thermal interface resistance is achieved with thin solder layers or silver-filled epoxics. In addition to the materials shown in Fig. 6.14, also ductile metal foils like indium foil are also used. Depending on the temperature range, the size of the TEG and the surface properties of the substrates the type, the thickness of the interface material as well as the contact pressure has to be optimized.

The next issue in integrating TEGs into energy-harvesting application is the design of the heat sink. On one hand, miniaturized systems are desirable from the user's perspective. On the other hand, the thermal resistance to ambient have to be matched with the TEG, sufficient heat has to be dissipated and a high temperature difference has to be maintained over the TEG. Figure 6.15 gives an overview of typical heat transfer coefficients α_w. In most cases, air cooling will be used at the cold side of energy harvesters. For natural air convection of an upright plate, thermal transfer coefficients between 3 and 30 W/m^{2}K are realistic, while a strong

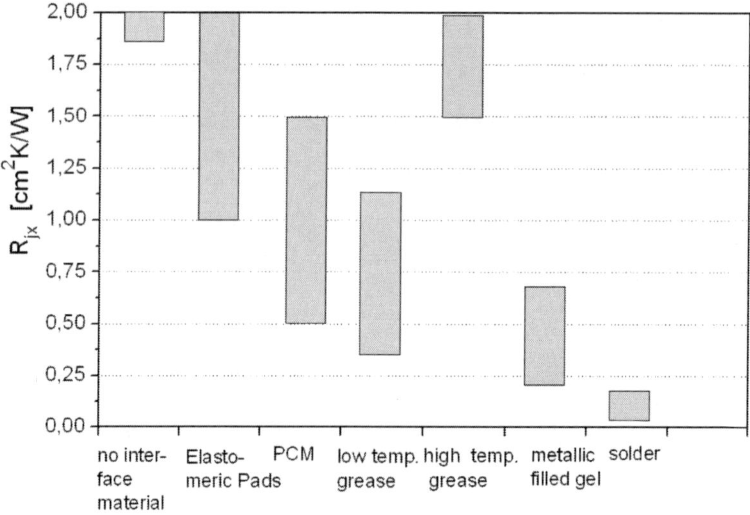

Figure 6.14   Thermal resistivity of several thermal interface materials.

forced air movement can increase the heat transfer coefficient up to $300 \ W/m^2 K$. Many different types and designs of heat sinks which were in most cases developed for electronics cooling can be used for thermal harvesters. Their heat transfer coefficients as function of air velocity and mounting orientation can be found in the data sheets. As can be seen from Fig. 6.15 the heat transfer coefficient can be increased significantly if a fan is used instead of natural convection air. Since the output power of the TEG is a power function of the temperature difference, it can be easily increased by a factor of 5 to 10 with help of an air mover. However, in most cases, this is not an option for energy-harvesting application since the air movement consumes more energy than the generator can produce. Therefore, passive cooling solutions have to be optimized, such as large surface heat sinks, heat pipes and thermosiphon cooling.

Table 6.1 gives an overview of typical commercial thermoelectric modules, with V_o the open circuit voltage, I_s short circuit current, R_i internal resistance, G the thermal conductance of the module, G/A the thermal conductance per module area and P1 the electrical power with respect to module area (power density).

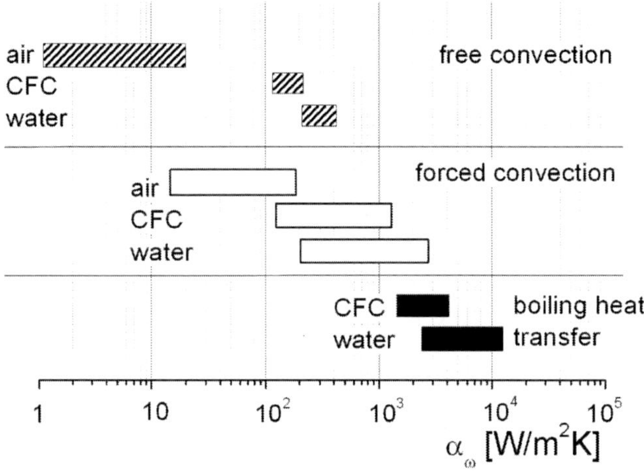

Figure 6.15   Typical heat transfer coefficients. (CFC-chlorofluorocarbon).

6.5 Microgenerators

The development of thermoelectric microconverters, particularly those compatible with standard silicon IC technology, is anticipated to provide many promising applications in energy-harvesting devices. The design of miciro-generators can basically be divided into two types: the vertical and the horizontal configuration.

In the vertical structure, a thermoelectric converter is fabricated with its thermoelement length aligned in the same direction as the thin-film growth and represents the configuration shown in Fig. 6.9. The temperature gradient is applied or generated perpendicular to the substrate surface. The difference to traditional generators here is that all dimensions are much smaller. Since the thermocouples are deposited and patterned with help of lithography, a large number of thermocouples can be fabricated at no additional cost and high module voltages can be achieved even in small modules.

In the horizontal structure, thermoelements are "lying" on the substrate surface and the temperature gradient is applied or generated along the substrate surface.

Relatively long thermoelement can be fabricated using the horizontal structure. This helps to ease the difficulty in obtaining

Table 6.1 Data sheet parameters of typical thermoelectric generators

Company	Type	$l \times w \times h$	V_o	I_s	R_i	α	G	G/A	P1*	T_{max}
		(cm $\times$ cm $\times$ mm)	(V)	(A)	(Ω)	(V/K)	(W/K)	(mW cm^2/K)	(mW/cm^2)	(°C)
T	199-150-6	$4 \times 4 \times 3.6$	8.20	3.15	2.6	0.082	1.136	71	400	200
T	287-200-14	$4 \times 4 \times 4.8$	11.5	0.97	9.0	0.100	0.518	32	174	200
T	127-250-32	$4 \times 4 \times 3.4$	10.8	7.0	1.5	0.054	0.696	43	590	225
T	097-300-33	$10 \times 5 \times 120$	4.2	1.05	4	0.021	0.021		ca. 50	1000
M	240-100-50	$0.5 \times 0.7 \times 0.5$	1.20	0.15	8.0	0.024	1.388	4.1	110	150
N	$N_x 2$	$0.5 \times 0.4 \times 0.7$	0.39	1.23	0.3			15	1500	
Hi-Z	HZ-2	$2.9 \times 2.9 \times 5.1$	3.2	0.8	4	0.032	0.4	47	148	250
Hi-Z	HZ-14	$6.27 \times 6.27 \times 5$	1.6				1.9	47	148	250

*At dT $=$ 100 K; supplier: T, Thermalforce; M, Micropelt; N, Nextreme.

very small contact parameters required in the vertical structure. Due to the high length of the thermocouples, the thermal impedance is better matched to passive low power air cooling. However, a thermal bypass is introduced due to the substrate in close proximity underneath the thermoelements in the horizontal structure.

6.5.1 *Microgenerators in Vertical Configuration*

The equations described in Section 6.4 can be applied directly to the vertical structure without need for modification. In this case, the length of the thermoelements is very short (10–20 μ m). Thus, very small contact parameters n and r are essential. This is demonstrated in Figs. 6.16 to 6.18. In these figures, voltage and power are plotted according to Eqs. 6.21 and 6.25 as function of λ_c the thermal conductivity of the contact and r_c the specific contact resistivity. To demonstrate the higher sensitivity of micro-TEGs compared to bulk (macro) TEGs, a comparison is made for two typical examples. The parameters used are summarized in Table 6.2.

As can be seen in Figs. 6.16 and 6.17, voltage and power reductions of the micro-TEG occur for a thermal conductivity of the

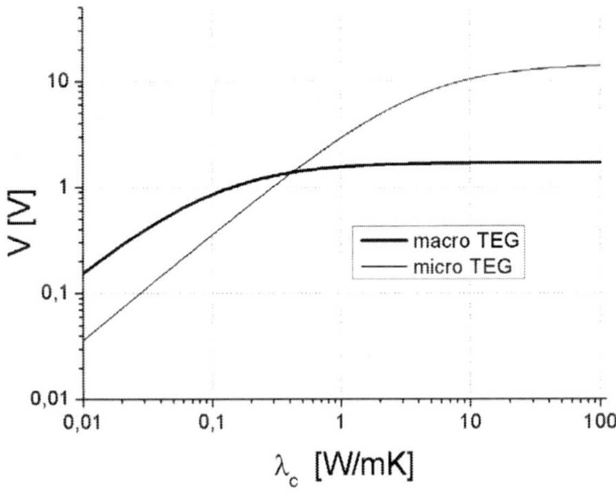

Figure 6.16 Comparison of bulk (macro) TEG and micro TEG with parameters according to Table 6.2: Influence of the thermal conductivity of the contact material on the TEG voltage.

Table 6.2 Parameters used for the performance comparison of micro- and macro TEG

Parameter	Unit	Bulk TEG	Micro-TEG
α	V/K	0.005	0.0027
l	Mm	1	0.02
l_c	μm	20	16
λ	W/mK	2.4	2.4
ρ_c	Ωm^2	10^{-10}	10^{-10}
ρ	Ωm	10^{-4}	10^{-5}
N	—	34	540
A-TEG	—	3×3 cm^2	3×3 mm^2
A-junction		2×4 mm^2	40×40 μm^2
ΔT	K	10	10

contact material below ca. 10 W/mK, while a visible influence on the bulk TEG can only be seen for λ values below ca. 1 W/mK.

The electrical contact resistance is even more critical for micro-TEGs. As demonstrated in Fig. 6.18, the specific contact resistance should be lower than $10^{-11}\Omega$m^2 while large TEGs can tolerate contact resistances of above $10^{-8}\Omega$m^2.

The thickness and thermal conductivity of the module substrates influences the TEG performance in a similar way like the thermal properties of the joints of the thermolegs. This means that very high thermal conductivity and thin substrates are a prerequisite for vertical micro-TEGs. Diamond and AlN ceramic substrates have the highest thermal conductivity but high cost. Silicon is a widely used material for micro-TEGs as alternative to the normal alumina ceramics. In addition, the thermal interface to the heat think (external interface in Fig. 6.12) and the thermal conductivity of the heat sink are of special importance. As demonstrated in Fig. 6.19 if a micro-TEG is used with higher power per module area than a bulk TEG than higher heat spreading is required in both, the heat source and the heat sink. The resistance of the thermal interface between TEG and heat source and heat sink must be lower as well.

Technology for vertical micro-TEGs

The simplest way is to start from bulk materials which are fabricated with conventional technologies of ca. 1 to 2 mm thickness and

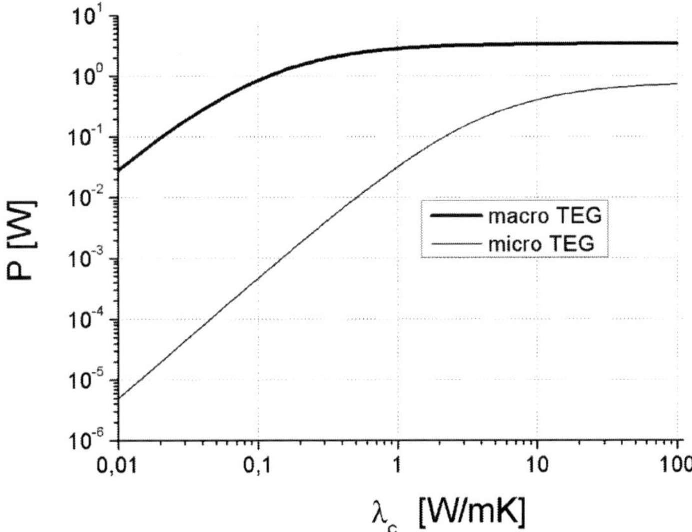

Figure 6.17 Comparison of macro TEG and micro-TEG with parameters according to Table 6.2: Influence of the thermal conductivity of the contact material on the TEG power.

thinning them down to ca. 200 µm. At this thickness, a high specific power output can be achieved (Fig. 6.11). Reducing the thickness of the brittle thermoelectric legs below ca. 200 µm is not an option, since the influence of the contact resistance between the thermoelectric material and the diffusion barrier and solder results in dramatic performance reductions (Figs. 6.17 and 6.18).

Another approach for further miniaturization is the electrochemical deposition of the p- and n-type materials. Relatively high growth rates of 5 to 10 µm/h can be achieved. Due to the anisotropic properties of Bi_2Te_3-based materials (electrical conductivity is about a factor of four higher along the "high-ZT" direction than along the "low-ZT" direction), electrodeposition allows the growth of Bi_2Te_3 with the crystalline high ZT direction perpendicular to the conducting substrate [5]. Even though different experimental results of a complete technology procedure have been presented, no commercial device has been fabricated to date. The mechanisms of electrodeposition of Bi_2Te_3 and related compounds are still being investigated. Research is now focussing on the electroplating of

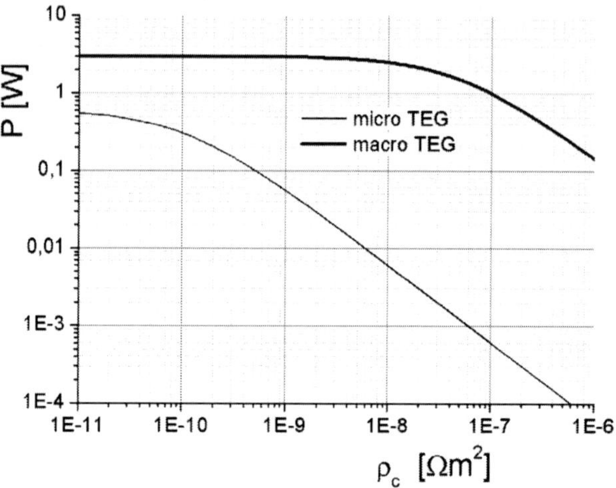

Figure 6.18 Comparison of macro TEG and micro TEG with parameters according to Table 6.2: Influence of the electrical contact resistivity on the TEG power ($\lambda_c = 10$ W/mK).

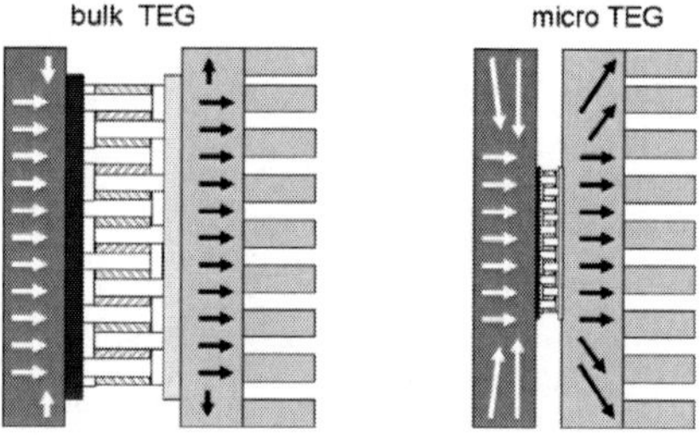

Figure 6.19 Illustration of increased heat spreading in the heat source and the heat sink in case of micro TEG implementation in comparison to the application of bulk TEGs.

thermoelectric nanowires but practical device fabrication strategies are not available [6]. A robust process has been published for the fabrication of flexible polymer-based micro-thermogenerators with electroplated Cu and Ni thermocouples [7]. Due to the low thermoelectric material properties, a power density of only ca. 20 nW/cm^2 was achieved.

The most successful approach to date for the miniaturization of TEGs is the vacuum deposition of the Bi_2Te_3-based materials either by means of sputtering [8, 9] or by vacuum evaporation [10].

Micro-TEGs based on a wafer level sputter technology of the company MicroPelt are commercially available for several years [9]. The high-rate sputter deposition (5 µm/h) from 99.995% element sources produces polycrystalline single phase p- and n-type materials. Effective ZT values of 0.75 have been achieved by annealed polycrystalline materials. n-type Bi_2Te_3 or $Bi_2(Se_{0.05}Te_{0.95})_3$ and p-type $(Bi_{0.25}Sb_{0.75})_2Te_3$ layers with a thickness of ca. 20 µm can be deposited on Si wafers. The issue of large thermal expansion coefficients mismatch (the CTE of Si and V–VI compounds differ by a factor of 5 to 6) was solved. The n- and p-type materials are separately produced and optimized on two different wafers (Fig. 6.20). The process sequence is as follows:

- fabrication of electrical contact/interconnection structure on oxidized Si-wafers
- sputtering of diffusion barrier and p- or n-type material and solder layer
- application of etch resist and photlithography for patterning
- dry etching of the thermocouple elements
- annealing
- singulation of n- and p-type dies with help of wafer saw
- soldering n- and p dies together

Figure 6.20 Cross section of a sputtered and etched thermoelectric leg (left) and patterned chip (right) [11], © MicroPelt.

Table 6.3   Parameters of MicroPelt TEGs [11]

Parameter	Unit	MPG-D651	MPG-D751
l × w	mm^2	3.3 × 2.5	4.2 × 3.3
Thickness	mm	1.09	1.09
Electrical resistance at 23°C	Ω	185	300
Thermal resistance at 85°C	K/W	22	12,5
Net Seebeck voltage at 23°C	mV/K	75	140

A patterning resolution of ca. 100 thermoelectric pairs/mm^2 has been demonstrated. Since the thermoelectric material is directly deposited onto a contact layer on one side, and a defined, very thin soldering layer on the other side can be used, the contact resistance can be kept at a minimum. A Ti/Pt backside metallization can be applied to facilitate the thermal contact to the heat source/ heat sink. Some data sheet information of typical MicroPelt TEGs is summarized in Table 6.3. The electrical power and heat flow as function of load resistance and temperature difference for the 4 × 3 mm^2 TEG is shown in Fig. 6.21. A device with 450 leg pairs achieves per pair $\alpha = 220\ \mu V/K$ [12].

6.5.2 *Microgenerators in Horizontal Configuration*

Relatively long thermoelements can be fabricated using the horizontal structure. This results in better contact parameters but also a

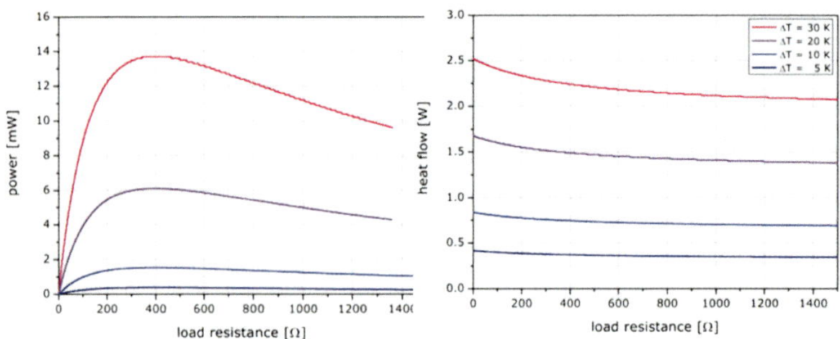

Figure 6.21   Electrical output power and heat flow as function of load resistance and temperature difference for MicroPelt TEG MPG-D751 [11].

thermal bypass due to substrate underneath the thermoelements. Due to the difference in design compared to vertical TEGs, Eqs. 6.21–6.28 cannot be used for the horizontal configuration. A modified theory has been developed which takes into account the influence of the substrate thermal bypass and radiation loss due to large surface areas of thermoelement which can be found in the literature [13].

The technology involves membrane-based thin-film patterning and sacrificial layers, which are already widely used for different sensor applications such as the measurement of IR radiation, or gas flow.

The fabrication and design optimization of a horizontal thermoelectric generator with optimized heat flow path is described in [14]. The meandering thermocouples (n-poly-Si/Al) are located on a Si wafer and are fabricated in thin-film technology to achieve a high integration density. The heat flux is guided perpendicular to the substrate plane to the planar thermocouple junctions by metal stripes with high thermal conductivity. The heat flux from the ambient area is introduced vertically over the footprint area of the chip, guided through the thermocouples in a planar direction, and released vertically over the footprint area as shown in Fig. 6.22. The device consists of several modules, the first with the thermocouples, the second and third modules include the heat conductive structures that are necessary to yield a high in-plane temperature gradient, while the heat flux is perpendicular (cross-plane) through the device. That way, the thermal contact areas are maximized. A prototype generator delivered 9.5 mV/K.

Thin-film SiGe thermocouples were used for another horizontal micro-TEG approach [15, 16]. The focus of this work was to increase the number of thermocouples as much as possible to obtain a sufficient high voltage (1–2 V) at very low temperature differences; 6 μm high thermolegs were patterned at a lateral size of 3–10 μm and only 2–3 μm distance using stepper technology for lithography. On a die of size 1×2.5 mm^2, up to 2500 thermocouples have been fabricated. The process sequence is as follows:

- fabrication of sacrificial 6 μm high SiO$_2$ bumps with tapered side walls
- deposition of 150 nm Si$_3$N$_4$ isolation on the substrate

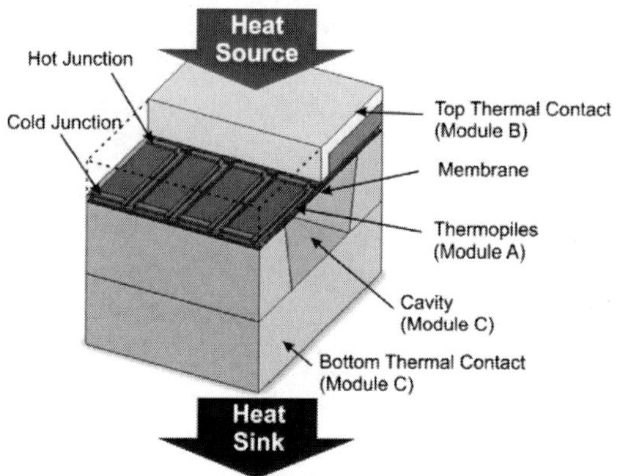

Figure 6.22 Horizontal micro-TEG configuration with deposited n-poly-Si and Al thermolegs [14].

- deposition of SiGe, doping for n- and p-type legs and patterning
- fabrication of Al-interconnects (top and bottom simultaneously)
- etching of the sacrificial SiO_2
- dicing and bonding the top-chip with $1\mu m$ polymer adhesive

Output voltages of up to 0.9 V have been obtained. A power density of 0.3 $\mu W/cm^2$ was measured at $\Delta T_g = 3.3$ K. Further improvements are possible by reducing the contact resistance between Al and SiGe. Figure 6.23 shows the principle set-up and a micrograph of the thermocouples.

6.6 System-Level Design and TEG Integration into Energy-Harvesting Applications

6.6.1 *Model at System Level*

The previous sections described the electrical performance of TEGs for various temperature differences. In the design of actual energy-harvesting systems, when the TEG is connected thermally with a small natural convection heat sink, one cannot assume that the

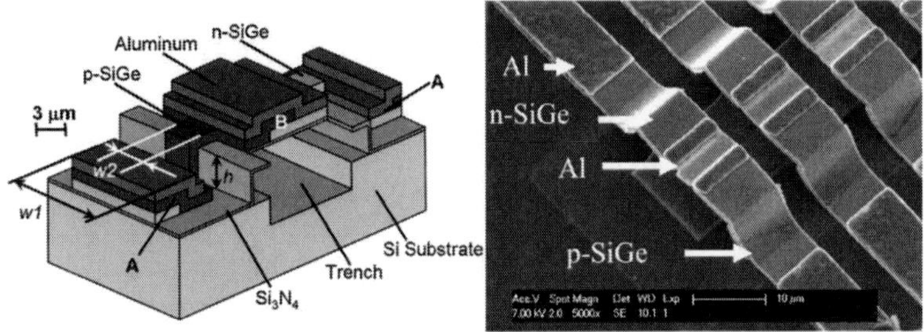

Figure 6.23 Principle set-up and SEM picture of thin-film SiGe thermocouples in horizontal configuration [15, 16].

desired temperature difference can be achieved across the TEG. The TEG must be matched thermally to the available heat sink in order to maximize the performance. A system optimization can be done for maximum power or for maximum voltage. Since recently high efficiency voltage converters have been developed which accept rather low input voltages (see Chapter 8), power optimization is of interest in the first place for energy-harvesting devices.

A physical model for TEGs was developed, which includes the thermal resistances of the heatsink and heat source as well as Thomson-, Peltier- and Joule heat [17] (Fig. 6.24). If the Thomson effect is neglected, the temperature difference over the thermolegs can be expressed as function of the overall temperature difference

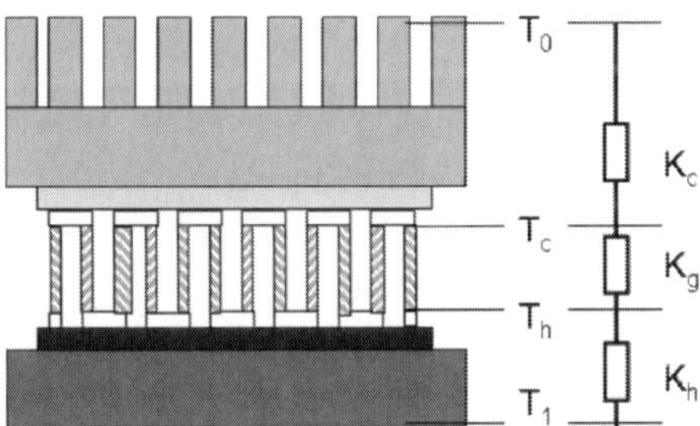

Figure 6.24 Equivalent circuit model for the thermal resistances of a TEG integrated into an application.

$\Delta T = T_1 - T_0$ in simplified terms:

$$\Delta T_g \approx \frac{\Delta T}{1 + (K_c + K_h)\left(\frac{1}{K_g} + \frac{\alpha^2 T_0}{R_l + R_g}\right)} \tag{6.29}$$

with K_c, K_g and K_h the thermal resistance of the cold side, the TEG-legs and the hot side, respectively. R_l and R_g are the electrical resistances of the load and the TEG. The output power is

$$P_{out} = (\Delta T \alpha)^2 \left(\frac{K_g}{K_g + K_c + K_h}\right)^2 \frac{R_l}{(R_l + R_{g,eff})^2} \tag{6.30}$$

The effective inner resistance $R_{g,eff}$ is definded as

$$R_{g,eff} = R_g + T_0 \alpha^2 (K_c + K_h)\frac{K_g}{K_g + K_c + K_h} \tag{6.31}$$

The maximum output power is achieved in case

$$R_l = R_{g,eff} \tag{6.32}$$

and thus depends on the thermal impedance of the heat sink and heat source as well. This was already demonstrated in Fig. 6.21, where the load resistance at maximum power is higher than the TEG-resistance shown in Table 6.3. In most energy-harvesting applications, the thermal resistance of the heat sink is relatively high and fixed due to size constraints, and the use of natural air convection. Thus, the TEG design must be adapted to that condition. This means high thermal resistance of the thermolegs (high aspect ratio) which is difficult to achieve. Most of today's traditional bulk TEGs will only be applicable in energy-harvesting applications when large temperature differences exist and when means of heat sinking beyond natural convection are available.

In case of very low temperature differences, for example, in human body energy harvesting, new TEGs with high thermal resistance are required. The thermal impedance of the surrounding air has also to be taken into consideration as well. The equivalent circuit model has to be extended according to Fig. 6.25.

6.6.2 *Human Body Integration of TEGs for Wearable Electronics*

The human body self-regulates its temperature at a constant ca. 37°C. Harnessing this against the ambient air temperature offers

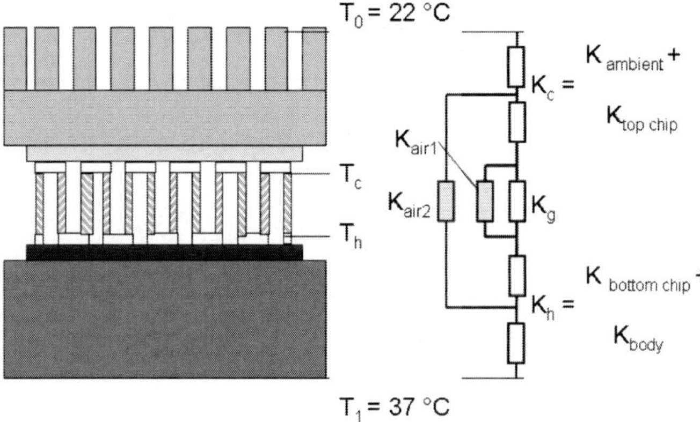

Figure 6.25 Equivalent circuit model of a TEG deployed on a human body.

a source for thermal harvesting for sensor nodes applied on the human body. Some companies producing body-worn products, for example watches, have already developed devices which utilize the small difference between our body heat and the ambient temperature, generating power on the order of microwatts demonstrating that similar techniques can be applied to sensor nodes.

According to Fig. 6.25, the TEG is connected thermally in series with human body and the ambient. Because of the low thermal conductivity of human skin, it constitutes a large thermal resistance. The thermal resistance of the ambient is also large owing to the normally inefficient heat dissipation under a limited temperature difference. To increase the heat dissipation, a relatively large heat sink has to be used. If the area of the heat sink is large and the distance to the skin is low, the thermal resistance K_{air2} which constitute a parasitic heat flow, becomes significant. For the TEG with optimized K_g (Fig. 6.23), it was shown that by increasing the distance between the heatsink and the skin, K_{air2} can be reduced significantly resulting in a higher output power. A maximum power density of ca. 25 $\mu W/cm^3$ (20°C) was achieved at a distance of 5 mm [18]. For practical estimates, some thermal resistances of the body and possible heat flows are summarized in Table 6.4. The values for the heat flow represent maximal values which are limited by the sensation of cold. The power generation rises with falling

Table 6.4 Thermal resistance of the human body and possible heat flow according to [18]

Location	Indoors		Outdoors (−4 to +2°C)	
	K_{body} (cm^{2}K/W)	q (mW/cm^2)	K_{body} (cm^{2}K/W)	q (mW/cm^2)
Forehead	200–400	10–20	—	<45
Leg	400–900	4–15	300–500	15–60
Wrist (watch location)	440	15	—	65
Wrist (radial artery)	160	20	—	120

ambient temperature but falls to zero if the ambient temperature approaches 37°C. Thus, hybrid systems consisting of a TEG and a small photovoltaic module have been proposed.

Thermal energy harvesting is not only applicable on the outer surface of the human body but can also be utilized internally to power implantable medical devices and sensors. In this case the temperature differences are typically as low as 0.3–1.5 K but as the TEG is in direct contact to body liquids, the thermal transfer coefficient is higher compared with an external air heat sink (see Fig. 6.15). These ideas can be extended to cases involving other warm blooded animals.

6.6.3 *Exploitation of Temperature Changes and Transient TEG Behaviour*

For some applications thermal energy is harvested not in the steady state but only during changes of the temperature. This may be due to changes in ambient conditions or changes of the operational mode of a device. During these temperature changes significant temperature gradients can be established over an integrated TEG while in the steady state there is only a minor temperature difference. Figure 6.26 shows output power and temperature difference over time of a commercial TEG which was equipped with metallic pin fin heat sinks on both sides and heated with a candle.

The metallic heat sink has a large heat capacity. Thus, during the start phase, a much higher temperature gradient is maintained than in the steady state when the heat sink has heated up.

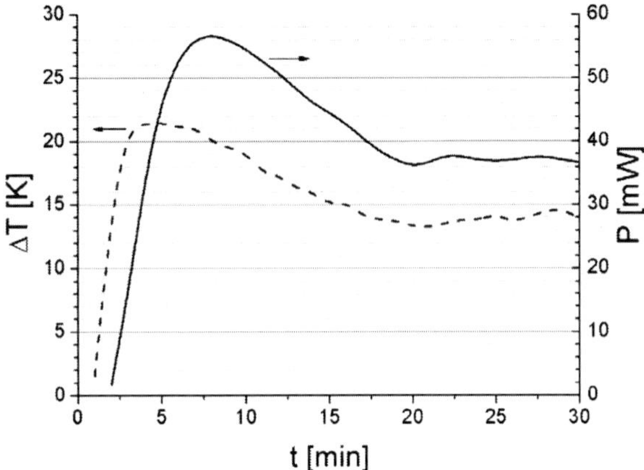

Figure 6.26 Generated electrical power and temperature difference over the TEG sandwiched between metallic heatsinks as function of time.

The transient thermal gradients which develop during an aircraft's flight were investigated to power sensor nodes with help of TEGs [19].

In this case the temperature decreases with the increase of the aircraft's altitude. To enhance the value of the thermal gradient and its duration water was used as a thermal energy storage inside the generator due to its high specific heat and its fusion point (0°C) being located within the aircraft's operating range allowing high energy storage. To give a rough upper bound on the available energy from this method it was considered that changing 1 ml of water from a ground temperature of 15°C at take-off, to a cruising temperature of −60°C, and then back to ground temperature at landing requires 1.3 kJ of energy. Tests showed that 34 J of energy could be extracted with an integrated TEG.

6.7 Conclusions

Further improvements of thermoelectric materials and the commercialization of TEGs which are optimized for low-temperature-difference energy-harvesting applications would be desirable. Nevertheless, if properly adapted and integrated the energy from the thermogenerators is sufficient to supply various types of sensors as well as radio transmitters or other energy autarkic systems. Numerous sources of temperature gradients can be exploited like solar heat, ground soil to ambient air, water to ambient air in outdoor environments, engine and combustion heat or motor waste heat in the transport sector, aircraft compartments, industrial waste heat and human body temperature. The deployment of TEGs as electrical power sources for autarkic systems like sensor nodes is facilitated by the continuous decrease in power requirements of digital and analog electronics and the developments of high efficiency step-up converters with low input voltages. Several evaluation units of thermoelectric harvesters and wireless sensors with a maximum power of ca. 20 mW are on the market [11].

References

1. Rowe, D. M. (2006) *Thermoelectrics Handbook: Macro to Nano*, CRC Press, New York, 2006.

2. Rowe, D. M., and Min, G. (1996) Design theory of thermoelectric modules for electrical power generation, *IEE Proc. Sci. Meas. Technol.*, **143**(6), 351–356.

3. Min, G., Rowe, D. M., Assis, O., and Williams, S. G. K. (1992) Determining the electrical and thermal contact resistances of a thermoelectric module, *Proceedings of the 11th International Conference on Thermoelectrics, Arlington*, TX, USA, pp. 210–212.

4. Min, G., and Rowe, D. M. (1995) Peltier device as a generator, in *CRC Handbook of Thermoelectric*, D. M. Rowe, CRC Press, New York, pp. 479–488.

5. Magri, P., Boulanger, C., and Lecuire, J. M. (1994) Electrodeposition of Bi_2Te_3 Films, *Proceedings of the 13th International Conference on Thermoelectrics*, Kansas City, Kansas, USA, 277–281.

6. Xiao, F., Hangarter, C., and Yoo, B. (2008) Recent progress in electrodeposition of thermoelectric thin films and nanostructures, *Electrochim. Acta*, **53**(28), 8103–8117.

7. Glatz, W., and Muntwyler, S. (2006) Optimization and fabrication of thick flexible polymer based micro thermoelectric generator, *Sens. Actuators A*, **132**, 337–345.

8. Beyer, H., Nurnus, J., and Böttner, H. (2002) High thermoelectric figure of merit ZT in PbTe and Bi_2Te_3-based superlattices by a reduction of the thermal conductivity, *Phys. E: Low-Dimens. Syst. Nanostruct.*, **13**(2–4), 965–968.

9. Böttner, H., Nurnus, J., Gavrikov, A., Kühner, G., and Jägle, M. (2004) New thermoelectric components using microsystem technologies, *J. Microelectromechan. Syst.*, **13**(3) 414–420.

10. Goncalves, L. M., Couto, C., Alpuim, P., and Rolo, A. G. (2010) Optimization of thermoelectric properties on Bi_2Te_3 thin films deposited by thermal co-evaporation, *Thin Solid Films*, **518**(10), 2816–2821.

11. www.micropelt.com, Datasheet 0025DSPG6&70210v1e.

12. Böttner, H., Nurnus, J., Schubert, A., and Volkert, F. (2007) New high density micro structured thermogenerators for stand alone sensor systems, Jeju Island, *Proc. ICT 2007*, 306–309.

13. Min, G. and Rowe, D. M. (1999), Cooling performance of integrated thermoelectric microcooler, *Solid-State Electron.*, **43**, 923–929.

14. Huesgen, T., Woias, P., and Kockmann, N. (2008) Design and fabrication of MEMS thermoelectric generators with high temperature efficiency, *Sens. Actuators A: Phys.*, **145–146**(7–8), 423–429.

15. Su, J., Goedbloed, M., van Andel, Y., and Leonov, V. (2009), Micromachined thermoelectrric energy harvester fabrication on 6-inch wafer and characterization, *Proceeding of 9th International Workshop on Micro and Nanotechnology for Power Generation and Energy Conversion Applications (PowerMEMS 2009)*, Washington, DC, USA.

16. Wang, Z., Leonov, V., and Fiorini, P. (2009) Realization of a wearable miniaturized thermoelectric generator for human body applications, *Sens. Actuators A: Phys.*, **156**(1), 95–102.

17. Freunek, M., Müller, M., Ungan, T., Walkner, W., and Reindl, L. M. (2009) New physical model for thermoelectric generators, *J. Electron. Mater.*, **38**(7), 1214–1220.

18. Leonov, V., Su, J., and Vullers, R. J. M. (2010) Calculated performance characteristics of micromachined thermopiles in wearable devices, *Proceedings of DTIP 2010*, 391–396.

19. Bailey, N., Dilhac, J., and Escriba, C. (2008) Scavenging based on transient thermal gradients: Applications to structural health monitoring of aircrafts, *Proceedings of 8th International Workshop on Micro and Nanotechnology for Power Generation and Energy Conversion Applications (PowerMEMS 2008)*, Sendai, Japan.

20. Böttner, H. (2008) Thermoelectrics for high temperature differences may complement renewable energies: A survey about state-of-the-art of so-called high temperature thermoelectric materials, *Proc. Int. Conf. Thermoelectrics* (ICT2008), Corvallis, USA.

21. Sootsman, J., Chung, D., and Kanatzidis, M. (2009) New and old concepts in thermoelectric materials, *Angew. Chem. Int. Ed.*, **48**, 8616–8639.

22. Vining, C. B. (2007) *ZT* ∼ 3.5: Fifteen Years of Progress and Things to Come, *European Conference on Thermoelectrics, ECT2007, Odessa, Ukraine, Sept. 2007.*

Chapter 7

Solar Cells

Monika Freunek Müller,[a] Birger Zimmermann,[b] and Uli Würfel[b]

[a] *ZHAW Zürcher Hochschule für Angewandte Wissenschaften, Technoparkstrasse 2, 8401 Winterthur, Switzerland*
[b] *Fraunhofer Institute for Solar Energy Systems ISE, Heidenhofstr. 2, 79110 Freiburg, Germany*

This chapter provides an overview of the working principles of photovoltaic cells and reviews them in terms of micro energy harvesting.

7.1 Photovoltaic Devices

Photovoltaic devices convert light into electric energy. There are three fundamental processes needed to achieve this goal. The first step is the absorption of a photon by which an electron is excited from a lower state to a higher (less tightly bound) state. The remaining unoccupied lower state from where the electron was excited is called a hole. Next, the electron–hole pair has to be spatially separated and finally the electron and the hole have to be transported to and collected at their respective electrodes. In the first step, i.e., the creation of an electron–hole pair through the absorption of a photon, the energy of the photon is converted

Handbook of Energy Harvesting Power Supplies and Applications
Edited by Peter Spies, Loreto Mateu, and Markus Pollak
Copyright © 2015 Pan Stanford Publishing Pte. Ltd.
ISBN 978-981-4241-86-1 (Hardcover), 978-981-4303-06-4 (eBook)
www.panstanford.com

into chemical energy as an electron–hole pair is uncharged. Subsequently, the chemical energy is converted into electric energy by separation of the charge carriers.

Photovoltaic devices are generally made from materials that exhibit states for electrons only at certain energy levels. In a solid, these levels are broadened and form so-called bands with gaps between them. At room temperature, the highest and almost entirely occupied energy range is called valence band. The next higher, largely unoccupied energy range is called conduction band. In organic materials, these states are referred to as HOMO (highest unoccupied molecular orbital) and LUMO (lowest unoccupied molecular orbital). The two energy ranges are separated by the energy gap E_G.

In a very general way, a photovoltaic device can be described as a two-level system, with energies E_V as the upper edge of the valence band and E_C as the lower edge of the conduction band. It is $E_G = E_C - E_V$. For sake of simplicity, we will neglect the fact that the bands have a distribution of states with respect to energy.

Generation of charge carriers is caused by absorption of photons with energies $h\nu = \hbar\omega \geq E_G$, which leads to transitions of electrons from E_V to E_C. The opposite process, i.e., the transition of an electron from E_C to E_V is a loss mechanism in a photovoltaic device and it is called recombination. It can also be described as a chemical reaction (annihilation) of an electron in E_C with a hole in E_V. The generation is a function of the incident photons, the occupied states in E_V and the unoccupied states in E_C, as only electrons occupying states in the valence band can be excited to (free) states in the conduction band. The recombination rate depends on the concentration of electrons in the upper energy level and the concentration of the holes in the lower energy level.

For convenience, we will assume in the following that the states at E_V are largely occupied and the ones at E_C largely unoccupied, which is true for photovoltaic devices under unconcentrated solar irradiation.

Figure 7.1 shows schematically a semiconductor with the energy levels E_C (lower edge of the conduction band) and E_V (upper edge of the valence band), separated by the energy gap E_G. Depicted are the generation of an electron–hole pair (G) via absorption of a

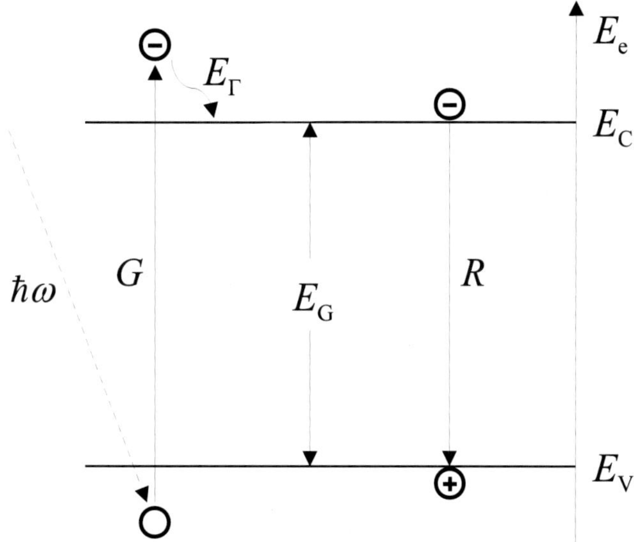

Figure 7.1 Scheme of the energy of an electron E_e in a semiconductor with bandgap E_G and the processes of generation G by absorption of a photon with energy $\hbar\omega$ and recombination R. Also depicted is the thermalization, i.e., the transformation of a part of the electron energy into heat by emission of a phonon with energy E_Γ. See text for details.

photon (with energy $\hbar\omega$) and the reverse process, the annihilation of an electron–hole pair, which is called recombination (R). Note that if $\hbar\omega > E_G$ the excess energy is dissipated by the emission of phonons (lattice vibrations, i.e., heat) on a very short timescale (in the range of 10^{-12} s). This heat will not be converted to electric energy by a photovoltaic device and is thus lost. This process is called thermalization.

All different types of photovoltaic devices have in common that they show a diode-like behavior in terms of their current–voltage characteristics. This is a consequence of the fact that the photovoltaic device contains one phase in which electrons are transported and another phase for the transport of holes. These two phases can be two chemically different materials like in organic or dye photovoltaic devices for example or rather from the same material as in silicon solar cells.[a] Applying a reverse bias to the photovoltaic device in the dark, i.e., positive polarity of the n-region with respect to the p-region, makes electrons and holes flow away from the junction of the two phases. This means that only those electrons and holes can contribute to the current that are generated inside the cell. The generation rate resulting from the 300 K radiation from the environment is very small and does not depend on the applied bias voltage. Therefore the reverse current density, called saturation current density j_S, is also very small. In forward direction, electrons are injected in the electron conducting phase and holes in the hole conducting phase; they are transported in their respective phases toward the junction of the two phases where they recombine.[b] Thus, the current in forward direction is a recombination current.

The recombination is a chemical reaction between an electron and a hole. It depends on the product of the concentration of the charge carriers and no more of them can flow toward the junction than disappear there. In a semiconductor, the concentration of electrons and holes depends exponentially on the applied bias voltage and therefore the current in forward direction increases exponentially with increasing voltage. An ideal diode is a hypothetical device with vanishing transport resistance of electrons and holes and purely radiative recombination. The corresponding equation describing the charge current density j_Q of such an ideal diode is given by

$$j_Q(V) = j_S \left[\exp\left(\frac{e_0 V}{k_B T} \right) - 1 \right] + j_{sc} \qquad (7.1)$$

Therein, j_S is the saturation current density, e_0 the elementary charge, V the voltage, k_B Boltzmann's constant, T the temperature and j_{sc} the short circuit current density. As already depicted above, j_S includes all the charge carriers that are generated not more than a diffusion length from the junction in the dark whereas j_{sc}

---

[a] It should be noted that materials such as n-doped and p-doped Si—although chemically nearly identical—are very distinct materials regarding their *electrical* properties.

[b] To be more exact, electrons and holes move as minority carriers into the respective other phase and recombine within an average path length that is called diffusion length.

represents the current density arising from the absorbed photons under illumination. In this notation, the short circuit current density is negative as electrons and holes flow away from the junction.

It can be shown[1] that this equation can also be written as

$$j_Q(V) = e_0 \int_{E_G}^{\infty} \phi_\gamma^0(\hbar\omega)\,\mathrm{d}\hbar\omega \times \left[\exp\left(\frac{e_0 V}{k_B T} \right) - 1 \right]$$
$$- e_0 \int_{E_G}^{\infty} \phi_\gamma^{\text{source}}(\hbar\omega)\,\mathrm{d}\hbar\omega \tag{7.2}$$

with $\phi_\gamma^0(\hbar\omega)$ being the infinitesimal photon current density of the 300 K background radiation for the photon energy $\hbar\omega$ and $\phi_\gamma^{\text{source}}(\hbar\omega)$ the one of the illumination source. Equation (7.2) means that under thermal equilibrium in the dark (no bias voltage applied) the generation rate at a particular photon energy $\hbar\omega$ resulting from the radiation of the environment, $\phi_\gamma^0(\hbar\omega)$, is compensated by an equal recombination rate. This is called the principle of detailed balance. The term $\exp(e_0 V / k_B T)$ contains the exponential dependence of the concentration of electrons and holes on the applied voltage and thus their recombination rate. Finally, the integration limits express the fact that in an ideal case, every photon with $\hbar\omega \geq E_G$ is absorbed by the device.

The spectrum of the 300K background radiation can be calculated with Planck's law of radiation (Eq. 2.14). If the spectrum of the light source is known, the photovoltaic conversion efficiency can be determined:

$$\eta = \frac{P_{\text{el}}}{P_{\text{rad}}} = \frac{j_{\text{mpp}} V_{\text{mpp}}}{\displaystyle\int_0^{\infty} \hbar\omega\,\phi_{\gamma,\text{source}}(\hbar\omega)\mathrm{d}\hbar\omega} = \frac{j_{\text{sc}} V_{\text{oc}} FF}{\displaystyle\int_0^{\infty} \hbar\omega\,\phi_{\gamma,\text{source}}(\hbar\omega)\mathrm{d}\hbar\omega} \tag{7.3}$$

j_{mpp} and V_{mpp} are the current density and the voltage at the so-called maximum power point, i.e., the point on the current–voltage characteristics where the electrical output power per unit area $P_{\text{el}} = jV$ has its maximum. The so-called fill factor FF is the ratio between $j_{\text{mpp}} V_{\text{mpp}}$ and $j_{\text{sc}} V_{\text{oc}}$. In Fig. 7.2 it corresponds to the area of the smaller rectangle (defined by j_{mpp} and V_{mpp}) divided by the area of the larger rectangle (defined by j_{sc} and V_{oc}) Fill factors of good solar cells are in the range of 0.8–0.9.

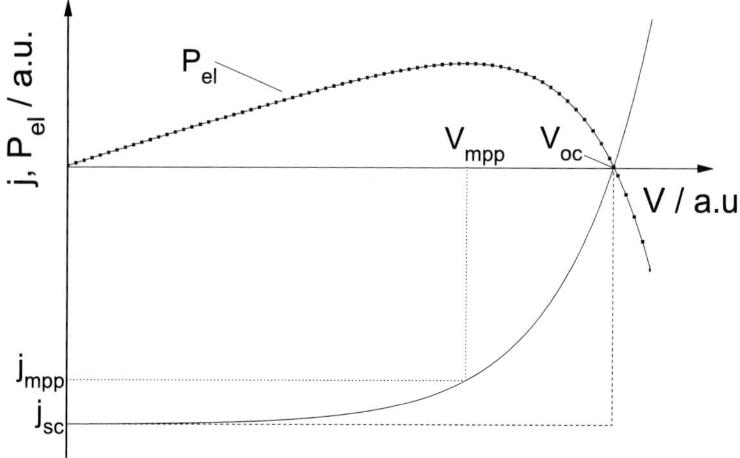

Figure 7.2 Current–voltage characteristics and electrical power of a solar cell showing the maximum power point. The fill factor FF is the area of the smaller rectangle, defined by j_{mpp} and V_{mpp}, divided by the area of the larger rectangle, defined by j_{sc} and V_{oc}.

7.1.1 *Maximum Efficiency of Solar Cells*

As we will see later, the photovoltaic conversion efficiency depends strongly on the spectrum of the light source and for every spectrum, there is an optimal bandgap for single junction solar cells. [a] Materials with smaller bandgaps absorb a higher fraction of the incoming photons but more energy is lost due to thermalization of the electron–hole pairs. Materials with larger bandgaps will convert the energy of the absorbed photons more efficient but absorb less of them. Figure 7.3 shows the results of calculations carried out according to Eq. (7.2), i.e., only radiative recombination is considered and all photons with energies $\hbar\omega \geq E_{\mathrm{G}}$ are absorbed. This approach was introduced by Shockley and Queisser.[2] It means to neglect possible losses due to a non-ideal transport of charge carriers, their recombination at surfaces or due to impurities. The maximum attainable efficiencies are almost 30% for the AM0 and the 5800 K black body spectrum with $\Omega = 6.8 \times 10^{-5}$, respectively,

---

[a] and an optimum combination of several bandgaps in the case of multi-junction solar cells.

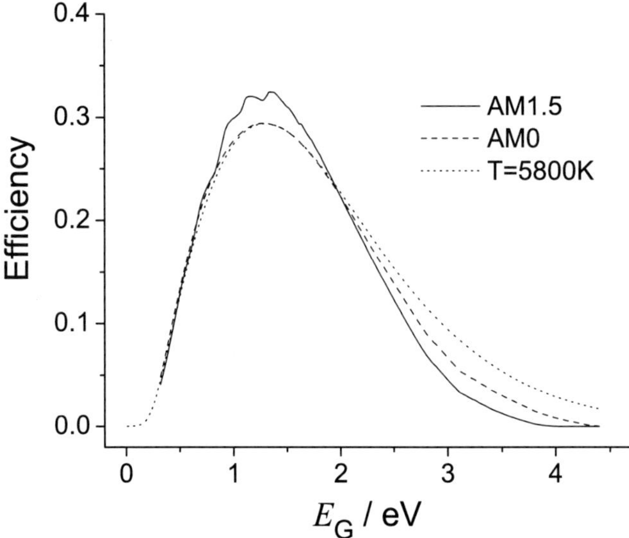

Figure 7.3 Conversion efficiency of solar cells with radiative recombination only as a function of their energy gap for three different spectra, AM1.5, AM0, and black body radiation with $T = 5800\,\mathrm{K}$ and $\Omega = 6.8 \times 10^{-5}$.

and approx. 32.5% for the AM1.5 spectrum. The reason is that in the AM1.5 spectrum there are less photons at lower wavelengths thus reducing the thermalization losses in the solar cell. The optimum bandgap is ca. 1.3 eV for the AM0 and the 5800 K spectrum. For the AM1.5 spectrum, there are two maxima at 1.16 eV and 1.36 eV.

Under indoor conditions, the intensities available for photovoltaic energy conversion are much lower compared to those of AM0 or AM1.5. They are typically in the range of 0.1–10 Wm^{-2}. For this reason, Fig. 7.4 shows efficiencies vs. bandgap energies for the spectrum of a fluorescent tube, an incandescent lamp (simulated as a black body with $T = 2856$ K) and the AM1.5 spectrum, all adjusted to an intensity of 1 Wm^{-2}.

The maximum efficiency for the spectrum of the fluorescent lamp exceeds 51.5% for a bandgap energy of $E_\mathrm{G} = 1.95$ eV. This high value is a consequence of the very narrow spectrum that minimizes thermalization losses.

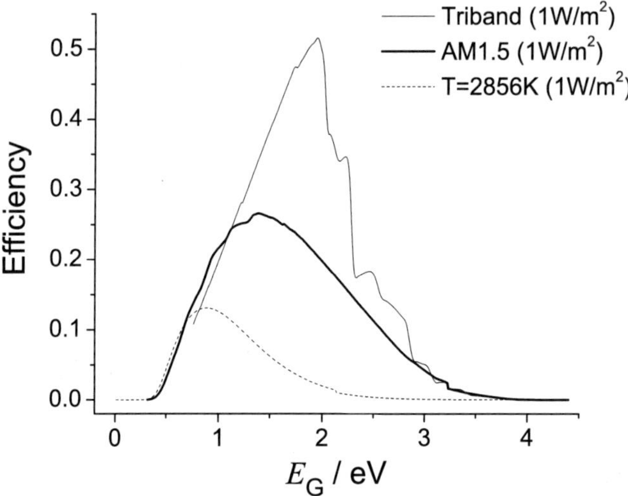

Figure 7.4 Efficiencies of ideal single junction photovoltaic cells for three different spectra (fluorescent bulb, incandescent light and AM1.5, all adjusted to an intensity of $1\,\mathrm{Wm^{-2}}$).

A very important aspect of an application of photovoltaic devices for energy harvesting systems is their performance under lower illumination intensities such as those that can be found indoors. In a real solar cell, the recombination of electrons and holes is not solely radiative. There are non-radiative recombination processes like impurity recombination, Auger recombination and surface recombination.[1,30] In addition, the different recombination processes show distinct dependencies on the light intensity. For high efficiency silicon solar cells for example, the limiting process under AM1.5 is usually Auger recombination whereas impurity recombination limits their performance at low light intensities. For detailed modeling of the performance under different light intensities of a particular solar cell, more parameters such as doping concentrations, purity of the photoactive material, surface passivation, etc., have to be considered.

To ensure a certain amount of power output from a solar cell, it is necessary to connect several cells to form a module. This is done to minimize the ohmic losses due to the limited conductivity of the current collectors. Usually the cells are connected

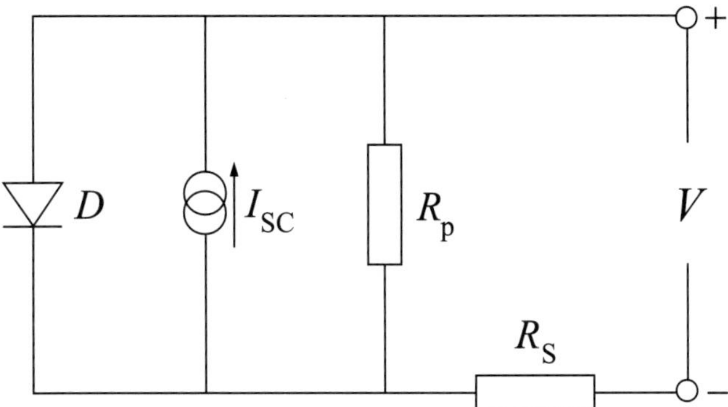

Figure 7.5 One-diode model with parallel resistance R_p and series resistance R_s.

electrically in series, which leads to a higher voltage leaving the current unchanged. Apart from series resistances, real solar cells and modules always have a parallel shunt resistance as well. This is illustrated in Fig. 7.5, which shows the one-diode model with parallel resistance R_p and series resistance R_s. The corresponding equation becomes

$$j_Q(V) = j_S \left[\exp\left(\frac{e_0\,(V - I_Q R_s)}{k_B T} \right) - 1 \right] + j_{sc} + \frac{V - I_Q R_s}{R_p} \quad (7.4)$$

Here $I_Q = j_Q A$ is the current with A being the area. To identify the impact of series and parallel resistances on the performance of photovoltaic devices we again applied the Shockley–Queisser limit formalism. Figure 7.6 shows the maximum attainable efficiency of a solar cell for the ideal case and the case where different resistances are involved. The input for the calculations was black body radiation of $T = 5800\,\mathrm{K}$ and $\Omega = 6.8 \times 10^{-5}$. It can be seen from Fig. 7.6 that for the ideal case ($R_s = 0$; $R_p = \infty$) the efficiency increases logarithmically with the illumination intensity. A series resistance leads to a decrease in efficiency growing larger with increasing light intensity and thus current. For lower light intensities (such as $I_e \leq 10\,\mathrm{W/m^2}$ in the example shown here), there is hardly a detrimental effect of the series resistance visible due to the low corresponding current. The consequence of the parallel resistance is the greater the

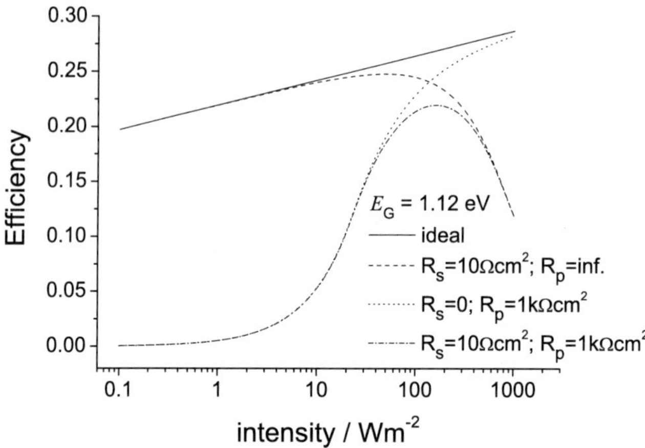

Figure 7.6 Maximum efficiency as a function of the illumination intensity for an ideal solar cell with a bandgap of $1.12\,\text{eV}$, and the influence of series and parallel resistances. The source of illumination was black body radiation with $T = 5800\,\text{K}$ and $\Omega = 6.8 \times 10^{-5}$.

lower the light intensity. The worst is obviously a combination of the two types of parasitic resistances.

As a second example, we modeled a solar cell optimized for the spectrum of the triband fluorescence bulb shown in Fig. 7.7. It has a bandgap of $1.95\,\text{eV}$ and as parallel resistance we chose $R_p = 5\,\text{M}\Omega\text{cm}^2$. Good solar cells can have even higher values of R_p. It can be seen in Fig. 7.7 that at intensities above $20\,\text{W/m}^2$, the series resistance limits the performance of the solar cell. For a range roughly between 2 and $20\,\text{W/m}^2$ there is no difference visible between the four curves. The reason is that the current is already low enough that the voltage drop over the series resistance can be neglected. On the other hand, it is still high enough that the loss due to the parallel resistance hardly plays a role. Decreasing the light intensity further leads to a reduction in efficiency due to the parallel resistance but this loss is still rather small for the lowest intensity of $0.1\,\text{W/m}^2$. Hence, if care is taken to avoid shunt resistances solar cells can maintain their good performance down to very low light intensities.

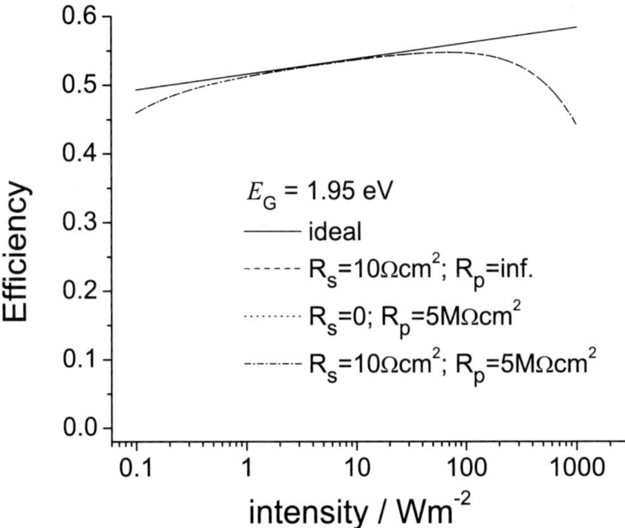

Figure 7.7 Maximum efficiency as a function of the illumination intensity for an ideal solar cell with a bandgap of 1.95 eV, and the influence of series and parallel resistance. The source of illumination was the triband fluorescence bulb (see Fig. 2.9) and its intensity was adjusted.

7.2 Photovoltaics in Micro Energy Harvesting Applications

In most applications, light from natural and/or artificial sources is available with comparable high power densities and a good predictability. Solar light on earth reaches maximum intensities in the range of 1000 Wm^{-2}. In office buildings, a minimum irradiation of 0.1 to 4 Wm^{-2} from both natural and artificial light can be expected.[3]

The designer of a photovoltaic system needs to choose a technology and to dimension the required area A of the photovoltaic module. Hence, the designer requires the knowledge of the efficiency η of the specific module technology and the incoming spectral photon current density ϕ_γ, as shown in Eq. (7.3). Therefore, the knowledge of the incoming light is essential for modeling the efficiency of a photovoltaic cell. The modeling of the output power P_{el} depends on the used photovoltaic cell technology. Design failures due to wrong assumptions of the environment lead to lacking user acceptance and unsatisfying market introductions of photovoltaic products. This is especially important for stationary indoor products, where a mistake in design cannot be compensated by an intervention of the user, such as orientating a calculator to a window.

7.2.1 *Demonstrated Efficiencies at Standard Test Conditions*

For outdoor applications, the efficiency η_{IEC} is generally measured following *Standard Test Conditions* (STC) based on *IEC 60904-3, 2nd ed.* (2008).[4] At STC, outdoor conditions are approximated with a light intensity of $1000\,\mathrm{Wm^{-2}}$, a cell temperature of $25\,°\mathrm{C}$, and the spectral distribution of solar light after passing 1.5 earth atmospheres, that is an Air Mass Factor (AM) of 1.5, see Eq. 2.16.

The current record efficiencies demonstrated at an independent laboratory are $\eta_{IEC} = 25.6\%$ for a mono crystalline silicon solar cell, $\eta_{IEC} = 10.2\%$ for a stabilized amorphous silicon cell, $\eta_{IEC} = 21.0\%$ for CdTe,[5] and $\eta_{IEC} = 11.0\%$ for an organic cell. The current record efficiency of a dye-sensitized cell measured by an independent laboratory is $\eta_{IEC} = 11.9\%$ demonstrated by Sharp.[5]

Table 7.1 summarizes commercially available small modules measured following conditions from *IEC 60904-3*.

7.2.2 *Demonstrated Efficiencies and Measurements at Indoor Conditions*

Photovoltaics in indoor applications is often referred to as ipv.[28] There is no measurement standard for ipv devices. A common practice is to refer to an illumination I_V of 200 Lux, usually from a fluorescent tube. A photometric illumination I_V in Lux or $\mathrm{lm/m^2}$ has the following relation to a radiometric spectral irradiance $I_{e,\lambda}$:[a]

$$I_V = K_m \int_{380nm}^{780nm} I_{e,\lambda}(\lambda)V(\lambda)\mathrm{d}\lambda. \tag{7.5}$$

---

[a]Again, I_V and $I_{e,\lambda}$ can be also found as E_V and $E_{e,\lambda}$.

Table 7.1 Commercially available small and flexible photovoltaic modules, characterization data referring to an irradiance around 1000 Wm^{-2}

Company	Product	Area cm^2	Power density Wcm^{-2}
EPS Soltec[6]	SM 02-700	108.36	12.9[1]
Flexcell[7]	Sunpack 7 W	3150	2.1[2]
IXYS[8]	IXOLAR TM XOB17-12x1	1.54	12.9[2]
Konarka[9]	KT 25	201.24	1.1[3]
Plastecs[10]	SPMIN-2.5	0.97	+
PowerFilm[11]	SP3-37	23.68	2.8
Solaronix[12]	Mini High Efficiency Cell	0.28	7.0[3]

[1] Sunny day, bright sky.
[2] Standard test conditions.
[3] At 1000 Wm^{-2}.
+ Information currently not available.

Photometry evaluates a lighting situation with the visibility function of the human eye $V(\lambda)$ in the visible range from about 380 to 780 nm. The factor $K_m = 683\,\text{lm/W}$ denotes the maximum spectral luminous efficiency for phototopic vision. From Eq. (7.5), a conversion between a photometric and a radiometric value is possible within the range of visible light, provided the spectral irradiance $I_{e,\lambda}(\lambda)$ of the light source is known.

A photometric illuminance $I_V(\lambda)$ can be measured with luxmeters, which usually consist of an Si-photodiode and a $V(\lambda)$-filter. Luxmeters are optimized for a Planckian radiator with a temperature of 2856 K (also known as CIE (*Commission Internationale de l'Eclairage*) Standard illuminant A $= I_{e,\lambda}^{\text{cal}}(\lambda)$), such as an incandescent lamp. For measurements of a deviating light source, the mismatch factor MMF needs to be calculated:

$$\text{MMF} = \frac{\int_0^\infty I_{e,\lambda}^{\text{cal}}(\lambda) S_{\text{cal}}(\lambda) \, d\lambda}{\int_0^\infty I_{e,\lambda}^{\text{act}}(\lambda) S(\lambda) \, d\lambda}. \tag{7.6}$$

$I_{e,\lambda}^{\text{act}}(\lambda)$ is the actual spectral irradiance to be measured with the detector having a spectral response $S(\lambda)$, and $I_{e,\lambda}^{\text{cal}}(\lambda)$ and $S_{\text{cal}}(\lambda)$ the spectral irradiance and the spectral response to which the measurement instrument has been calibrated. While the deviation of $S(\lambda)$ to $S_{\text{cal}}(\lambda)$ usually can be neglected, the MMF due to the differing spectra is essential for luxmeter measurements of light sources as LED or fluorescent light tubes (see Figs. 2.9 and 2.12).

The main radiometric methods to determine indoor irradiance include

- irradiance sensors based on photodiodes
- pyranometers based on thermoelectric arrays
- ray tracing programs, such as *Radiance* [13]

Also for radiometers, the MMF needs to be considered. Due to their wide spectral response function, pyranometers are most suitable for determination of absolute values. For studies with many variables, ray tracing programs are a reliable alternative to measurements. A more detailed, comparative study of these methods can be found in.[14] Currently, for indoor applications, mainly amorphous modules are implemented, which are well-known from calculators. Historically, the performance of an amorphous module under the irradiance of an incandescent bulb could be estimated using photometric measurements with luxmeters. This is due to the spectral response of amorphous solar cells, which is close to the human visibility function $V(\lambda)$.

Other technologies include crystalline, organic and dye-sensitized modules, where crystalline modules need to be optimized for indoor conditions. Standard production line cells can show a voltage drop at low intensities, depending on the adjustment of their doping level. High-efficiency crystalline cell for low intensities have been demonstrated by Glunz et al.[29]

Table 7.2 summarizes commercial amorphous silicon small modules measured under photometric conditions. At the current state of measurement standards, for indoor applications the information from the data sheets in Tables 7.1 and 7.2 need to be completed by own measurements.

7.2.3 *Outside and Standard Conditions*

For outside applications, an estimation of available solar radiant power P_S can be calculated by describing the sun as a Planckian radiator with a surface temperature of $T = 5800$ K and a radius $r_S = 6.96 \times 10^8$ m. From the Stefan–Boltzmann law, the emitted solar

Table 7.2 Commercially available small photovoltaic modules, characterization data referring to a photometric illuminance of 200 Lux

Company	Product	Area [cm^2]	Power density [μWcm^{-2}]
Sanyo Solar[15]	AM-1411	3.5	3.4[1]
Schott Solar[16]	ASI20i06/ 025/0020 JJF	5.0	3.1 [2]
Sinonar[17]	SS8223A-BY11	18.9	5.0 [3]

[1] At 200 Lux, fluorescent tube, not stabilized.
[2] At 200 Lux, fluorescent tube.
[3] At 200 Lux, electric light, 25°C.

power P_S is

$$P_S = \sigma T^4 4\pi r_S^2 = 3.91 \times 10^{26}\,\text{W}, \tag{7.7}$$

where σ is the Stefan–Boltzmann constant of 5.67×10^{-8} Wm^{-2}K^{-4}. With a mean radius of the earth's orbit $r_{\bar{e}} = 1.496 \times 10^{11}$ m, the extraterrestrial solar constant I_e^{AM0} per unit area becomes[a]

$$I_e^{\text{AM0}} = \frac{P_S}{4\pi r_{\bar{e}^2}} = 1390\,\text{Wm}^{-2}. \tag{7.8}$$

For an estimation of the spectral distribution, Eq. 2.14 can be applied.

The received irradiance on earth is determined by the incident angle of the light passing the atmosphere. Due to interactions of the photons with atmospheric gases, the received light changes. Standard test conditions following *IEC 60904-3, 2nd ed.* refer to AM1.5 with an incident angle of 48.19° and an irradiance of $I_e^{\text{AM1.5}} = 1000$ W/m^2. The current reference spectral distribution can be found at the *National Renewable Energy Laboratory*.[18] The extraterrestrial solar spectral irradiance $I_{e,\,\lambda}^{\text{AM0}}$ and the terrestrial distribution $I_{e,\,\lambda}^{\text{AM1.5}}$ are depicted in Fig. 2.9.

Assuming a uniform atmospheric mass distribution, the AM can be calculated depending on the latitude and the season. The AM for Freiburg, Germany, for example, with a northern latitude of 48° and an eastern longitude of 7.8° varies between 1.12 on summer and 3.15 on winter solstice.

Other variables to be considered include the albedo of the environment, the circumsolar radiation, meteorological conditions, such

[a] As Eqs. (7.7) and (7.8) show, the result depends on the assumed astronomical data. The current official solar constant is 1366 Wm^{-2}.

as clouds, obstacles, module orientation, temperature, and other environmental influences such as dirt and water. An introduction on this subject using analytical and numerical approaches is given in.[19]

7.2.3.1 Summary outdoor conditions

For most available datasheets, modules are characterized referring to the solar spectral distribution of *IEC 60904-3, 2nd ed.* with an irradiance around $1000\,\mathrm{Wm}^{-2}$ and AM1.5. Although in most real applications, the effective irradiance will remain below this value, characterization data from these conditions provide a reasonable orientation from repeatable measurement conditions for the design process.

7.2.4 *Indoor Conditions*

Indoor light conditions are typically based on an artificial light source with a spectral distribution optimized for the human visibility function, and daylight transmitted through a window. The irradiance typically ranges below $10\,\mathrm{Wm}^{-2}$, although it can exceed $500\,\mathrm{Wm}^{-2}$ at certain conditions (depending on factors as distance to a window, solar zenith angle, etc.). For indoor irradiance, interactions of the light with an object becomes more likely.

The interaction is described by the object's coefficient of absorbance a, its coefficient of transmission t and its coefficient of reflection r:

$$1 = a + t + r. \tag{7.9}$$

For the case of an ideal black body, every photon is absorbed and transferred to heat, with $a = 1$, and $r = 0$. For an ideal white body, every photon is reflected, with $a = 0$, and $r = 1$. Real bodies are in between these limits.

The transmission coefficient of window glass can be measured following[13] or obtained from manufacturers. Depending on the specific technology and materials, t is generally close to zero in the far infrared range and is optimized for the visible spectrum. For heat protection glass, t typically ranges from 0.6 to 0.95 in the visible spectrum.

Table 7.3 Reflection coefficients from *EN12464-1*[20]

Element	EN 12464
Ceiling	0.6–0.9
Wall	0.3–0.8
Work space	0.2–0.6
Floor	0.1–0.5

Table 7.3 shows typical indoor reflection coefficients from the *EN12464-1*.[20]

Figure 7.8 provides a schematic of the composition of indoor light. In addition to the variables from the calculation of solar light, the reflection and transmission parameters and the geometry of the particular room need to be considered, as well as user behavior concerning the use of electric light and the use of blinds.

Due to the many variables, an economic and reliable approach is the use of ray tracing programs combined with user models. Especially in office buildings, the main spatial parameters can be obtained from existing CAD models. These models can also be used in ray tracings programs as *Radiance*. *Radiance* is a validated backward ray tracer that enables the calculation of both electric and natural light, and to model complex material properties.

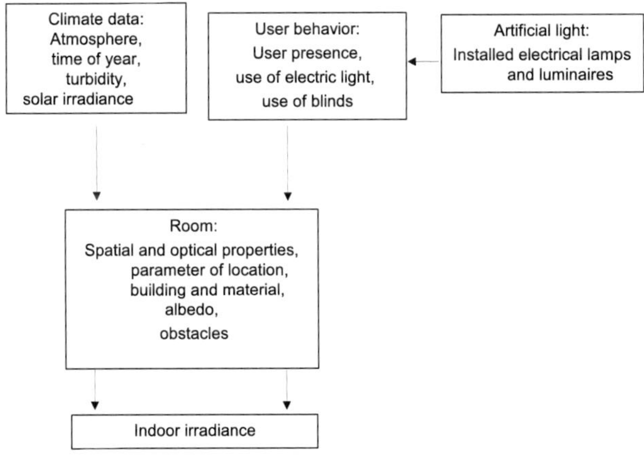

Figure 7.8 Main influences determining indoor irradiance.

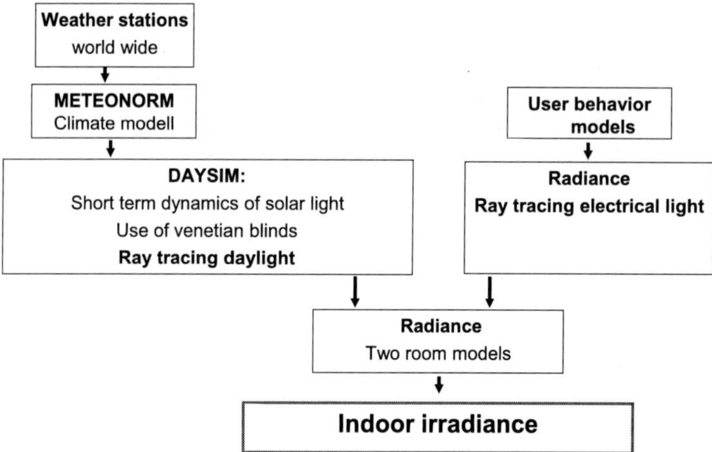

Figure 7.9 Simulation pathways for the ray tracing approach.

Figure 7.9 shows simulation pathways for such a ray tracing approach and relevant measurements for indoor irradiance. The irradiance data can be obtained from climate simulation programs as *METEONORM*, which is based on measurement data from weather stations worldwide.[21] The electric light distribution is simulated with *Radiance*. The daylight contribution is modeled with *DAYSIM*, using the same room models as in Radiance. *DAYSIM* is based on *Radiance*, and calculates daylight on a time step basis using the daylight coefficient method.[22]

Figures 7.10 and 7.11 show the mean irradiance contribution of daylight from an exemplary simulation study in Freiburg, Germany.[3] The investigated points N1-N3 and S1-S3 are exemplary installation points with different orientations and distance to the window in two office rooms in Freiburg, Germany. The obtained annual mean irradiance from natural light ranged from 0.8 to 50.1 Wm^{-2}. The peak values ranged from 7 to 609 Wm^{-2}.

Figure 7.12 depicts the result for the simulation of a worst-case scenario using the combined simulation approach from Fig. 7.9. This worst-case scenario consisted of a room with a length of 10 m and a north-window, many pieces of dark furniture, December, and more than 10 days without use of electric light due to holidays.

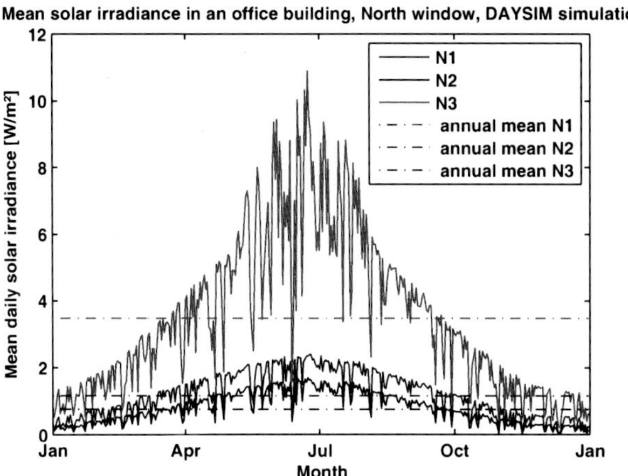

Figure 7.10 Indoor received solar irradiance for virtual sensors in an office with a north window. *DAYSIM*-simulation, location Freiburg, Germany.

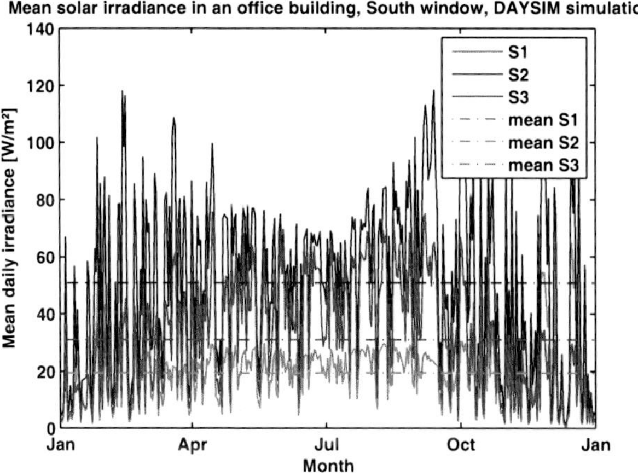

Figure 7.11 Indoor received solar irradiance for virtual sensors in an office with a south window. *DAYSIM*-simulation, location Freiburg, Germany.

7.2.4.1 Summary indoor conditions

A minimum irradiance around 2–4 Wm^{-2} can be assumed, provided a regular use of electric lighting and a usual office room. A combined simulation approach consisting of a climate simulation program,

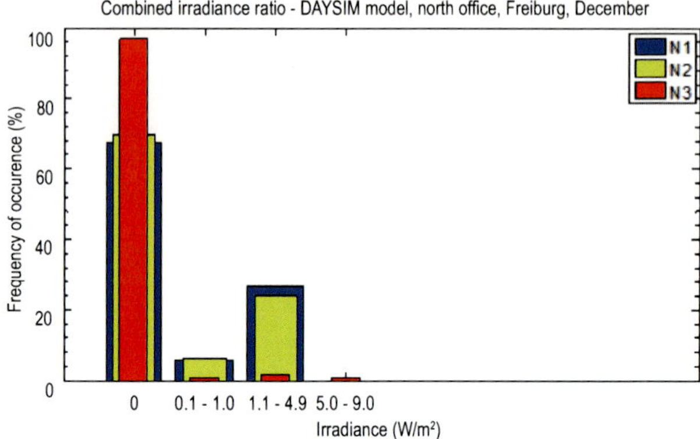

Figure 7.12 Received solar and artificial irradiance for virtual sensors in an office with a north window, worst-case scenario. *Radiance* and *DAYSIM* simulation, and user model for Freiburg, Germany, December.

such as *METEONORM*, ray tracing programs, such as *Radiance* and *DAYSIM*, and a user model is suggested for ipv-design. For measurement data referring to photometric values, the mismatch factor MMF depending on the light source needs to be considered.

7.3 Tailoring the Current-, Voltage-, and Power Output of Photovoltaic Cells

As described in the previous sections, the output of a simple photovoltaic cell is mainly determined by the bandgap of the semiconductor material used to fabricate the photovoltaic cell and the irradiance in the spectral region, which the photovoltaic cell can absorb light. The output voltage is logarithmically dependent on irradiance and cannot exceed the bandgap energy in eV divided by the elementary charge e_0. In principle, the output voltage and the FF are area independent, while the current- and power-output scale with the area of the photovoltaic cell. The output voltage can be multiplied by connecting several photovoltaic cell elements in series. Therefore, the photovoltaic cell is patterned into individual cell elements by isolating lines in the electrodes. Neighbored cell elements are connected in series by overlapping the negative

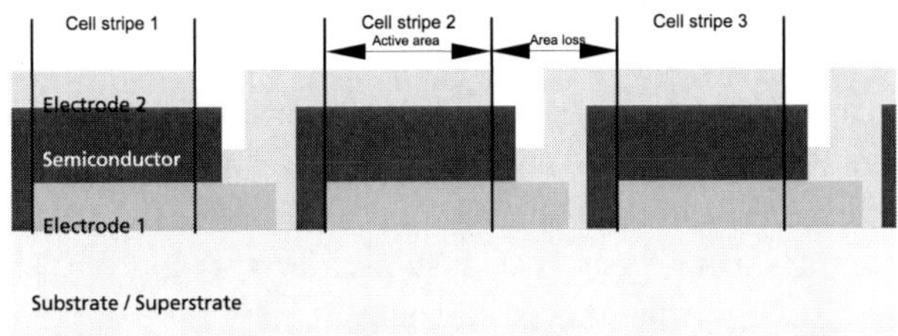

Figure 7.13 Monolithic interconnection of single cell elements. In this way the voltage of the cell elements sum up and resistive losses can be minimized.

electrode of one element with the positive electrode of the next cell element (Fig. 7.13).

If such a series circuitry is employed, the device is called a module. The ideal module output and size can be calculated as follows:

$$V_{\text{oc, module}} = n_{\text{cells}} \cdot V_{\text{oc}} \tag{7.10}$$

$$I_{\text{sc}} = A_{\text{cell}} \cdot J_{\text{sc}} \tag{7.11}$$

$$A_{\text{module}} = n_{\text{cells}} \cdot A_{\text{cell}} \tag{7.12}$$

For the real layout, some technical constraints have to be taken into account. The most important factors for the module design are the area loss induced by the circuitry and the resistivity of the contact materials, which introduces a voltage drop ΔV upon current flow, i.e., a loss of power $\Delta P = I \Delta V = R I^2$. The relative area loss can be reduced if the cell elements are made wider, while the ohmic losses due to the resistance of the electrode materials demand narrow cell elements.

7.3.1 Optimization of the Circuitry Geometry

In large-area photovoltaic modules, the dimensions of the electrodes are of utmost importance to avoid unnecessary losses. photovoltaic cells require at least one transparent electrode to allow light to enter the photoactive layer. Typical materials for this electrode are transparent conducting oxides (TCOs), such as indium tin oxide, fluorinated tin oxide or insulator/metal/insulator systems[31-39] deploying only several nanometer thin silver layers sandwiched by semiconducting oxides. Novel approaches also include layers of doped conducting polymers[23–25,40] or films of carbon nanotubes.[41–51] An overview is given in references.[52,53] The limited conductivity of these materials induces restrictions on the design of photovoltaic modules. To model the performance of photovoltaic modules, we assume the transparent electrode to be resistive, while the intransparent electrode is considered not to contribute to resistive losses of the converted incident radiation power. We justify this simplification by a significantly smaller sheet resistance of typically used intransparent metallic electrodes $(1\,\Omega/\square)$ yielding negligible losses compared to TCOs $(10$–$100\,\Omega/\square)$ or even organic electrodes $(100$–$1000\,\Omega/\square)$. The I–V characteristics of a photovoltaic cell of width b with one resistive electrode can be calculated by solving the two coupled first-order differential equations for current $\mathbf{I}(\mathbf{x})$ and voltage $V(\mathbf{x})$ at the point in space $\mathbf{x}$:

$$\nabla \cdot \mathbf{I} = \mathbf{j}_\perp(V) \tag{7.13}$$

$$\nabla V = \rho_\square \mathbf{I} \tag{7.14}$$

Equation (7.13) is the continuity equation for the current $\mathbf{I}$ in the transparent electrode. Further $\mathbf{j}_\perp(V)$ is the current density of an infinitesimal small photovoltaic cell depending on the voltage V at a specific working point. Equation (7.14) is Ohm's law with the surface resistance $\rho_\square$ of the transparent contact.

The series circuitry induces a certain area loss due to the isolation lines and overlapping electrodes. Depending on the structuring method, this loss is typically in the range of several $100\,\mu$m-1 mm. The trade-off between area loss and resistive loss can be optimized with a numerical model or, in good approximation especially at the optimum, with an analytical model. As the JV curve of the photovoltaic cell, especially J_{sc}, depends on the irradiance, this optimum will vary for different light intensities. Furthermore, production-related parameters may alter the results, e.g., contact resistances at the series circuitry and parallel resistances due to, for instance, imperfect isolation between cell elements or defects on the active area of the module. The parallel resistances can be a major issue for low-light performance as shown for single photovoltaic cells in Section 7.1.

Example: organic photovoltaic module

As mentioned in Section 7.1 for energy-efficient indoor light sources, the optimum bandgap is in the range of 1.5–2.0 eV. Organic semiconductors used for organic photovoltaic cells have bandgaps in this range and also the low-light performance is potentially very high for defect free organic photovoltaic cells[25] making them a promising potential low-cost option for indoor applications. The printability of such photovoltaic cells[54?-57] makes a specific layout for any new application very easy to realize, as printing machines are designed to be easily adapted to new motifs. An area loss of 1 mm is assumed taking the typical accuracy of high speed printing technologies on flexible substrates into account. Details about the chosen algorithm are found elsewhere.[27] Two illumination intensities were considered, AM1.5 photovoltaic irradiation with an intensity of $1000\,\mathrm{Wm^{-2}}$ and an indoor illumination of $5\,\mathrm{Wm^{-2}}$. Current–voltage characteristics were derived from a 4.5% efficient organic photovoltaic cell with a short circuit current density of $10\,\mathrm{mA\,cm^{-2}}$ under AM1.5 illumination at $1000\,\mathrm{Wm^{-2}}$. The calculations were made for four types of transparent electrodes, ITO on glass with a sheet resistance of $15\,\Omega/\square$, $60\,\Omega/\square$ ITO on polymer substrates, and a transparent polymer anode (PEDOT) with $500\,\Omega/\square$ and $2000\,\Omega/\square$.

AM1.5 solar irradiation with intensity of $1000\,\mathrm{Wm^{-2}}$ is considered in Fig. 7.14a. As expected, the highest relative module efficiency of 84% of the elementary cell (4.5% solar efficiency) is obtained for the lowest sheet resistance ($15\,\Omega/\square$ ITO on glass). The optimum width of a single cell is 9 mm. The relative efficiency decreases down to 76% (of the elementary cell) at an optimum cell width of 6mm when choosing ITO on a flexible substrate ($60\,\Omega/\square$). The resistive losses are considerably reduced at low light intensities ($5\,\mathrm{Wm^{-2}}$). 96% of the single cell efficiency is obtained with a low sheet resistance of $15\,\Omega/\square$ and a cell width of 40 mm. A relative efficiency of 89% can be obtained with a polymer anode ($500\,\Omega/\square$) and a width of 13 mm. Even for a resistivity of $2000\,\Omega/\square$ over 80% of the small area efficiency can be obtained, showing that under low-light conditions fully organic photovoltaic cells become an interesting option with high potential for cost reduction.

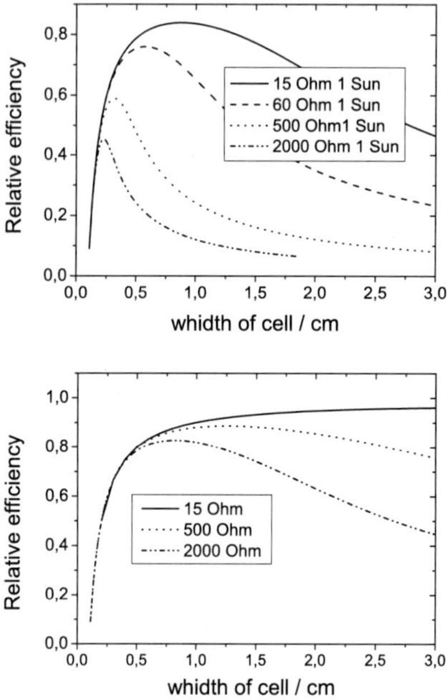

Figure 7.14 Simulated relative photovoltaic module efficiency depending on the width of the photoactive layer for different sheet resistances. (a): Simulated AM1.5 photovoltaic irradiation with intensity of $1000\,\mathrm{Wm^{-2}}$. (b): Illumination with $5\,\mathrm{Wm^{-2}}$ (same spectral distribution).

7.3.2 Layout for Specific Applications

A specific energy harvesting application implies certain boundary conditions for the layout of the photovoltaic module. As the output of a photovoltaic cell is highly dependent on the lighting conditions as irradiance and spectrum of the light, the first input into a specific module layout is the typical expected illumination scenario and the power and energy consumed by the application. While the spectra of the different light sources encountered during use of the system will influence which bandgap and thus which semiconductor material is the most favorable, the expected light intensities and their typical percentage during use will determine the size and layout of the module.

At this point, one has to decide if an energy storage system like a capacitor or accumulator will be part of the system.

This is mandatory if a chronological discrepancy between power generation by the photovoltaic cell and power consumption by the application is anticipated. Explicitly under highly fluctuating lighting conditions with phases of high irradiance resulting in an oversupply of energy and phases in which the irradiance is too low to deliver enough power on the maximum allowable area of the photovoltaic module while the system has to run during those periods of time. An example for such a system might be a sensor system designated to collect data continuously 24 h per day. Another example is a solar bag charging a battery, which then can deliver power to consumer electronic devices on demand. A storage system can be omitted, if there is enough light available during the operating time of the application. One example could be advertisement signature in a supermarket, where the lighting conditions are constant during the opening hours. This is the simplest case, as the photovoltaic module can be tailored to one specific spectrum and intensity of light to deliver the requested power. From Fig. 7.15, one can derive that for a system without energy storage one will optimize the module to worst-case conditions, while under fluctuating conditions one has to optimize the module to generate a maximum energy during a typical

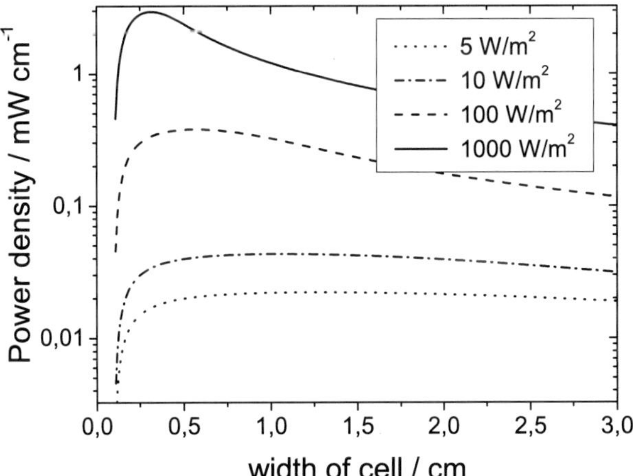

Figure 7.15 Power density available for different light intensities against width of the module cell elements. The sheet resistance is $500\,\Omega/\square$, which is a good value for an organic electrode.

day of use, i.e., maximizing the integral in Eq. 7.15.

$$E = \int_{t_0}^{t1} \eta(P_{rad}(t)) \times P_{rad}(t)dt \qquad (7.15)$$

As the loss for non-optimal dimensions at high intensities is larger as compared to the loss of a module optimized for high intensities at lower irradiance, a best case optimization is a simple and well performing approach to harvest a maximum amount of energy. For energy autonomous systems the balance between battery capacitance and photovoltaic cell area might favor an optimization for a day with worst-case illumination, as the battery would become extremely large, if the seasonal differences in the available radiative energy shall be equalized. This decision will depend on the affordable battery capacity and area of the photovoltaic module.

7.3.3 *Module Layout for a System without Energy Storage*

If no energy storage system is employed, the module must be able to power the application under worst-case illumination. Under higher light intensities the efficiency of the module will be lower, but the output power will nevertheless be higher (Fig. 7.15). The corresponding spectrum determines the optimum bandgap of the photovoltaic cell, while the lowest expected irradiance during operation mainly influences the circuitry dimensions of the module. To calculate the appropriate module dimensions and geometry, the JV-curve of the photovoltaic cell must be measured under these lighting conditions. With this JV-curve the circuitry dimensions are optimized as described in Section 7.3.1 to allow for highest efficiency under worst-case conditions, where the available power is lowest. With the resulting efficiency η_{module} one can easily calculate the area A_{module} needed to power the device from the power consumed by the application P_{appl} and the irradiance P_{rad} according to the following equation:

$$A_{module} = \frac{\eta_{module} \cdot P_{el}}{P_{rad}} > \frac{\eta_{module} \cdot P_{appl}}{P_{rad}} \qquad (7.16)$$

The operating voltage of the application determines the number of cells connected in series thus the length of the module l (and implicitly the width w) is determined by number of cells connected in series n_{cells} times optimum width b_{opt} of the cell stripe for the series circuitry according to equations 7.17 to 7.19. It has to be considered that V_{mpp} of the photovoltaic cell and the optimum width of the cell stripes b_{opt} depend on the irradiance.

$$n_{\text{cells}} > \frac{V_{\text{system}}}{V_{\text{mpp}}(P_{\text{rad}})} \tag{7.17}$$

$$l = n_{\text{cells}} \cdot b_{\text{opt}}(P_{\text{rad}}) \tag{7.18}$$

$$w = \frac{A}{l} \tag{7.19}$$

If one cannot exclude that the system is exposed to significantly higher light intensities from time to time, e.g., natural daylight shining through a window, one should consider to include an overvoltage protection to avoid damage to the application electronics.

7.3.4 Layout of a System with Energy Storage

Energy storage is mandatory, if there is a chronological mismatch between energy generation and consumption. One question to be answered before dimensioning the area of the module is if the system shall work energy autonomous or if the photovoltaic charging is only used to prolong the intervals of charging from the grid. In the case of a supplementary system, the area of the module will be determined by the application, for example the size of a solar bag and the preferences of the customer in case of external charging units. As explained in Section 7.3.1, the dimensioning of the series circuitry will be optimized to the highest irradiances that will be regularly encountered during operation.

Generally the area should be at least large enough to generate enough power for a significant extension of the device operation. Therefore it must deliver a certain percentage of the energy consumed during daily use. The accumulator must be dimensioned to power the device at least for the longest period of darkness that will be encountered during the use of the system. From Fig. 7.16b one can conclude that typical consumer applications can be run for a substantial fraction of the day (24 h) by a 4% efficient photovoltaic module with the size of a Din A4 page during summer. In the winter significantly less radiative energy is available, thus the area needed

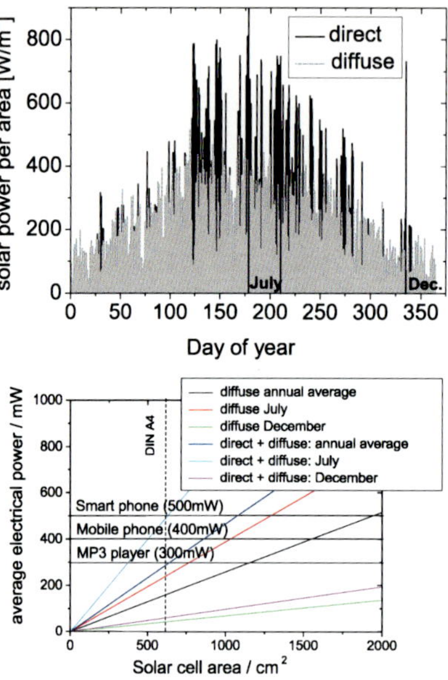

Figure 7.16 Irradiance distribution of a reference year (DWD TRY-Dataset, Region Oberrheingraben unteres Neckartal, Repräsentanzstation Mannheim) (a) and average amount of energy that can be collected by a 4% efficient photovoltaic cell during 24 h against size of the photovoltaic cell for different lighting scenarios (b).

to power those devices becomes very large (ca. 1 m^2). Furthermore, for this simple estimation no reduction of the efficiency under low irradiance was taken into account, which in reality will further increase the needed area. As people will be more time outdoors during summer and on sunny days in the winter, the demand for remote energy will be significantly larger on those days too, mitigating this obvious shortcoming. In Fig. 7.17 A Din A4 page sized photovoltaic cell with approximately 4 % power conversion efficiency under AM 1.5 is used to charge a battery of 10 Wh and power a mobile phone with a power consumption of ca. 3 W in peak and 400 mW in average, which means that on an average, the battery needs to be fully recharged every 30 h. A logarithmic decrease of the efficiency under lower light intensities is included for the calculations.

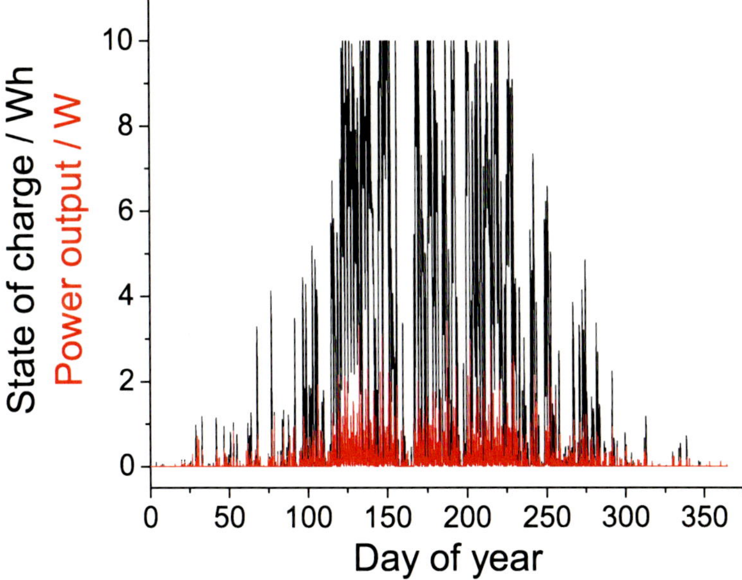

Figure 7.17   State of charge of and power drain from a 10 Wh battery charged with a Din A4 page sized photovoltaic module with 4% efficiency under daylight and discharged with a frequently used cell phone as electrical load. During summer the system works well and a significant amount of the used energy can be supplied by the photovoltaic cell. In winter only on sunny days the energy supplied by the PV module is sufficient to prolong the use of the cell phone significantly.

It is clearly visible that the system performs very well in summer but only on very sunny days during winter. As mentioned earlier the typical user will need the remote power with higher probability on the sunny days. Thinking of direct electrical heating of gloves or shoes on cloudy cold winter days, nevertheless, will not work well with solar power as one would need several square meters of photovoltaic cells. A second scenario is shown in Fig. 7.18, where a sensor with average energy consumption of 100 µW shall be powered by an organic photovoltaic cell and a battery in a north facing office in Freiburg. The available light is taken from the DAYSIM calculations as shown in Fig. 7.10 scenario N3. An efficiency of 4% is assumed under the occurring light intensities and spectrum.

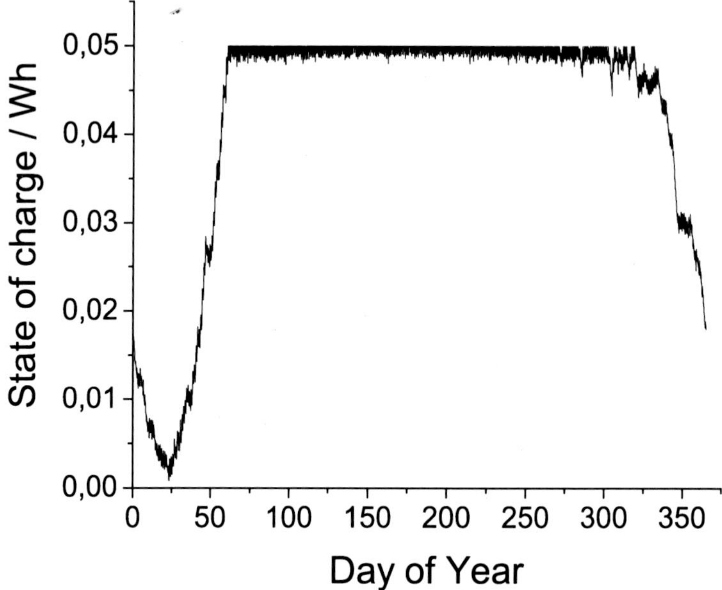

Figure 7.18 Example of an energy autonomous sensor system mounted under worst-case conditions as described in Fig. 7.10 scenario N3. A sensor with average energy consumption of 100 µW is powered by a 4% efficient 100 cm^2 large photovoltaic cell and the energy is stored in a 50 mWh battery for seasonal energy transfer. Depicted is the state of charge of the battery showing that the photovoltaic cell can supply enough energy to power the sensor during the year even for this worst-case scenario without artificial lighting.

The area of the photovoltaic cell is 100 cm^2 and the capacity of the battery is 50 mWh, which corresponds to a button cell. From these calculations one can conclude that for the operation of energy autonomous sensor systems with a low power requirement in the µW to low mW range photovoltaic cells can provide enough power even under typical indoor lighting conditions with acceptable areas.

7.4 Concluding Remarks

Photovoltaic energy is a powerful, but dilute and highly fluctuating (orders of magnitude) energy source. Even more complexity is added by the fact that also the spectral distribution is changing with different light sources and weather conditions, which influences even the choice of the semiconductor material. Any solar cell only generates electrical power if it is exposed to light; therefore, the placement of this energy harvester is very critical for the usefulness of such a system. The direct integration into the housing of an electronic device such as a cell phone or mp3 player is in often not the best option, as they are most of the time in pockets or bags and have a small surface area. In order to have a working system with added value for the customer, many specific questions have to be answered and careful dimensioning of the components is needed to prevent frustration. But facing this challenge and optimizing the system to the needs and habits of the customers by knowing the relevant lighting conditions can make a significant difference and result in an abundant, reliable and powerful energy source.

References

1. P. Würfel, *Physics of Solar Cells*. Wiley-VCH (2009).
2. W. Shockley and H. J. Queisser. Detailed balance limit of efficiency of P-N junction solar cells. *J. Appl. Phys.* **32**, 510 (1961).
3. M. Müller, J. Wienold, W. D. Walker, and L. M. Reindl. Characterization of indoor photovoltaic devices and light. *Proc. 34th IEEE Photovoltaic Specialist Conference*, June 7–12, 2009, Philadelphia, USA (2009), 738–743.
4. *Standard Test Conditions*, following *IEC 60904-3*, Ed. 2 (2008).
5. M. A. Green, K. Emery, Y. Hishikawa, and W. Warta. Short communication solar cell efficiency tables (version 45). *Prog. Photovoltaics Res. Appl.*, **17**(5), 320–326, (2009).
6. EPS soltec, http://www.eps-soltec.com, December 10, 2009.
7. Flexcell, http://www.flexcell.com, December 10, 2009.
8. IXYS, http://www.ixys.com/Product_portfolio/solar.asp, December 10, 2009.
9. Konarka, http://www.konarka.com/, December 10, 2009.
10. Plastecs, http://www.plastecs.com/, December 10, 2009.
11. PowerFilm, http://www.powerfilmsolar.com/index.htm, December 10, 2009.

12. Solaronix, http://www.solaronix.com/, December 10, 2009.

13. G. W. Larson and R. A. Shakespeare. *Rendering With Radiance: The Art and Science of Lighting Visualization*. Morgan Kaufmann Publishers, S. 536 (1998).

14. M. Müller, W. D. Walker, and L. M. Reindl. Simulations and measurements for indoor photovoltaic devices. In: *Proc. 24th European Photovoltaic Specialist Conference*, September 21–25, 2009, Hamburg, (2009).

15. Sanyo Solar, http://us.sanyo.com/solar/, December 10, 2009.

16. Schott Solar, http://www.schottsolar.com, December 10, 2009.

17. Sinonar, http://www.sinonar.com.tw/, December 10, 2009.

18. National Renewable Energy Laboratory: *Reference Solar Spectral Irradiance: Air Mass 1.5*, http://rredc.nrel.gov/solar/spectra/am1.5/, 10.12.2009

19. A. Wagner. *Photovoltaik Engineering: Handbuch für Planung, Entwicklung und Anwendung*. Berlin Heidelberg, Springer-Verlag, 2nd ed. (2006).

20. EN12464-1, *Light and lighting - Lighting of work places - Part 1: Indoor work places*, 2003.

21. *Metenorm—Global Solar Radiation Database*, http://www.meteonorm.ch, December 10, 2009.

22. C. F. Reinhart. *Daylight Availability and Manual Lighting Control in Office Buildings: Simulation Studies and Analysis of Measurements*. Karlsruhe, Universität Karlsruhe, Fakultät für Architektur, Dissertation (2001).

23. M. Glatthaar, M. Niggemann, B. Zimmermann, P. Lewer, M. Riede, A. Hinsch, and J. Luther. Organic solar cells using inverted layer sequence, *Thin Solid Films* **491** (1–2), 298–300 (2005). ISSN 00406090 (ISSN). URL http://www.scopus.com/scopus/inward/record.urlfieid=2-s2.0-25144508569&p% artner=40&rel=R5.0.4.

24. B. Zimmermann, M. Glatthaar, M. Niggemann, M. K. Riede, A. Hinsch, and A. Gombert, ITO-free wrap through organic solar cells: A module concept for cost-efficient reel-to-reel production, *Solar Energy Mater Solar Cells* **91** (5), 374–378 (March 2007).

25. C. Lungenschmied, G. Dennler, H. Neugebauer, S. N. Sariciftci, M. Glatthaar, T. Meyer, and A. Meyer. Flexible, long-lived, large-area, organic solar cells, *Solar Energy Mater. Solar Cells* **91** (5), 379–384 (March 2007).

26. C. Wang, M. Waje, X. Wang, J. Tang, R. Haddon, and Y. Yan, Proton exchange membrane fuel cells with carbon nanotube based electrodes, *Nano Lett.* **4** (2), 345–348, (2004). URL ISI:000188965700031. 182.

27. Glatthaar. *Zur Funktionsweise organischer Solarzellen auf der Basis interpenetrierender Donator/Akzeptor-Netzwerke*. PhD thesis, Albert-Ludwigs-Universität Freiburg im Breisgau, (2007).

28. J. F. Randall. On the use of photovoltaic ambient energy sources for powering indoor electronic devices. p. 67, These N° 2806, EPFL, Lausanne (2003).

29. S. W. Glunz, et al. High-efficiency silicon solar cells for low-illumination applications. *Conference Record of the Twenty-Ninth IEEE Photovoltaic Specialists Conference*, pp. 450-453 (2002).

30. M. A. Green. *Silicon Solar Cells: Advanced Principles and Practice*. Sydney, Bridge Printery (1995).

31. M. Bender, W. Seelig, C. Daube, H. Frankenberger, B. Ocker, and J. Stollenwerk, Dependence of film composition and thicknesses on optical and electrical properties of ITO-metal-ITO multilayers, *Thin Solid Films* **326** (1–2), 67–71, (1998). ISSN 0040-6090. DOI: 10.1016/S0040-6090(98)00520-3. URL http://www.sciencedirect.com/science/article/B6TW0-3TMPJBG-8/2/af98d448% 535d7d91b5b7c3 909be49686.

32. X. Liu, X. Cai, J. Mao, and C. Jin. ZnS/Ag/ZnS nano-multilayer films for transparent electrodes in flat display application, *Appl. Surf. Sci.* **183** (1–2), 103–110, (2001). ISSN 0169-4332. DOI: 10.1016/S0169-4332 (01)00570-0. URL http://www.sciencedirect.com/science/article/ B6THY-44C84C2-J/2/cc53068e898b7c78b926385040e261e7.

33. D. Sahu and J.-L. Huang. High quality transparent conductive ZnO/Ag/ZnO multilayer films deposited at room temperature, *Thin Solid Films* **515** (3), 876–879 (November 2006). URL http://www. sciencedirect.com/science/article/B6TW0-4KSD819-8/2/ebbd0f9f% 4d47f85813c40469895abf56.

34. D. Sahu, S.-Y. Lin, and J.-L. Huang. ZnO/Ag/ZnO multilayer films for the application of a very low resistance transparent electrode, *Appl. Surf. Sci.* **252** (20), 7509–7514 (2006). ISSN 0169-4332. DOI: 10.1016/ j.apsusc.2005.09.021. URL http://www.sciencedirect.com/ science/article/B6THY-4H9YBXD-B/2/c14423f9% 55213585b974a 98d5b02913e.

35. H. Pang, Y. Yuan, Y. Zhou, J. Lian, L. Cao, J. Zhang, and X. Zhou. ZnS/Ag/ZnS coating as transparent anode for organic light emitting diodes, *J. Luminescence* **122–123**, 587–589 (2007). ISSN 0022-2313. DOI: 10.1016/j.jlumin.2006.01.232. URL http://www.sciencedirect.com/ science/article/B6TJH-4JGJGY5-S/2/c26459886e333966e958974e99 b6a575.

36. H. Park, J. Park, J. Choi, J. Lee, J. Chae, and D. Kim. Fabrication of transparent conductive films with a sandwich structure composed of ITO/Cu/ITO, *Vacuum* **83** (2), 448–450, (2008). ISSN 0042-207X. DOI: 10.1016/j.vacuum.2008.04.061. URL http://www.sciencedirect.com/science/article/B6TW4-4S9P5M4-D/2/b6253a9eb9f868c15b4de8779ecf5c3c.

37. D. Sahu, S.-Y. Lin, and J.-L. Huang. Investigation of conductive and transparent Al-doped ZnO/Ag/Al-doped ZnO multilayer coatings by electron beam evaporation, *Thin Solid Films* **516** (15), 4728 – 4732, (2008). ISSN 0040-6090. doi: DOI:10.1016/j.tsf.2007.08.089. URL http://www.sciencedirect.com/science/article/B6TW0-4PGPVP5-M/2/1b4c18dc% 4322a85650d23db7eec1b61d.

38. L. Cattin, F. Dahou, Y. Lare, M. Morsli, R. Tricot, S. Houari, A. Mokrani, K. Jondo, A. Khelil, K. Napo, and J. C. Bernede, MoO_3 surface passivation of the transparent anode in organic solar cells using ultrathin films, *J. Appl. Phys.* **105**(3), 034507 (2009). doi: 10.1063/1.3077160. URL http://link.aip.org/link/?JAP/105/034507/1.

39. B. Szyszka, P. Loebmann, A. Georg, C. May, and C. Elsaesser. Development of new transparent conductors and device applications utilizing a multi disciplinary approach, *Thin Solid Films* **518**(11), 3109–3114 (2010), (2009). ISSN 0040-6090. doi: DOI:10.1016/j.tsf.2009.10.125. URL http://www.sciencedirect.com/science/article/B6TW0-4XP37PM-1/2/a47ad59fbb1384ec5613e70fac468b83.

40. S. K. Hau, H.-L. Yip, J. Zou, and A. K.-Y. Jen. Indium tin oxide-free semi-transparent inverted polymer solar cells using conducting polymer as both bottom and top electrodes, *Organic Electron.* **10** (7), 1401–1407 (2009). ISSN 1566-1199. doi: DOI:10.1016/j.orgel.2009.06.019. URL http://www.sciencedirect.com/science/article/B6W6J-4WNRK3J-2/2/aa33c95aed73638d725ff9d190aa5886.

41. D. L. Carroll, R. Czerw, and S. Webster. Polymer-nanotube composites for transparent, conducting thin films, *Synthetic Met.* **155** (3), 694–697 (2005). ISSN 0379-6779. doi: DOI:10.1016/j.synthmet.2005.08.031. URL http://www.sciencedirect.com/science/article/B6TY7-4HH81HY-G/2/744c1958810cbddde1b288aec42fd15e.

42. J. Moon, J. Park, T. Lee, Y. Kim, J. Yoo, C. Park, J. Kim, and K. Jin. Transparent conductive film based on carbon nanotubes and PEDOT composites, *Diamond Relat. Mater.* **14** (11–12), 1882–1887 (2005). ISSN 0925-9635. doi: DOI:10.1016/j.diamond.2005.07.015. URL http://www.sciencedirect.com/science/article/B6TWV-4GYH7MY-1/2/38bff34faa2e1edc36c77585122311bc.

43. M. Kaempgen, G. Duesberg, and S. Roth. Transparent carbon nanotube coatings, *Appl. Surf. Sci.* **252** (2), 425–429 (2005). ISSN 0169-4332. DOI: 10.1016/j.apsusc.2005.01.020. URL http://www.sciencedirect.com/science/article/B6THY-4FFN4Y7-2/2/04482628% 96f1fead23f754d 602c15bba.

44. R. Ulbricht, S. B. Lee, X. Jiang, K. Inoue, M. Zhang, S. Fang, R. H. Baughman, and A. A. Zakhidov. Transparent carbon nanotube sheets as 3-d charge collectors in organic solar cells, *Solar Energy Mater. Solar Cells* **91** (5), 416–419 (2007). ISSN 0927-0248. doi: DOI:10.1016/j.solmat.2006.10. 002. URL http://www.sciencedirect.com/science/article/B6V51-4MKV2M0-1/2/f9a11f5adae760e05bdb729f2f3567e6.

45. X. Yu, R. Rajamani, K. Stelson, and T. Cui. Fabrication of carbon nanotube based transparent conductive thin films using layer-by-layer technology, *Surf. Coatings Technol.* **202** (10), 2002–2007 (2008). ISSN 0257-8972. doi: DOI:10.1016/j.surfcoat.2007.08.064. URL http://www.sciencedirect.com/science/article/B6TVV-4PKXBJK-2/2/d254da00e6b24f56d1a76dd187292d1d.

46. L. Valentini, M. Cardinali, D. Bagnis, and J. M. Kenny, Solution casting of transparent and conductive carbon nanotubes/poly(3,4-ethylenedioxythiophene)-poly(styrenesulfonate) films under a magnetic field, *Carbon* **46** (11), 1513–1517 (2008). ISSN 0008-6223. DOI: 10.1016/j.carbon.2008.05.025. URL http://www.sciencedirect.com/science/article/B6TWD-4SPC0S8-2/2/c7764b67% c67a12c7d87eed 0b8b124c39.

47. S. Paul and D.-W. Kim. Preparation and characterization of highly conductive transparent films with single-walled carbon nanotubes for flexible display applications, *Carbon* **47** (10), 2436–2441 (2009). ISSN 0008-6223. DOI: 10.1016/j.carbon.2009.04.045. URL http://www.sciencedirect.com/science/article/B6TWD-4W7J12T-1/2/8719852 2% f2cb50c831ea87fee03176ad.

48. T. Kitano, Y. Maeda, and T. Akasaka. Preparation of transparent and conductive thin films of carbon nanotubes using a spreading/coating technique, *Carbon* **47** (15), 3559–3565 (2009). ISSN 0008-6223. DOI: 10.1016/j.carbon.2009.08.027. URL http://www.sciencedirect. com/science/article/B6TWD-4X24VR9-1/2/8a527864% cd53ada53d 2be68f5632041d.

49. G. Xiao, Y. Tao, J. Lu, and Z. Zhang. Highly conductive and transparent carbon nanotube composite thin films deposited on polyethylene terephthalate solution dipping, *Thin Solid Films* **518**(10), 2822–2824 (2010), (2009). ISSN 0040-6090. doi: DOI:10.1016/j.tsf.2009.

11.021. URL http://www.sciencedirect.com/science/article/B6TW0-4XR5N0B-2/2/f399170714471f34c2a8a03978cef9bf.

50. H. S. Ki, J. H. Yeum, S. Choe, J. H. Kim, and I. W. Cheong, Fabrication of transparent conductive carbon nanotubes/polyurethane-urea composite films by solvent evaporation-induced self-assembly (EISA), *Composites Sci. Technol.* **69** (5), 645–650 (2009). ISSN 0266-3538. DOI: 10.1016/j.compscitech.2008.12.012. URL http://www.sciencedirect.com/science/article/B6TWT-4V88FT1-1/2/b504efc0% 8a29ead4835 a48518d544a55.

51. R. A. Hatton, N. Blanchard, L. W. Tan, G. Latini, F. Cacialli, and S. R. P. Silva. Oxidised carbon nanotubes as solution processable, high work function hole-extraction layers for organic solar cells, *Org. Electron.* **10** (3), 388–395 (2009). ISSN 1566-1199. doi: DOI:10.1016/j.orgel.2008.12.013. URL http://www.sciencedirect.com/science/article/B6W6J-4V94X0D-5/2/b15d581f% 48fa01954fd6a4ba9f012811.

52. Z. Spitalsky, D. Tasis, K. Papagelis, and C. Galiotis. Carbon nanotube-polymer composites: Chemistry, processing, mechanical and electrical properties, *Prog. Polym. Sci.* **35** (3), 357–401 (2010). ISSN 0079-6700. doi: DOI:10.1016/j.progpolymsci.2009.09.003. URL http://www.sciencedirect.com/science/article/B6TX2-4X9NV3D-2/2/d2c2b1b4% d38ceb49265f68f55aace839.

53. C. G. Granqvist. Transparent conductors as solar energy materials: A panoramic review, *Solar Energy Materials and Solar Cells* **91** (17), 1529–1598 (2007). ISSN 0927-0248. doi: DOI:10.1016/j.solmat.2007.04.031. URL http://www.sciencedirect.com/science/article/B6V51-4P3TYF4-1/2/33b862d8% b208b7dd3ca821e723081cf1.

54. S. Shaheen, R. Radspinner, N. Peyghambarian, and G. Jabbour, Fabrication of bulk heterojunction plastic solar cells by screen printing, *Appl. Phys. Lett.* **79** (18), 2996–2998 (2001). URL ISI:000171726300045. 120.

55. F. C. Krebs, H. Spanggard, T. Kjaer, M. Biancardo, and J. Alstrup. Large area plastic solar cell modules, *Mater. Sci. Eng. B* **138** (2), 106–111 (March 2007).

56. F. C. Krebs. Fabrication and processing of polymer solar cells: A review of printing and coating techniques, *Solar Energy Mater. Solar Cells* **93**(4), 394–412 (2009), (2008). ISSN 0927-0248. doi: DOI:10.1016/j.solmat.2008.10.004. URL http://www.sciencedirect.com/science/article/B6V51-4V0VBXJ-1/2/c660b684% cbf8c104f1dd085e7f1ec1f4.

57. J. M. Ding, A. de la Fuente Vornbrock, C. Ting, and V. Subramanian. Patternable polymer bulk heterojunction photovoltaic cells on plastic by rotogravure printing, *Solar Energy Mater. Solar Cells* **93** (4), 459–464 (2009). ISSN 0927-0248. doi: DOI:10.1016/j.solmat.2008.12.003. URL http://www.sciencedirect.com/science/article/B6V51-4VF0XTX-5/2/a88f80d7% 05e2c2e26e76b3fb6ed3c0b0. Processing and Preparation of Polymer and Organic Solar Cells.

Chapter 8

DC–DC Converters

Markus Pollak

Fraunhofer Institute for Integrated Circuits IIS, Nordostpark 93,
90411 Nuremberg, Germany
markus.pollak@iis.fraunhofer.de

In energy harvesting applications, DC–DC converters play an essential role in powering electrical loads. Energy transducers such as thermogenerators, piezoelectric and electrodynamic generators provide varying output voltages, which are in general too high or too low for supplying electrical loads directly. Therefore, DC–DC converters are required to provide stable supply voltages. In the following sections, first the linear regulators working like regulated Ohmic resistors are described and afterward the switching regulators that transport the electrical energy in at least two periodic steps with the help of energy storage devices, that is, capacitors or inductors. In the last section, the matching of a load to an energy transducer employing DC–DC converters is considered in detail, where the output of the converter is controlled in order to provide maximum power to the electrical load.

Handbook of Energy Harvesting Power Supplies and Applications
Edited by Peter Spies, Loreto Mateu, and Markus Pollak
Copyright © 2015 Pan Stanford Publishing Pte. Ltd.
ISBN 978-981-4241-86-1 (Hardcover), 978-981-4303-06-4 (eBook)
www.panstanford.com

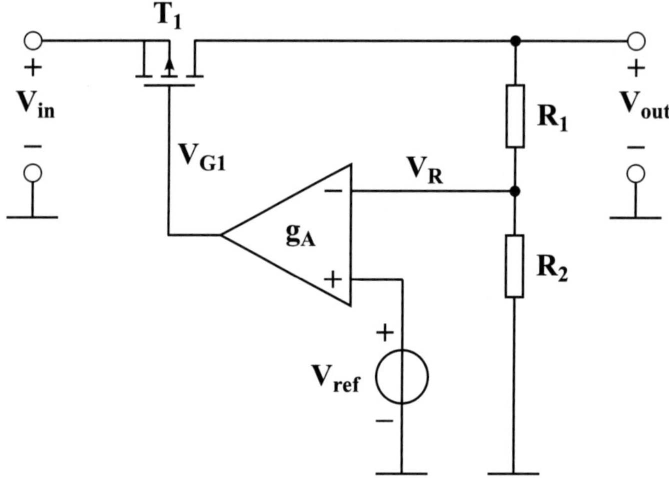

Figure 8.1 Block diagram of a linear regulator.

8.1 Linear Regulators

The task of linear regulators is to establish a constant output voltage independently from output current and input voltage. Therefore, a regulation element consisting of one or more transistors is needed. To control the regulation element, a loop amplifier, a network of several resistors as well as a voltage reference is used.

8.1.1 *Electrical Circuit*

Generally, an operational amplifier fulfills the task of the loop amplifier that compares a part of the output voltage of the regulator with a reference voltage. Fig. 8.1 illustrates the topology of a linear regulator. The regulation loop adjusts itself until the voltage (V_{fb}) deviates as little as possible (depending on the open circuit amplification and the offset voltage of the amplifier) from the reference voltage (V_{ref}). The disadvantage of such a regulation concept is the power loss due to the voltage drop between input and output voltage. That means the efficiency is high when the difference between input and output voltage is low. In addition, a linear regulator cannot have a bigger output voltage than the input

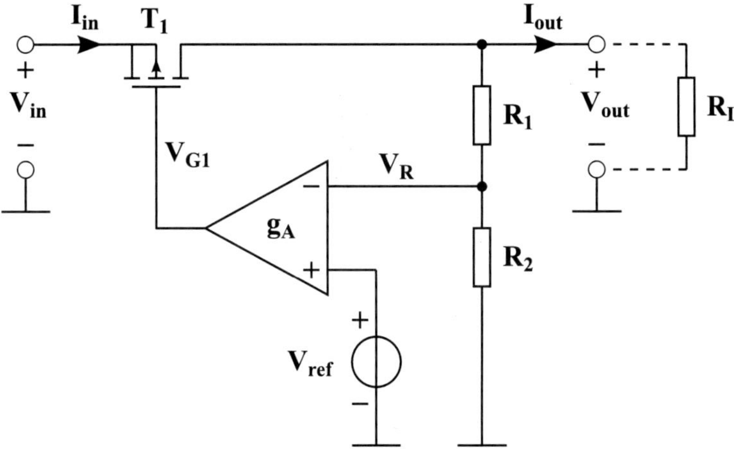

Figure 8.2 Signals in a linear regulator.

voltage. The advantage of such a concept is—in comparison to the switching regulators described in the following subsection—that the power path contains no switching elements. Therefore, there is no ripple generated at the output.

8.1.2 *Analytical Model*

Referring to Fig. 8.2, the gate voltage of T_1 is determined as follows:

$$V_{G1} = g_a \left(V_{\text{ref}} - V_{\text{out}} \frac{R_2}{R_1 + R_2} \right). \tag{8.1}$$

Rearranging terms leads to

$$V_{\text{out}} = \left(V_{\text{ref}} - \frac{V_{G1}}{g_a} \right) \frac{R_1 + R_2}{R_2}. \tag{8.2}$$

Assuming that the amplifier has an infinite DC gain or at least a very high DC gain—which is generally the case with operational amplifiers—it can be found that

$$V_{\text{out}} = V_{\text{ref}} \frac{R_1 + R_2}{R_2}. \tag{8.3}$$

In this case, the output voltage V_{out} of the regulator can be considered independent of its input voltage V_{in} and of the output load R_{L}. Furthermore, V_{out} is directly determined by the relation of resistors R_1 and R_2 and the reference voltage V_{ref}.

8.1.3 *Efficiency Calculation*

The output current I_{out} of the regulator (see Fig. 8.2) is calculated using Eq. (8.3):

$$I_{\text{out}} = I_{\text{in}} - \frac{V_{\text{out}}}{R_1 + R_2} = I_{\text{in}} - \frac{V_{\text{ref}}}{R_2}. \tag{8.4}$$

The overall efficiency of the circuit can be stated as follows:

$$\eta = \frac{P_{\text{out}}}{P_{\text{in}}} = \frac{V_{\text{out}} I_{\text{out}}}{V_{\text{in}} I_{\text{in}}} \tag{8.5}$$

and therefore using Eqs. (8.3) and (8.4):

$$\eta = \frac{V_{\text{out}}(I_{\text{in}} - \frac{V_{\text{ref}}}{R_2})}{V_{\text{in}} I_{\text{in}}} = \frac{V_{\text{out}}}{V_{\text{in}}} - \frac{V_{\text{out}} V_{\text{ref}}}{V_{\text{in}} I_{\text{in}} R_2} = \frac{V_{\text{out}}}{V_{\text{in}}} \left(1 - \frac{V_{\text{ref}}}{I_{\text{in}} R_2}\right). \tag{8.6}$$

8.1.4 *Design Optimization*

From Eq. 8.2, it can be noticed that the linear regulator of Fig. 8.2 works the more accurately, the higher the gain g_A of its loop amplifier is. Referring to Eq. (8.6), the highest possible efficiency is achieved minimizing the current across resistors R_1 and R_2. Consequently, for fixed output voltages V_{out}, highest possible resistor values have to be used. Furthermore, the efficiency depends on the voltage ratio between output and input. For example, for an output voltage that is half of the input voltage, 50% of the input power is lost in form of heat in transistor T_1 because the output current of the regulator is equal or less than the input current. Otherwise the regulator can work very efficiently, if the output voltage is a little below the input voltage. Detailed considerations on high-efficiency linear regulators can be found, for example, in an application note by Linear Technology [1].

8.2 Switching Regulators

Switching regulators are an alternative to linear regulators. They also fulfill the task of generating a constant output voltage independent of input voltage and output current. Here the regulation element works like a switch that can be either in on- or in off-state. Moreover, for switching regulators the output voltage can be higher than the input voltage depending on the type of converter used. Besides, one or more energy storage elements—like inductors or capacitors—are needed. The types of switching regulators can be divided into four main subtypes.

A buck converter (Section 8.2.1) delivers an output voltage lower and a boost converter (Section 8.2.2) an output voltage higher than the input voltage. These two types have in common that they use an inductor as an energy storage element. For these two types of converter two different modes of operation are defined concerning the inductor current. In continuous-conduction mode (CCM), the current is always greater than zero, whereas in discontinuous-conduction mode (DCM), it is zero for a certain time period. This will be discussed later on in more detail.

Flyback- and forward-converters use a transformer instead of an inductor. The flyback converter—like the boost converter—transfers the input energy during off-state of the switch to the output capacitor. The forward converter instead transfers the input energy directly in on-state. For high output powers, it has at least two switching transistors in a "push-pull" configuration. However, this type of converter is mainly used at output powers greater than 100 W, so it is not of importance in state-of-the-art energy harvesting applications. One last important type of converter described in Section 8.2.5 is the charge pump. Here, several switching transistors are used as well and only capacitors are present as energy storage elements. Charge pumps are mostly useful for low output currents determined by frequency and size of the capacitors.

Finally, a less common type of converter based on the Meissner oscillator is explained. It can be useful for energy harvesting applications, where the energy transducer delivers output voltages below 500 mV. Like in flyback converters, a transformer is employed, but in this case the secondary winding is used to control the switching transistor.

8.2.1 *Buck Converter*

8.2.1.1 Physical principles

The step-down (buck) converter delivers an output voltage lower than its input voltage. Basically, the input voltage is periodically

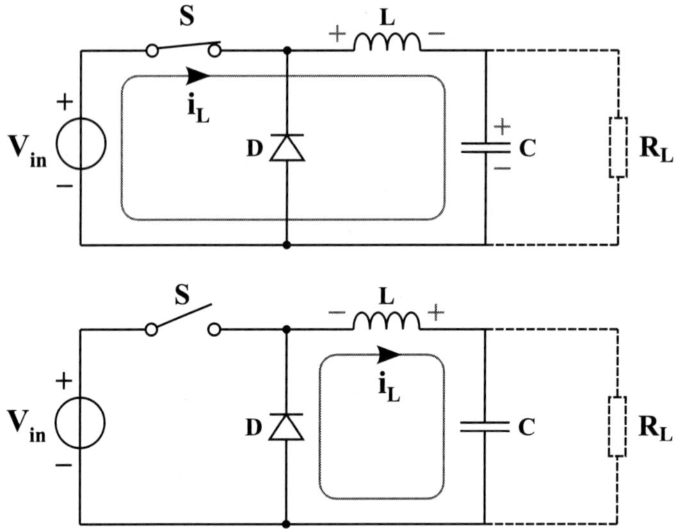

Figure 8.3 Physical principle of a step-down DC–DC converter.

switched on and off generating a rectangular voltage and afterward a low pass filter transfers the mean value to the output of the converter. The basic topology of the converter is shown in Fig. 8.3. After the switch is opened, the diode is conducting due to the current still flowing in the inductor. This way it is redirected to the load.

8.2.1.2 Electrical circuit

A typical electrical circuit of a buck converter is shown in Fig. 8.4. A voltage proportional to the output voltage is compared to a voltage reference according to the desired output voltage. The difference between these two signals v_{err} is amplified and again compared to a sawtooth voltage v_{st}. Thus, the comparator output v_{ctrl} is a pulse-width-modulated signal controlling the switching transistor T_1. The output voltage V_{out} can only be smaller than or—at a duty cycle of 1—equal to the input voltage V_{in}, where switch T_1 is closed at any time.

Figure 8.5 shows the generation of the pulse-width-modulated (PWM) signal in more detail [2]. It can be noticed that if the amplified error signal v_{err} goes up—meaning that the output voltage

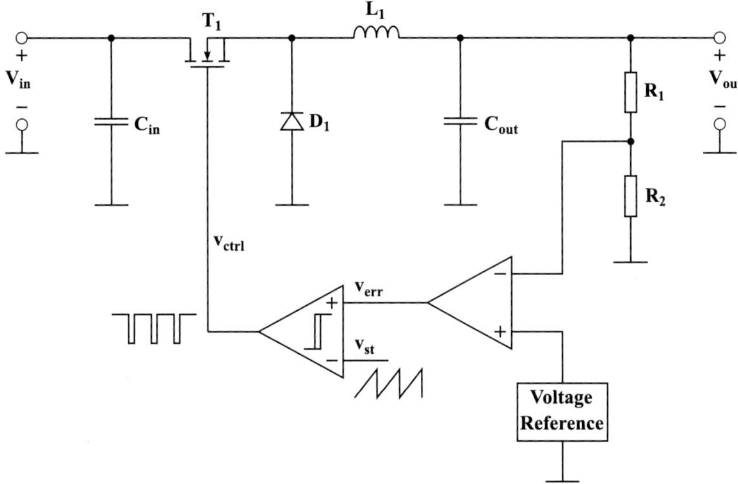

Figure 8.4 Block diagram of a step-down converter.

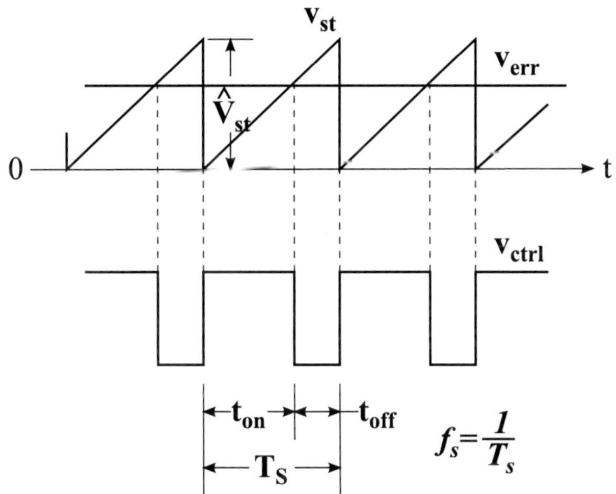

$$f_s = \frac{1}{T_s}$$

Figure 8.5 Generation of a pulse-width-modulated (PWM) signal [2].

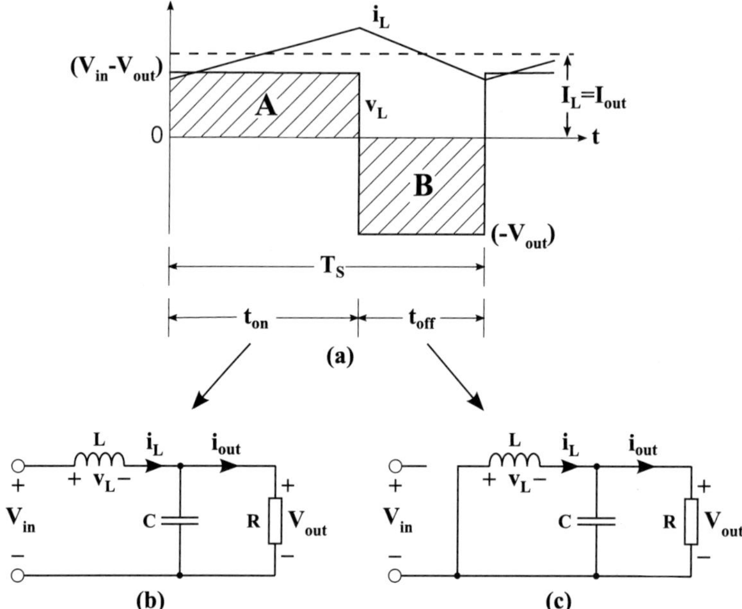

Figure 8.6 Buck converter circuit for closed (b) and opened (c) switch and corresponding waveforms (a) in continuous conduction mode [2].

drops below the desired value—then the on-time of the switching transistor is increased. In that way, the output voltage of the converter increases, until the desired value is reached. The opposite of this happens when v_{err} goes down.

8.2.1.3 Analytical model

Mohan et al. [2] did a comprehensive analysis of a buck converter, which is explained in this paragraph. For the circuit in Fig. 8.3, the corresponding waveforms of inductor current and voltage in steady-state for the continuous conduction mode are shown in Fig. 8.6. When the switch is closed (Fig. 8.6b), the inductor current i_L rises linearly resulting in a positive voltage $v_L = V_{in} - V_{out}$ across the inductor. When the switch is opened (Fig. 8.6c) the inductor current continues flowing across the diode transferring its stored energy to the output. The voltage v_L across the inductor now becomes negative that is $-V_{out}$. Assuming a lossless inductor and steady-state operation, the integral of v_L over each period T_s has to be zero. With $T_s = t_{on} + t_{off}$ it can be stated:

$$\int_0^{T_s} v_L \, dt = \int_0^{t_{on}} v_L \, dt + \int_{t_{on}}^{T_s} v_L \, dt = 0 \qquad (8.7)$$

For an ideal diode D and an ideal switch S as well as an infinite output capacitor C (see Fig. 8.3), this equation can be simplified to

$$(V_{in} - V_{out})t_{on} = V_{out}(T_s - t_{on}) \Leftrightarrow \frac{V_{out}}{V_{in}} = \frac{t_{on}}{T_s} = D, \qquad (8.8)$$

where D is the duty cycle of the switching period of T_1 (Fig. 8.4). Taking into account Fig. 8.5, it is concluded:

$$D = \frac{t_{on}}{T_s} = \frac{V_{out}}{V_{in}} = \frac{v_{err}}{\hat{V}_{st}}. \qquad (8.9)$$

Finally, the output voltage V_{out} can be expressed as

$$V_{out} = \frac{V_{in}}{\hat{V}_{st}} v_{err}. \qquad (8.10)$$

Thus, the output voltage V_{out} is directly proportional to the amplified error signal v_{err}. If no power is lost in the circuit elements, the assumption $P_{in} = P_{out}$ can be made and for the current relationship between input and output of the converter it follows:

$$V_{in} I_{in} = V_{out} I_{out} \Leftrightarrow \frac{I_{out}}{I_{in}} = \frac{V_{in}}{V_{out}} = \frac{1}{D}. \qquad (8.11)$$

From the foregoing equations, it can be observed that the circuit works like a DC transformer where the winding ratio between primary and secondary side can be adjusted linearly between 0 and 1 changing the duty cycle. Due to the switching process, the voltage across the inductor is stepping each cycle with an amplitude V_{in}. Therefore, it can be useful to connect a filter at the input of the converter to avoid undesired current harmonics.

The considerations up to this point were made for the continuous-conduction mode (CCM) of a buck converter, where the inductor current is always positive and never reaches zero. Now, the boundary between continuous- and discontinuous-conduction mode is examined, where the inductor current reaches zero at the end of every time period T_s. Fig. 8.7a shows the relevant waveforms. Subscript B stands for the boundary condition of the mean inductor

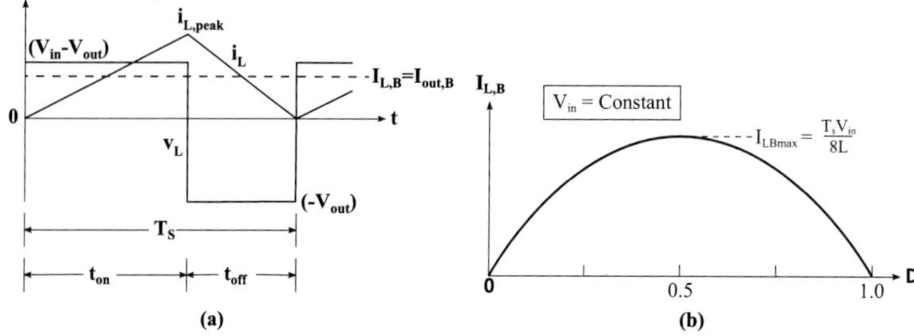

Figure 8.7 Buck converter in boundary between discontinuous-(DCM) and continuous-conduction mode (CCM). (a) Inductor voltage and current profile (b) $I_{L,B}$ as a function of D with $V_{in} = const$ [2].

current I_L. Since the waveform of the inductor current i_L is triangular it can be immediately followed that $I_{L,B} = \frac{1}{2} i_{L,peak}$. With $i_L = \frac{1}{L} \int v_L \, dt$ it is further calculated:

$$I_{L,B} = \frac{t_{on}}{2L}(V_{in} - V_{out}) = \frac{DT_s}{2L}(V_{in} - V_{out}) = I_{out,B} \qquad (8.12)$$

If the input voltage V_{in} is kept constant, V_{out} can be substituted in Eq. (8.12) with DV_{in} from Eq. (8.11). This results in

$$I_{L,B} = \frac{T_s V_{in}}{2L} D(1 - D), \qquad (8.13)$$

which is plotted in Fig. 8.7b as a function of the duty cycle D. The maximum of $I_{L,B}$ is reached at

$$I_{L,Bmax} = \frac{T_s V_{in}}{8L}, \qquad (8.14)$$

as displayed in the diagram. If the output current I_{out} drops below $I_{L,B}$, the inductor current will become discontinuous, that is, it will stay zero for a certain time before rising again.

The waveforms in continuous-conduction mode (CCM) are shown again in Fig. 8.8a, whereas the waveforms in discontinuous-conduction mode (DCM) are shown in Fig. 8.8b. In DCM, each period is divided into three time intervals DT_s, where the switch is closed, $D_1 T_s$, where the switch is opened and $D_2 T_s$, where the switch is still open and the inductor current stays at zero. At first, it should be considered that the input voltage V_{in} is constant. This is the typical case in a motor speed control for example.

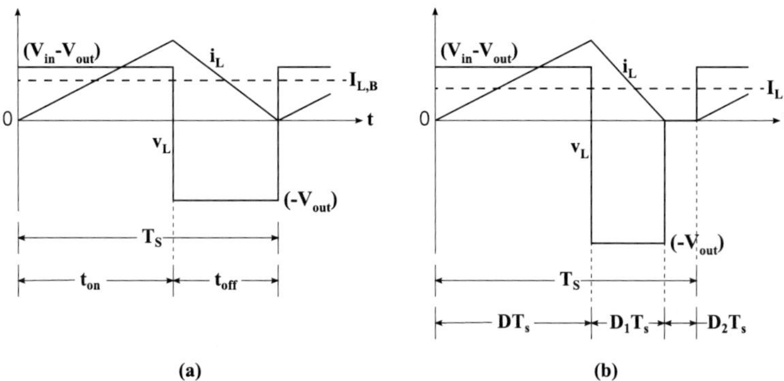

Figure 8.8 Waveforms of buck converter in boundary between CCM and DCM (a) compared to waveforms in DCM (b) [2].

At the boundary between CCM and DCM, $V_{out} = DV_{in}$ and Eq. (8.14) can be introduced into Eq. (8.12), leading to

$$I_{L,B} = \frac{DT_s}{2L}(V_{in} - V_{out}) = \frac{T_s D}{2L}(1 - D)V_{in} = 4I_{L,Bmax}D(1 - D). \quad (8.15)$$

In discontinuous mode again the integral over one time period (see Fig. 8.8) must be zero:

$$\int_0^{DT_s} v_L\, dt + \int_{DT_s}^{D_1 T_s} v_L\, dt + \int_{D_1 T_s}^{D_2 T_s} v_L\, dt = 0. \quad (8.16)$$

Simplifying the last equation

$$DT_s(V_{in} - V_{out}) - D_1 T_s V_{out} + 0 = 0 \Rightarrow DT_s V_{in} - V_{out} T_s(D + D_1) = 0$$

leads to

$$\frac{V_{out}}{V_{in}} = \frac{D}{D + D_1} \quad (8.17)$$

with $D + D_1 < 1$ (DCM). Looking at Fig. 8.8, it can be derived that

$$i_{L,peak} = V_{out}\frac{D_1 T_s}{L}. \quad (8.18)$$

Combining Eq. (8.18), Eq. (8.17) and Eq. (8.14), the calculation of the mean output current I_{out} is done:

$$I_{out} = i_{L,peak}\frac{D + D_1}{2} = \frac{V_{out} T_s}{2L}(D + D_1)D_1$$

$$= \frac{V_{in} T_s}{2L}DD_1 = 4I_{L,Bmax}DD_1 \quad (8.19)$$

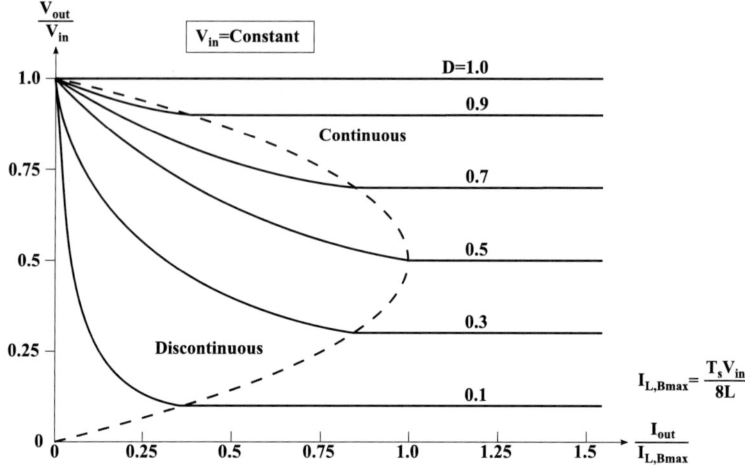

Figure 8.9 Limits between CCM and DCM of a buck converter with constant input voltage V_{in} [2].

Rearranging terms gives

$$D_1 = \frac{I_{out}}{4 I_{L,Bmax} D} \tag{8.20}$$

Furthermore, using Eq. (8.17) it can be finally stated that

$$\frac{V_{out}}{V_{in}} = \frac{D^2}{D^2 + \frac{I_{out}}{4 I_{L,Bmax}}}. \tag{8.21}$$

In Fig. 8.9, various curves of the output voltage V_{out} are plotted versus the output current I_{out} for different duty cycles. V_{out} is normalized to the fixed input voltage V_{in} and I_{out} to the maximum mean inductor current $I_{L,Bmax} = \frac{T_s V_{in}}{8L}$. The dashed line marks the boundary between CCM and DCM.

Now, the situation is analyzed for a constant output voltage of the buck converter V_{out}. That would be the desired case if, for example, the rectified AC output of a piezoelectric generator is used as input for the regulator. With $V_{in} = \frac{V_{out}}{D}$ introduced into $I_{L,B} = \frac{D T_s}{2L}(V_{in} - V_{out})$ from Eq. (8.12), the following is obtained:

$$I_{L,B} = \frac{T_s V_{out}}{2L}(1 - D) \tag{8.22}$$

and with a constant V_{out} the maximum mean inductor current $I_{L,B}$ is reached at

$$I_{L,Bmax} = \frac{T_s V_{out}}{2L}. \tag{8.23}$$

In this way, Eq. (8.22) can be stated as follows:

$$I_{L,B} = (1 - D)I_{L,Bmax}. \tag{8.24}$$

It has to be mentioned that the operation at a duty cycle D equal to zero—whereas V_{out} is greater than zero—is only possible in theory since that would require an infinite input voltage V_{in}. For the functionality of the converter at constant V_{out}, it is important to express the duty cycle D as a function of $I_{out}/I_{L,Bmax}$. For DCM, Eqs. (8.17) and (8.19) can also be used for a constant V_{out}. Combining Eq. (8.23) and Eq. (8.19), I_{out} can be reduced to the following:

$$I_{out} = I_{L,Bmax}(D + D_1)D_1 \tag{8.25}$$

and rearranging terms in Eq. (8.17), D_1 can be expressed as follows:

$$D_1 = D\left(\frac{V_{in}}{V_{out}} - 1\right). \tag{8.26}$$

Inserting Eq. (8.26) into Eq. (8.25) and rearranging terms gives the following expression for D:

$$D = \frac{V_{out}}{V_{in}}\sqrt{\frac{I_{out}/I_{L,Bmax}}{1 - V_{out}/V_{in}}}. \tag{8.27}$$

This function is plotted in Fig. 8.10 for the ratio between V_{out} and V_{in} of 1.25, 2.0 and 5.0. The dashed line again shows the boundary condition between DCM and CCM.

8.2.1.4 Efficiency calculation

A detailed analysis of the efficiency of a buck converter has been done by Gildersleeve et al. [3]. In Fig. 8.11, a typical design of a synchronous step-down converter with a load is shown. It is similar to the circuit in Fig. 8.3, but the diode is replaced by a second switching transistor that diminishes the losses due to the diode forward voltage. Additionally, the parasitic elements are displayed that account for the power consumed by the converter itself. The specific losses in the electrical elements of the converter are considered in the next passages to estimate the efficiency of a step-down converter.

The load current I_{Load} (see Fig. 8.11) in a buck converter is identical with the mean inductor current I_L. It produces losses due to

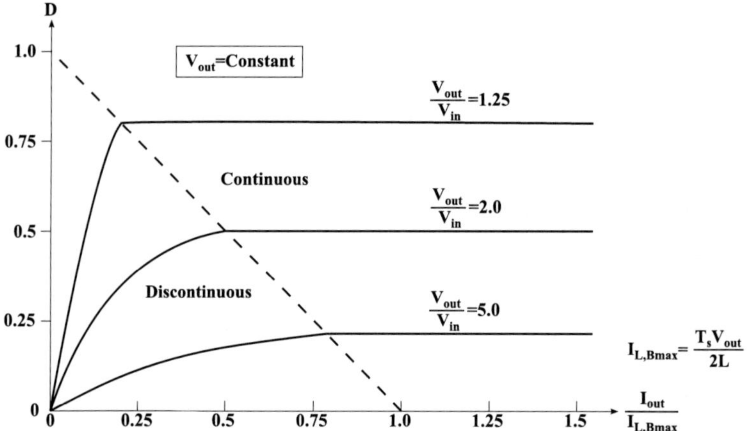

Figure 8.10 Duty cycle as a function of the ratio of I_out to $I_{LB,max}$ for different values of $\frac{V_\text{out}}{V_\text{in}}$ [2].

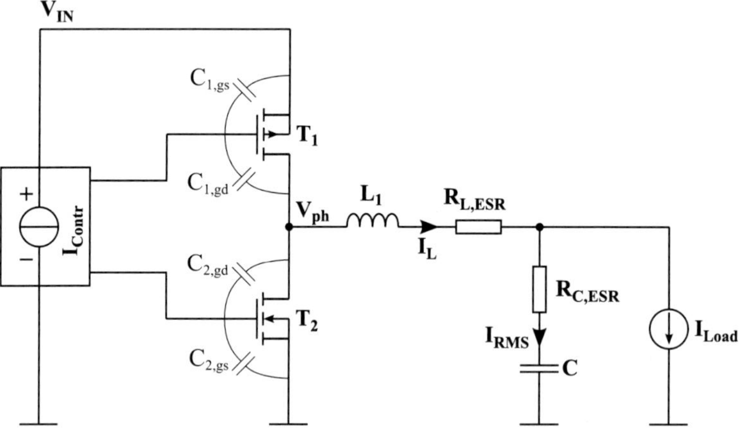

Figure 8.11 Synchronous step-down converter with parasitic elements.

the equivalent series resistance (ESR) of inductor L and the on-state resistances of the switches. The resulting power dissipation can be expressed as follows:

$$P_{\text{I,Load}} = (R_{\text{L,ESR}} + R_{T\,1/2}) \cdot I_{\text{Load}}^2 \tag{8.28}$$

For simplicity, the on-state resistances are considered equally for both transistors (either T_1 or T_2 is conducting).

The RMS value I_{RMS} of the ripple inductor current—which is the RMS value of the AC component of the transient inductor current i_{L}—produces additional power losses. These power losses are dissipated in the capacitor, the inductor, and the switches. Also, the dead times of the switches create power losses through their body diodes, but this is not taken into consideration here. Therefore, the RMS losses can be summarized in the following equation:

$$P_{\text{I,RMS}} = (R_{\text{L,ESR}} + R_{\text{C,ESR}} + R_{T1/2}) \cdot I_{\text{RMS}}^2. \tag{8.29}$$

Additionally, there are power losses due to the current consumption of the controller for the switching transistors. One part is due to its quiescent current and the second part is due to the current for charging and discharging the parasitic gate capacitors C_{gs} and C_{gd} (see Fig. 8.11). The quiescent current loss of the controller together with the gate drive switching losses can be estimated as [4]:

$$P_{\text{Contr}} \approx I_{\text{Contr}} \cdot V_{\text{in}} + (Q_{\text{g,T1}} + Q_{\text{g,T2}}) \cdot V_{\text{in}} \cdot f_{\text{s}} \tag{8.30}$$

with the switching frequency f_{s} and the gate charge $Q_{\text{g,T1}}$ and $Q_{\text{g,T2}}$ required to charge and discharge C_{gs} and C_{gd}. Furthermore, there are switching losses due to the voltage and current overlap of the transistors T_1 and T_2. This can also be approximated as follows:

$$P_{\text{SW}} \approx V_{\text{in}} \cdot I_{\text{Load}} \cdot t_{\text{sum}} \cdot f_{\text{s}}, \tag{8.31}$$

where the rise and fall times of current and voltage at V_{ph} in Fig. 8.11 are summarized to t_{sum}. Finally, adding the above equations to $P_{\text{loss}} = P_{\text{I,Load}} + P_{\text{Q,Contr}} + P_{\text{SW}}$ the total power loss of a step-down converter can be estimated. It can be noticed that at light loads the losses due to controller quiescent current as well as gate charging and discharging become significant, whereas for heavy load currents the dominant losses are due to voltage and current overlap as well as ESR resistances. Besides, there is, of course, a dependence on the switching frequency of the converter. At high frequencies, the losses because of voltage and current overlap and gate drive increase.

8.2.1.5 Design optimization

Optimizing a buck converter is in general a trade-off between geometrical size and efficiency because the inductor is usually the biggest part of the circuit. If it is big in size, then the inductance can be high with a low ESR at the same time. Therefore, the switching frequency can be low leading to a high efficiency due to the assumptions made in the last paragraph. If the inductor is small, the inductance also has to be small to avoid degrading the ESR. Therefore, the switching frequency has to be chosen higher and something has to be done to limit the frequency dependent losses. Important techniques to achieve this—which may not be easy to implement though—are zero-voltage switching (ZVS) and zero-current switching (ZCS) architectures. These designs try to eliminate the voltage-current overlap during switching transitions because the transistors are switched when either the voltage or the current is zero. In the easiest case, only a capacitor is added between V_{ph} and ground in Fig. 8.11. Each transition is controlled by the inductor, which acts like a current source that charges and discharges the parasitic capacitances of the switching transistors, but this is not considered in detail here. For example, Bill Andreycak [5] describes a zero-voltage-switching method in a buck converter design and also extends this technique to other converter topologies. Another article of A. K. Panda et al. [6] can be found, were zero-voltage-switching is implemented using additional active components.

8.2.2 Boost Converter

8.2.2.1 Physical principles

A step-up (boost) converter is a useful circuit, if a bigger voltage from an input source is needed. Concerning energy harvesting transducers, it is an important device for thermogenerators and inductive generators, where the output voltage is typically lower than 1 V. Fig. 8.12 shows the basic function of a boost converter.

In the first phase (upper circuit in Fig. 8.12), the switch is closed and the inductor current is rising ideally linear and therefore, the energy stored in the inductor is also rising. In the second phase (lower circuit in Fig. 8.12), the switch is opened and the energy stored in the inductor is transferred to the output where a second energy storage element—namely capacitor C—is connected. The current change due to the opened switch S induces an inductor voltage that is added to the input voltage. The two phases alternate

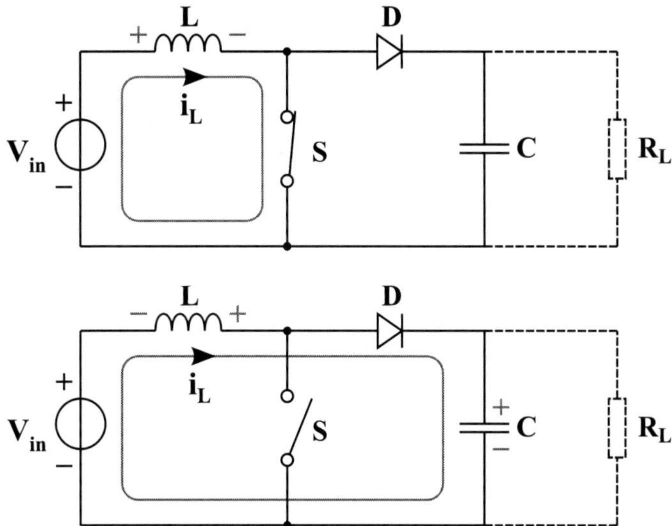

Figure 8.12   Physical principle of a step-up converter.

periodically and in that way, the output voltage is always higher than or equal to the input voltage of the converter.

8.2.2.2 Electrical circuit

A typical block diagram of the step-up converter is shown in Fig. 8.13, which is very similar to the one of the buck converter in Fig. 8.4. The regulation loop works in the same manner as the one described for the buck converter in Section 8.2.1.

8.2.2.3 Analytical model

The following calculations for a boost converter are based on the analysis of Mohan et al. [2]. For the circuits of Fig. 8.12, the waveforms in steady-state for the continuous conduction mode of a boost converter are displayed in Fig. 8.14.

As with the step-down converter the areas A and B have to be equal assuming a lossless circuit. Therefore,

$$V_{in}(T_s - t_{off}) + (V_{in} - V_{out})t_{off} = 0 \Leftrightarrow \frac{V_{out}}{V_{in}} = \frac{T_s}{t_{off}} = \frac{1}{1 - D}. \quad (8.32)$$

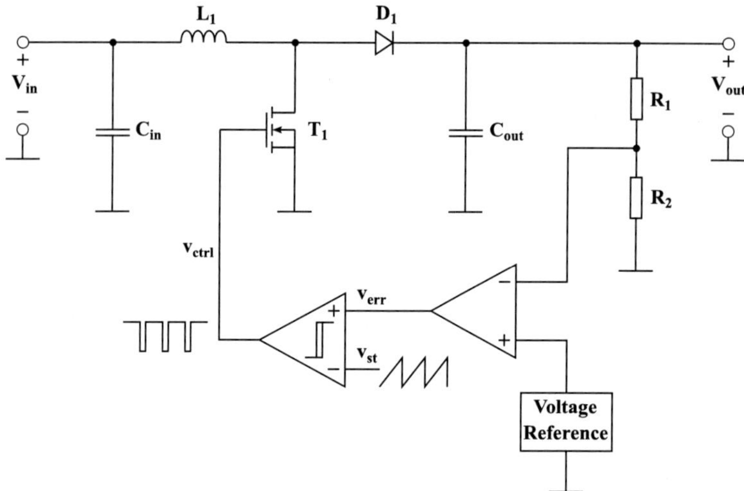

Figure 8.13   Block diagram of a step-up converter.

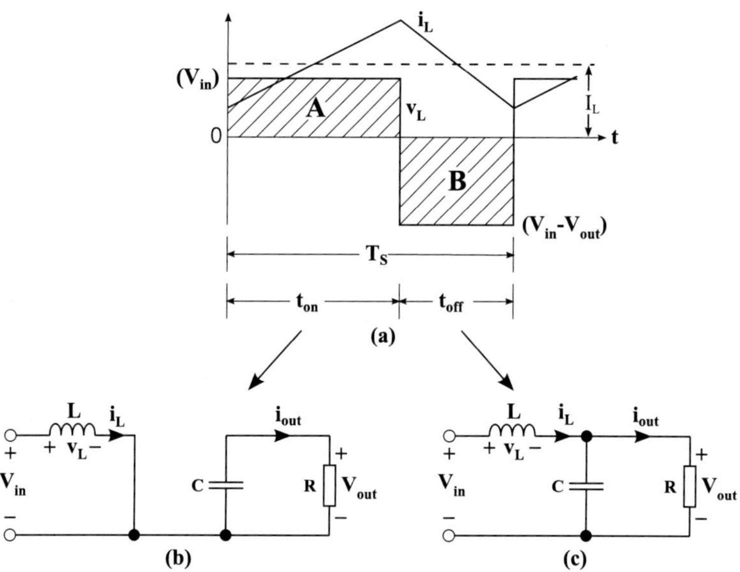

Figure 8.14   Boost converter circuit for closed (a) and opened (b) switch and corresponding waveforms (c) in continuous conduction mode [2].

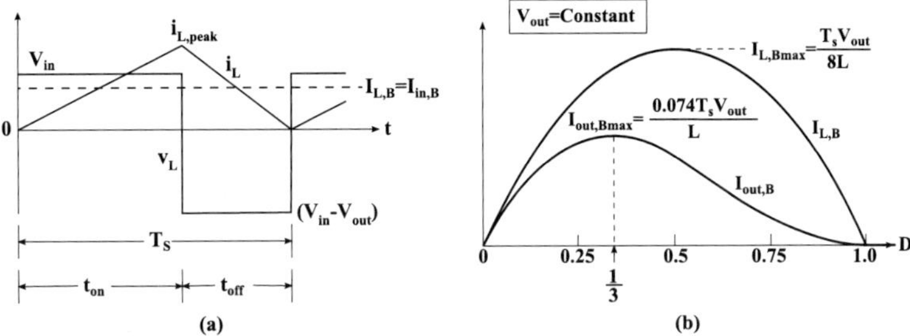

Figure 8.15 Boost converter in boundary between discontinuous-(DCM) and continuous-conduction mode (CCM). (a) Inductor voltage and current profile (b) $I_{L,B}$ as a function of D with (V_{out} = const) [2].

Since a lossless circuit is assumed, the input power is equal to the output power, $P_{in} = P_{out}$. Therefore, with Eq. (8.32) it can be stated that

$$V_{in}I_{in} = V_{out}I_{out} \Leftrightarrow \frac{I_{out}}{I_{in}} = 1 - D. \tag{8.33}$$

For the boundary condition between CCM and DCM, Fig. 8.15a shows the waveforms in the steady state of inductor current i_L and voltage v_L. The mean inductor current $I_{L,B}$ can be calculated as follows:

$$I_{L,B} = \frac{1}{2}i_{L,peak} = \frac{1}{2}\frac{V_{in}}{L}t_{on} = \frac{T_s V_{out}}{2L}D(1 - D). \tag{8.34}$$

Since in a boost converter the inductor current equals the input current ($i_L = i_{in}$), it can be furthermore stated for the output current $I_{out,B}$ with Eq. (8.33):

$$I_{out,B} = \frac{T_s V_{out}}{2L}D(1 - D)^2. \tag{8.35}$$

An important case for practical boost converter designs is that the output voltage should be constant. Therefore, in Fig. 8.15b the graphs are illustrated for $I_{L,B}$ and $I_{out,B}$ versus the duty cycle D for a constant V_{out}. The maximum values for $I_{L,B}$ and $I_{out,B}$ are found at

$$D = 0.5 \Leftrightarrow I_{L,Bmax} = \frac{T_s V_{out}}{8L}, \tag{8.36}$$

$$D = \frac{1}{3} \Leftrightarrow I_{out,Bmax} = \frac{2}{27}\frac{T_s V_{out}}{L} = \frac{0.074 T_s V_{out}}{L}. \tag{8.37}$$

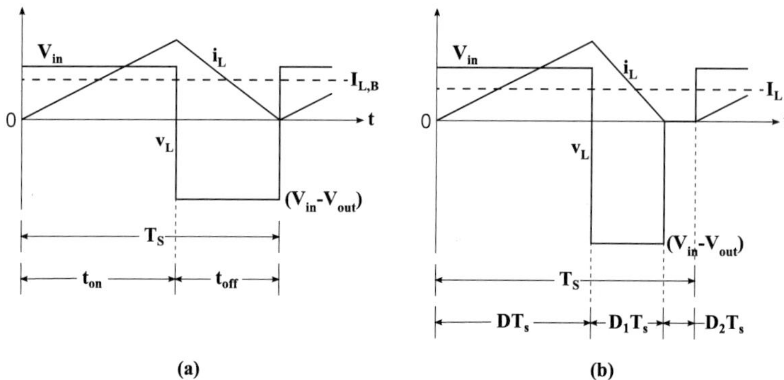

Figure 8.16 Inductor voltage and current profile for a boost converter in (a) CCM and (b) DCM with $(V_{in} = \text{const})$ [2].

Finally, Eqs. (8.34) and (8.35) can be written as

$$I_{L,B} = \frac{1}{4} I_{L,Bmax} D(1 - D), \tag{8.38}$$

$$I_{out,B} = \frac{27}{4} I_{out,Bmax} D(1 - D)^2. \tag{8.39}$$

From Fig. 8.15, it can be furthermore observed that if the output current drops below $I_{out,B}$, the inductor current i_L will become discontinuous. The discontinuous-conduction mode of a step-up converter is considered in more detail in the next passages.

In Fig. 8.16 the waveforms of inductor current and voltage for CCM (a) and DCM (b) are shown. For both, the input voltage V_{in} and the duty cycle D are constant, but the output power in (b) is lower. That happens, for example, when the load resistance R is rising. Therefore, in (b) $I_L = I_{out}$ is decreasing and the inductor current becomes discontinuous. In other words, after going down to zero the inductor current remains at zero for the time $D_2 T_s$ before rising again in the next cycle. At the same time the output voltage V_{out} increases due to the faster transition to zero of the inductor current during the time interval $D_1 T_s$. Integrating again the inductor voltage i_L of Fig. 8.16b, which has to be zero for one cycle, leads to

$$DT_s V_{in} + D_1 T_s (V_{in} - V_{out}) = 0 \Leftrightarrow \frac{V_{out}}{V_{in}} = \frac{D_1 + D}{D_1} \tag{8.40}$$

and with $P_{in} = P_{out}$, the ratio between I_{out} and I_{in} is found:

$$\frac{I_{out}}{I_{in}} = \frac{D_1}{D_1 + D}. \tag{8.41}$$

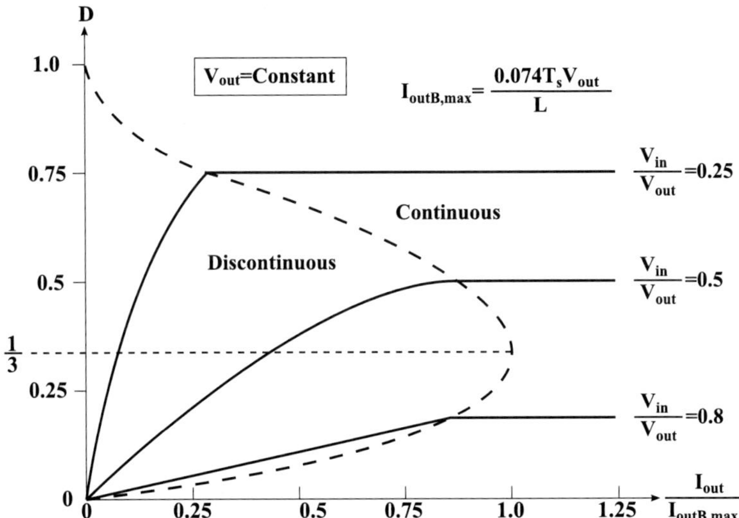

Figure 8.17 Limits between DCM and CCM (dashed line) for a boost converter with ($V_{\text{out}} = \text{const}$) [2].

Since $I_{\text{in}} = \frac{1}{2}i_{\text{L,peak}}(D + D_1)$ (from Fig. 8.16) and $i_{\text{L,peak}} = \frac{1}{2L}\int_0^{DT_s} V_{\text{in}}\, dt$ it follows that

$$I_{\text{in}} = \frac{V_{\text{in}}}{2L} DT_s(D + D_1). \tag{8.42}$$

Introducing Eq. (8.41) into the previous equation yields

$$I_{\text{out}} = \frac{V_{\text{in}} T_s}{2L} DD_1. \tag{8.43}$$

For practical boost converter designs it is generally desired to have a constant output voltage V_{out} at a varying input voltage V_{in} and output current I_{out}. Therefore, it is necessary to adjust the duty cycle. Combining Eqs. (8.37), (8.40), and (8.43), the following is obtained:

$$D = \sqrt{\frac{4}{27}\frac{V_{\text{out}}}{V_{\text{in}}}\left(\frac{V_{\text{out}}}{V_{\text{in}}} - 1\right)\frac{I_{\text{out}}}{I_{\text{out,Bmax}}}}. \tag{8.44}$$

The duty cycle D is plotted in Fig 8.17 versus the current ratio $I_{\text{out}}/I_{\text{out,Bmax}}$ for different ratios of input to output voltage ratios ($V_{\text{in}}/V_{\text{out}} = 0.25, 0.5, 0.8$).

If the duty cycle D would be fixed and the load would be very light in DCM then the output voltage would become very high due to

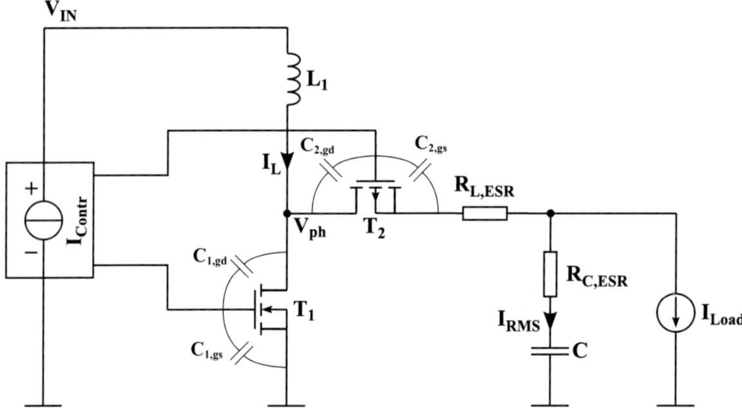

Figure 8.18 Synchronous step-up converter with parasitic elements.

$\frac{V_{out}}{V_{in}} = \frac{I_{in}}{I_{out}}$. In a practical design, this case could lead to an overvoltage on the output capacitor.

8.2.2.4 Efficiency calculation

The efficiency calculation of Gildersleeve et al. [3] can be used also for a synchronous boost converter design of Fig. 8.18. It is obvious that the same parasitic elements as in the buck converter design of Fig. 8.11 are present. Moreover, the occurring gate-source and gate-drain voltages are basically the same. The difference for the calculation of the efficiency of a step-up converter compared to the calculation of a buck converter is that the average input current I_{in} determines the ESR losses $P_{I,In}$ and the switching losses P_{SW}.

Similarly, as shown in Section 8.2.1, the power losses for the step-up converter from Fig. 8.18 can be stated as follows:

$$P_{I,In} = (R_{L,ESR} + R_{T1/2}) \cdot I_{In}^2 \tag{8.45}$$

$$P_{I,RMS} = (R_{L,ESR} + R_{C,ESR} + R_{T1/2}) \cdot I_{RMS}^2 \tag{8.46}$$

$$P_{Q,Contr} \approx I_{Contr} \cdot V_{In} + 16 \cdot C_{gs} \cdot V_{In}^2 \cdot \frac{f_s}{3} \tag{8.47}$$

$$P_{SW} \approx V_{In} \cdot I_{In} \cdot t_{sum} \cdot f_s \tag{8.48}$$

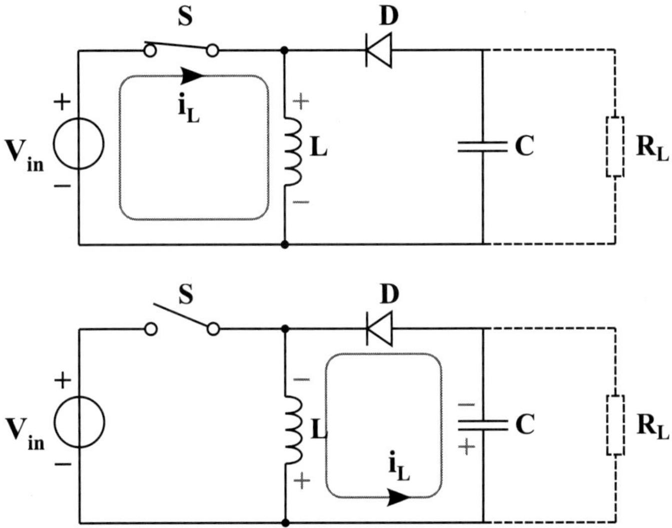

Figure 8.19 Basic circuitry and operation principle of a buckboost converter.

8.2.3 *Buck-Boost Converter*

8.2.3.1 Physical principles

The physical principle of a buck-boost converter is shown in Fig. 8.19. When switch S is closed, the inductor is connected to the input voltage V_{in} and the inductor current i_l is rising linearly. Afterward, when the switch is opened, i_L is falling and the inductor voltage changes its polarity. Now diode D is conducting and capacitor C is charged to the negative inductor voltage. Thus, the buck-boost converter is useful when a negative voltage compared to an input voltage is needed. Additionally, this topology is capable of generating any output voltage bigger or smaller than the input voltage.

8.2.3.2 Electrical circuit

The block diagram of a buck-boost converter is shown in Fig. 8.20 similar to the ones of Sections 8.2.1 and 8.2.2. In the regulation loop, the supply connections of error amplifier and comparator are shown

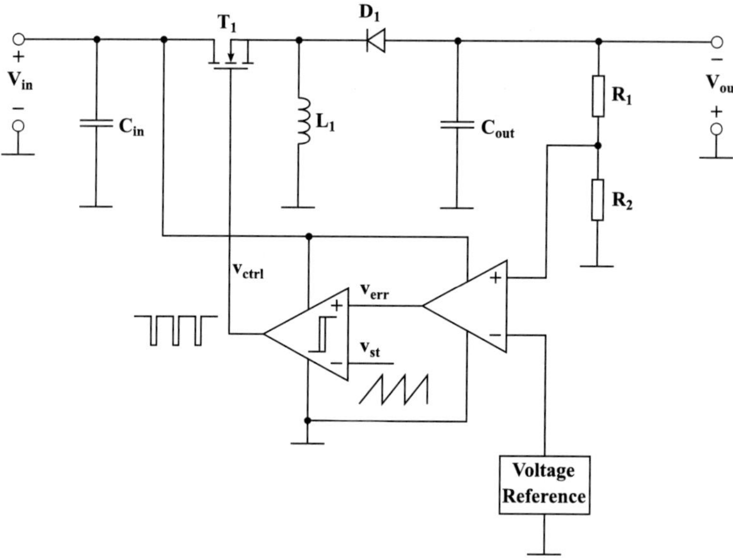

Figure 8.20 Block diagram of a buck-boost converter.

as well, since they have to be connected to the input voltage in this case. In fact, the supply of error amplifier and comparator cannot be connected to the output voltage V_{out}, because it is negative compared to ground (see Fig. 8.20).

8.2.3.3 Analytical model

In this section, an analysis of the buck-boost converter is done as shown by Mohan et al. [2]. The buck-boost converter can be figured as a cascade connection between buck and boost converter (see Sections 8.2.1 and 8.2.2). Therefore, the voltage conversion ratio of a buck-boost converter can be obtained by multiplying Eq. (8.8) and Eq. (8.32):

$$\frac{V_{out}}{V_{in}} = \frac{D}{1 - D}. \tag{8.49}$$

In consequence, the output voltage V_{out} can be higher or lower than the input voltage V_{in} depending on the duty cycle D. The circuit in Fig. 8.19 is the simplification of the cascade connection between buck and boost converter. Fig. 8.21 shows the waveforms

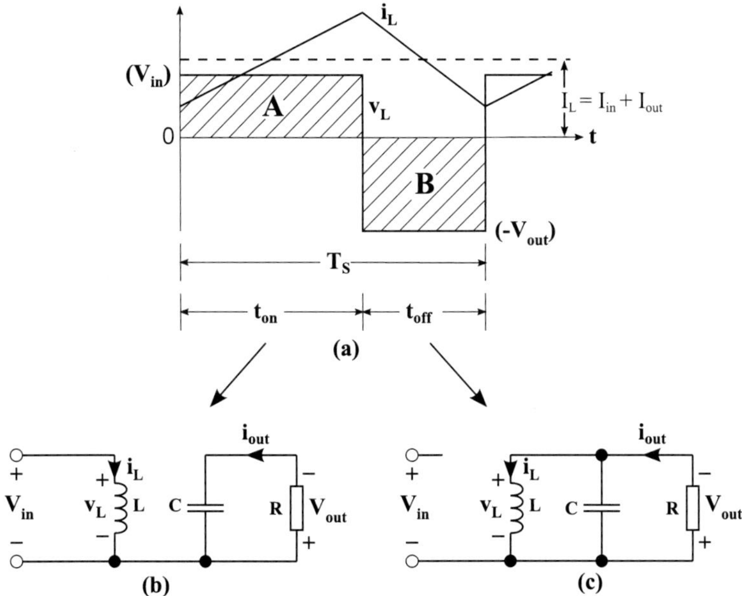

Figure 8.21 Model of a buck-boost converter in CCM [2].

in continuous-conduction mode. The integral of the inductor voltage v_l of one switching period has to be zero, resulting in Eq. (8.49):

$$V_{in} D T_s + (-V_{out})(1 - D)T_s = 0 \Leftrightarrow \frac{V_{out}}{V_{in}} = \frac{D}{1 - D}. \tag{8.50}$$

For the current conversion ratio, the following is obtained (power losses are neglected, $P_{in} = P_{out}$):

$$\frac{I_{out}}{I_{in}} = \frac{1 - D}{D}. \tag{8.51}$$

The waveforms of a buck-boost converter at the boundary of CCM and DCM are displayed in Fig. 8.22a. Here, the inductor current goes to zero at the end of every period. The mean inductor current $I_{L,B}$ (see Fig. 8.22a) can be calculated as follows:

$$I_{L,B} = \frac{1}{2} i_{L,peak} = \frac{T_s V_{in}}{2L} D. \tag{8.52}$$

Due to the assumption that the average capacitor current is zero (capacitor C has no Ohmic losses) (see Fig. 8.21), it can be stated that

$$I_{out} = I_L - I_{in}. \tag{8.53}$$

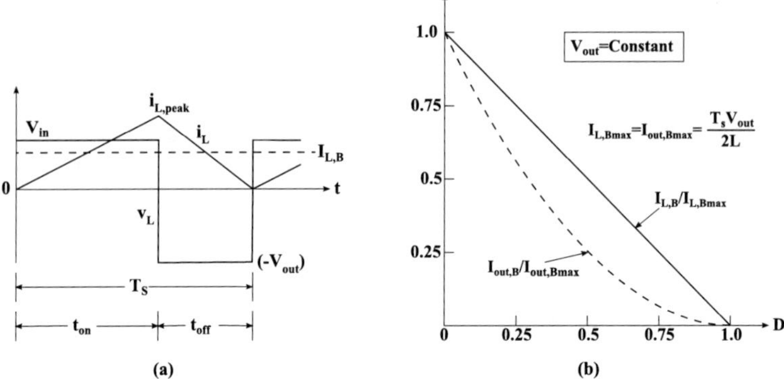

Figure 8.22 Buck-boost converter in boundary between discontinuous-(DCM) and continuous-conduction mode (CCM). (a) Inductor voltage and current profile (b) $I_{L,B}$ versus D with (V_{in}, $D = $ const) [2].

Introducing Eq. (8.50) into Eq. (8.52) yields

$$I_{L,B} = \frac{T_s V_{out}}{2L}(1 - D) \tag{8.54}$$

and furthermore using Eq. (8.51) and (8.52) results in the average output current of the converter $I_{out,B}$:

$$I_{out,B} = \frac{T_s V_{out}}{2L}(1 - D) - \frac{D I_{out,B}}{1 - D}$$

$$\Rightarrow \frac{I_{out,B}(1 - D) + D I_{out,B}}{1 - D} = \frac{T_s V_{out}}{2L}(1 - D)$$

$$\Rightarrow \frac{I_{out,B}}{1 - D} = \frac{T_s V_{out}}{2L}(1 - D)$$

$$\Rightarrow I_{out,B} = \frac{T_s V_{out}}{2L}(1 - D)^2. \tag{8.55}$$

In many applications, it is necessary that the buck-boost converter deliver a constant output voltage V_{out}. Equations (8.54) and (8.55) show that average inductor and output current $I_{L,B}$, $I_{out,B}$ in CCM are maximum when the duty cycle D is equal to zero:

$$I_{L,B} = I_{out,B} = \frac{T_s V_{out}}{2L}. \tag{8.56}$$

Therefore, $I_{L,B}$ and $I_{out,B}$ can be normalized as follows to their maximum values using Eq. (8.54) through (8.56):

$$I_{L,B} = I_{L,Bmax}(1 - D), \tag{8.57}$$

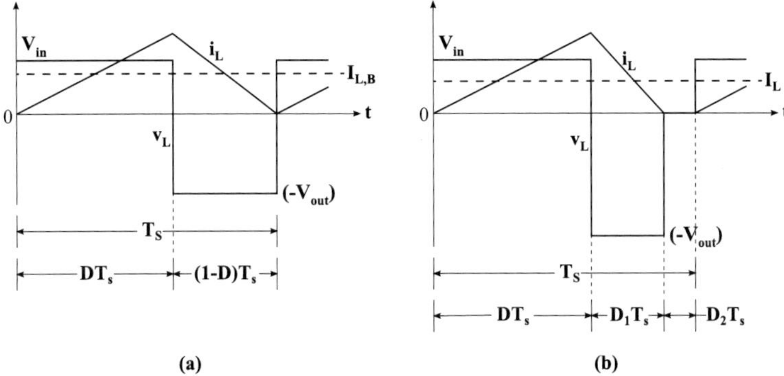

Figure 8.23 Waveforms of buck-boost converter in boundary between CCM and DCM (a) compared to waveforms in DCM (b) [2].

$$I_{out,B} = I_{out,Bmax}(1 - D)^2. \tag{8.58}$$

$I_{L,B}$ and $I_{out,B}$ are displayed as a function of the duty cycle D in Fig. 8.22b, when the output voltage V_{out} is held constant. Fig. 8.23b now shows the waveforms in discontinuous-conduction mode compared to the ones in Fig. 8.23a in continuous-conduction mode.

The integral of the inductor voltage v_L (see Fig. 8.23b) is equal to zero and therefore it can be calculated:

$$V_{in}\,DT_s + (-V_{out})\,D_1\,T_s = 0 \Leftrightarrow \frac{V_{out}}{V_{in}} = \frac{D}{D_1} \tag{8.59}$$

and

$$\frac{I_{out}}{I_{in}} = \frac{D_1}{D}. \tag{8.60}$$

When V_{out} is kept constant again and using the formulas from above, D as a function of $\frac{V_{out}}{V_{in}}$ can be expressed as

$$D = \frac{V_{out}}{V_{in}}\sqrt{\frac{I_{out}}{I_{out,Bmax}}}, \tag{8.61}$$

which is plotted in Fig. 8.24 as a function of $\frac{I_{out}}{I_{out,Bmax}}$ for different values of $\frac{V_{out}}{V_{in}}$.

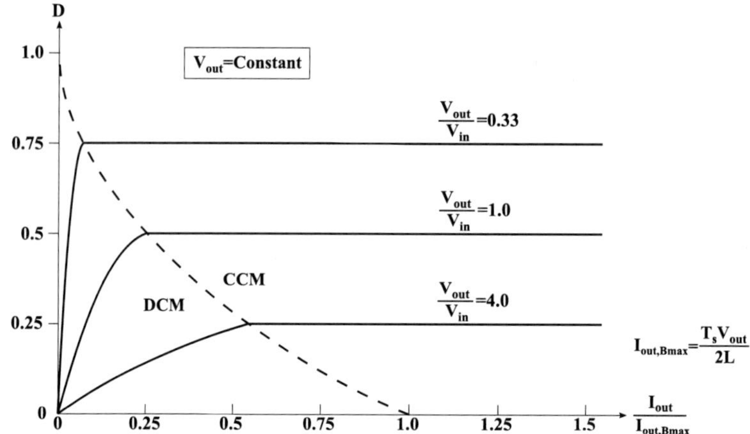

Figure 8.24 Limits between DCM and CCM (dashed line) for a buck-boost converter with (V_{out} = const) [2].

8.2.4 *Flyback Converter*

The flyback converter topology is a buck-boost converter that employs a transformer instead of an inductor. Therefore, it is suitable for switched mode power supplies, for example, where an electrical isolation between input and output is needed. In Fig. 8.25, a basic circuit and its operation is shown. In the first phase switch S is closed and the energy of the input is stored on primary inductor L_1. In the second phase, the energy is passed via the secondary inductor L_2 to output capacitor C and from there to the load R_L.

Like the buck-boost converter, the flyback converter can be used for both up- and down-conversion of the input voltage. Nevertheless, a flyback converter is less useful for energy harvesting applications, because generally no isolation from the energy source to the output is needed. A transformer would even increase the board space of the circuit, because a good coupling between the windings is necessary in form of a closed core. Additionally, two windings—instead of one in case of a single inductor—have to be designed for acceptable low Ohmic losses. The flyback converter can be considered similarly to the buck-boost converter and is not discussed in more detail here. For further information the book "Power Electronics" of Mohan et al. [2] can be helpful.

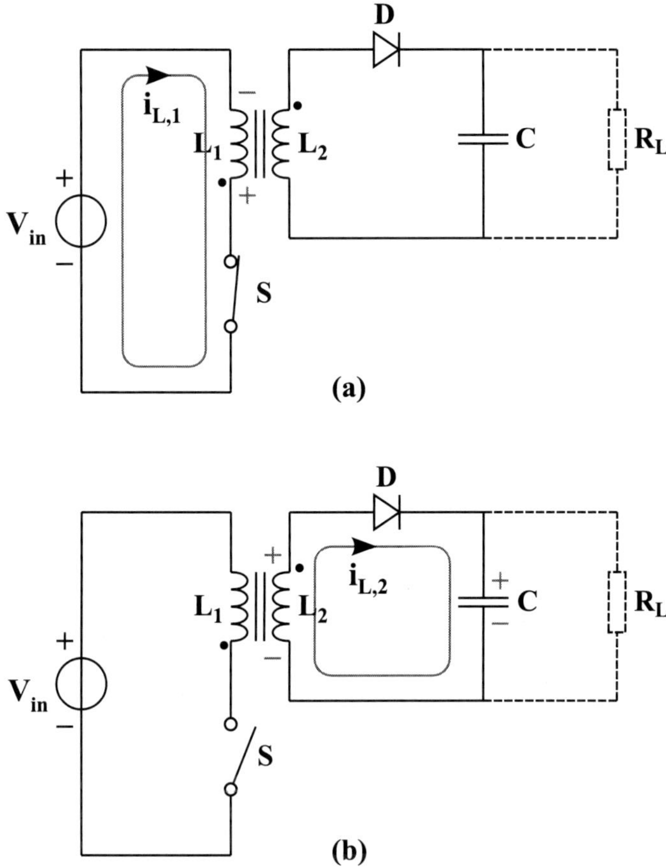

Figure 8.25   Basic circuitry and operation principle of a flyback converter.

8.2.5 *Charge Pump*

8.2.5.1 Physical principles

A charge pump circuit is an alternative to the previously described switching converter topologies that use an inductor or a transformer as an energy storage element. In charge pumps, only those capacitors are used for the task that can be built with a lower ESR when assuming the same switching frequency. That makes them also easier to be integrated on a chip. There are many different concepts

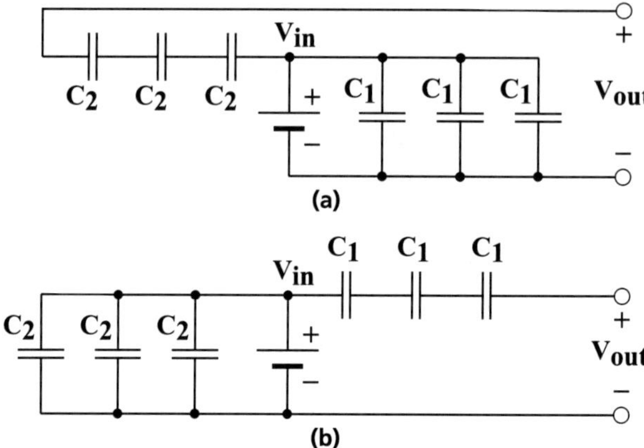

Figure 8.26 General operation principle of a charge pump.

of charge pumps, two of them are described in the following paragraphs.

Fig. 8.26 illustrates a general operation principle of an efficient charge pump design [7] that is divided into two working phases. In phase 1 (a) (see Fig. 8.26), three capacitors C_1 are connected in parallel to a voltage source and charged up to the voltage V_{in}. In phase 2 (b), another three capacitors C_2 are charged up to V_{in}, while the other three are connected in series to the voltage source. This series chain is also connected to the output at that point. Therefore, the output voltage V_{out} is four times the voltage V_{in} of the input source. Although this circuit is not a practical approach of a charge pump—six capacitors for an output voltage of $4 \cdot V_{in}$ are needed—it works efficiently, because the current of the input source is flowing continuously.

Figure 8.27 shows the operation of a four-stage Dickson charge pump that is also divided into two working phases. In phase (a) (see Fig. 8.27), capacitor C_1 is charged by the input source and C_3 is charged by C_2. In phase (b), C_2 is charged by C_1 while C_4 is charged by C_3. In this charge pump concept, only four capacitors are needed to achieve $4 \cdot V_{in}$. Additionally, the stages can be simply cascaded to achieve higher output voltages. Here the input current is only flowing in phase (a) (see Fig. 8.27).

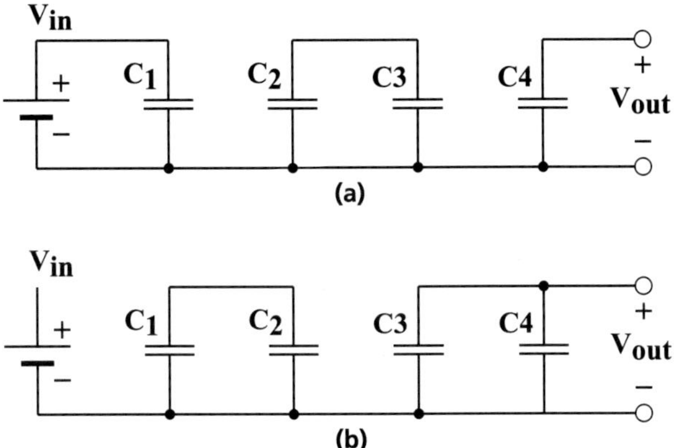

Figure 8.27 Operation principle of a Dickson charge pump.

8.2.5.2 Electrical circuit

In Fig. 8.28, a block diagram of a three-stage Dickson charge pump [8] is shown. It delivers an output voltage V_{out} three times the input voltage V_{dd} and is divided into two separate schematics illustrating the two states at steady state.

In the first phase (see Fig. 8.28a), the output of inverter U_1 is low ($= V_{ss}$) and therefore the output of inverter U_2 is high (V_{dd}). In that way, C_1 is charged via diode D_1 and C_{out} is charged by C_2 via diode D_3. D_2 is not conducting (reverse-biased) at that time. In the second phase (see Fig. 8.28b), the output of inverter U_1 is high ($= V_{dd}$). Thus, the output of inverter U_2 is low (V_{ss}) and C_2 is charged by C_1 via diode D_2 while diode D_1 and D_3 are not conducting. The clock signal V_{clk} is in general provided by a rectangular oscillator supplied by V_{dd}.

Diodes have a certain forward voltage (depending on the type of diode), which degrade the output voltage and consequently the efficiency of a charge pump. Using switched transistors instead of diodes improves the efficiency and the output voltage gets closer to the theoretical value ($3 \cdot V_{in}$ in the concept of Fig. 8.28). In literature, a lot of possibilities to implement this can be found, which lead, of course, to a higher bill of material for the design. For example, Jong-Min Baek et al. [9] published an upconverter for EEPROMs containing an efficient Dickson charge pump. Alternatively, Mensi

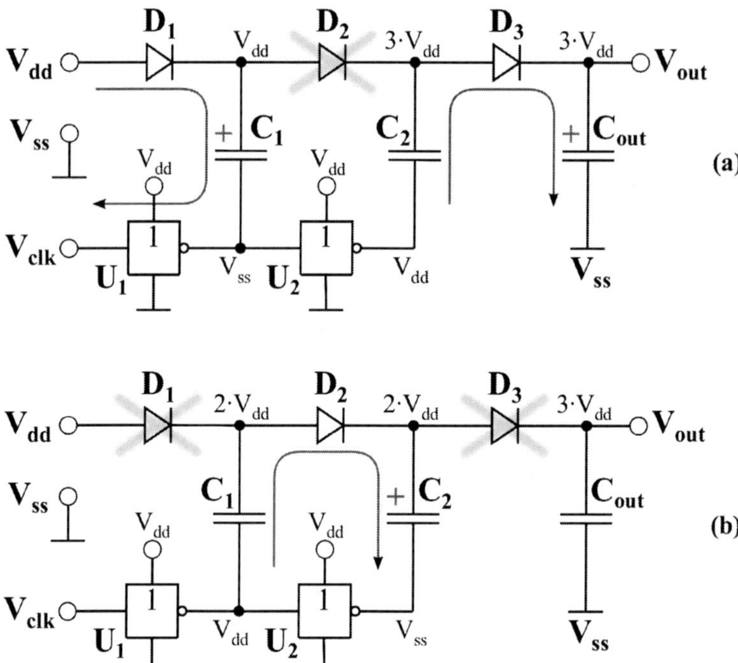

Figure 8.28 Operation of a Dickson charge pump.

et al. [10] designed a charge pump with a high voltage efficiency, integrated on a 90 nm CMOS process.

8.2.5.3 Analytical model

A detailed analysis of a charge pump converter is done by Steensgard et al. [7]. It is started with a simple model of stacked voltage sources—batteries for instance—with an output voltage V_{cell} that typically have an internal resistance R_{cell} (see Fig. 8.29).

With this topology a relative voltage drop β across each (battery) cell can be defined:

$$\beta = \frac{R_{cell} I_{cell}}{V_{cell}}. \tag{8.62}$$

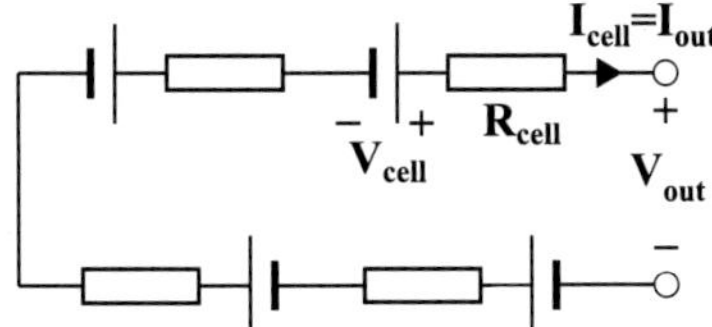

Figure 8.29 Simple model of series connected voltage sources with internal resistance.

The efficiency (voltage source plus internal resistance R_{cell}) is defined as follows:

$$\eta = \frac{P_{load}}{P_{loss} + P_{load}} = \frac{(V_{cell} - R_{cell}I_{cell})I_{cell}}{R_{cell}I_{cell}^2 + (V_{cell} - R_{cell}I_{cell})I_{cell}}$$

$$= \frac{V_{cell} - R_{cell}I_{cell}}{R_{cell}I_{cell} + V_{cell} - R_{cell}I_{cell}}$$

$$= \frac{V_{cell} - R_{cell}I_{cell}}{V_{cell}} = 1 - \frac{R_{cell}I_{cell}}{V_{cell}} \qquad (8.63)$$

Introducing Eq. (8.62) into Eq. (8.63) yields

$$\beta = 1 - \eta. \qquad (8.64)$$

Considering a charge pump circuit, in which capacitors are stacked in series to a voltage source, a maximum relative voltage drop β_{max} can be defined. Fig. 8.30 shows such a constellation at the state, in which the pumping capacitors C_{pp} are fully charged to the voltage V_{dd} across the input source. A voltage regulator is connected to the output of the charge pump to achieve a constant output voltage V_{out} independent from the load connected (see Fig. 8.30). Additionally, a buffer capacitor C_{buf} is present for the state, in which the pumping capacitors are charged and not connected to

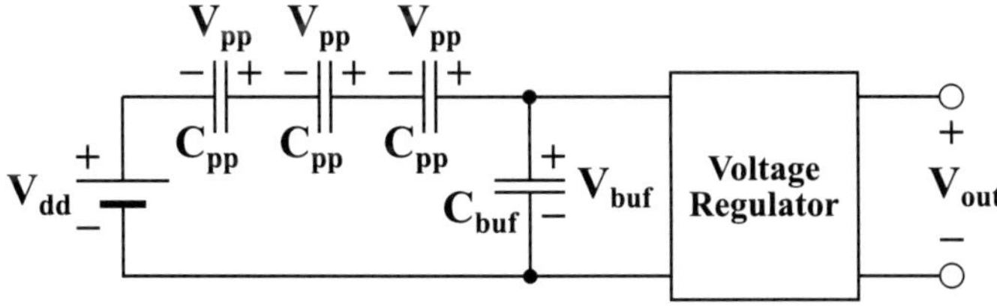

Figure 8.30 Charge pump in cascade with a linear regulator at the state, in which the capacitors C_{pp} are charged to V_{dd}.

the voltage regulator. A linear regulator (see Section 8.1) needs a minimum input voltage to keep the desired output voltage stable. Consequently, the output voltage V_{buf} of the charge pump itself has to be always higher than the output voltage V_{out} of the linear regulator. Therefore, β_{max} is defined as the point, where the output voltage of the charge pump V_{buf} has dropped to the minimum input voltage of the regulator.

The transient voltage V_{pp} across each capacitor C_{pp} (see Fig. 8.30), is now defined as follows:

$$V_{pp}(t) = V_{dd} \cdot (1 - \beta(t)). \tag{8.65}$$

Considering one period of time while the pumping capacitors C_{pp} are discharged with the current

$$I_{pp}(t) = C_{pp}\frac{dV_{pp}(t)}{dt} = C_{pp} \cdot V_{dd} \cdot \left(-\frac{d\beta(t)}{dt}\right), \tag{8.66}$$

the maximum available output current I_{max} is

$$I_{max} = C_{pp} \cdot V_{dd} \cdot \frac{V_{dd} - V_{pp,min}}{T} = C_{pp} \cdot \beta_{max} \cdot V_{dd} \cdot f_s, \tag{8.67}$$

with the switching frequency f_s. The efficiency of a charge pump is calculated in the next paragraphs employing the equations obtained in this subsection.

8.2.5.4 Efficiency calculation

Steensgard et al. [7] calculated the efficiency of a charge pump, employing the voltage doubler of Fig. 8.31, that works in "push-pull" operation. This means that in each phase one pumping capacitor

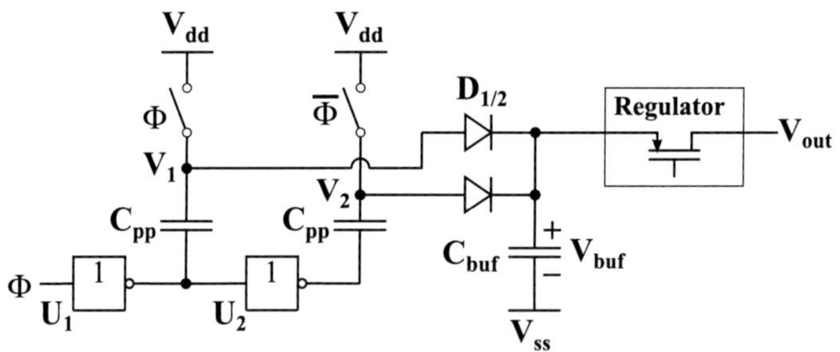

Figure 8.31 Simple charge pump: voltage doubler.

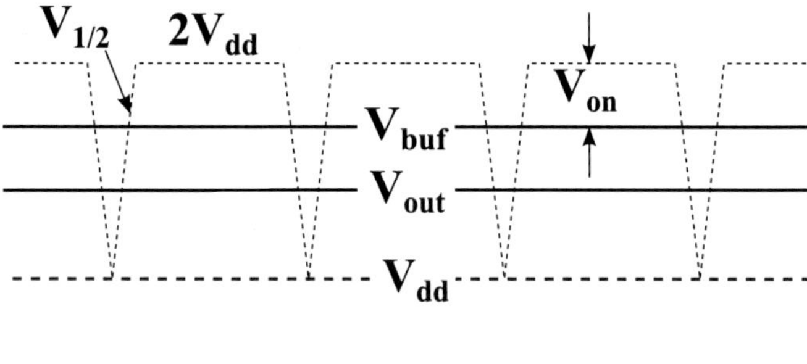

Figure 8.32 Waveforms of the voltage doubler in Fig. 8.31 with output in open-circuit.

C_{pp} is charged while the other one is discharged to the output buffer capacitor C_{buf}. Therefore, the circuit can be understood as an implementation of the concept in Fig. 8.26. ϕ is a rectangular clock signal to control the inverters and switches of the voltage doubler (see Fig. 8.31).

The corresponding waveforms with the output of the regulator in open-circuit are shown in Fig. 8.32. It can be discovered that there is a voltage drop between the maximum of the voltages V_1, V_2 and V_{buf}. The reason is that in reality there is typically a voltage drop V_{on} due to the diodes D_1 and D_2. Also, it can be noticed that V_{buf} is greater than V_{out}. This can be understood regarding Fig. 8.33, where the maximum load is connected at the output of the linear regulator. The output current causes periodic ripples on V_{buf} due to the charge transport from C_{buf} to the output. Additionally, in a real linear regulator, there is always a minimum voltage drop between its input and output. Therefore, the output voltage V_{out} of the regulator is reduced by the forward voltage of the diodes, the maximum voltage ripple on V_{buf} and the minimum voltage drop of the regulator. This implies that the voltage multiplication factor of the charge pump has to be higher than the theoretical value to compensate these effects.

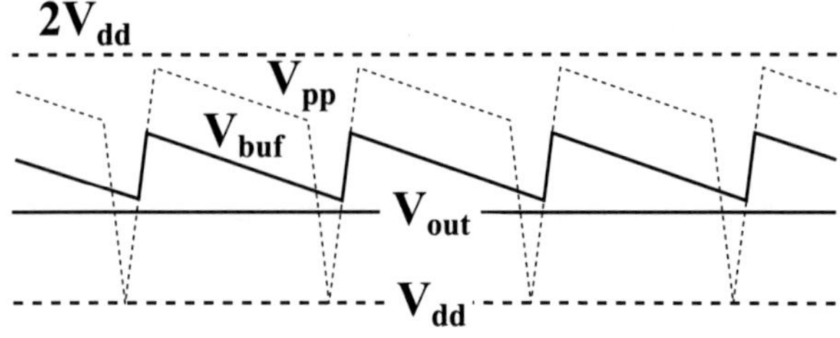

Figure 8.33 Waveforms of the voltage doubler in Fig. 8.31 with maximum load at the output.

For every clock cycle, with $E_{load} = Q \cdot V_{out}$, $E_{supply} = 2 \cdot Q \cdot V_{dd}$ and $Q = C_{pp} \cdot \beta \cdot V_{dd}$, the efficiency is given by:

$$\eta = \frac{E_{load}}{E_{supply}} = \frac{V_{out} \cdot Q}{2 \cdot V_{dd} \cdot Q} = \frac{V_{out}}{2V_{dd}}. \tag{8.68}$$

From this equation it can be derived that the efficiency increases, if the forward voltage drop V_{on} (see Fig. 8.32) of the diodes is eliminated or at least decreased. This can be done exchanging the diodes by switches in form of transistors, for example. In the following calculations, the regulator is considered ideal and switches are used instead of diodes for the voltage doubler of Fig. 8.31. The switches are controlled by the clock signal ϕ (see Fig. 8.34). The output voltage V_{out} as a function of the maximum relative voltage

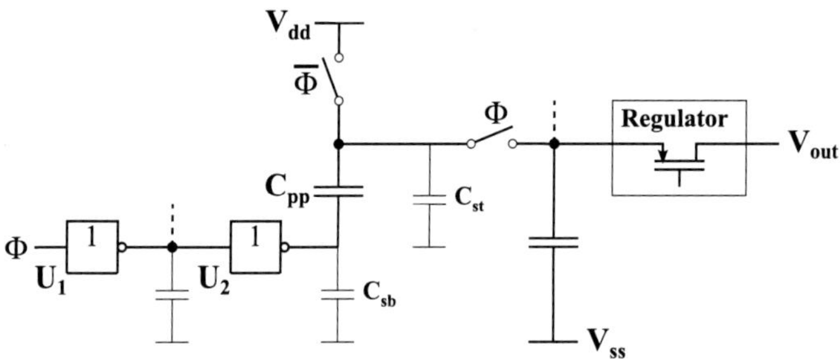

Figure 8.34 Voltage doubler with parasitic elements.

drop β_{max} is

$$V_{out} = V_{buf} = V_{dd} + V_{dd} \cdot (1 - \beta_{max}) = V_{dd} \cdot (2 - \beta_{max}). \qquad (8.69)$$

Introducing this equation into Eq. (8.68) yields

$$\eta = \frac{V_{dd} \cdot (2 - \beta_{max})}{2V_{dd}} = 1 - \frac{\beta_{max}}{2} \qquad (8.70)$$

In conclusion, the efficiency is only a function of β_{max}—considering an ideal regulator, ideal switches, inverter, and capacitors. In Fig. 8.34, stray capacitances C_{sb} are illustrated. These are generally found in integrated circuits, formed by the metal paths and the substrate. It is assumed that C_{sb} includes the parasitic capacitances of the transistors in the inverters. A similar parasitic capacitance (top-plate stray capacitance C_{st}) can be found between the upper supply rail and the common node of the switches (see Fig. 8.34).

In the following equations, the top- and the bottom-plate stray capacitances C_{st} and C_{sb} are expressed as a fraction of the pumping capacitance C_{pp}:

$$C_{st} = \alpha_{st} \cdot C_{pp}, \qquad (8.71)$$

$$C_{sb} = \alpha_{sb} \cdot C_{pp}. \qquad (8.72)$$

There are power losses occuring in the inverters (U_1 and U_2, (see Fig. 8.34) that mainly depend on the driver "strength." The driver strength is determined by the maximum output current of digital gates. Since the transistors of the output stages of digital gates work like a resistor in series to the capacitances present at their output, the driver strength determines the current and thus the time it takes until the output capacitance is charged or discharged. This means that at high frequencies, "strong" drivers have to be used in order to charge and discharge the output capacitor fast enough to reach the supply rails. The strength of the drivers in the voltage doubler of Fig. 8.34 can also be expressed as a factor α_{dr} of the pumping capacitors C_{pp}. With $E = \frac{1}{2}CV^2$, the switching energy losses can be summarized to

$$E_{sw} = (\alpha_{st} + \alpha_{sb} + \alpha_{dr}) \cdot C_{pp} \cdot V_{dd}^2, \qquad (8.73)$$

while the losses due to the voltage drop of the pumping capacitors are expressed by

$$E_{\beta} = \beta_{max} \cdot V_{dd} \cdot Q_{out} = \beta_{max} \cdot \beta \cdot V_{dd}^2 \cdot C_{pp} \qquad (8.74)$$

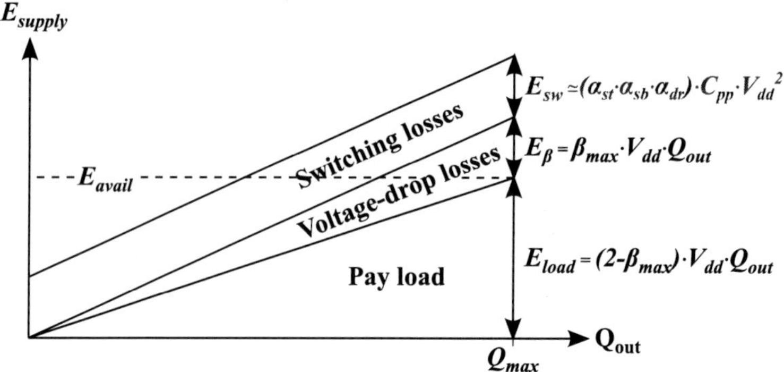

Figure 8.35 Voltage doubler with parasitic elements.

with $Q_{out} = C_{pp} \cdot V_{dd} \cdot \beta$. The load energy left, employing Eq. (8.69) is

$$E_{load} = (2 - \beta_{max}) \cdot V_{dd} \cdot Q_{out} = (2 - \beta_{max}) \cdot \beta_{max} \cdot V_{dd}^2 \cdot C_{pp}. \quad (8.75)$$

The previous energy equations are plotted as a function of the available output charge in Fig. 8.35. It can be noticed that the voltage-drop losses E_{beta} are rising if the output load current is increasing while the switching losses E_{sw} stay constant. Thus, for light loads it is recommended to decrease the frequency to reduce switching losses.

The overall efficiency employing Eqs. (8.73), (8.74), and (8.75) is calculated as follows:

$$\eta = \frac{E_{load}}{E_{load} + E_{\beta} + E_{sw}} = \frac{1}{1 + \frac{\beta_{max}}{2 - \beta_{max}} + \frac{E_{sw}}{E_{load}}}. \quad (8.76)$$

If this function is plotted as a function of the normalized load energy $\frac{E_{load}}{C_{pp} V_{dd}^2}$, the result is an array of curves for different values of β_{max}. In Fig. 8.36, the efficiency η is plotted for β_{max} from 0.05 to 1.00. The total constant $\alpha = \alpha_{st} + \alpha_{sb} + \alpha_{dr}$ of the switching losses $E_{sw} = \alpha C_{pp} V_{dd}^2$ is assumed to be 0.25, which is 25% of the supply energy. At the endpoint of the curves, E_{load} is equal to the available (output) energy E_{avail} and β is equal to β_{max}. It can be observed that an optimum of the efficiency is achieved at $\beta_{max} = 0.35$ corresponding to around 61%.

If we now assume the switching losses to be only 1% of the supplied energy, that is, $\alpha_{st} + \alpha_{sb} + \alpha_{dr} = 0.01$, the graphs of the

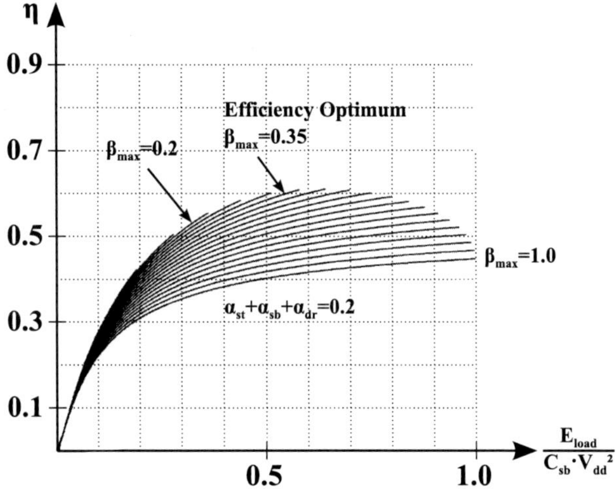

Figure 8.36 Efficiency of a voltage doubler with $\alpha_{st} + \alpha_{sb} + \alpha_{dr} = 0.25$.

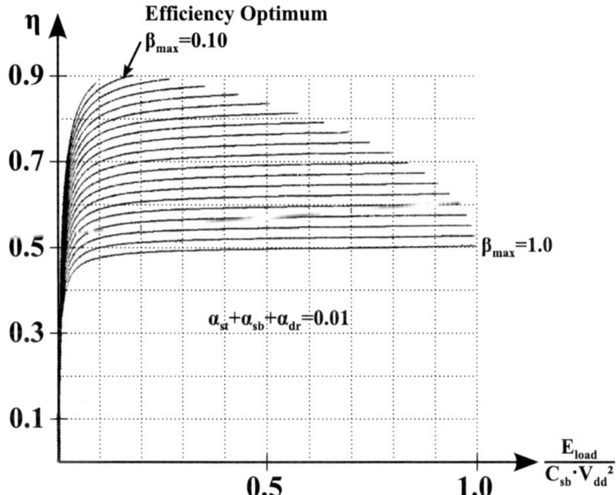

Figure 8.37 Efficiency of a voltage doubler for $\alpha_{st} + \alpha_{sb} + \alpha_{dr} = 0.01$.

efficiencies η as a function of the load energy E_{load} for different values of β_{max} look like in Fig. 8.37. Here, it can be noticed that a maximum efficiency of 90% is reached at a β_{max} of 0.1.

Figure 8.37 illustrates that an efficiency of 90% in a charge pump design is possible. Nevertheless, the efforts for this are generally large capacitors and a low switching frequency. This is obviously a challenge when implementing a charge pump circuit on an integrated circuit design, because big capacitances can only be achieved off-chip. Since the number of capacitors needed for a charge pump multiplies with the number of stages, the geometrical size can make a charge pump design unpractical. But there is still room for optimizations, which is shown in the following paragraph.

8.2.5.5 Design optimization

Steensgard et al. [7] show various examples of optimizing the efficiency of a charge pump. Following are the main possibilities:

1. varying the size of pumping capacitance C_{pp} depending on load current
2. varying the clock frequency dependent on load current to reduce switching losses
3. employing large buffer capacitor C_{buf} and switching on charge pump only if voltage of C_{buf} drops below certain value

The first method is illustrated in Fig. 8.38, in which several stages of a charge pump are connected in parallel. The charge pump stages are switched on and off depending on the power required at the output. In that way, the pumping capacitance C_{pp} is adapted to the load. Since there are a lot of capacitors needed for this design, it is only useful in an integrated circuit. The advantage of this solution

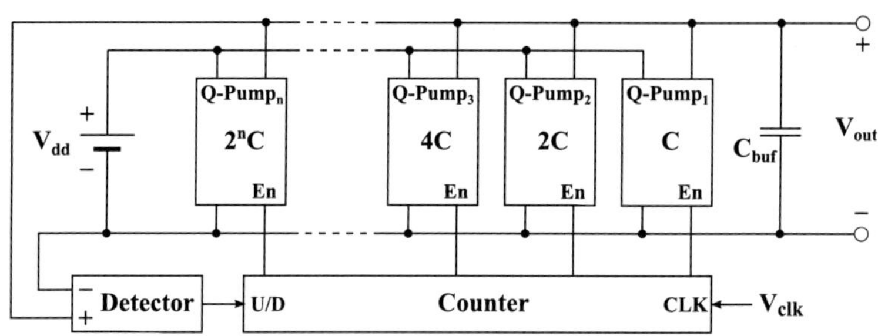

Figure 8.38   Charge pump solution with frequency regulation.

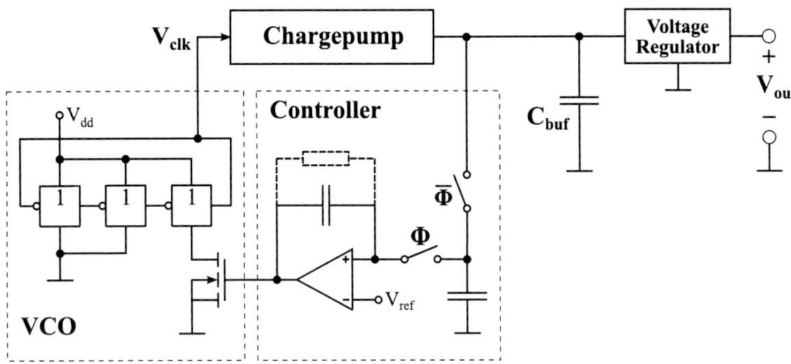

Figure 8.39 Charge pump solution with variable output power.

is that the switching losses can be reduced at low loads without changing the frequency of the V_{clk} signal. This charge pump scheme is also utilized in a patent of Bedarida et al. [11].

Another optimizing method for charge pumps is shown in Fig. 8.39. Here, the frequency is regulated by a switched-capacitor controller in order to reduce switching losses at low load conditions. Therefore, a voltage-controlled oscillator is required. A switched-capacitor error amplifier is used in this circuit to prevent power losses due to Ohmic voltage dividers. One last alternative to optimize the efficiency of a charge pump is illustrated in Fig. 8.40. A toggle flip-flop, which generates the clock pulses for the charge pump is controlled by a voltage regulation loop. The goal of this design is to

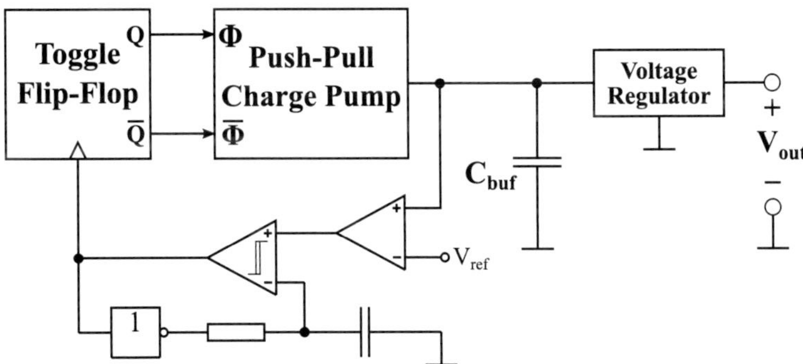

Figure 8.40 Charge pump solution with duty cycle regulation.

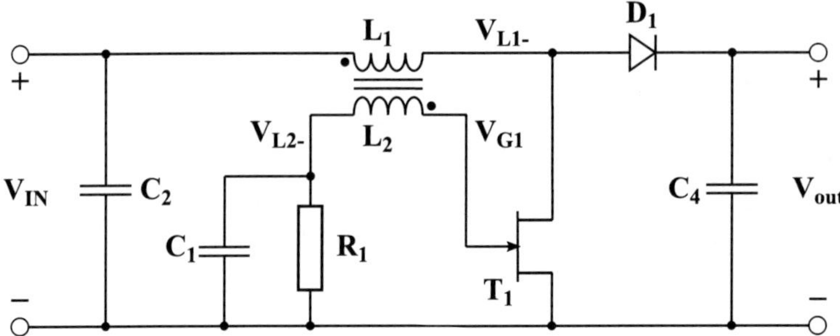

Figure 8.41   Circuit example of a Meissner oscillator based converter.

reduce the number switching transitions per time to a minimum. In this way, the duty cycle of the clock signal ϕ is varied between almost 0 and 0.5. Therefore, the efficiency is optimal for a wide range of loads.

8.2.6 *Meissner Oscillator Based Converter*

8.2.6.1 Physical principles

For energy harvesting applications, it can be an advantage to use oscillator topologies like the Meissner or Armstrong oscillator [12] to built up a switching converter. This way a self-oscillating circuit can be created, where the switching transistor is driven by a winding on a transformer. Consequently, the regulation circuit can be simplified, because no extra clocking circuit with a driver to control the switching transistor is needed. This helps also to save power in low-load conditions. In the following sections, a Meissner-based converter concept is explained in detail according to the work of Pollak et al. [13].

8.2.6.2 Electrical circuit

Figure 8.41 shows a simple solution of the Meissner-based converter concept described earlier. In this case, a junction-field-effect transistor (JFET) is used, which is a normally-on active element. Thus, its threshold voltage is negative, or, in other words, a negative voltage is needed to switch it off. The circuit is working like a boost converter, see Section 8.2.2, just the control of the transistor is done with a coupled inductor. This is in contrast to the Flyback topology in Section 8.2.4, where the transformer is used to convert the power and not to control the switching transistor. This can be noticed as an advantage of the circuit of Fig. 8.41, because the secondary winding does not need to be optimized for a low ESR, which would otherwise increase the size of the transformer.

The operation of the circuit is as follows: since the N-JFET T_1 is conducting when the input voltage V_{IN} is rising from zero, the current through the transformer inductance L_1 is rising, too. Thus, a voltage is induced across the secondary inductance L_2, which is negative counting from V_{L2-} to V_{G1}. Furthermore, the voltage V_{L2-}—referred to the negative supply rail—also becomes negative because the N-JFET has a p-n-junction from its gate to its source that is conducting at that time. In fact, the current through L_1 is not rising anymore at some point. The reason is that either the core of the transformer gets into saturation or the JFET limits the current, because of its on-state resistance, which is generally in the area of 50 Ω for standard types of N-JFETs. Nevertheless, at that point the voltage over L_2 reaches zero and the negative voltage V_{L2-} stored on the capacitor C_1 becomes equal to the JFET's gate voltage V_{G1} resulting in the JFET to switch off fast. The capacitor is now discharged via resistor R_1 until the JFET is conducting again and an oscillation starts. The frequency of the oscillation is just determined by the time constant of R_1 and C_1.

8.2.6.3 Simulation results

This paragraph shows some simulation results of a self-oscillating converter described in the previous section. The "Spectre" simulator of the "Cadence" design tool is used for that task [14]. The corresponding circuit diagram is shown in Fig. 8.42. It is the same circuit as in Fig. 8.41, but additionally the parasitic DC resistances R_{L1} and R_{L2} of the transformer are inserted. For the simulation, an input voltage of 300 mV is present and a load resistance of 100 KΩ as well as an output capacitor of 100 nF.

The waveforms of Fig. 8.43 show the simulation results at start-up. The voltages on nets $V L1-$, $V L2-$ and $V G1$ are plotted as well

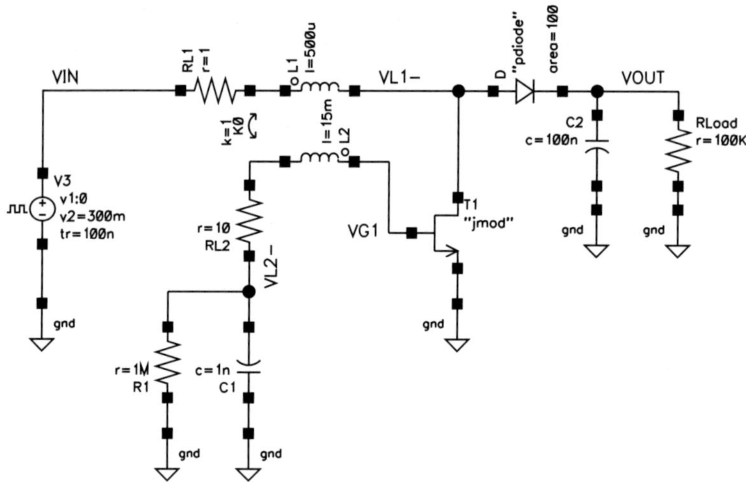

Figure 8.42 Simple simulation circuit of a Meissner oscillator based converter.

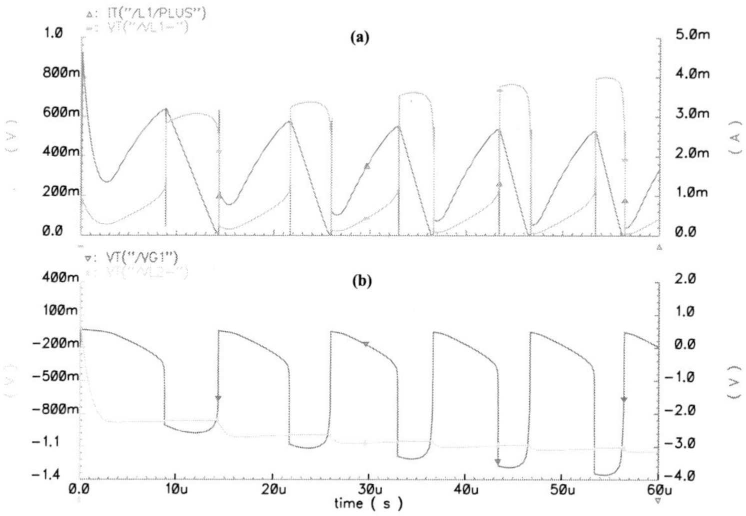

Figure 8.43 Waveforms of the simulation of the coupled inductor converter during start-up phase.

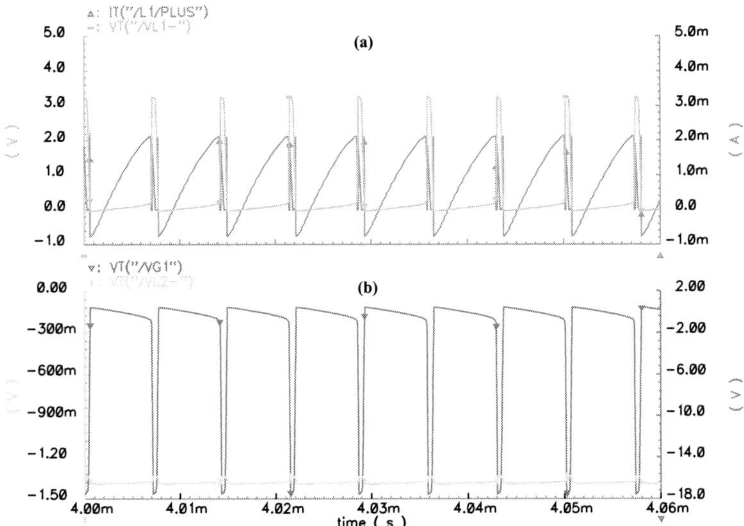

Figure 8.44 Waveforms of the simulation of the coupled inductor converter during steady state.

as the transient current at the node $/L1/PLUS$ of the inductor $L1$. At the moment where the supply voltage on net VIN is switched on, the current flowing through the inductor increases because of the coupling of the primary inductor $L1$ to the secondary inductor $L2$ on which the voltage cannot be higher than the forward voltage of the pn-junction of the N-JFET. The reason for this is that capacitor $C1$ is still not charged, which can be noticed regarding the waveform $VT("/VL2-")$ of Fig. 8.43a. After the supply is switched on, $C1$ is charged more and more to a negative voltage because of the rising current of inductor L_1. The waveform $VT("/VG1")$ of Fig. 8.43b shows that the voltage never reaches more than 500 mV due to the conducting pn-junction at the gate of $T1$.

The equivalent waveforms of Fig. 8.43 at steady-state operation can be seen in Fig. 8.44. When observing the waveform $IT("/L1/PLUS")$ of the inductor current a negative part can be noticed. This is caused by charging and discharging the parasitic capacitances at the Gate of $T1$. Since the energy for this is only exchanged between the Gate capacitances and the transformer, the

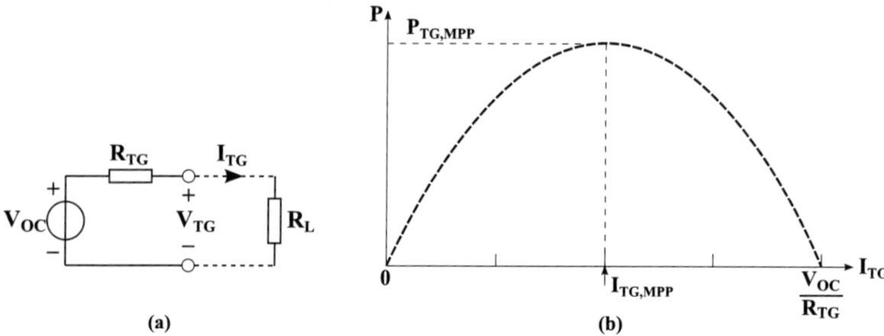

Figure 8.45 Matching a load to an energy transducer directly connected (a), after a voltage converter (b).

current of $L1$ has a negative component and therefore the energy is not lost.

8.2.7 Matching Loads

Since some energy transducers—like a thermoelectric generator (TEG) based on the Seebeck effect or a solar cell—have an internal parasitic Ohmic resistance, it is useful to think about matching the load to this resistance in order to get the maximum power out of the transducer. The adaptive matching of a load is considered in this chapter for the thermogenerator transducer. Nevertheless, this method can be used in an analogue way for equivalent energy transducers.

8.2.7.1 Analytical model

In Fig. 8.45a the model of a thermoelectric generator (TEG), which is composed of a voltage source V_{OC} and an internal resistor R_{TG}.

The output power P_{TG} of the TEG can be stated as follows:

$$P_{TG} = (V_{OC} - R_{TG}I_{TG})I_{TG} = V_{OC}I_{TG} - R_{TG}I_{TG}^2, \tag{8.77}$$

which is a quadratic function and equals to zero when there is no output current I_{TG} or when it is equal to $\frac{V_{OC}}{R_{TG}}$. P_{TG} is plotted versus I_{TG} in Fig. 8.45b, where a maximum current I_{MPP} can be found differentiating P_{TG} respect to I_{TG} and equaling to zero:

$$\frac{\Delta P_{TG}}{\Delta I_{TG}} = V_{OC} - 2R_{TG}I_{TG} = 0 \Leftrightarrow I_{TG,MPP} = \frac{V_{OC}}{2R_{TG}}. \tag{8.78}$$

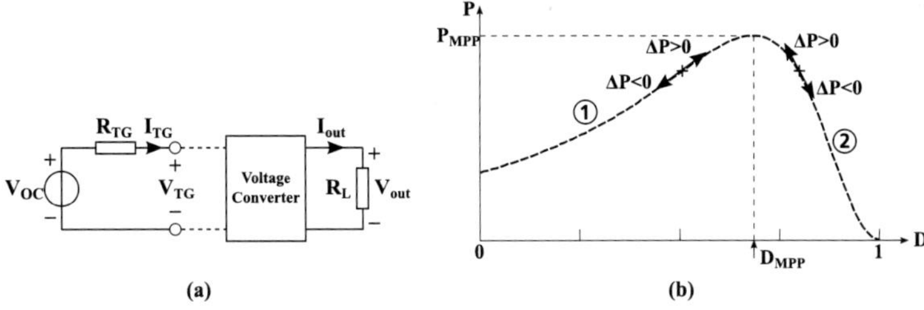

Figure 8.46 Matching a load to an energy transducer: Output power versus (a) output current, (b) duty cycle.

Employing Eq. (8.77), the maximum power $P_{TG,MPP}$ is

$$P_{TG,MPP} = V_{OC}\frac{V_{OC}}{2R_{TG}} - R_{TG}\frac{V_{OC}^2}{(2R_{TG})^2} = \frac{V_{OC}^2}{2R_{TG}} - \frac{V_{OC}^2}{4R_{TG}}$$

$$= \frac{2V_{OC}^2 - V_{OC}^2}{4R_{TG}} = \frac{V_{OC}^2}{4R_{TG}}. \tag{8.79}$$

The resulting output voltage $V_{TG,MPP}$ at the maximum power point $I_{TG,MPP}$ is consequently

$$V_{TG,MPP} = V_{OC} - R_{TG}I_{TG,MPP} = V_{OC} - R_{TG}\frac{V_{OC}}{2R_{TG}} = \frac{V_{OC}}{2} \tag{8.80}$$

with an equivalent load resistance $R_{L,MPP}$:

$$R_{L,MPP} = \frac{V_{TG,MPP}}{I_{TG,MPP}} = \frac{V_{OC}}{2} \cdot \frac{2R_{TG}}{V_{OC}} = R_{TG}. \tag{8.81}$$

Concluding, the TEG provides maximum power if the condition $R_L = R_{TG}$ and $V_{TG} = \frac{V_{OC}}{2}$ is fulfilled. In general, state-of-the-art TEGs provide output voltages less than 1 V for temperature gradients lower than 10°C, which is in general too low for supplying electrical circuits. Employing a voltage converter (see Fig. 8.46a), the resistive load R_L connected to the converter creates an equivalent resistance R_{eq} connected to TEG. Concerning the following analysis, it is assumed that a boost converter is used. Therefore, Eq. (8.32) from Section 8.2.2 is helpful:

$$V_{TG} = V_{out}(1 - D). \tag{8.82}$$

In reality, a boost converter is not ideal and has a certain efficiency η. Simplifying, it is assumed that the efficiency is equal for any load current. Therefore, it can be calculated with $P_{out} = \eta P_{TG}$:

$$V_{out}I_{out} = \eta V_{TG}I_{TG} \Rightarrow I_{TG} = \frac{I_{out}}{\eta(1 - D)}. \tag{8.83}$$

The equivalent resistance R_{eq} at the output of the TEG (see Fig. 8.45a), can now be expressed as

$$R_{eq} = \frac{V_{\text{TG}}}{I_{\text{TG}}} = \frac{V_{\text{out}}(1-D)}{\frac{I_{\text{out}}}{\eta(1-D)}} = \eta(1-D)^2 \frac{V_{\text{out}}}{I_{\text{out}}} = \eta(1-D)^2 R_{\text{L}}. \quad (8.84)$$

The output current I_{TG} of the TEG can be stated as follows:

$$I_{\text{TG}} = \frac{V_{\text{OC}}}{R_{\text{TG}} + R_{eq}} = \frac{V_{\text{OC}}}{R_{\text{TG}} + \eta(1-D)^2 R_{\text{L}}}. \quad (8.85)$$

Introducing I_{TG} into Eq. (8.77) leads to

$$\begin{aligned}
P_{\text{TG}} &= \frac{V_{\text{OC}}^2}{R_{\text{TG}} + \eta(1-D)^2 R_{\text{L}}} - \frac{R_{\text{TG}} V_{\text{OC}}^2}{(R_{\text{TG}} + \eta(1-D)^2 R_{\text{L}})^2} \\
&= \frac{R_{\text{TG}} V_{\text{OC}}^2 + \eta(1-D)^2 R_{\text{L}} V_{\text{OC}}^2 - R_{\text{TG}} V_{\text{OC}}^2}{(R_{\text{TG}} + \eta(1-D)^2 R_{\text{L}})^2} \\
&= \frac{\eta(1-D)^2 R_{\text{L}} V_{\text{OC}}^2}{(R_{\text{TG}} + \eta(1-D)^2 R_{\text{L}})^2}
\end{aligned} \quad (8.86)$$

for the output power of the TEG. If this function is plotted versus the duty cycle D (see in Fig. 8.46b), it can be observed again that there exists a global maximum at D_{MPP}. This means that we can find an optimum power point for any resistive load adapting the duty cycle. The same can be done with any other converter type, only Eq. (8.82) has to be exchanged. Considering Eq. (8.84), R_{L} always has to be greater than R_{eq} since a boost converter is used. In case of employing a buck-boost converter, R_{L} can be greater or smaller than R_{eq}. Nevertheless, for finding the maximum power point the duty cycle D has to be adjusted, which results in varying the output voltage of the converter, see Eq. (8.82), in case of a resistive load R_{L}. In general, loads—like wireless transceivers or sensors—are not resistive and need to be supplied by a fixed voltage. In this case, it is not possible to draw the maximum power out of the energy transducer, since the current and thus the power is fixed by the load itself. Nevertheless, if an energy storage device like a battery or a capacitor is used between the load and the output of the voltage converter, it is still possible to operate the energy transducer in the maximum power point. How that works is shown in the following paragraphs.

Figure 8.47 illustrates the configuration of Fig. 8.46(a) with a battery as the load for the voltage converter. Under this condition

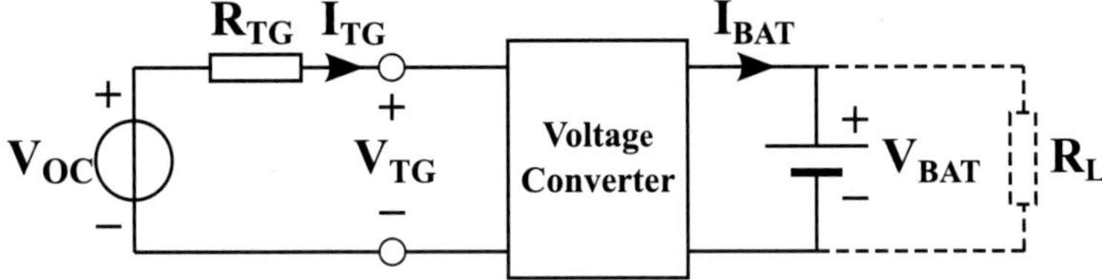

Figure 8.47   Matching a battery load to an energy transducer employing a voltage converter.

it can be assumed that the maximum power is supplied to the load, when the output current I_{BAT} of the converter is maximized since the output voltage V_{BAT} changes slowly. In fact, for the following analysis, V_{BAT} is assumed to be constant. As an example, now a buck-boost converter is used for voltage conversion. Therefore, the voltage conversion ratio from Eq. (8.59) is written down again for the configuration of Fig. 8.47:

$$\frac{V_{BAT}}{V_{TG}} = \frac{D}{1-D}.$$

(8.87)

Consequently, I_{TG} can be calculated as follows:

$$I_{TG} = \frac{V_{OC} - V_{TG}}{R_{TG}} = \frac{1}{R_{TG}}\left(V_{OC} - V_{BAT}\frac{1-D}{D}\right)$$

(8.88)

If this result is introduced in Eq. (8.77), the output power of the TEG can be calculated:

$$\begin{aligned}
P_{TG} &= V_{OC}I_{TG} - R_{TG}I_{TG}^2 \\
&= \frac{V_{OC}}{R_{TG}}\left(V_{OC} - V_{BAT}\frac{1-D}{D}\right) - \frac{1}{R_{TG}}\left(V_{OC} - V_{BAT}\frac{1-D}{D}\right)^2 \\
&= \frac{V_{OC}^2}{R_{TG}} - \frac{V_{OC}V_{BAT}}{R_{TG}}\cdot\frac{1-D}{D} \\
&\quad - \frac{1}{R_{TG}}\left(V_{OC}^2 - 2V_{OC}V_{BAT}\frac{1-D}{D} + V_{BAT}^2\cdot\frac{(1-D)^2}{D^2}\right) \\
&= \frac{V_{OC}V_{BAT}}{R_{TG}}\cdot\frac{1-D}{D} + \frac{V_{BAT}^2}{R_{TG}}\left(\frac{1-D}{D}\right)^2.
\end{aligned}$$

(8.89)

A maximum of the output power of the TEG can be achieved by differentiating P_{TG} respect to the duty cycle D:

$$\frac{\Delta P_{TG}}{\Delta D} = \frac{V_{OC} V_{BAT}}{R_{TG}} \cdot \frac{-D - (1 - D)}{D^2}$$

$$-2 \cdot \frac{V_{BAT}^2}{R_{TG}} \cdot \frac{1 - D}{D} \cdot \frac{-D - (1 - D)}{D^2}$$

$$= \frac{V_{OC} V_{BAT}}{R_{TG}} \cdot \frac{-1}{D^2} + 2 \cdot \frac{V_{BAT}^2}{R_{TG}} \cdot \frac{1 - D}{D^3}$$

$$= 2 \cdot \frac{V_{BAT}^2}{R_{TG}} \cdot \frac{1 - D}{D^3} - \frac{V_{OC} V_{BAT}}{R_{TG} D^2} \tag{8.90}$$

and equaling the resulting function to zero:

$$\frac{\Delta P_{TG}}{\Delta D} = 2 \cdot \frac{V_{BAT}^2}{R_{TG}} \cdot \frac{1 - D}{D^3} - \frac{V_{OC} V_{BAT}}{R_{TG} D^2} = 0$$

$$\Rightarrow 2 V_{BAT} \cdot \frac{1 - D}{D} - V_{OC} = 0 \Rightarrow \frac{1}{D} - 1 = \frac{V_{OC}}{2 V_{BAT}} \Rightarrow \frac{1}{D} = \frac{V_{OC}}{2 V_{BAT}} + 1$$

$$\Rightarrow D_{MPP} = \frac{1}{\frac{V_{OC}}{2 V_{BAT}} + 1} = \frac{2 V_{BAT}}{V_{OC} + 2 V_{BAT}}. \tag{8.91}$$

It can be summarized that for maximizing the output power of the TEG using a battery at the output of the voltage converter (see Fig. 8.47), the duty cycle D_{MPP} is only dependent on V_{OC} and V_{BAT}. In reality, the open-circuit voltage V_{OC} and the internal resistance R_{TG} cannot be measured without disconnecting the TEG from the rest of the circuit. Therefore, it is more feasible to find an electrical structure that measures the output power of the energy transducer finding the maximum by itself. Only one general solution is shown in the next paragraphs, since there is a lot of work on this topic already done in literature. For more information, the work of Sullivan et al. [15] or Koutroulis et al. [16], for example, can be helpful.

8.2.7.2 Physical principles

A general algorithm for a "Maximum-Power-Point-Tracker" (MPPT) can be found using Fig. 8.46b. In the graph of P_{MPP}, two areas are marked (1 and 2). Depending on which side of P_{MPP} the regulation

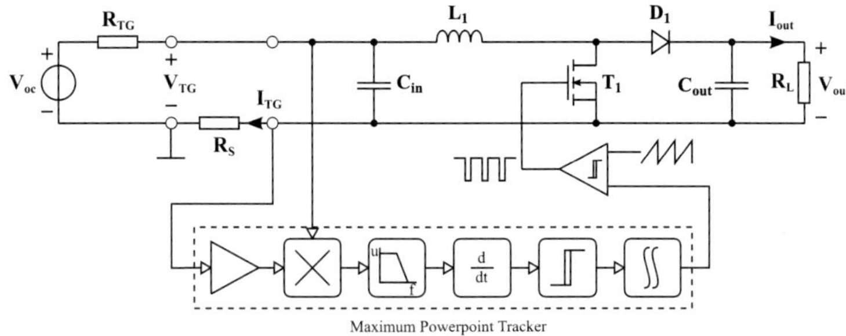

Figure 8.48   Design example of a maximum-powerpoint-tracker.

starts, the expressions for the algorithm are as follows:

$$Area(1) : \Delta P > 0 \Rightarrow \Delta D \uparrow \tag{8.92}$$

$$\Delta P < 0 \Rightarrow \Delta D \downarrow \tag{8.93}$$

$$Area(2) : \Delta P > 0 \Rightarrow \Delta D \downarrow \tag{8.94}$$

$$\Delta P < 0 \Rightarrow \Delta D \uparrow . \tag{8.95}$$

From these statements, it can be deduced that the starting point has to be chosen according to the implemented solution. Figure 8.48 shows a block diagram of a possible design employing the boost converter from Fig. 8.13. Output current and voltage of the energy transducer, V_{TG} and I_{TG}, are used as an input to the MPPT and are multiplied to deliver a signal proportional to the output power of the transducer. Generally, a shunt resistor R_S transforms the output current I_{TG} of the TEG into a voltage signal, which is amplified afterward (see Fig. 8.48). The signal after the multiplier is fed through a low-pass filter to remove the current ripple due to the switching of the converter and to prevent the MPPT to regulate on the resulting ripple of the power signal. The output signal of the filter is then differentiated to know if the power is rising or falling. A comparator is connected to the differentiator in order to increase or decrease the duty cycle of the converter. This is realized with an integrator afterward that sums up the comparator signal as a function of time. The output of the integrator is the control signal for the boost converter, which corresponds to the error signal v_{err} in the voltage-regulated converter from Fig. 8.13.

Whether the output of the comparator is inverted or not related to its input determines which algorithm—either Eqs. (8.92)/(8.93) or Eqs. (8.94)/(8.95)—is used. In fact it has to be decided to start the operation of the circuit either at duty cycle $D = 0$ or at $D = 1$. The proposed MPPT can only work properly when the output power of the transducer does not change faster than the reaction of the MPPT, otherwise it could happen that the maximum powerpoint changes to a point in the wrong area (see diagram in Fig. 8.46b) and the algorithm fails to lead the duty cycle in the right direction. Therefore, this design is more practical for TEGs than for solar cells, for example, since in general the temperature gradient applied to a TEG does not change as fast as the light intensity applied to a solar cell. In literature other solutions can be found, for example, Sullivan et al. [15] work with the help of a flip-flop for having the ability to switch between algorithm of Eq. (8.92)/(8.93) and Eq. (8.94)/(8.95).

8.2.7.3 Electrical circuit

For the MPPT loop of Fig. 8.48, there are both analog and digital solutions possible, whereas for a digital solution at least one analog-to-digital and one digital-to-analog converter is necessary as well as a microcontroller. For energy harvesting applications, it is in general a choice of which architecture consumes more power.

For the analog solution in Fig. 8.48, a circuit is presented in Fig. 8.49, where one operational amplifier is needed per stage. The other parts are standard circuits with operational amplifiers for differentiator, comparator, and integrator. Sometimes it makes sense to use an amplifier stage instead of a comparator, because its gain is lower for a possible DC-component from the output of the differentiator. The output of the integrator serves as the control signal for the PWM of the converter.

For the multiplier, it is not given a typical design here. For the multiplier task commercial integrated circuits like the AD633 [17] from Analog Devices can be used, for example. Nevertheless, especially the AD633 consumes some mA of current, which can be too much for energy harvesting applications. However, it is possible to get around the multiplier measuring only current instead of power. For instance, if a battery at the output of the voltage converter is connected—like in the previously described solution in Fig. 8.47— the output current can be considered to be constant in relation to the

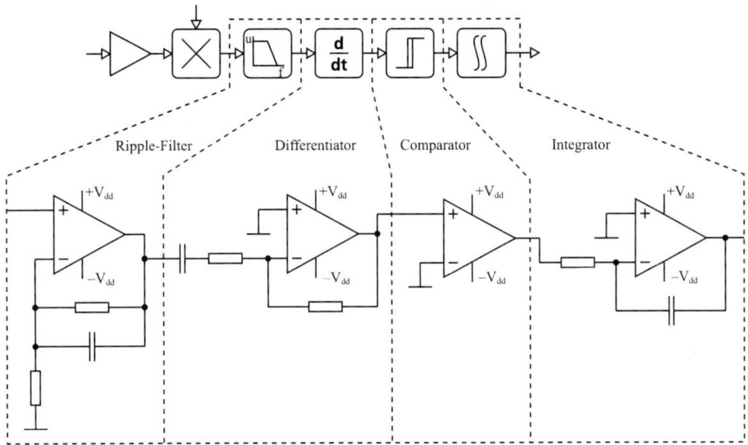

Figure 8.49 Analog implementation of a maximum-powerpoint-tracker.

output voltage. In this case, the MPPT tracks the maximum output power of the converter, which is additionally more accurate than tracking the output power of the transducer. This can be understood considering that a typical boost converter has a higher efficiency at higher input voltages, which means that the maximum output power of the transducer is at a different point than the maximum power delivered to the battery.

8.2.7.4 Efficiency considerations

The benefit in efficiency using a maximum-power-tracking loop can be estimated using a diagram where the output power is plotted as a function of the output current of a thermoelectric generator. Therefore, the theoretical model of Fig. 8.45 and Eq. (8.77) is used. Using a TEG, V_{OC} is directly proportional to the applied temperature gradient ΔT:

$$V_{\mathrm{OC}} = \alpha_m \cdot \Delta T \tag{8.96}$$

with the Seebeck coefficient α_m. For a typical TEG, for example, the 127-150-26 [18] from the company "thermalforce.de" or the PKE-128-A-1027 [19] from the company Peltron GmbH, with $\alpha_m = 0.05$ the graph of the output power P_{TG} as a function of the output current I_{TG} is shown in Fig. 8.50 for open-circuit voltages V_{oc} of 100, 200,

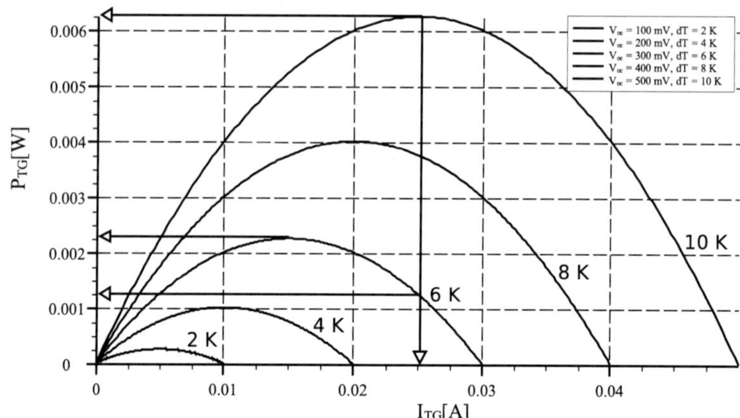

Figure 8.50 Output power versus output current of a typical TEG for different open-circuit voltages V_{oc}.

300, 400 and 500 mV. Beside these voltages, the corresponding temperature gradients ΔT of the TEG are noted.

For example, at a temperature gradient of 10 K the TEG works at its maximum power point of $P_{TG} = 6.2$ mW at an output current of $I_{TG} = 25$ mA. Now it is considered that a reduction of the applied temperature gradient from 10 K to 6 K occurs and that the output current remains with 25 mA (see Fig. 8.50). In this case, the output power of the TEG drops to $P_{TG} = 1.2$ mW, whereas the maximum power point would be at $P_{TG} = 2.5$ mW and $I_{TG} = 15$ mA. This example shows that $100 \cdot \frac{2.5 \text{ mW} - 1.2 \text{mW}}{1.2 \text{ mW}} = 108\%$ of output power of the TEG could be gained using an adaptive method for working always in an optimum power point of an energy transducer.

References

1. J. Williams, High efficiency linear regulators, *Linear Technology Application Note.* **32**, 1–12 (1989).

2. N. Mohan, T. M. Undeland, W. P. Robbins, *Power Electronics: Converters, Applications and Design* (Wiley, 1995).

3. M. Gildersleeve, H. P. Forghani-zadeh, G. A. Rincón-Mora. A comprehensive power analysis and a highly efficient, mode-hopping dc-dc converter. In *Proc. 2002 Asian-Pacific Conference on ASICs*, pp. 153–156 (2002).

4. Y. Chen. Resonant gate drive techniques for power mosfets. Master's thesis, Virginia Polytechnic Institute and State University (May, 2000).

5. B. Andreycak, *Zero Voltage Switching Resonant Power Conversion* (Texas Instruments Incorporated, 2001).

6. A. Panda, H. N. Pratihari, B. Prasad, Panigrahi, and L.Moharana, A zero voltage transition synchronous buck converter with an active auxiliary circuit, *DSP Journal.* **Volume 9**, 41–49, (2009).

7. J. Steensgaard, V. Ivanov, Switched-capacitor power supplies. In *Advanced Engineering Course on Power Management Lausanne, Switzerland*, pp. 1–89 (Sept., 2007).

8. Charge pump (11, 2011). URL http://en.wikipedia.org/wiki/Charge_pump.

9. J.-M. Baek, J.-H. Chun, and K.-W. Kwon, A power-efficient voltage upconverter for embedded eeprom application, *IEEE Transactions on Circuits and Systems II: Express Briefs.* **57** (2010).

10. L. Mensi, A. Richelli, L. Colalongo, and Z. M. K. Vajna, A voltage efficient PMOS charge pump architecture, *Research in Microelectronics and Electronics* (2006).

11. L. Bedarida. Modular charge pump architecture (09, 2004).

12. Armstrong oscillator. URL http://en.wikipedia.org/wiki/Armstrong_oscillator.

13. M. Pollak, L. Mateu, and P. Spies. Step-up dc-dc converter with coupled inductors for low input voltages. In *Procedings of the PowerMEMS Conference*, pp. 145–148 (November 2008).

14. Cadence design systems. URL http://en.wikipedia.org/wiki/Cadence_Design_Systems.

15. C. R. Sullivan and M. J. Powers, A high-efficiency maximum power point tracker for photovoltaic arrays in a solar-powered race vehicle, *Power Electronics Specialists Conference.* pp. 574–580 (1993).

16. E. Koutroulis, K. Kalaitzakis, and N. C. Voulgaris, Development of a microcontroller-based, photovoltaic maximum power point tracking control system, *IEEE Transactions on Power Electronics.* **16**, 46–54 (01, 2001).

17. Low cost analog multiplier. URL http://www.analog.com/static/imported-files/data_sheets/AD633.pdf.

18. Thermogenerator (12, 2011). URL http://www.thermalforce.de/de/product/thermogenerator/TG127-150-26e_.pdf.

19. Thermogenerator pke 128 a 1027 (12, 2011). http://www.peltron.de/elemstdt.html

Chapter 9

AC–DC Converters

Loreto Mateu and Peter Spies

Fraunhofer Institute for Integrated Circuits IIS, Nordostpark 93,
90411 Nuremberg, Germany
loreto.mateu@iis.fraunhofer.de, peter.spies@iis.fraunhofer.de

Vibrations are a ubiquitous ambient source employed in energy harvesting systems. Piezoelectric, electrostatic, and electrodynamic transducers provide AC power from ambient vibrations. Thus, AC–DC converters are necessary in order to convert their AC power into the DC power that is required by the load of the energy harvesting system. The AC–DC converter employed for those transducers is normally a two-stage power converter consisting of the following:

- AC–DC rectifier. The rectification is usually done with a full-wave or half-wave rectifier using diodes. Another solution consists of using voltage multipliers for the case of electrodynamic transducers to increase the low-output voltages and rectify them at the same time. Current multipliers are employed in piezoelectric transducers for increasing the low output currents and rectifying the output power.
- DC–DC converter. After the rectification of the AC power, it is necessary to adapt the voltage levels of transducer and

Handbook of Energy Harvesting Power Supplies and Applications
Edited by Peter Spies, Loreto Mateu, and Markus Pollak
Copyright © 2015 Pan Stanford Publishing Pte. Ltd.
ISBN 978-981-4241-86-1 (Hardcover), 978-981-4303-06-4 (eBook)
www.panstanford.com

load with a DC–DC converter. Sometimes the objective of the DC–DC converter is not to set a certain voltage but to maximize the power harvested from the transducer what is of special interest in energy harvesting systems where the output power is in the microwatt or milliwatt range.

There are also direct AC–DC converters presented in the section of electrodynamic converters where a single converter rectifies the AC power and adapts the voltage to the requirements of the load.

The voltage and current levels as well as the conversion principle associated with each one of the three transducers pointed out is different. Therefore, the chapter is divided in three main sections, one for each of the transducers where the specific AC–DC converters are presented and analyzed. This chapter is a compendium of published papers of different authors about this topic. For each one of the AC–DC convertors, the electrical circuit is introduced as well as the physical principles behind the converter. A theoretical analysis and efficiency data are also presented when there are available.

9.1 AC–DC Converters for Piezoelectric Transducers

A rectifier is necessary when piezoelectric transducers are employed in an energy harvesting power supply. Voltage or current multiplier rectifiers are an alternative to full-wave rectifiers when it is required to increase the voltage or the current, respectively, and rectify the signal.

9.1.1 *Voltage Doubler*

Han et al. [1] present three different AC–DC rectifiers: a diode-resistor pair rectifier, a diode-pair rectifier, and a synchronous rectifier (see Fig. 9.1). The rectifiers are employed for converting piezoelectric AC power into DC power. Afterward, a charge pump is used to provide a regulated output voltage for a filter capacitor and resistive load.

Electrical Circuit   The piezoelectric element is modeled as a sinusoidal voltage source v_p in series with capacitor C_1 and resistor R_p

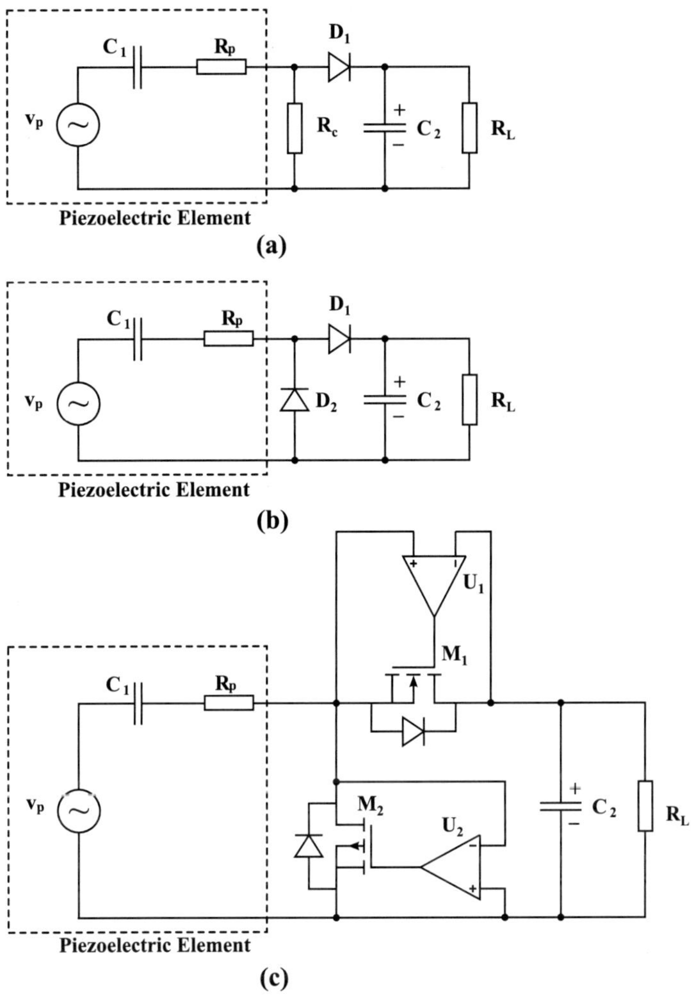

Figure 9.1 AC–DC rectifiers (a) diode-resistor pair rectifier (b) passive full-wave rectifier (c) synchronous full-wave rectifier [1].

(see Fig. 9.1). The circuit shown in Fig. 9.1a is a half-wave rectifier with a parallel connection of resistor R_c to the piezoelectric element before diode D_1. The resistor is placed in this position since it reduces the charging time of the output capacitor. The optimum value of R_c is the one that makes possible to reach the final voltage on capacitor C_2 without load in the minimum time. Additionally, more energy is extracted with the parallel connection of resistor R_c [1].

The circuit shown in Fig. 9.1b is a full-wave rectifier that works as a voltage doubler. At resonant frequency, the capacitance of the piezoelectric element dominates its internal resistor, and therefore the rectifier works as a voltage doubler with a capacitor fixed by the piezoelectric element. The circuit shown in Fig. 9.1c has the same topology as the circuit shown in Fig. 9.1b but the diodes are replaced by transistors driven synchronously. The body diodes of the MOSFETs shown in Fig. 9.1c make possible that when there is no power for supplying the comparators, the circuit shown in Fig. 9.1c becomes the circuit shown in Fig. 9.1b. Therefore, when no initial power is available in the system, Fig. 9.1c can start up and afterward supply power to the operational amplifiers to begin synchronous operation. When the input voltage has a negative value, comparator U_2 provides a high signal that turns on transistor M_2. The operation principle of circuit shown in Fig. 9.1c is as follows. When the input voltage of the rectifier is higher than the output voltage, comparator U_1 provides a positive signal to the gate of transistor M_1 turning it on.

The efficiency obtained employing a load of 80 kΩ in the circuits shown in Fig. 9.1a,b,c is 34%, 57%, and 92%, respectively [1].

9.1.2 *Half-Wave Rectifier with Voltage Doubler*

Le et al. [2] present the same designs shown previously in Fig. 9.1b,c and an additional circuit that is shown in Fig. 9.2a. This circuit is a half-wave rectifier with a voltage doubler block that is displayed in Fig. 9.2b. Therefore, only one half of the piezoelectric wave is rectified and afterward multiplied by a factor 2.

9.1.2.1 Electrical circuit

A synchronous rectifier composed by a PMOS transistor, an inverter, and a comparator is connected after the piezoelectric element. When the output voltage of the piezoelectric element is higher than the voltage after the PMOS, this is turned on and clock signal ϕ_1 has a

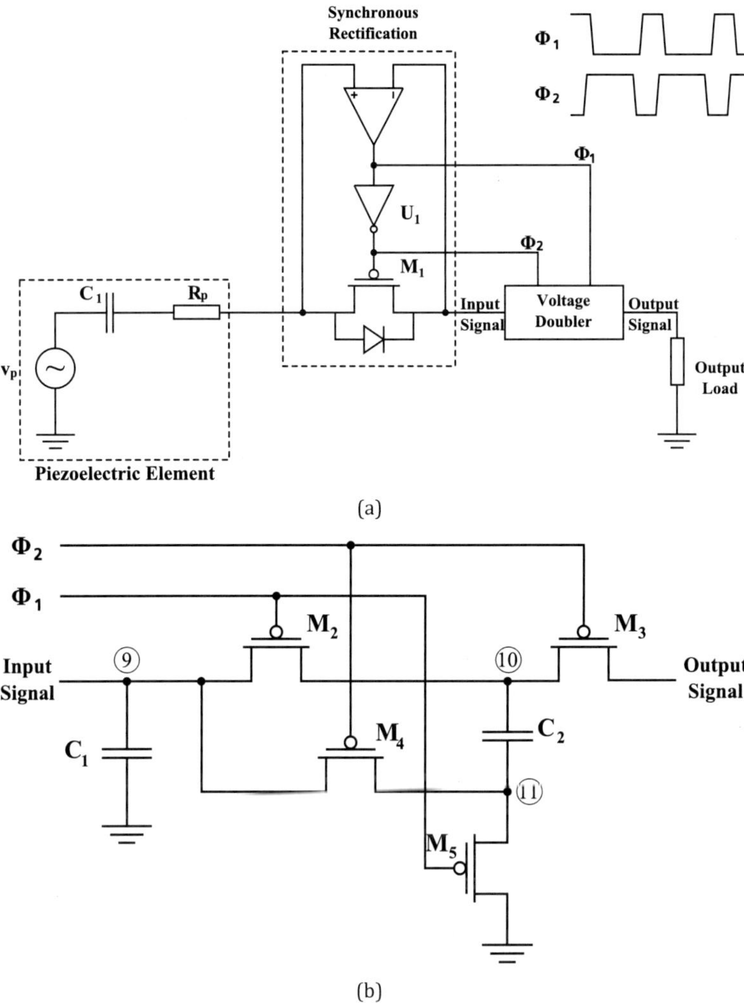

Figure 9.2 (a) Half-wave synchronous rectifier and voltage doubler (b) Transistor-level circuit of the voltage doubler [2].

high level. Then, capacitor C_1 shown in Fig. 9.2b is charged with the input voltage. Moreover, the two PMOS transistors M_2 and M_5 are turned on and capacitor C_2 is charged with the output voltage of the piezoelectric element. When the output voltage of the piezoelectric element becomes lower than the voltage after the PMOS transistor

M_1, signal ϕ_2 has a high level and the PMOS transistors connected to it in the voltage doubler circuit are turned on. Then, the capacitor connected between nodes 10 and 11 is connected in series with the capacitor connected between node 9 and ground and so, the output voltage obtained is twice the piezoelectric output voltage.

9.1.2.2 Efficiency measurement

Le et al. [2] built an ASIC with the passive full-wave rectifier, the synchronous full-wave rectifier and the half-wave synchronous rectifier, and voltage doubler shown in Fig. 9.2, Fig. 9.1b,c, respectively. The measured peak efficiency obtained for the passive full-wave rectifier was 65% for an output current between 5 and 10 µA. The synchronous full-wave rectifier has a measured efficiency between 70% and 85% for an output current between 5 and 30 uA. For the case of the half-wave synchronous rectifier and voltage doubler topology, the efficiency ranges from 65% to 88% for an output current between 58 nA and 5 µA. Consequently, the synchronous full-wave rectifier has the best combination of efficiency and output power since it can deliver up to 22 µW.

For the design of synchronous rectifiers, the power consumption of the comparator must be taken into the consideration. The comparator designed by Le et al. consumes only 165 nW.

9.1.3 *Direct Discharge Circuit*

9.1.3.1 Physical principles

The direct discharge circuit, also called standard circuit, is the simplest circuit that provides DC power to a load employing a piezoelectric transducer. The circuit rectifies the AC power and stores the harvested energy on a capacitor. The piezoelectric element is modeled in Fig. 9.3 as a sinusoidal current source i_p in parallel with capacitor C_1.

9.1.3.2 Electrical circuit

The circuit shown in Fig. 9.3 shows the topology of the direct discharge circuit. A diode bridge converts the AC power into DC

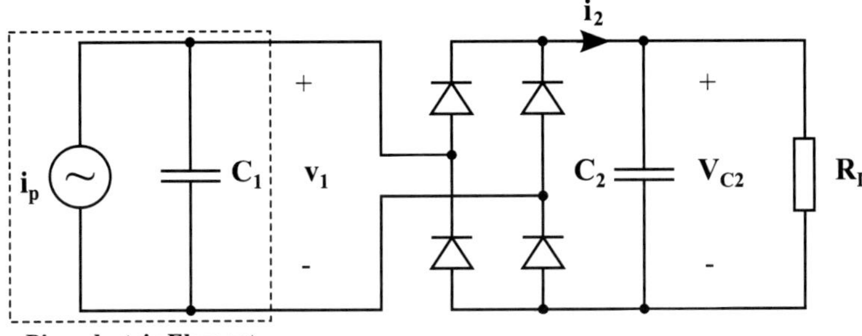

Figure 9.3 Piezoelectric power supply using a diode bridge and a storage capacitor as AC–DC power converter [3].

power and a second capacitor is employed to filter the DC power that supplies the resistive load.

The waveforms associated with this circuit are shown in Fig. 9.4. During time interval u, the piezoelectric current is charging its capacitance and no current flows from the piezoelectric element to the load through the diode bridge. During the rest of the half period, the voltage on the piezoelectric element is higher than the voltage on capacitor C_2, and therefore the diodes of the bridge are in conduction and the current flows from the piezoelectric element to filter capacitor C_2 and load.

9.1.3.3 Analytical model

In this analysis, it is assumed that C_2 is large enough to consider that the voltage on C_2, V_{C2}, is constant [4] and that the mechanical force that excites the piezoelectric element is a sine. The current that flows from the piezoelectric element to C_2 and R_L is

$$i_2(t) = \begin{cases} 0 & 0 \le t \le u/\omega \\ \frac{C_2}{C_2+C_1} I_p \, |\sin(\omega(t))| & u/\omega \le t \le \pi/\omega. \end{cases} \tag{9.1}$$

If it is assumed that $C_2 \gg C_1$, almost all the current generated by the piezoelectric material is transferred to the load. The mean output current is

$$\langle i_2(t) \rangle = \frac{2I_p}{\pi} - \frac{2V_{c2}\omega C_1}{\pi}. \tag{9.2}$$

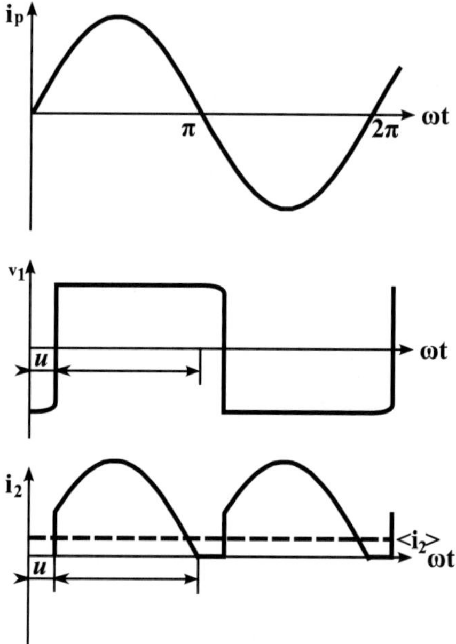

Figure 9.4 Waveforms of the current i_p generated by the piezoelectric element, voltage v_1 on the piezoelectric element and current i_2 through the rectifier bridge [4].

The mean output power delivered by the piezoelectric element is given by

$$\langle P\,(t)\rangle = \frac{2V_{C2}}{\pi}\left(I_p - V_{C2}\omega C_1\right).\tag{9.3}$$

The previous analysis for the direct discharge circuit assumes a sinusoidal mechanical excitation. However, other types of mechanical excitations are present in the environment [5, 6]. A generic mechanical excitation is going to be analyzed to find an expression for the value of the maximum voltage that can be accumulated in the storage capacitor and the choice of appropriate values for both the storage capacitor and the number of piezoelectric elements to connect in parallel. This number is especially important when certain energy requirements have to be met. The analysis is valid for any kind of mechanical excitation where the amount of charge generated by the piezoelectric is known.

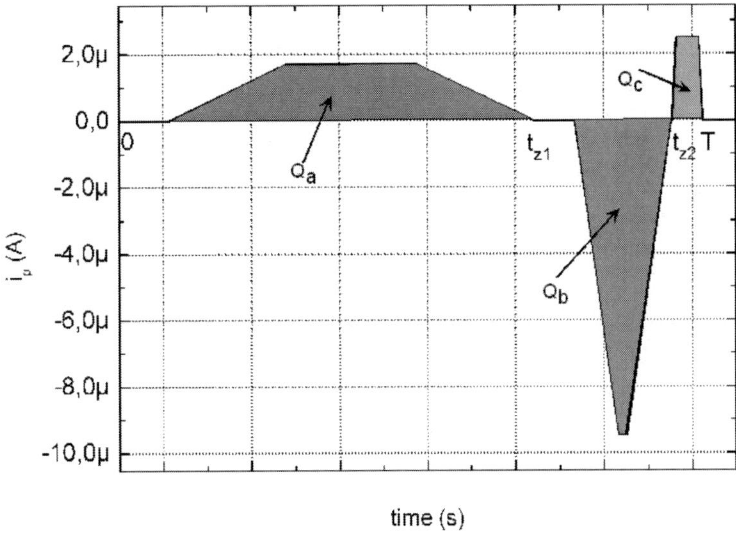

Figure 9.5 Current waveform i_p for a piezoelectric element of PVDF during walking activity [5].

Figure 9.5 displays the current waveform generated by a piezoelectric Polyvinylidene fluoride (PVDF) element during human walking activity. Three regions can be distinguished in this waveform. Region A goes from 0 to t_{z1} and corresponds to the first positive region of the current waveform. Region B goes from t_{z1} to t_{z2} and corresponds to the negative region of the current waveform. Region C is the last region and goes from t_{z2} to T. The integration of one period of the piezoelectric current is zero since no charge is accumulated on the piezoelectric material after each period. Hence, the sum of Q_a, charge from 0 to t_{z1}, Q_b, charge from t_{z1} to t_{z2}, and Q_c, charge from t_{z2} to T, is zero.

Parameters Q_a, Q_b, and Q_c are defined as the charge generated between the zero crossings of the current waveform:

$$Q_a = \int_0^{t_{z1}} i_p(t)\, dt \tag{9.4}$$

$$Q_b = \int_{t_{z1}}^{t_{z2}} i_p(t)\, dt \tag{9.5}$$

$$Q_c = \int_{t_{z2}}^{T} i_p(t)\, dt \tag{9.6}$$

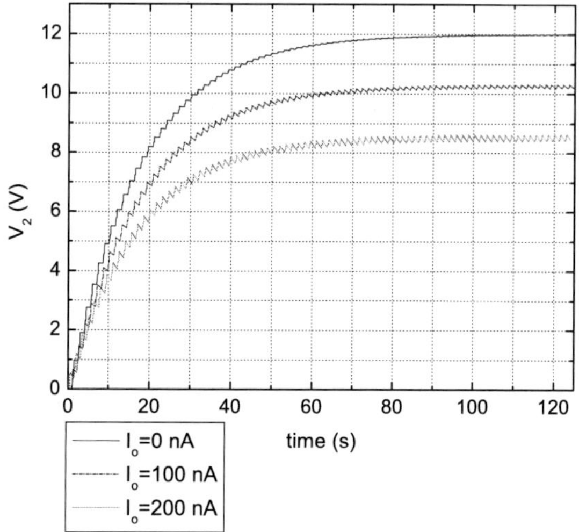

Figure 9.6 Simulated voltage waveform on C_2 for different current consumptions of the control circuit. The values employed for the HSPICE simulation are $C_1 = 22$ nF, and $C_2 = 1$ μF. The diode model selected is 1n4148.

If the piezoelectric element is cyclically excited and its current is rectified, the start-up voltage on C_2 has the profile shown in Fig. 9.6. Initially, when C_2 is completely discharged, all the current generated by the piezoelectric element i_p charges both C_1 and C_2. Therefore, the voltage on C_2, V_2 increases rapidly at first, but during the subsequent steps it becomes increasingly difficult to turn on the diodes to charge C_2. Finally, when the peak voltage on C_1 is equal to the voltage on C_2 plus $2V_d$, the diodes turn off definitively and V_2 no longer increases.

As shown in Fig. 9.7, between t_{z2} and T, the diodes are turned off since the maximum voltage on C_2 during one period is reached at t_{z2}. Taking this into account and assuming for simplicity that the forward voltage of the diodes is zero, the expression for the final output voltage on C_2 is [5, 6]

$$V_{\text{maxC2}}|_{I_o=0} = -\frac{Q_b}{2C_1} \tag{9.7}$$

As this last expression shows, the final voltage of V_2 is related to the piezoelectric capacitance C_1 and the mechanical excitation given by the generated charge.

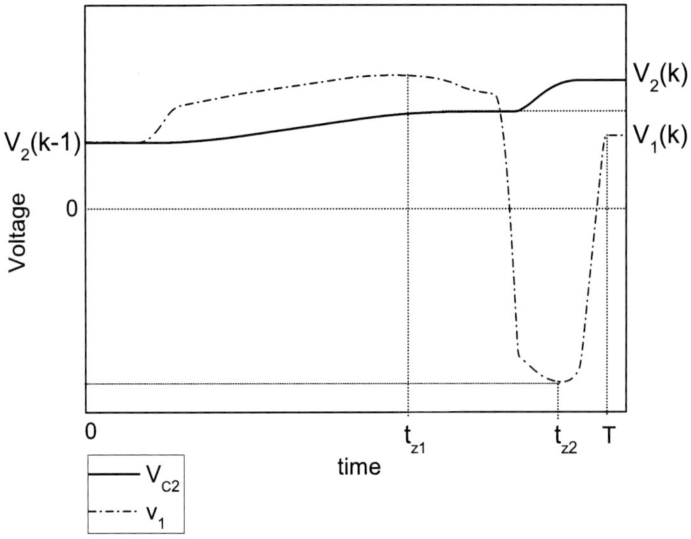

Figure 9.7 Simulated voltage waveform on C_1 (V_1) and on C_2 (V_2). The values employed for the HSPICE simulation are $C_1 = 22$ nF and $C_2 = 1$ μF. The diode model employed is 1n4148 [5].

The number of steps k needed to achieve a certain voltage $V_2(k)$ on C_2 is

$$k = \frac{1}{2} \frac{\ln\left(1 - \left(V_2\left(k\right) / V_{\text{maxC2}}\right)\right)}{\ln\left(\left(C_2 - C_1\right) / \left(C_1 + C_2\right)\right)} \tag{9.8}$$

The number of steps, or in a more general way, the number of mechanical excitations needed to achieve a certain voltage, is independent of the period of piezoelectric current source i_p. The time to achieve a certain voltage $V_2(k)$ can be obtained by multiplying the number of steps by the step period.

The term t_f is defined as the time required to achieve a voltage on C_2 that is equal to 95% of the final voltage. Thus, the number of mechanical excitations to achieve this voltage is

$$k_f = \frac{1}{2} \frac{\ln\left(0.05\right)}{\ln\left(\left(C_2 - C_1\right) / \left(C_1 + C_2\right)\right)} \tag{9.9}$$

and t_f is k_f times the period of one mechanical excitation T.

$$t_f = k_f T \tag{9.10}$$

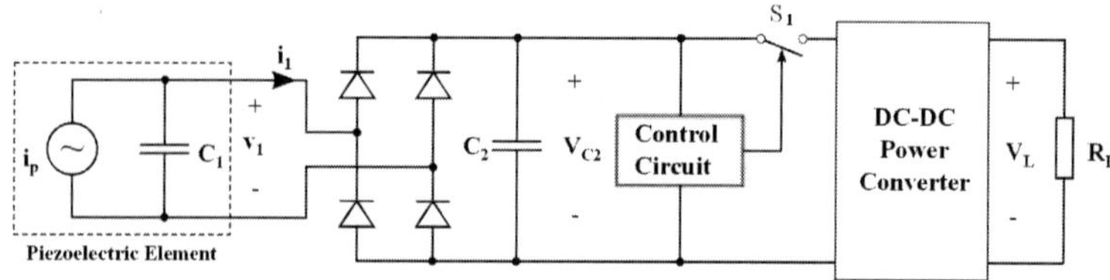

Figure 9.8 Direct discharge circuit with control and regulator circuit [5].

Similarly, it is possible to calculate the number of stress cycles needed to increase the voltage of C_2 from a given voltage V_{off} to another voltage V_{on}:

$$k_s = \frac{1}{2} \frac{\ln\left(\left(V_{maxC2} - V_{on}\right) / \left(V_{maxC2} - V_{off}\right)\right)}{\ln\left(\left(C_2 - C_1\right) / \left(C_1 + C_2\right)\right)} \tag{9.11}$$

Figure 9.8 shows the direct discharge circuit with a regulator circuit and a load. As the piezoelectric element is repetitively stressed, the voltage in the storage capacitor increases. Whenever this voltage reaches a certain limit V_{on}, the capacitor is connected to the regulator input and it is discharged to a lower voltage V_{off}, providing the energy E_{req} to the load (including the energy needed by the regulator). Capacitor C_2 is then disconnected from the regulator again, allowing it to be recharged. The control circuit has a current consumption I_o while it monitors the voltage on C_2.

In order to calculate the maximum voltage on C_2 for the case of a current consumption I_o, it has been considered that C_2 is not discharged in region B (see Fig. 9.5). In this case, the maximum voltage on C_2 is given by the following expression:

$$V_{maxC2} = V_{maxC2}\big|_{I_o=0} - \frac{I_o \left(C_2 - C_1\right)\left(t_{z1} - t_{z2p}\right)}{4C_1C_2} \tag{9.12}$$

where t_{z2p} corresponds to the second time that the current waveform crosses 0 A of the previous period.

9.1.3.4 Design optimization

The electronic load to be powered can be characterized by two parameters. First, the required energy E_{req} needed to perform its operation. Second, the power supply voltage V_{dd} of the device. It is assumed that this value is fixed by a voltage regulator.

The circuit operates as described previously, with the capacitor working between V_{off} and V_{on}. The number of stress cycles, given a certain piezoelectric element and type of excitation, depends on the capacitance C_2 as well as the values of V_{on} and V_{off} (see Eq. (9.11)). It is then desirable to choose the value of these three parameters to minimize the number of cycles for the recharge of C_2. The minimum allowed voltage V_{off} must be larger than V_{dd} to allow a proper operation of the voltage regulator.

$$V_{off} = V_{dd} + \epsilon \tag{9.13}$$

In general, the energy that must be provided by C_2 is

$$E_{req} = \frac{1}{2}C_2 \left(V_{on}^2 - V_{off}^2\right) = \frac{1}{2}C_2 \left(V_{on}^2 - (V_{dd} + \epsilon)^2\right) \tag{9.14}$$

From Eq. (9.14), C_2 can be obtained as a function of V_{on}, V_{dd}, and ϵ. Then, by substituting the expression of C_2 obtained into Eq. (9.8), the relation between number of steps and V_{on} for the required energy is obtained. k_{ini}, Eq. (9.15), gives the expression for the number of steps needed to increase the voltage on C_2 from 0 V to a value V_{on}, whereas k_s (Eq. (9.16)) gives the expression for the number of steps required to increase the voltage on C_2 from V_{off} to V_{on}.

$$k_{ini} = \frac{1}{2}\frac{\ln\left(1 - (V_{on}/V_{maxC2})\right)}{\ln\left(\left(2E_{req}/\left(V_{on}^2 - (V_{dd} + \epsilon)^2\right) - C_1\right)/\left(2E_{req}/\left(V_{on}^2 - (V_{dd} + \epsilon)^2\right) + C_1\right)\right)} \tag{9.15}$$

$$k_s = \frac{1}{2}\frac{\ln\left((V_{maxC2} - V_{on})/(V_{maxC2} - (V_{dd} + \epsilon))\right)}{\ln\left(\left(2E_{req}/\left(V_{on}^2 - (V_{dd} + \epsilon)^2\right) - C_1\right)\left(2E_{req}/\left(V_{on}^2 - (V_{dd} + \epsilon)^2\right) + C_1\right)\right)} \tag{9.16}$$

The optimum value for the capacitor C_2 and the voltage V_{on} is not the same for minimum k_{ini} (the number of cycles for the first initial charge) and for minimum k_s (the number of cycles for recharging). However, the most appropriate value for V_{on} and C_2 will be the one that minimizes k_s since the initial charge cycle k_{ini} takes place only once in the operation of the load. Magnitudes C_2, k_{ini}, and k_s from Eqs. (9.14)–(9.16) are represented in Fig. 9.9 as a function of V_{on}. It is observed that there is a minimum value $k_{s,opt}$ for the number of steps at a certain voltage $V_{on,opt}$ on capacitor C_2. The value of

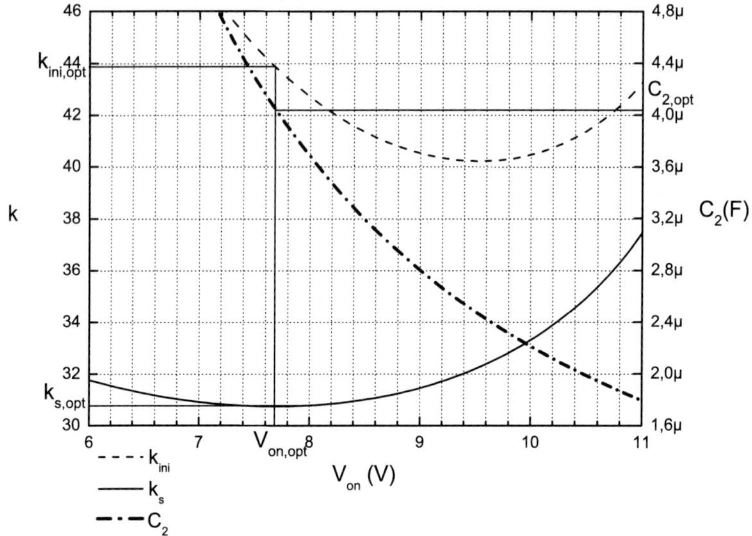

Figure 9.9 Optimum values of V_{on}, C_2, and k_{ini} to supply 100 μJ to a load powered at 3 V with $C_1 = 22$ nF with the minimum number of recharging cycles $k_{s,opt}$.

capacitor C_2 for $V_{on,opt}$ is $C_{2,opt}$. If another capacitance value of C_2 is selected, the number of steps needed to ensure 100 μJ of energy to supply the load increases. Therefore, the selection of the appropriate capacitance value is important in order to use fewer steps to supply power to the load.

9.1.4 *Direct Discharge Circuit in Conjunction with DC–DC Converters*

9.1.4.1 Physical principles

A general harvesting circuit is composed by the piezoelectric transducer, an AC–DC rectifier and a capacitor where the rectified energy is stored. Afterward, a switching converter is used to transfer the energy from the storage capacitor to the battery (see Fig. 9.10). There is an optimum voltage level on the storage capacitor C_2 for which the power flowing from the piezoelectric transducer

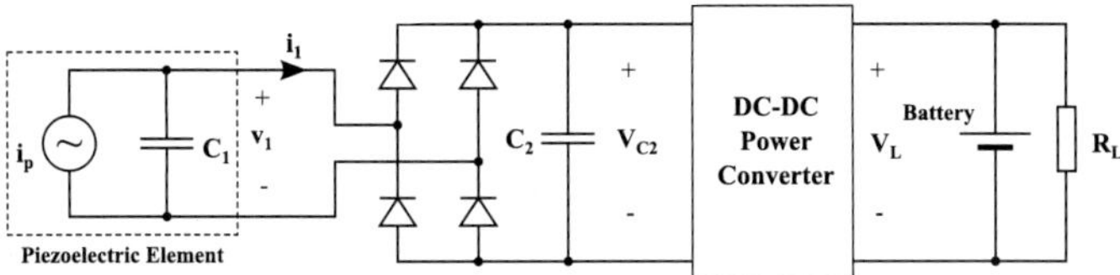

Figure 9.10 Piezoelectric power supply with a direct discharge circuit and a DC–DC converter as power management unit [3].

is maximum. Thus, the DC–DC converter must work with an appropriate duty cycle to achieve this optimum voltage.

9.1.4.2 Efficiency optimization

Ottman et al. use a step-down converter as DC–DC converter to regulate the input voltage V_{C2} of the converter to the value that maximizes the power provided by the piezoelectric element (see Eq. (9.17)). This concept is developed in two different ways for a sinusoidal mechanical excitation. The first approach is an adaptive circuit that maximizes dynamically the current that charges the battery [4], whereas the second approach includes the calculation of the optimum duty cycle and the design of a circuit with a fixed duty cycle equal to the optimum [7].

The voltage of capacitor C_2 is calculated as follows:

$$V_{C2} = \frac{I_\mathrm{p}}{2\omega C_1} = \frac{V_\mathrm{oc}}{2} \tag{9.17}$$

where I_p is the peak current, C_1 is the internal capacitance, and V_oc is the open circuit voltage of the piezoelectric element.

9.1.4.3 Design optimization employing a maximum power point tracking algorithm

The power delivered by the piezoelectric element is approximately equal to the input power of the step-down DC–DC converter. If it is assumed that output and input power of the converter are related by a constant value, the maximization of the power delivered by the piezoelectric element is equivalent to the maximization of the output power of the converter. Moreover, when a battery is connected at the output of the DC–DC converter, it can be considered that its voltage

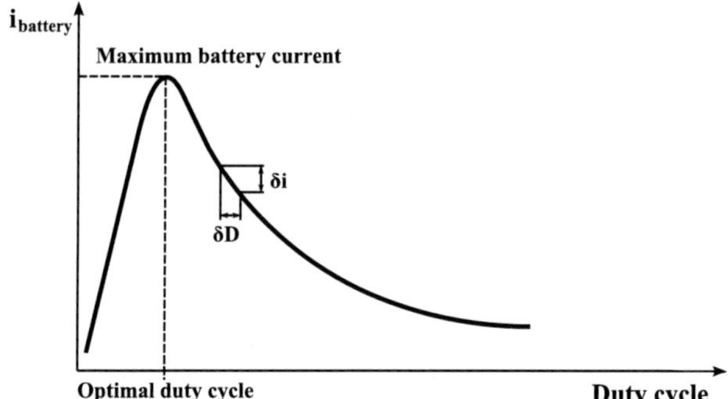

Figure 9.11 Current flowing into the battery versus duty cycle for a piezoelectric energy harvesting circuit with a step-down converter in steady state [4].

is constant. Then, the maximization of the output power of the DC–DC converter is equivalent to the maximization of its output current. This method is widely used in control algorithms for maximum power point trackers of, for example, solar cells [8]. Figure 9.11 displays the relation between battery current and duty cycle of the step-down converter in steady state and Eq. (9.18) describes the algorithm of the control circuit.

$$D_{i+1} = D_i + K \cdot \text{sgn}\left(\frac{\partial I}{\partial D}\right) \tag{9.18}$$

where D_{i+1} is the value of the duty cycle in the next iteration, D_i is the actual duty cycle of the converter, K is the multiplying coefficient of the sign function sgn that is applied to the partial derivative $\frac{\partial I}{\partial D}$.

The control algorithm changes the value of the actual duty cycle D_i depending on the slope of the curve ($\frac{\partial I}{\partial D}$) in order to achieve the point with the optimum duty cycle that provides the maximum battery current. If the slope is positive, it implies that the duty cycle is smaller than the optimum duty cycle and it has to be increased while if the slope is negative, the duty cycle must be reduced to achieve the optimum value. This algorithm was implemented in a digital signal processor (DSP) by Ottman et al. [4].

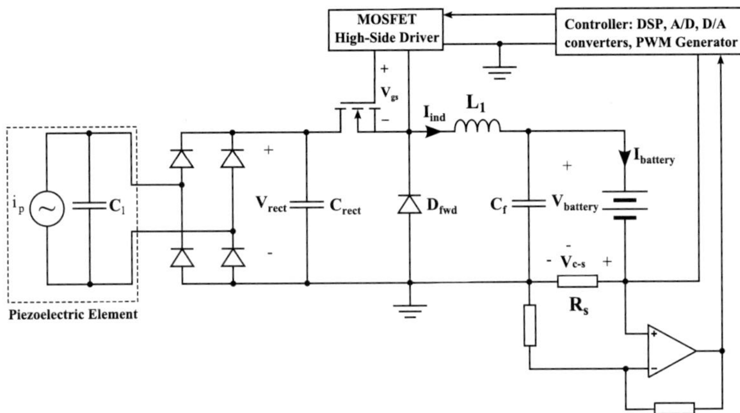

Figure 9.12 Piezoelectric energy harvesting circuit [4].

9.1.4.4 Electrical circuit employing a maximum power point tracking algorithm

Figure 9.12 shows the detailed energy harvesting circuit implemented. The step-down converter has a shunt resistor for sensing the battery current that is afterward amplified and sampled by an A/D converter. The controller generates the control signal for the duty cycle of the converter with the algorithm of Eq. (9.18).

The efficiency of the implemented converter, excluding the rectifier, is between 74% and 88%. [4].

Design Optimization employing a Fixed Duty Cycle Ottman et al. [7] developed a second circuit for maximizing the power harvested by a piezoelectric element with a sinusoidal mechanical excitation employing also a step-down converter. In this case, the optimum duty cycle is calculated and its value is fixed in the control circuit that drives the converter. Assuming that the step-down converter is working in discontinuous current conduction mode (DCM), the power produced by the piezoelectric element is [7]

$$P_{\text{in}} = \frac{D^2 \left(\frac{2I_{\text{p}}}{\pi} - \frac{2\omega C_1 V_{\text{out}}}{\pi} \right) \left(\frac{2I_{\text{p}}}{\pi} + \frac{D^2 V_{\text{out}}}{2Lf_{\text{s}}} \right)}{2Lf_{\text{s}} \left(\frac{2\omega C_1}{\pi} + \frac{D^2}{2Lf_{\text{s}}} \right)^2}, \qquad (9.19)$$

where D is the duty cycle of the switching converter, I_{p} is the piezoelectric peak current, C_1 is the piezoelectric capacitance, ω is

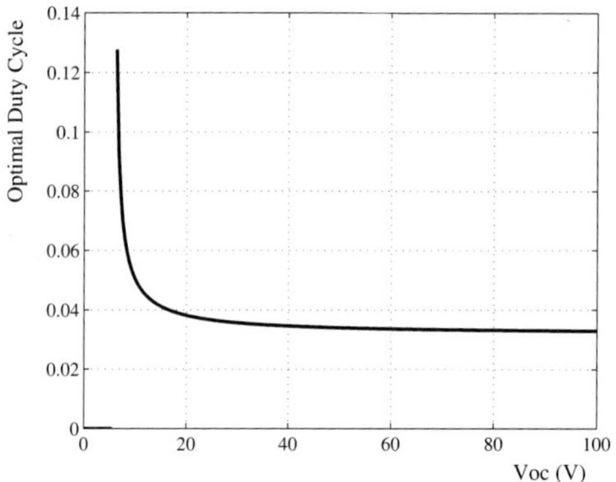

Figure 9.13 Optimal duty cycle for achieving the maximum power transfer from the piezoelectric transducer to the battery as a function of the piezoelectric open circuit voltage in a step-down converter working in discontinuous conduction mode with $L = 10$ mH, $C_p = 200$ nF, $\omega = 400$ rad/s, and $f_s = 1$ kHz.

the angular frequency of the piezoelectric element, V_{out} is the output voltage of the step-down converter, L is the inductance employed in the converter, and f_s is the switching frequency of the converter.

The optimum duty cycle D_{opt} that provides the maximum power is calculated with $\frac{\partial P_{in}}{\partial D} = 0$, resulting in

$$D_{opt} = \sqrt{\frac{4V_{C2}\omega LC_1 f_s}{\pi\left(V_{C2} - V_{battery}\right)}},\tag{9.20}$$

where V_{C2} is the voltage on C_2 and $V_{battery}$ is the voltage of the battery (see Fig. 9.12).

Figure 9.13 shows the optimum duty cycle as a function of the piezoelectric open circuit voltage. In Eq. (9.20), V_{C2} corresponds to half of the piezoelectric open circuit voltage V_{oc} and this voltage is much larger than $V_{battery}$. Thus, Eq. (9.20) can be approximated to

$$D_{opt} = \sqrt{\frac{4\omega LC_1 f_s}{\pi}}.\tag{9.21}$$

9.1.4.5 Electrical circuit employing a fixed duty cycle

Figure 9.14 shows the energy harvesting circuit which is a two-mode converter circuit. When the energy provided by the piezoelectric

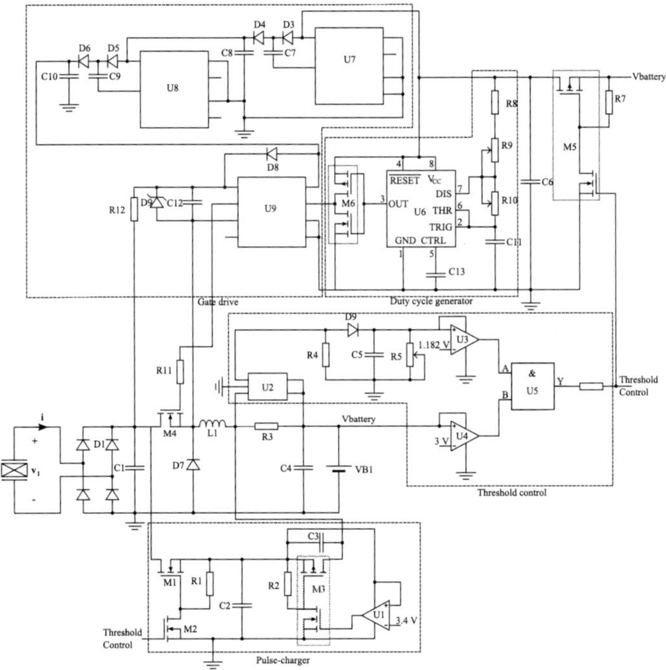

Figure 9.14 Energy harvesting circuit with fixed duty cycle in the step-down converter [7].

transducer is too low to power the switching converter, it is accumulated in the battery in a manner similar to the direct discharge method. The output current of the converter flows through R3 and is amplified by a low-power current amplifier U2. The output of the current amplifier is connected to a passive low-pass filter to average the value of the current taking into account several mechanical excitations. Afterward, the voltage proportional to the current is compared with a fixed voltage in an ultra low power comparator U3 with an internal reference voltage. A second comparator U4 is employed to decide if the battery voltage is over or under a certain voltage. The output of both comparators is connected to an AND gate U5. The output signal of the AND gate determines which one of the two converter circuits charges the battery with the power extracted from the piezoelectric transducer.

The control circuit of the step-down converter is activated when the battery is charged above a certain voltage and the battery current is high enough. A high battery current value implies that the mechanical excitation is enough to power the control circuit of the step-down converter. When one condition is not accomplished, the pulse charger circuit is activated.

The pulse-charger circuit charges capacitor C2 to a certain voltage that is fixed by the comparator U1. When the voltage on C2 is higher than the nominal battery voltage of, for example, 3.4 V, C2 is discharged to the battery through PMOS transistor M3. It can also be considered that the pulse-charger is the startup circuit of the converter.

The constant duty cycle signal that controls the step-down converter is generated by timer U6, working in astable operation [9]. The output signal obtained by the timer is inverted by M6 and used as input of high voltage driver U9. A voltage driver is necessary in order to operate switching transistor M4 of the step-down converter due to the high voltage values that can be obtained by the piezoelectric elements. Moreover, it is necessary to increase the battery voltage to power the voltage driver. Thus, two voltage doublers (U7 and U8) are employed.

9.1.4.6 Efficiency calculation of the circuit employing a fixed duty cycle

Figure 9.15 shows the output power as a function of the duty cycle of the step-down converter for different switching frequencies. The maximum power is achieved for different combinations of duty cycle and switching frequency. Nevertheless, higher switching frequencies provide a response curve with a broader range of the duty cycle to obtain the maximum power point. Consequently, the choice of the switching frequency is of special importance.

The efficiency of the converter, excluding the rectifier, is between 10% and 70%. [7]

9.1.5 *Non-linear Techniques*

Non-linear converters are employed with piezoelectric transducers to increase the harvested energy by active discharge. A part of the

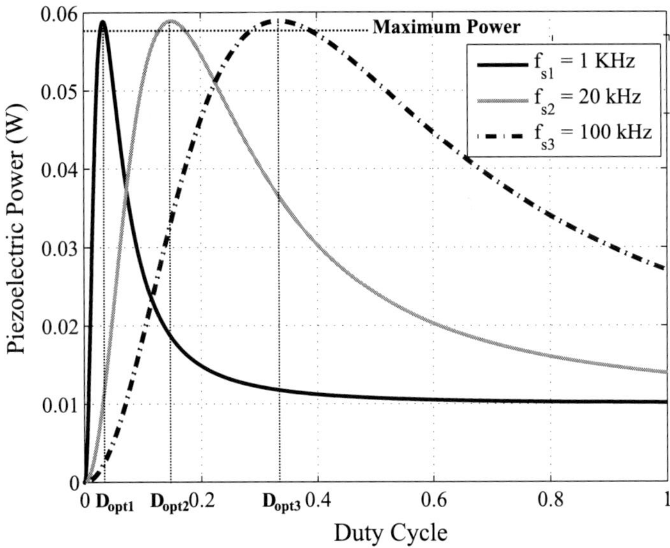

Figure 9.15 Output power of the piezoelectric transducer as a function of the step-down converter duty cycle working in discontinuous current mode for different switching frequencies with $L = 10$ mH, $C_\mathrm{p} = 200$ nF, $\omega = 400$ rad/s and $f_\mathrm{s} = 1$ kHz, $V_\mathrm{out} = 3$ V and $V_\mathrm{rect} = 34$ V [10].

generated charge by the piezoelectric element is employed to charge and discharge its internal capacitor. In non-linear techniques, when the maximum voltage on the piezoelectric element is achieved, the connection of an inductor with the piezoelectric element causes a resonance, due to the internal capacitor of the piezoelectric element, that results in the inversion of the piezoelectric voltage polarity in a very short time compared to the mechanical excitation. The non-linear techniques that are under consideration in this section are three: the parallel synchronized switch harvesting on inductor (parallel SSHI), the series synchronized switch harvesting on inductor (series SSHI) and the synchronous electric charge extraction (SECE).

Non-linear techniques and linear-techniques can be compared in two different scenarios. In the first scenario, piezoelectric transducers for which there is no vibration damping induced by the energy harvesting procedure are under consideration like weakly coupled systems, piezoelectric elements not excited at their

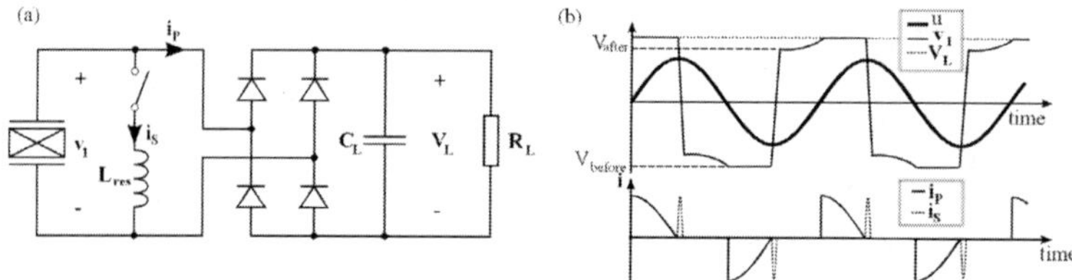

Figure 9.16 Parallel SSHI energy harvesting technique (a) circuit (b) waveforms [13].

resonant frequency, and structures where the displacement is fixed. In the second scenario, strongly coupled systems subjected to a sinusoidal force of constant magnitude are examined. Linear techniques achieve better results than non-linear techniques in the second scenario, whereas in the first scenario non-linear techniques harvest more power than linear techniques [11–13].

9.1.5.1 Parallel SSHI technique

Physical Principles   Figures 9.16a,b show the parallel SSHI circuit without its control circuit and its associated waveforms at steady state, respectively [13]. In this technique, the switch is open most of the time and the current flows through the diode bridge when the voltage on the piezoelectric element is equal to the voltage V_L on the load plus the voltage drop on the diode bridge. Nevertheless, when the maximum voltage is detected on the piezoelectric element, the switch is closed and the piezoelectric current flows through inductor L_{res}. At this moment, a resonant LC circuit is created with the piezoelectric internal capacitor and the voltage on the piezoelectric element changes its polarity in a time given by Eq. (9.22).

$$t_I = \pi \sqrt{L_{\text{res}} C_1} \tag{9.22}$$

Electrical Circuit   Figure 9.17 shows the parallel SSHI circuit including its control circuit implemented by Ben-Yaakov et al. [14]. The peak values of the piezoelectric voltage waveform are detected with a passive differentiator with hysteresis. After that, this signal is compared in an ultra-low-power comparator referenced to one side of the piezoelectric element.

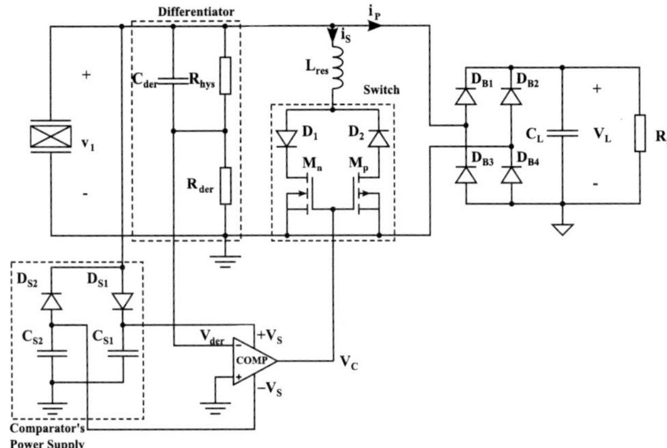

Figure 9.17 Parallel SSHI circuit including its control circuit [14].

The current generated by the piezoelectric material flows through the rectifier diode and the load only when the switch is open. When a maximum is reached in the piezoelectric voltage, the switch is closed. The ultra-low-power comparator is powered by the piezoelectric element through the circuit composed by diodes D_{S1} and D_{S2} and capacitors C_{S1} and C_{S2}. Capacitor C_{S1} is charged through D_{S1} with the positive peak voltage of the piezoelectric element and provides the positive supply voltage $+V_S$, whereas C_{S2} is charged through D_{S2} with the negative peak voltage of the piezoelectric element and provides the negative supply voltage $-V_S$.

The switch that connects the inductor in parallel with the piezoelectric element is composed by two diodes, D_1 and D_2, and two MOSFETs, M_n (NMOS) and M_p (PMOS). D_1 and M_n constitute the switch for the piezoelectric element when the piezoelectric voltage changes its polarity from positive to negative, whereas D_2 and M_p constitute the switch for the opposite change of polarity (negative to positive) in the piezoelectric element.

The differentiator with hysteresis is composed by components C_{der}, R_{der}, and R_{hys}. Resistor R_{hys} generates the hysteresis to avoid undesired commutations of transistors M_n and M_p. The frequency of the mechanical vibration that stimulates the piezoelectric element determines the value of the components of the differentiator. Thus, the parallel SSHI circuit has a limited operating bandwith [15].

Figure 9.18 shows measured waveforms of the parallel SSHI circuit shown in Fig. 9.17. When the piezoelectric element vibrates with a frequency lower than the cut-off frequency of the differentiator, the signal obtained at v_{der} corresponds to the derivative of signal v_1. When the piezoelectric current changes its polarity from positive to negative, the piezoelectric voltage starts to decrease and v_{der} is negative, whereas when the piezoelectric voltage is negative and starts to increase, v_{der} is positive. The derivative signal is compared with the reference voltage obtained at one side of the piezoelectric element. Thus, when v_{der} is negative, the output signal of the comparator v_c is positive and M_n is turned on. Then, the positive current of the piezoelectric element flows through L_{res}, D_1, and M_n. The internal capacitor of the piezoelectric element and the inductor form a resonant circuit that changes the polarity of the piezoelectric element in a time given by Eq. (9.22). When no current flows through the inductor anymore, the piezoelectric element has changed its polarity from positive to negative. In the same way, when v_{der} is positive, v_c is negative and M_p is turned on making it possible that the negative current flows through the inductor and reverses the piezoelectric voltage.

Efficiency Considerations The absolute value of the piezoelectric voltage after inversion is lower than before the inversion due to the losses in the switching circuit and inductor. The inversion is characterized by the ratio γ that relates the voltage before (V_{before}) and after (V_{after}) the polarity change of the piezoelectric transducer [13].

$$V_{after} = -V_{before}\gamma \tag{9.23}$$

When the switches are open and no current is flowing through the inductor, the current flows through the diode bridge and the load. From Fig. 9.19, it is deduced that there is an optimum load as in the case of the standard circuit that maximizes the power extracted from the piezoelectric element to work in the maximum power point. A DC–DC voltage regulator can be used after the diode bridge that adjusts the voltage after the SSHI to the value obtained with the optimum load.

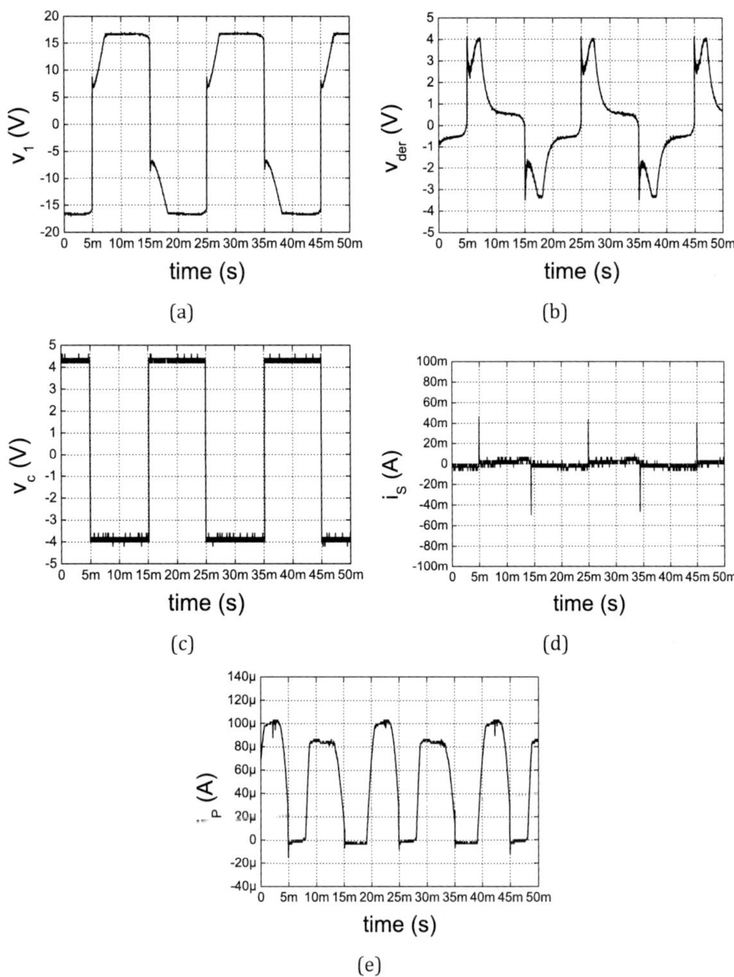

Figure 9.18 Steady-state waveforms for the SSHI technique: (a) Voltage on the piezoelectric element v_1; (b) voltage on the differentiator v_{der}; (c) voltage on the gate of the switching transistors v_c; (d) current flowing through the inductor i_s; (e) rectified current flowing at the output of the diode bridge i_p.

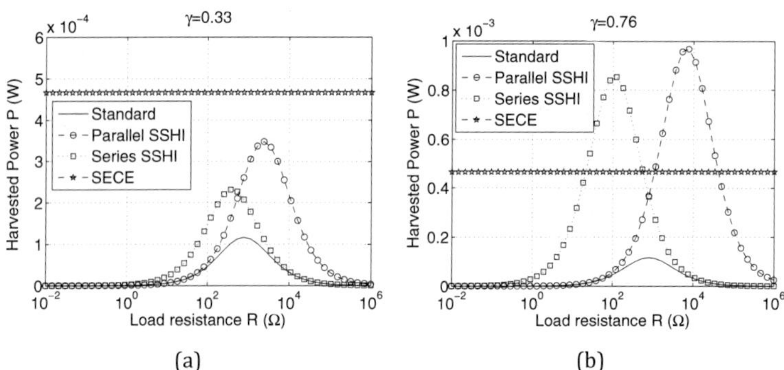

Figure 9.19 Simulated power as a function of the load for different energy harvesting methods with (a) $\gamma = 0.33$ and (b) $\gamma = 0.76$.

9.1.5.2 Series SSHI technique

Physical Principles　Figures 9.20a,b show the series SSHI harvesting circuit and the associated waveforms, respectively. In this case, the inductor is connected in series with the piezoelectric element. The current generated flows only through the inductor, the diode bridge, and the load when the switch is closed. Otherwise, there is no current flowing from the piezoelectric element. As in the parallel SSHI, the control circuit has to detect the peak voltage on the piezoelectric transducer and close the switch at this moment. Therefore, the working principle of the control circuit employed is the same in parallel and series SSHI.

The series and parallel SSHI techniques provide comparable harvested power but their optimum loads are different (see Fig. 9.19).

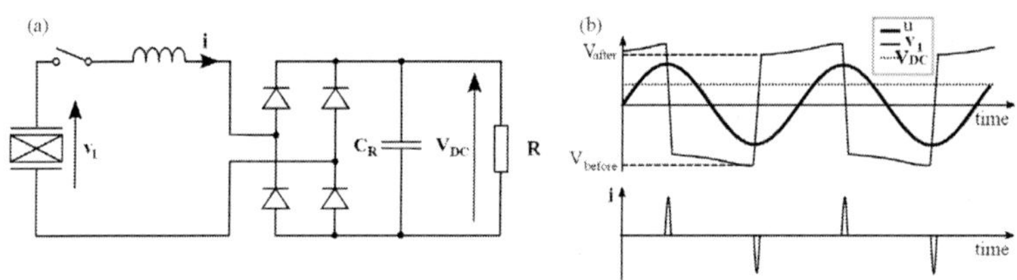

Figure 9.20 Series SSHI energy harvesting technique: (a) Circuit; (b) waveforms [13].

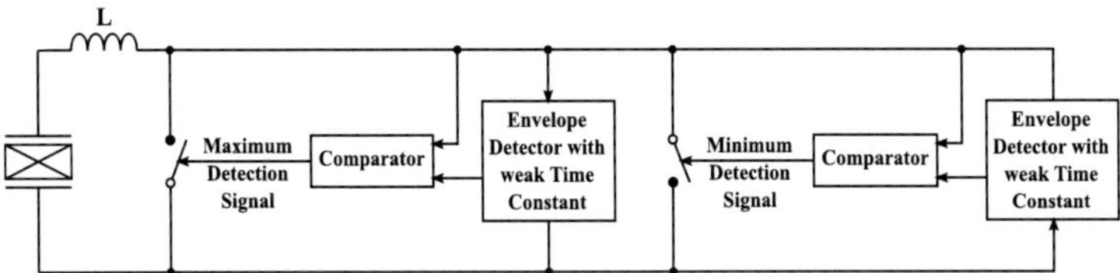

Figure 9.21   Block diagram for the SSDI [15].

Electrical Circuit   Figure 9.21 shows a block diagram for the synchronized switch damping on inductor (SSDI) technique [15]. The control circuit topology employed by the SSDI technique can be also employed by the parallel and series SSHI circuits instead of the one presented in Fig. 9.17. The control circuit shown in Fig. 9.21 that includes an envelope detector, a comparator, and a switch is displayed in Fig. 9.22 for finding the piezoelectric positive peak.

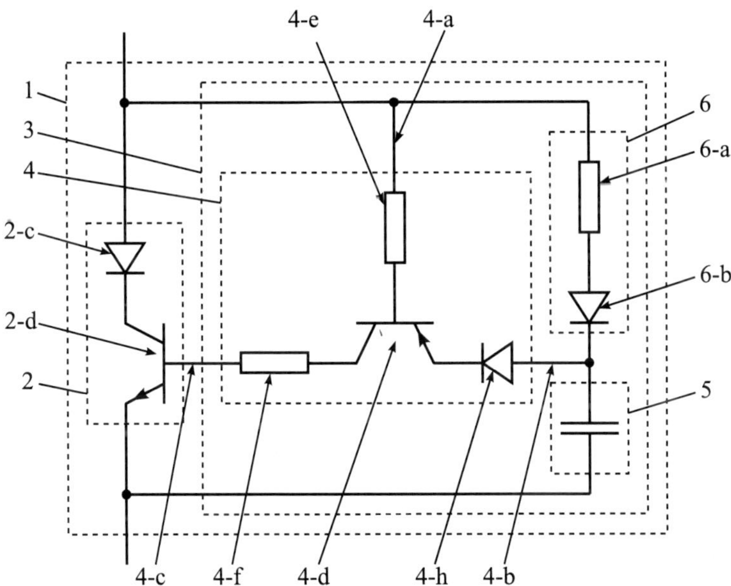

Figure 9.22   Circuit implementing the envelope detector, comparator, and switch in Fig. 9.21 for detecting the positive peak on the piezoelectric element [15].

The envelope detector consists of a differentiator and its components are resistor 6-a, diode 6-b, and capacitor 5. The comparator function is done by block 4. The PNP transistor implements the function of a comparator since it only allows that current flows when the voltage on the piezoelectric element is lower than the voltage on the differentiator output or, in other words, when a maximum is detected. The switch functionality corresponds to block 2 and is composed by diode 2-c and transistor 2-d like in the circuit shown in Fig. 9.17. It is turned on only when a maximum voltage on the piezoelectric element is detected.

9.1.5.3 SECE technique

Physical Principles   Figures 9.23a,b show the circuit for the synchronous electric charge extraction (SECE) technique with two possible topologies: flyback and buck-boost. In the case of the SECE circuit, the inductor and the switch are connected after the diode bridge instead of before as it occurs in the SSHI topologies.

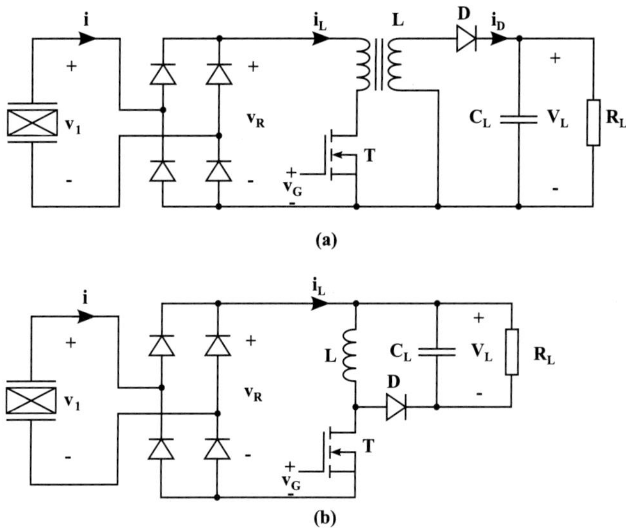

Figure 9.23   SECE energy harvesting circuits: (a) Flyback topology; (b) buck-boost topology [16].

When a maximum is detected at the rectified voltage V_R, the control circuit generates a positive voltage that is applied to the gate of transistor T. Once the transistor is turned on, the piezoelectric current flows through the inductor and the energy is transferred from the piezoelectric element to the inductor. When the current flowing through the inductor is maximum, the rectified voltage is zero and the transistor is turned off since the energy transfer from the piezoelectric element to the inductor has finished. Hence, the piezoelectric element is in open circuit. Afterward, the current starts to flow to storage capacitor C_L and resistive load R_L through diode D [16]. Figure 9.24 shows the different waveforms for the SECE circuit.

Electrical Circuit   Tan et al. [17] designed a practical realization of the SECE circuit that is shown in Fig. 9.25. This circuit is implemented with the flyback topology and includes a startup circuit. The startup circuit has a second diode bridge (D_6, D_7, D_8, and D_9) that rectifies the AC power of the piezoelectric transducer, a resistive load R_{10} and a depletion-mode NMOS M_3 that is normally conducting. During the startup phase, M_3 connects the second diode bridge to the output of the flyback converter. The startup circuit charges C_L that is able to power the control circuit which operates the transistor of the flyback converter during the initial phase. Afterward, transistor M_3 is turned off since the gate-source voltage becomes negative.

The control circuit for the SECE harvesting technique is displayed in Fig. 9.26. The signal obtained with the second diode bridge (D_6-D_9) and R_{10} as load is the input for the control circuit. The control circuit is composed of an active differentiator, a comparator, and a pulse generator. The on-time of the control signal for the NMOS transistor T of the flyback converter is fixed by the pulse generator. The pulse generator output voltage v_G goes from 0 to VDD when the differentiator detects a peak voltage on the piezoelectric transducer. The output of the pulse generator stays at VDD the time fixed by the RC network composed by R_{21}, R_{22}, and C_4.

The two graphs in Fig. 9.19 show the simulated power converted with the standard circuit, the parallel SSHI, the series SSHI, and the SECE circuits as a function of the load. The standard and the SSHI circuits extract the maximum power for a certain load, whereas

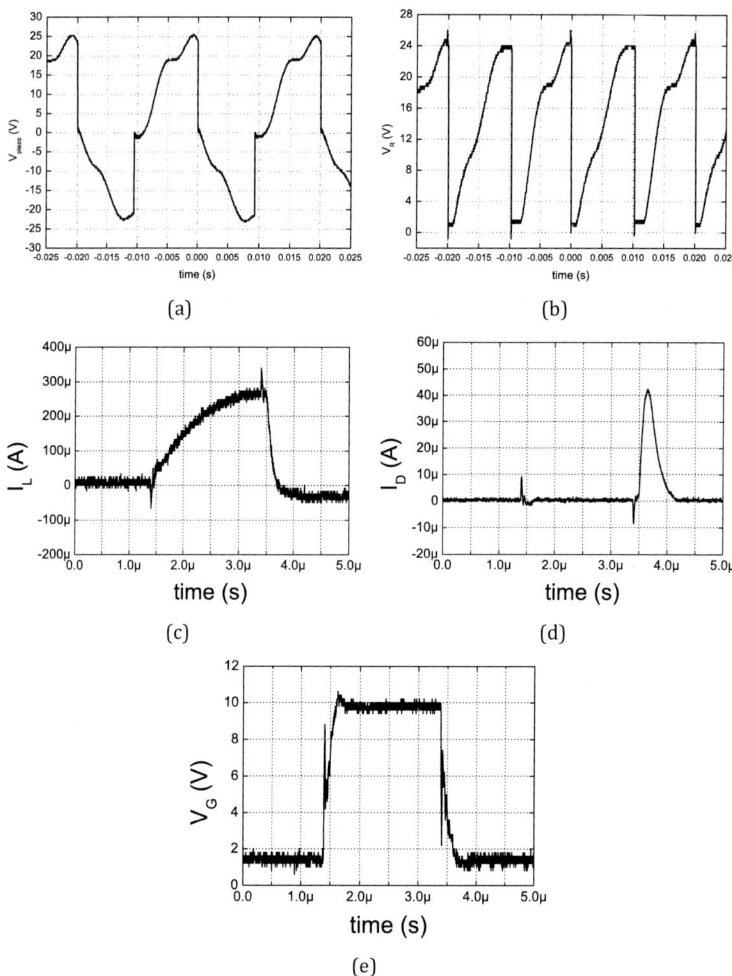

Figure 9.24 Steady-state waveforms for the SECE technique: (a) Voltage on the piezoelectric element V_{piezo}; (b) rectified voltage after the diode bridge V_{R}; (c) current flowing through the inductor I_{L}; (d) current flowing through the diode I_{D}; (e) gate voltage V_{G}.

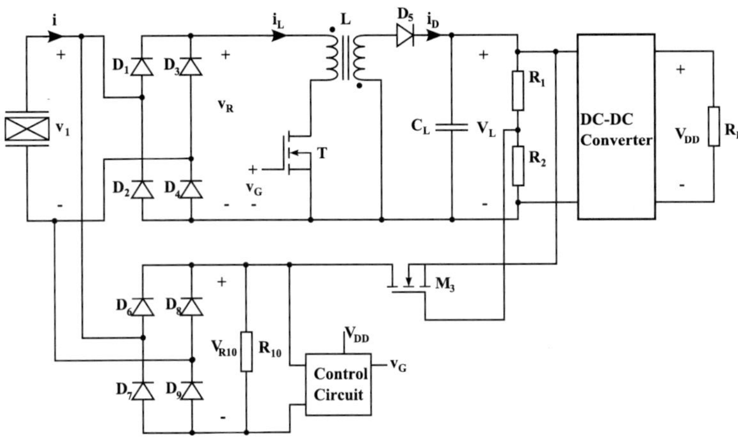

Figure 9.25 Synchronous electric charge extraction circuit (SECE) including startup circuit [17].

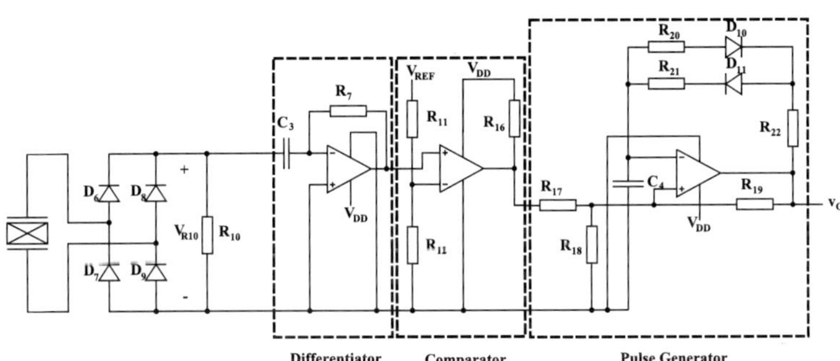

Figure 9.26 Control circuit for the synchronous electric charge extraction (SECE) technique [17].

the SECE technique provides a constant power for a wide range of resistances. Nevertheless, the output power extracted with the SECE technique is lower than the extracted power with the parallel and series SSHI technique when it is working with the optimum resistive load. The optimum load that provides the maximum power for the standard and the SSHI circuits is among others a function of the vibration frequency [18]. Hence, for using the SSHI circuits with different single excitation frequencies, it is necessary to readapt the input voltage value of the step-down converter in order to work in the maximum power point. However, in the case of the SECE circuit this is not necessary since the output power is almost independent of the resistive load.

9.1.5.4 Low-frequency pulsed resonant technique

Physical Principles The circuit presented in this section (see Fig. 9.27) is a modification of the SECE circuit. Like in the SECE circuit, the AC power delivered by the piezoelectric element is first rectified and when a maximum is detected in the rectified voltage, T_9 is turned on to connect the piezoelectric element to the inductor. Once the piezoelectric element has been discharged and the rectified

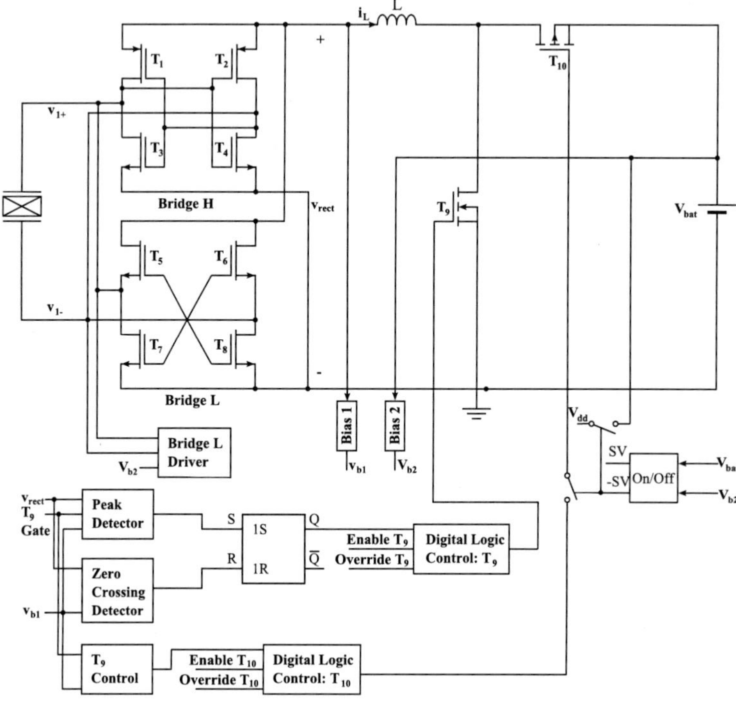

Figure 9.27 Pulsed-resonant converter including bridge H, full wave rectifier with P-MOS and N-MOS transistors, and bridge L, synchronous rectifier [19].

voltage reaches the zero value, T_9 is turned off. At this moment, current i_L, which flows through the inductor, has its maximum value. Afterward, T_{10} is turned on and the inductor current charges the battery. When i_L is zero, T_{10} is turned off to assure that the battery is not discharged.

The input frequency range of the converter is 10 Hz to 1 kHz which is in the application range of most of the vibration sources that are employed as mechanical excitation of piezoelectric elements.

Electrical Circuit Xu et al. [19] designed the low-frequency pulsed resonant converter shown in Fig. 9.27. The converter is composed by two bridge rectifiers: bridge H and bridge L. Bridge H consists of n-channel and p-channel MOSFETs, whereas bridge L consists only of n-channel MOSFETs. The gate voltages of bridge L are controlled by the bridge L driver.

Bridge H is a synchronous bridge rectifier that employs the diodes associated with the n-channel MOSFETs. The body diodes are located between drain and source for the case of the p-channel MOSFETs and between source and drain for the case of the n-channel MOSFETs. Thus, body diodes of MOSFETs T_1 and T_4 are forward biased when the piezoelectric voltage is positive while body diodes of MOSFETs T_2 and T_3 are forward biased when the piezoelectric voltage is negative.

During the positive half cycle of the piezoelectric voltage, transistors T_1 and T_4 are turned on allowing that the current flows without the power losses associated to their body diodes. In a similar way, when there is a negative voltage on the piezoelectric element that turns on transistors T_2 and T_3, the half-wave current is rectified without the power losses associated to their body diodes. Thus, bridge H has low power losses for voltages above a certain limit since the gate threshold voltage of the MOSFETs limits the value of the input voltage that can be rectified employing the conducting transistors and not the body diodes.

Bridge L has the same working principle as bridge H based on the associated body diodes of the MOSFETs, but in this case only n-channel MOSFETs are employed. Bridge L is employed when the

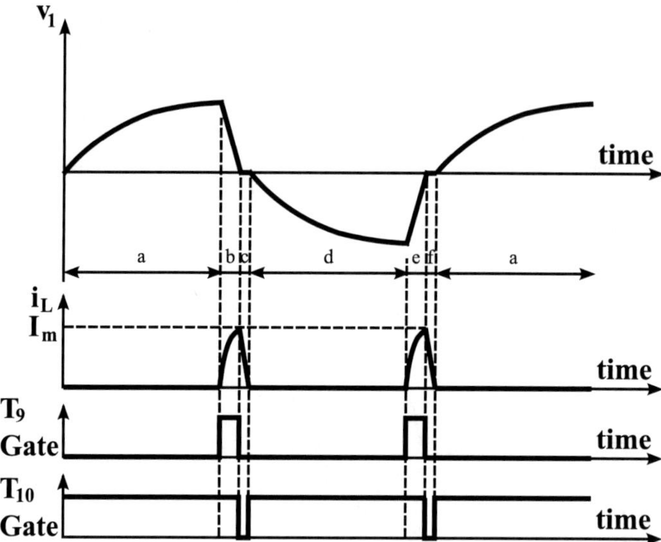

Figure 9.28 Timing waveforms of the piezoelectric voltage and NGate and PGate signals [19].

piezoelectric voltage is close to zero. However, bridge H can operate also when the battery is discharged.

Figure 9.28 shows the timing diagram of the converter circuit for the control signals T_9 Gate and T_{10} Gate. During time a, the piezoelectric voltage increases its value from 0 to its peak value. At this moment, the peak detector turns on T_9. The peak detector circuit is shown in Fig. 9.29. While the voltage on the piezoelectric element does not reach a peak, T_{11} is turned on and capacitor

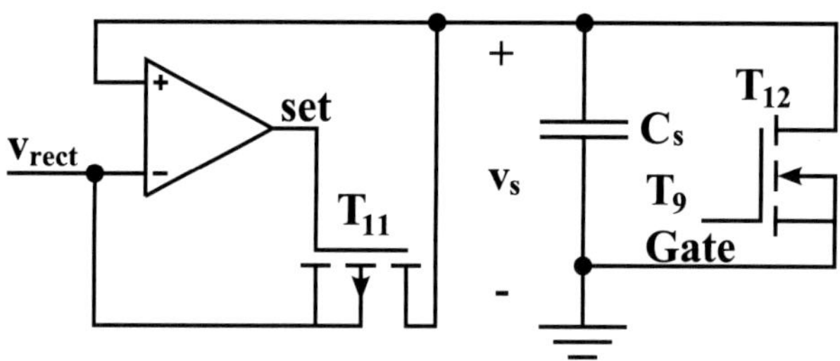

Figure 9.29 Peak detector circuit [19].

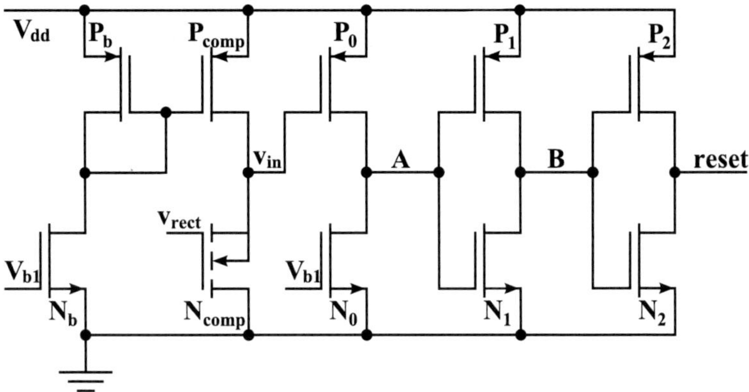

Figure 9.30 Zero voltage crossing detector circuit [19].

C_s is charged to the piezoelectric voltage. Once the voltage on the piezoelectric element is lower than v_s, a peak voltage has been reached and the set signal is high to turn off transistor T_{11}. Consequently, T_9 Gate gate signal goes high to turn on T_9 and to discharge capacitor C_s through transistor T_{12} in order to prepare the circuit for the detection of the next peak. When T_9 is turned on, the piezoelectric element is connected to inductor L composing an LC resonant circuit. When the piezoelectric element is discharged, the inductor current reaches its maximum value I_m.

The zero-crossing detector circuit (see Fig. 9.30) is in charge of finding when the piezoelectric voltage is zero. The zero-crossing detector circuit is activated when signal V_{b1} is high. Transistors P_b and P_{comp} act as a current mirror since gate and drain are connected. The reset signal is high when v_{rect} is lower than the threshold voltage of the circuit. The reset signal generates a low value on NGate signal that turns off T_9 and starts phase c.

The T_{10} control block of the converter generates the control signal T_{10}Gate of T_{10}. This control signal has a high value at the beginning of phase c and lasts the time needed by i_L for going from I_m to zero connecting the inductor to the battery for extracting the stored energy of the inductor. Similar operation takes place during the negative half-cycle of the piezoelectric element.

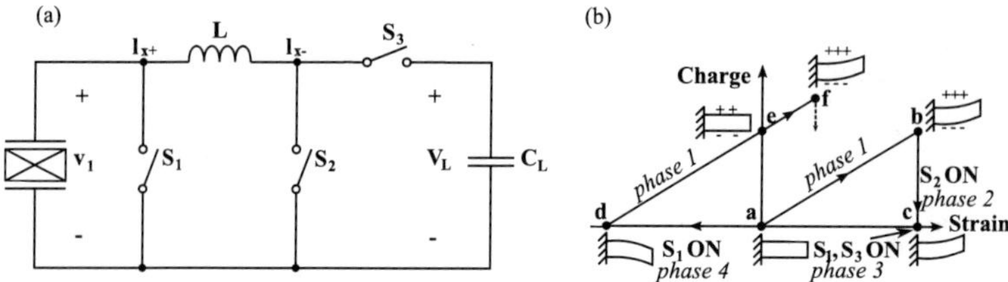

Figure 9.31 (a) Schematic of the non-linear technique, (b) Working cycle of the piezoelectric element [20].

9.1.5.5 AC–DC inductive step-up converter

The non-linear circuit presented in this section does not include a classical AC–DC converter like the full bridge or voltage doubler employed in other energy harvesting circuits.

Physical Principles Figure 9.31a shows a simplified schematic of the AC–DC inductive step-up converter designed by Dallago et al. [20]. The converter leaves the piezoelectric element in open circuit until it reaches its maximum strain, which occurs when the piezoelectric voltage is maximum. At this moment, the energy stored in the piezoelectric element is extracted by the AC–DC converter. The working principle of the converter in conjunction with the piezoelectric element, which is mechanically stressed, is displayed in Fig. 9.31b.

The converter has three switches: S_1, S_2, and S_3 that are controlled by a driving circuit. The operation of the circuit starts in point **a** when the piezoelectric element is at rest. During *phase 1*, the piezoelectric is strained (path **a-b**) and all the switches are open. When the piezoelectric material is at its maximum deflection, the piezoelectric voltage is maximum and *phase 2* starts by switching on S_2 (path **b-c**). Thus, the energy available in the piezoelectric cantilever is transferred into inductor L. All the energy is stored on the inductor when the piezoelectric voltage is zero. At this moment, *phase 3* (point **c**) starts and S_2 is opened while S_1 and S_3 are closed, allowing the energy transfer from inductor L to output capacitor C_L. Once all the energy has been transferred to the output capacitor, *phase 4* (path **c-d**) begins and S_2 and S_3 are opened while only S_1 remains closed. During *phase 4* the piezoelectric transducer is bended in the opposite direction and the piezoelectric voltage is almost zero since S_1 is closed. At point **d**, the piezoelectric transducer suffers its maximum deflection and it starts to be strained in the opposite direction initiating again *phase 1* (path **d-f**) with all the switches opened. During *phase 1*, the piezoelectric voltage increases its value from 0 to $|-V_1| + V_1$, where V_1 is the voltage peak of the equivalent Thevenin piezoelectric voltage source of the piezoelectric cantilever, since the piezoelectric cantilever is bended in both directions during path **d-f**.

The principle of operation of the present converter does not require the use of a classical AC–DC topology since the piezoelectric element only provides positive voltages. During *phase 4*, when the piezoelectric cantilever is bended in the opposite direction, the piezoelectric transducer is short-circuited. Hence, no energy is extracted during *phase 4*.

An advantage of the AC–DC inductive step-up converter over the standard circuit and even the voltage doubler rectifiers is the fact that the piezoelectric voltage has not to be higher than the output voltage V_L to be able to charge the output capacitor since the energy is transferred to the output capacitor via an inductor [20]. Thus, the mechanical energy provided by all the strains applied to the piezoelectric transducer is transferred to the output capacitor.

Electrical Circuit A more detailed circuit shown in the AC–DC inductive step-up converter including its driving circuit for controlling the switches is presented in Fig. 9.32. In Fig. 9.31a, S_1 is replaced by two n-channel MOSFETs, S_1', and S_1''. S_1' is closed during *phase 4*, whereas S_1'' is closed during *phase 3*.

During the startup phase, the parasitic body diodes of transistors S_1', S_1'', and S_3 provide the topology of a voltage doubler, and therefore the voltage on C_L starts to increase.

During *phase 1*, the driving circuit maintains all the switches of the converter opened. When the peak detector circuit detects that there is a peak on the piezoelectric voltage, *phase 2* starts. The output signal of the peak detector is connected to the set input of a flip-flop and the flip-flop output is connected to the driver of switch S_2. This phase finishes when all the energy stored in the piezoelectric element has been transferred to the inductor. At

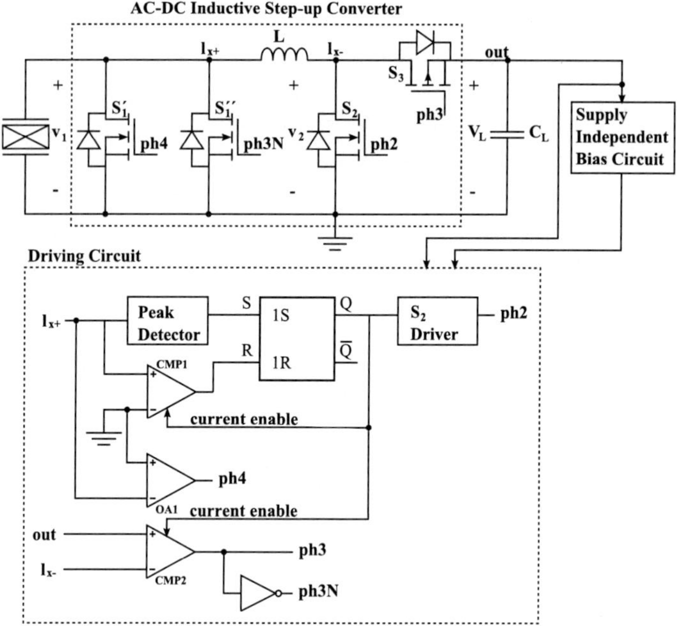

Figure 9.32 AC–DC inductive step-up converter including its driving circuit [20].

this moment, the inductor current is maximum, which means that the piezoelectric voltage is zero. Comparator CMP1 evaluates this condition and resets the flip-flop output to open S_2. When voltage V_2 is higher than the output voltage V_L, CMP2 turns on switches S_1'' and S_3 and phase 3 begins. When voltage V_2 is lower than V_L, the inductor current is zero, and therefore switches S_1'' and S_3 are opened. Phase 4 begins when the piezoelectric current i_p becomes negative which causes that OA1 turns on switch S_1'. Operational amplifier OA1 and switch S_1' limit the drain to source voltage of the transistor to the input offset voltage of the operational amplifier that is 20 mV.

9.1.5.6 Design optimization for non-linear techniques

A major concern is the operation of the control circuit of the non-linear techniques when it has to deal with different excitation frequencies applied to the piezoelectric transducer. There are two possibilities:

- Make a parallel or series connection of different piezo-electric elements with resonant frequencies close to each other in order to obtain a broadband piezoelectric energy harvesting device like in Ref. [21]. In this case, the range of frequencies is small and the differentiator included in the control circuits of the non-linear techniques will operate correctly in the entire frequency range.
- Use different resonant modes of the piezoelectric elements at the same time. In this case, there is only one type of piezoelectric transducer that is employed at different bending modes. The frequencies are far away and the differentiator of the control circuit cannot respond to all of them because of bandwidth limitations.

Figure 9.33 shows four different graphs for the output power obtained with the SECE technique where $SECE_{in}$ represents the power obtained after the diode bridge and $SECE_{out}$ is the power obtained after the buck-boost converter. The figures (a), (b), and (c) display the output power of the SECE circuit for the first, second, and third natural frequency, respectively, of the piezoelectric transducer, whereas (d) shows the output power obtained when the three first natural frequencies are applied at the same time to the piezoelectric transducer [3]. Figures 9.33a–c show a measured value of $SECE_{in}$ of three times the maximum output power obtained with the standard circuit. Nevertheless, the graph in Fig. 9.33d shows a power gain of only 1.7 for the SECE technique compared with the standard technique due to the combination of three different frequencies. Lefeuvre et al. recommended in this case to select the appropriate local extrema of the piezoelectric voltage.

The same problem is analyzed by Lallart et al. [15] for the SSDI technique when broadband frequencies are applied to the piezoelectric element. It is concluded that the most effective method is to close the switch and connect the piezoelectric to the inductor only when global maximums are achieved on the piezoelectric element and not during local maximums. This switching method requires a new control technique that is detailed in Fig. 9.34, where

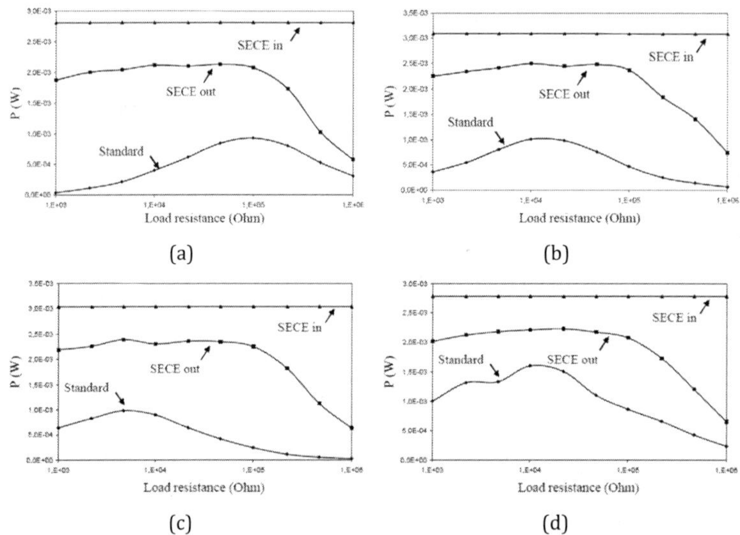

Figure 9.33 Power as a function of load for the piezoelectric: (a) First natural frequency at 56 Hz; (b) second natural frequency at 334 Hz; (c) third natural frequency at 915 Hz; (d) for the three modes mixed [3].

two envelope detectors: A first one for large time constants and a second one for weak time constants are used. The only difference between those envelope detectors is the value of the resistor and capacitor of the differentiator. The control diagram in Fig. 9.34 is also valid for achieving better results for the series and parallel SSHI technique.

9.2 AC–DC Converters for Electrostatic Transducers

In electrostatic generators, the movable plate of the capacitor converts the mechanical energy into electrical energy by the movement against the Coulomb force present between the two plates of the capacitor.

Meninger et al. [22] presented an electrostatic generator with two different designs. The first design had two parallel capacitors operated with a constant charge where one of them was a variable capacitor with moving plates. The second design was a variable

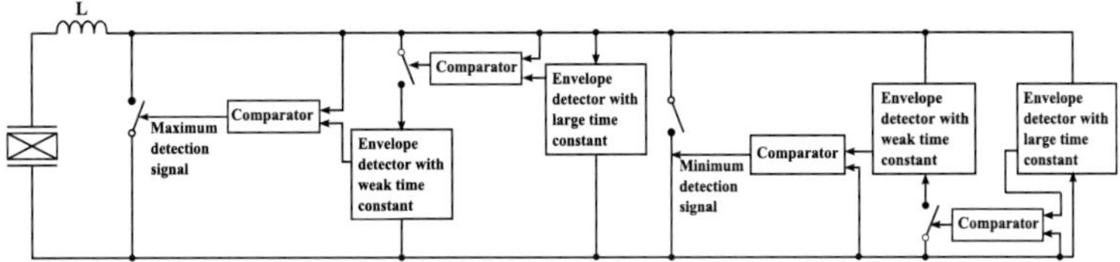

Figure 9.34 Control block diagram for the adaptative SSDI technique [15].

capacitor operated with a constant voltage. These generators are also called Coulomb-damped resonant generators (CDRGs) because they are based on electrostatic damping.

The change in the capacitance of the variable capacitor can be implemented by a change of the distance between the capacitor plates or the overlap area of the capacitor plates.

9.2.1 *Physical Principles*

If the charge on the variable capacitor is maintained constant while the capacitance decreases (e.g., reducing the overlap area of the plates or increasing the distance between them), the voltage will increase. On the other hand, if the voltage on the capacitor is maintained constant while the capacitance decreases, the charge will decrease. Figure 9.35 illustrates the process of charging and discharging the capacitor following constant charge (path A-B-D-A) or constant voltage (path A-C-D-A) approaches. The energy enclosed by the total path is the energy extracted in the mechanical to electrical conversion process.

The charge-constrained conversion cycle starts when the capacitance (given by the slope of the Q–V curve) is maximum. At this moment, a voltage source charges the variable capacitor with an initial voltage V_{start} and the conversion cycle goes from point A to point B. During the path B-D the plates are moving from maximum capacitance C_{max} to minimum capacitance C_{min} with constant charge Q_a. As the capacitance decreases and the charge is maintained constant, the voltage increases its value until it reaches its maximum value V_{max} at point D. The mechanical vibrations are converted into electrical energy during the path B-D. The discharge of the

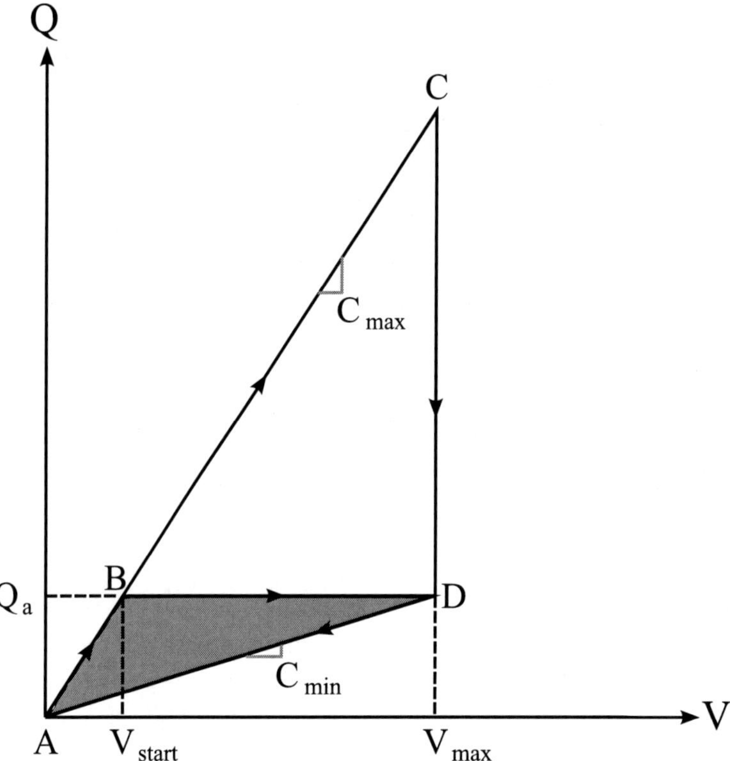

Figure 9.35 Diagram explaining electrostatic energy conversion [22]. The area enclosed by the points ABD corresponds to the energy converted in one charge-constrained conversion cycle while the area enclosed by the points ACD is the energy converted in one voltage-constrained conversion cycle.

capacitor is shown in path D-A. The energy converted in one charge-constrained conversion cycle is given by Eq. (9.24) and corresponds to the shaded area A-B-D-A in Fig. 9.35.

$$E_{charge} = \frac{1}{2}\left(C_{min}V_{max}^2 - C_{max}V_{start}^2\right) \tag{9.24}$$

As there is no variation in the charge,

$$C_{max}V_{start} = C_{min}V_{max} \tag{9.25}$$

Then, Eq. (9.24) can be rewritten as:

$$E_{\text{charge}} = \frac{1}{2} V_{\text{start}} V_{\text{max}} \left(C_{\text{max}} - C_{\text{min}} \right) \tag{9.26}$$

The voltage-constrained conversion cycle also starts at point A, when the capacitance is maximum. At this moment, a voltage source charges the variable capacitor with an initial voltage V_{max}, and therefore the cycle goes from point A to point C. The path C-D corresponds to the plates moving from maximum capacitance C_{max} to minimum capacitance C_{min}. Path D-A shows the discharge of the capacitor. The mechanical movement that takes place in path C-D is converted to electrical energy with a constant voltage. The energy converted in one voltage-constrained conversion cycle is given by Eq. (9.27) and it is equal to the area enclosed by the points ACD in Fig. 9.35.

$$E_{\text{voltage}} = \frac{1}{2} \left(C_{\text{max}} - C_{\text{min}} \right) V_{\text{max}}^2 \tag{9.27}$$

The energy gained in the conversion process is transferred from the variable capacitor to the energy storage capacitor C_{res} shown in Fig. 9.36a, along path D-A for both charge and voltage-constrained cycle. Figure 9.35 shows graphically that the mechanical energy converted into electrical energy is larger if the voltage across the capacitor is constrained than if the charge across the capacitor is constrained. The analytical comparison of the converted energies is done in Eq. (9.28).

$$\frac{E_{\text{voltage}}}{E_{\text{charge}}} = \frac{V_{\text{max}}}{V_{\text{start}}} \tag{9.28}$$

The charge-constrained conversion cycle employs only the voltage source V_{start} to power the control electronics and to charge the variable capacitor with an initial charge. Nevertheless, the voltage-constrained conversion cycle requires two voltage supplies, one to power the control electronics and a second one V_{max}, with a higher value to charge the variable capacitor. Thus, the ideal alternative would be to use only one voltage supply but obtaining the energy of the voltage-constrained energy conversion cycle. This solution is achieved adding a second capacitor C_{par} connected in parallel to the variable capacitor.

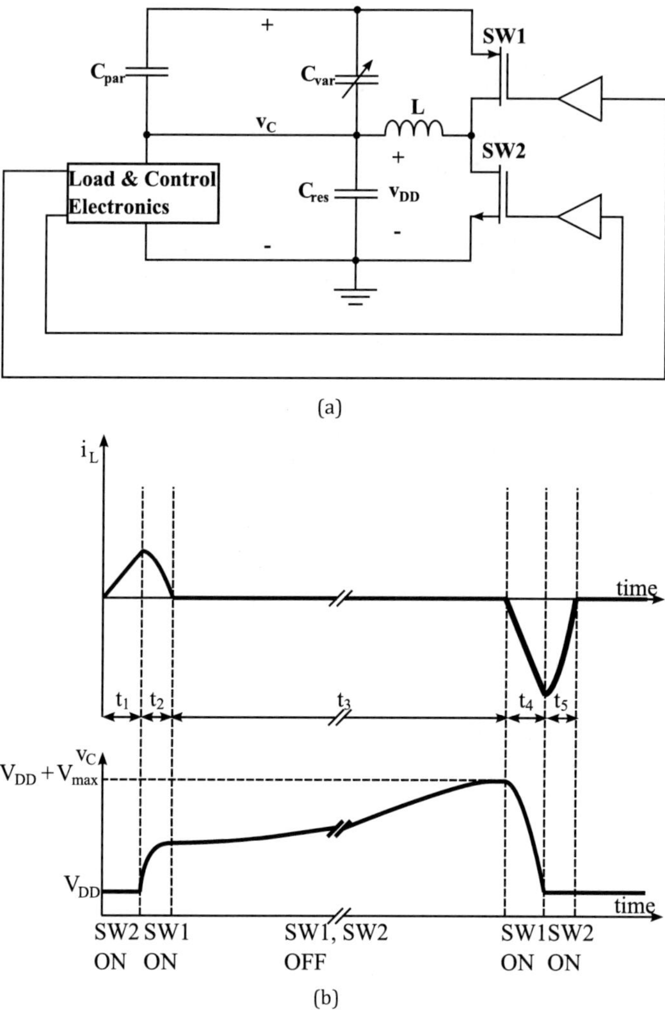

Figure 9.36 (a) AC–DC circuit for the charge-constrained conversion cycle; (b) time waveforms for V_c and i_L [22].

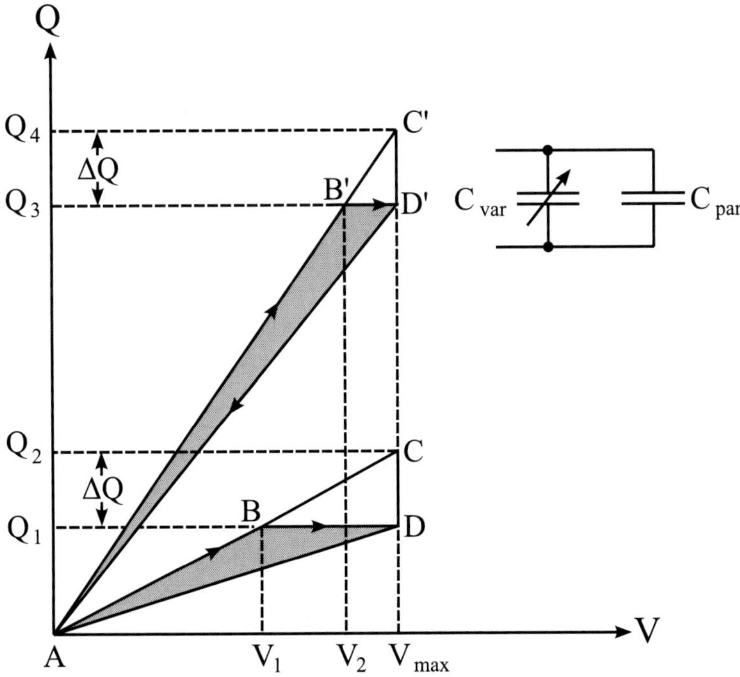

Figure 9.37   Energy conversion cycle with C_{par} in parallel with C_{var} [23].

Figure 9.37 shows the energy conversion cycles with and without parallel capacitor. In the case of voltage-constrained energy conversion cycle, the area A-C-D-A (without C_{par}) is equal to the area A-C'-D' (with C_{par}). Nevertheless, for the charge-constrained energy conversion cycle there is more energy converted when C_{par} is connected (area A-B'-D'-A) than when C_{par} is not in use (area A-B-D-A). Now, comparing again the charge and the voltage-constrained energy conversion cycles, the following is obtained:

$$E'_{charge} = E'_{voltage} - \frac{(\Delta Q)^2}{2\left(C_{par} + C_{max}\right)} \tag{9.29}$$

From the previous equation, it is deduced that the electrical energy obtained for the charge-constrained method is increased by adding a capacitor C_{par}. When C_{par} approaches infinity, the energy of both cycles is equal. Nevertheless, the disadvantage of adding C_{par} is that the required initial voltage is higher for the charge-constrained cycle.

9.2.2 AC–DC Electrical Circuit for the Charge-Constrained Conversion Cycle

Figure 9.36 shows the converter required for the charge-constrained energy conversion cycle when a capacitor C_{par} is connected in parallel to the variable capacitor C_{var}. It also displays the current waveform i_L on the inductor and the voltage waveform V_C. The converter circuit has two switches SW1 and SW2 that are turned on and off by the control circuit.

At the beginning, both switches are turned off and only C_{res} is charged to the supply voltage V_{dd}, and therefore $V_C = V_{dd}$. When a maximum capacitance of the parallel connection of C_{par} and C_{var} is detected, the conversion cycle starts. During time interval t_1, SW2 is turned on and C_{res} is connected in parallel with inductor L. Therefore, there is a transfer of energy from C_{res} to L and the inductor current i_L increases. During t_2, SW2 is turned off and SW1 is turned on. Then, the parallel capacitor $C_{par} + C_{var}$ is charged by the inductor and its current decreases. At t_3, both switches are open and the variable capacitor C_{var} connected in parallel to C_{par} changes its capacitance from the maximum value to the minimum value due to the mechanical vibration. Thus, the voltage across capacitors C_{par} and C_{var} reaches its maximum value and the mechanical-electrical energy conversion cycle is done. During t_4, SW1 is turned on and the electrical energy is totally transferred from C_{par} and C_{var} to the inductor. Time interval t_5 starts when the voltage across the capacitors is zero and the current through the inductor reaches its maximum absolute value. At this moment, SW1 is turned off, SW2 is turned on and the inductor is connected in parallel to C_{res}, charging the capacitor.

Figure 9.38 displays the different electrical connections of the electronic circuit shown in Fig. 9.36 depending on the time interval under consideration [23].

The electrical time constant LC of the power converting circuit in any of the time intervals has to be much lower than the vibration time constant. A synchronous rectification can be done when the

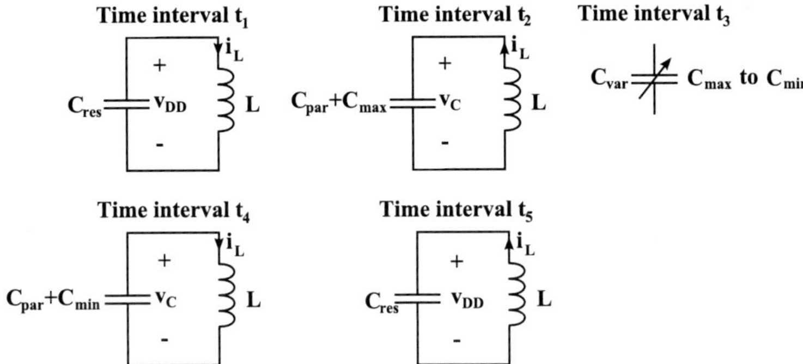

Figure 9.38 Electronic connections of the power electronic circuit for the electrostatic converter during the different time intervals [23].

values of the components L, C_{var}, C_{par}, and C_{res} are known. For synchronization, the vibration frequency must be considered to assure that $C_{par} + C_{var}$ is charged at its maximum capacitance and that it is discharged at its minimum capacitance. In this way, the conversion of vibrations into electrical energy is maximum [24].

Meninger et al. [25] report predicted data from their electrostatic generator, which provides 8.6 µW of output power without the ability to be synchronized with the capacitor motion.

Miyazaki et al. [26] present a circuit that generates the control signals for commuting the switches SW1 and SW2 in Fig. 9.36. Figures 9.39 and 9.40 show the schematic of the control circuit and the timing waveforms of the control signals, respectively. Figure 9.39 includes the detailed circuit for one of the delay blocks (del1) that is composed by a monostable multivibrator. The time interval of the delay signal at high state is given by $t = \ln(2)R_d C_d$ where R_d is a potentiometer, and therefore its value is adjusted manually. The clock signal f_{ref} corresponds to the resonance frequency of the electrostatic generator.

The upper delay blocks (del1, del2, and del3) of the control circuit switch during one half of the vibration cycle while the lower delay blocks (del4, del5, and del5) switch during the other half of the cycle. After the delay blocks, there is a two-input NOR gate connected to the gate of the PMOS transistor and a two-input OR gate connected to the NMOS transistor. The NOR gate has the output signals of

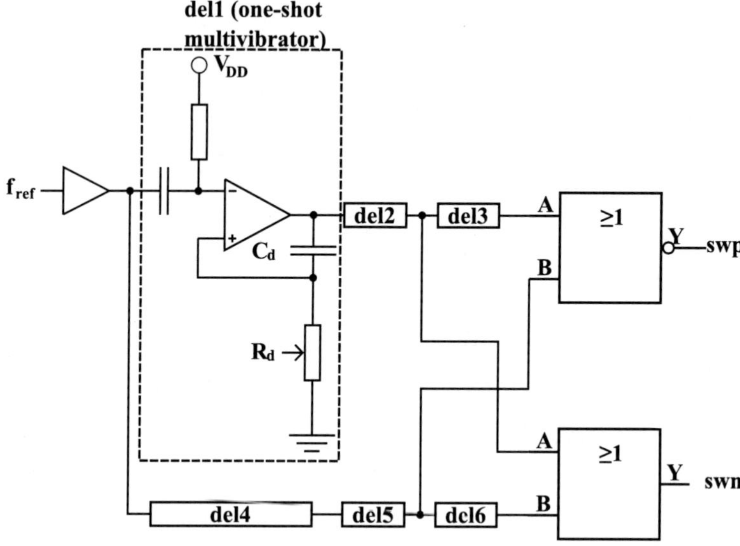

Figure 9.39 Control circuit for the electrostatic converter [26].

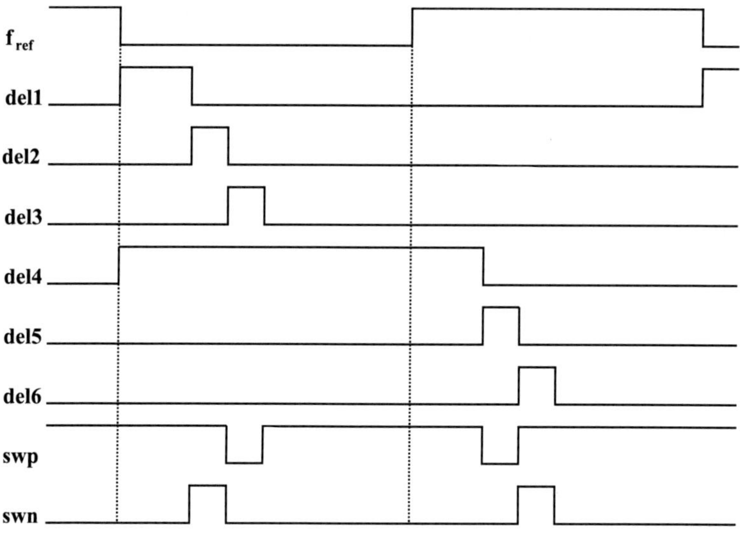

Figure 9.40 Timing waveforms of the control circuit shown in Fig. 9.39 [26].

the delay blocks del5 and del3 as input and therefore, the PMOS transistor is turned on when block del5 or del3 output is high. The OR gate has the outputs of the delay blocks del6 and del2 as input and consequently, the NMOS transistor is turned on when block del6 or del2 output is high.

Miyazaki et al. [26] provide an analysis of the efficiency of the electrostatic generator. The efficiency decreases due to the mechanical to electrical conversion losses of the variable capacitor, the losses in the charge transportation from the variable capacitor to the LC tank circuit, and the losses due to the non-ideal synchronization of the switches to the mechanical excitation. The mechanical to electrical conversion has a power-maximizing condition that determines the optimum design of the micro-generator. The charge transportation efficiency is analyzed as a function of the energy consumption in the inductance L, the capacitance C, and the parasitic resistance R. The timing-capture efficiency is assumed to be 100%. The mechanical to electrical conversion efficiency was 57% and the charge transportation efficiency was 37% in this approach, and so the total converter efficiency was 21%. The measured power of the micro-generator was 120 nW for an input vibration of 1 μm at 45 Hz.

Roundy et al. [27] propose an electrostatic converter for the charge-constrained conversion. Figure 9.41 [28] shows a simplified circuit for this electrostatic converter. V_{in} is the initial voltage needed to charge the variable capacitor. This voltage can be obtained from another capacitor or a battery. C_{var} represents the variable capacitor done with a MEMS structure. C_{par} is the parasitic capacitance associated with C_{var} and C_{stor} is the storage capacitor where the converted electrical energy is accumulated. SW1 and SW2 are the

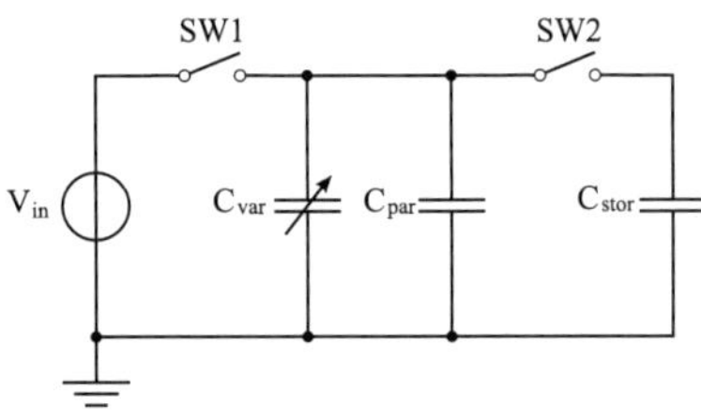

Figure 9.41 Circuit for the electrostatic converter [27].

two switches of the converter circuit. The principle of operation is as follows: When the variable capacitor has its maximum capacitance, SW1 is switched on and C_{var} is charged at V_{in}. Afterward, SW1 is opened again and C_{var} goes from its maximum to its minimum value. Therefore, the conversion of mechanical into electrical energy takes place and when C_{var} reaches its minimum value, SW2 is switched on and the charge is transferred to C_{stor}.

Sterken et al. [29] developed a new approach to electrostatic MEMS CDRG. The main improvement in their design is the employment of an electret for polarization. Thus, no voltage source is needed as occurring in the design of Meninger et al. [25]. The device consists of two micromachined capacitors connected in parallel and carrying a constant charge. The variable capacitors have opposite capacitance variations, when one increases, the other decreases. The systems from Roundy et al. [27] and Meninger et al. [25] have time periods where no electrostatic conversion takes place since the variable capacitor has to be charged and discharged when the converted energy increases. However, the working principle of the variable capacitor presented in this paper ensures a duty cycle of 100%. The designed micro-generator prototype is capable to produce 100 µW of electrical power at 1.2 kHz for a displacement of 20 µm.

Miao et al. [30] presented an electrostatic generator with a variable capacitance range from 100 pF to 1 pF operated in constant charge mode with an energy conversion rate of 2.4 µJ per cycle or 24 µW with a vibration frequency of 10 Hz. The power converter circuit is shown in Fig. 9.42. When one of the plates of the variable capacitor is connected to the terminal V_{in}, transistor Q1 is turned on to transfer the energy from battery $B1$ to inductor $L1$. Afterward, Q1 is turned off and the energy is transferred from the inductor to the variable capacitor. Once the variable capacitor is charged, its capacitance increases from its minimum to its maximum value and the variable capacitor is connected to the V_{out} terminal for extracting the electrical energy stored. First, the energy is transferred to inductor $L2$ and later on to the battery through diode $D2$.

Stark et al. [31] designed a non-resonant electrostatic generator for low-frequencies that can be employed for recovering energy from human body motion. The power converter circuit employed is

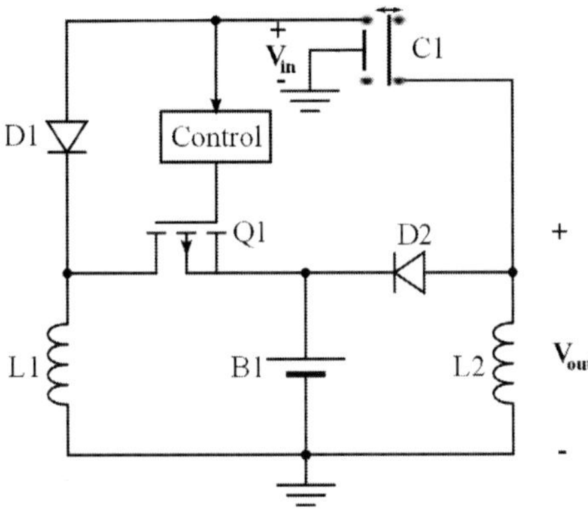

Figure 9.42 Schematic of the power converter proposed by Miao et al. [30].

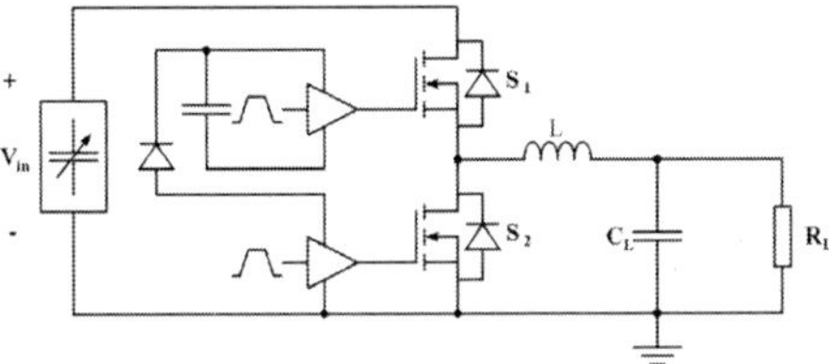

Figure 9.43 Half-bridge step-down converter [31].

a modified step-down converter (see Fig. 9.43). The voltage stored in the variable capacitor after the mechanical movement is around 250 V. Thus, the high side MOSFET has to block high voltages and the driver circuit for controlling its gate voltage is referenced to the source of the MOSFET.

The variable capacitor is connected in parallel to the parasitic capacitor of the depletion layer and the energy stored in this parasitic capacitor is lost when the high side MOSFET is turned on in order to step-down the voltage stored on the variable capacitor. The step-down converter has to deal with currents in the range of 0.1-1 A during short pulses of up to 1 μs. The increase in the inductance rises the pulse duration of the current; therefore the peak inductor current decreases, which reduces the on-state power losses of the MOSFET. Simulations with different inductances and numbers of MOSFETs connected in parallel were done and a

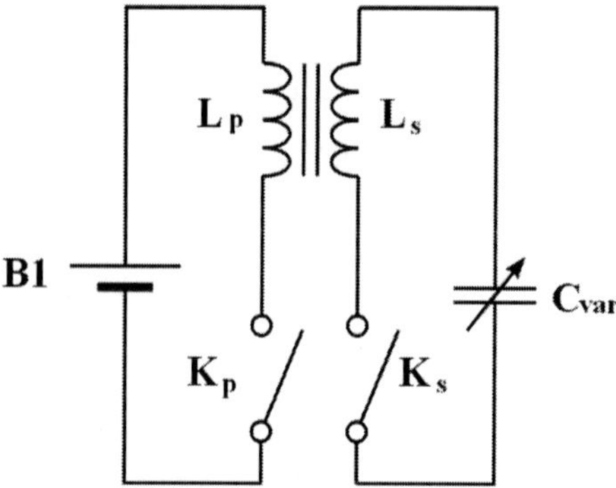

Figure 9.44 Electrostatic energy harvesting flyback converter [32].

conversion efficiency of 65% was obtained with 20 MOSFETs and 100 μH and 42% for 30 MOSFETs and 10 μH.

Despesse et al. [32] designed a modified flyback converter, shown in Fig. 9.44, that overpasses two problems of the modified buck converter from Stark et al.: the gate drive for the high side MOSFET and a shoot-through current which reverse biases the blocking junction of the low side MOSFET. The new converter has two transistors: K_p and K_s, an energy storage unit represented by $B1$, a variable capacitor C_{var} and a transformer with inductances L_p and L_s. Transistor K_p is switched on for transferring the energy from $B1$ to inductor L_p. Once K_p is switched off and K_s is switched on, the energy is transferred to the variable capacitor. The discharge of the variable capacitor starts switching on transistor K_s and afterward switching off and on K_s and K_p, respectively, to transfer the energy stored on the transformer to the energy storage element $B1$.

Mitcheson et al. [33] simulated the modified buck shown in Fig. 8.43 and a modified flyback converter. Higher efficiencies were achieved with the modified buck converter since the parasitic capacitance of the diode of the modified flyback converter reduces the generation efficiency.

Yen et al. [34] designed a converter circuit composed by a charge pump and a buck converter. Figure 9.45 shows the converter where C_{RES} represents the energy storage capacitor that powers the resistive load R_L. Diodes D_1 and D_2 with the variable capacitor C_{var} compose the charge pump that provides the initial voltage to

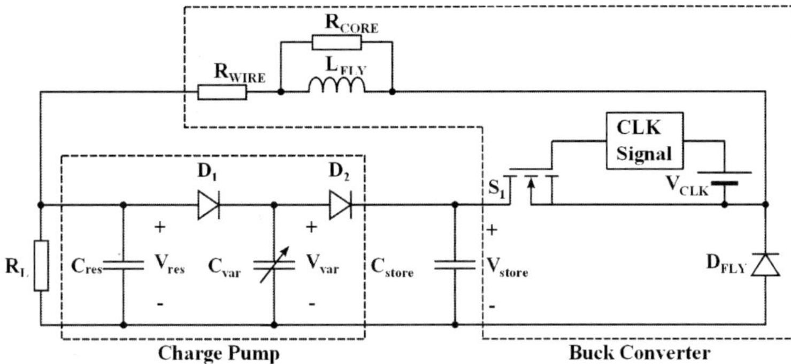

Figure 9.45 Electrostatic energy harvesting converter circuit constituted by a charge pump and a buck converter [34].

the variable capacitor through diode D_1 and transfers the energy harvested on C_{var} to C_{store} through diode D_2. Then, the buck converter decreases the voltage V_{store} to V_{res}.

The variable capacitor is charged through diode D_1 to the value V_{res}, point 4 of the energy conversion cycle shown in Fig. 9.46. Due to a mechanical vibration, the variable capacitor increases its capacitance to its maximum value at constant voltage, path 4-1 in Fig. 9.46. Afterward, a mechanical vibration reduces the variable capacitance at constant charge. Thus, V_{var} increases and diode D_1 is reverse-biased, path 1-2 in Fig. 9.46. When voltage V_{var} is higher than V_{store}, diode D_2 is forward-biased, path 2-3 in Fig. 9.46. While D_2 is forward-biased, energy is transferred from C_{var} to C_{store} while C_{var} continues decreasing, and therefore V_{var} continues increasing, until C_{var} reaches its minimum capacitance value. Then, a mechanical vibration forces an increment on C_{var} which causes diode D_2 is reverse-biased, path 3-4 in Fig. 9.46.

From Fig. 9.46, it is deduced that the first part of the energy conversion cycle is done at constant charge (path 1-2). Nevertheless, once D_2 is turned on, the energy conversion cycle is neither done at constant charge nor at constant voltage (path 2-3). The voltage on capacitor C_{store} is calculated as follows:

$$V_{store,n} = V_{res}\left[\left(1 - \frac{C_{max}}{C_{min}}\right)\left(\left(\frac{C_{store}}{C_{min} + C_{store}}\right)^n + \frac{C_{max}}{C_{min}}\right)\right], \quad (9.30)$$

where n is the number of cycles done.

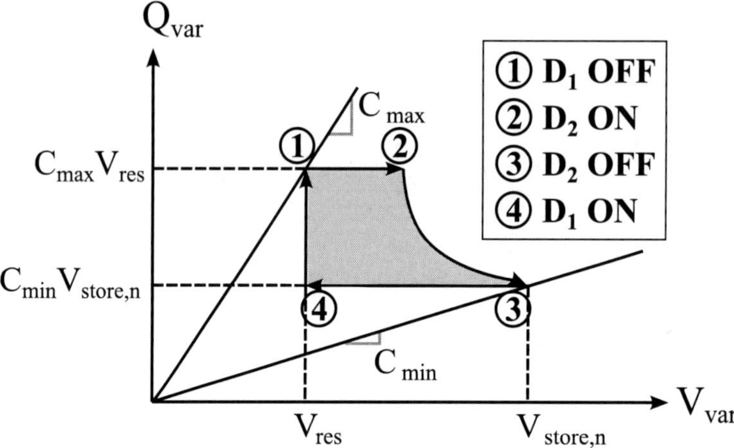

Figure 9.46 Energy conversion cycle for the electrostatic energy-harvesting circuit shown in Fig. 9.45 [34].

If no buck converter is placed after capacitor C_{store}, the voltage on this capacitor would reach the value:

$$v_{store,\infty} = \frac{C_{max}}{C_{min}} V_{res},\qquad(9.31)$$

where $n = \infty$.

It is not necessary that the turn-on time of the MOSFET is synchronized with the motion of the variable capacitor. It is enough to initiate the on-state when v_{store} is bigger than a threshold value. Thus, Yen et al. turned on the MOSFET once every four energy conversion cycles. The electronic converter has an efficiency of 19.1% when an output load of 20 MΩ is powered at 6 V.

9.2.3 Efficiency Calculation for the Charge-Constrained Conversion Cycle

The total efficiency of the electrostatic energy harvesting system is the combination of the mechanical effectiveness, the efficiency during the generation phase, and the conversion efficiency of the electronic converter to discharge the variable capacitor [35].

$$\eta_{eff} = \eta_{mech}\,\eta_{gen}\,\eta_{conv}\qquad(9.32)$$

9.2.4 *Electrical Circuit for the Voltage-Constrained Energy Conversion Cycle*

Torres et al. [36] presented a power converter circuit for the voltage-constrained energy conversion cycle. Figure 9.47 shows the operation principle of the converter that has three phases: (a) pre-charge, (b) harvest, and (c) recover. A Lithium-ion battery is the energy storage element employed for charging with a constant voltage V_{bat} the variable capacitor and for storing the converted energy during the voltage-constrained energy conversion cycle.

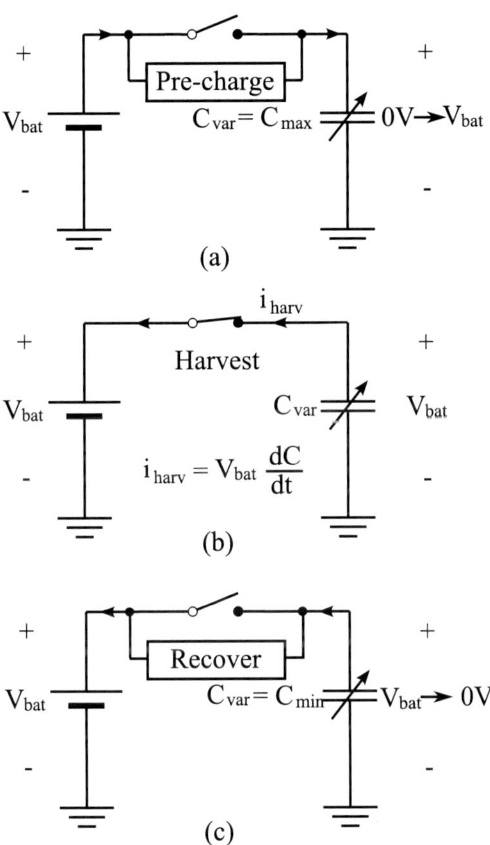

Figure 9.47 Electronic circuit for the voltage-constrained energy conversion cycle proposed by Torres et al. [36].

Thus, in phase (a), the variable capacitor, which has its maximum capacitance C_{max}, is charged by the Li-ion battery. During phase (b), the variable capacitor changes its capacitance from its maximum value C_{max} to its minimum value C_{min}. Therefore, current i_{harv} (see Eq. (9.33)) flows from the variable capacitor to the battery where the harvested energy is stored. Once the variable capacitor reaches its minimum value, the variable capacitor is discharged into the battery in phase (c).

A more detailed schematic of the converter circuit employed for the voltage-constrained energy conversion cycle is displayed in Fig. 9.48. In order to reduce the power losses, the variable capacitor is charged to its initial voltage through inductor L. Thus, the pre-charge phase comprises a first stage where the energy is transferred from the Lithium-ion battery to the inductor (step 1 with switches S_1 and S_3 closed) and a second stage where this energy is transferred from the inductor into the variable capacitor (step 2 with switches S_2 and S_4 closed). During the harvesting phase (step 3), switch

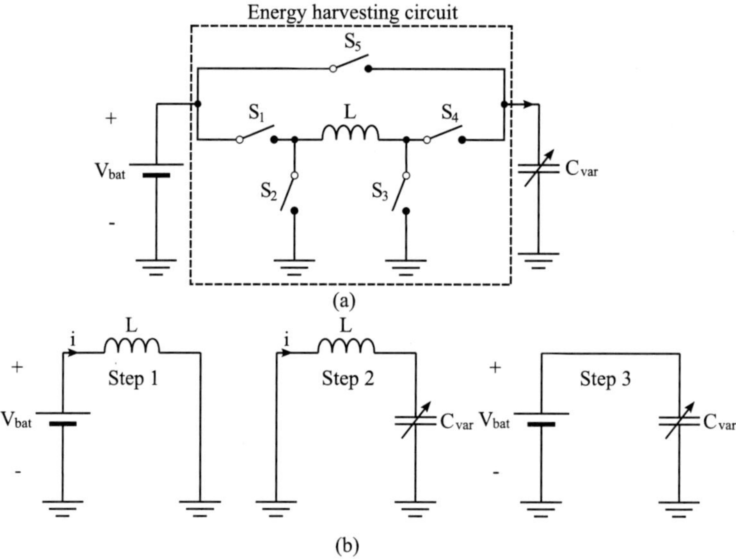

Figure 9.48 (a) Electronic circuit for the voltage-constrained energy conversion cycle; (b) connectivity of the power electronics circuit during different phases proposed by E.O. Torres et al. [36].

S_5 is closed connecting the variable capacitor to the battery. Thus, the variable capacitor goes from its maximum capacitance to its minimum capacitance while a constant voltage V_{bat} is applied to its terminals causing that current i_{harv} charges the battery. In this case, there is no recovery phase since the energy that is remaining in the capacitor is considered very low in relation to the harvested energy and the voltage on the variable capacitor is reduced under constant charge conditions.

The current generated during the harvesting phase is given by the following expression:

$$i_{harv} = V_{bat} \frac{\partial C}{\partial t} \tag{9.33}$$

An initial analysis of the energy gained with this harvesting circuit is done by Torres et al. assuming that there are no power losses. The net energy gained per cycle is

$$\Delta E_{Net} = -\Delta E_{Invested} + \Delta E_{Harvested} + \Delta E_{Recovered} = \frac{1}{2} \Delta C V_{bat}^2 \tag{9.34}$$

The invested energy corresponds to the energy needed to pre-charge the capacitor at its maximum capacitance:

$$\Delta E_{invested} = \frac{1}{2} C_{max} V_{bat}^2 \tag{9.35}$$

The energy stored in the battery during the harvesting phase is given by

$$\Delta E_{Harvested} = \int V_{bat} i_{harv}(t)\, dt = V^2 \int \frac{dC(t)}{dt}\, dt = V_{bat}^2 \Delta C \tag{9.36}$$

The energy that remains in the capacitor after the harvesting phase finishes and that is recovered is

$$\Delta E_{Recovered} = \frac{1}{2} C_{min} V_{bat}^2 \tag{9.37}$$

Figure 9.49 shows the complete electrostatic energy harvesting generator including its control signals. In this circuit, the battery is modeled as a large capacitor in series with a resistance R_{ESR_BAT}. The variable capacitor is modeled as a parasitic capacitor C_{PAR} in parallel with a variable capacitor C_{var} and a series resistance R_{ESR_VAR}. The inductor model includes a series resistance R_{ESR_L}. Switch S_1 is a PMOS transistor (MP_1), whereas S_2 and S_3 are NMOS transistors

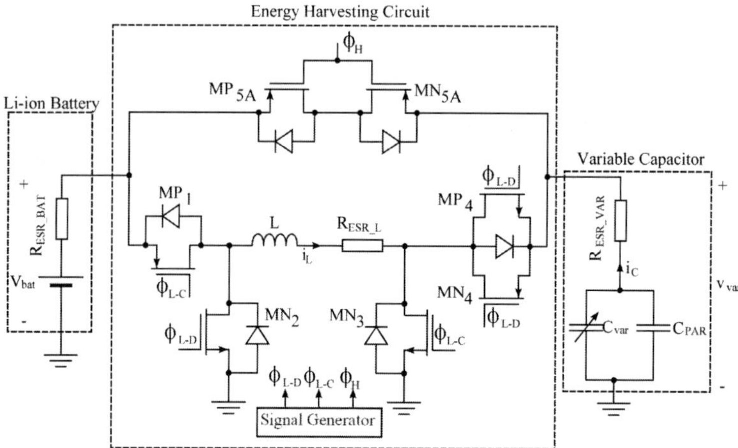

Figure 9.49 Detailed electrostatic energy harvesting circuit for the voltage-constrained energy conversion cycle by E.O. Torres et al. [36].

(MN_2 and MN_3). Switch S_4 is a CMOS transmission gate that consists of a parallel connection of transistors MN_4 and MP_4 allowing that the current flows in both directions through the switch. The control signals ($\phi_{L\text{-}C}$, $\phi_{L\text{-}D}$ and ϕ_H) connected to the gates of the CMOS switching transistors are generated by a low-power digital signal processing (DSP) unit.

Control signal $\phi_{L\text{-}C}$ is connected to the gate of transistors MP_1 and MN_3 to close the switches when the signal goes to a high value during the pre-charge phase. After each switching, a dead-time interval, where all the switches are opened, prevents short-circuits and high peak voltages. $\phi_{L\text{-}D}$ controls the switches MN_2, MN_4, and MP_4. Thus, the switches are closed when $\phi_{L\text{-}D}$ has a high value and the inductor charges C_{var} with the battery voltage. Once the variable capacitor is charged, the pre-charge phase ends, control signal $\phi_{L\text{-}D}$ goes to a low value and switches MN_2, MN_4, and MP_4 are opened. After a dead-time, control signal ϕ_H goes to a low state, closing switches MP_{5A} and MP_{5B} and starting the harvesting phase.

Figure 9.50 displays the control signals, the capacitance value of C_{var}, the voltage on the variable capacitor (v_{var}), the current through the inductor i_L and the current through the variable capacitor i_C as a function of time highlighting the different phases.

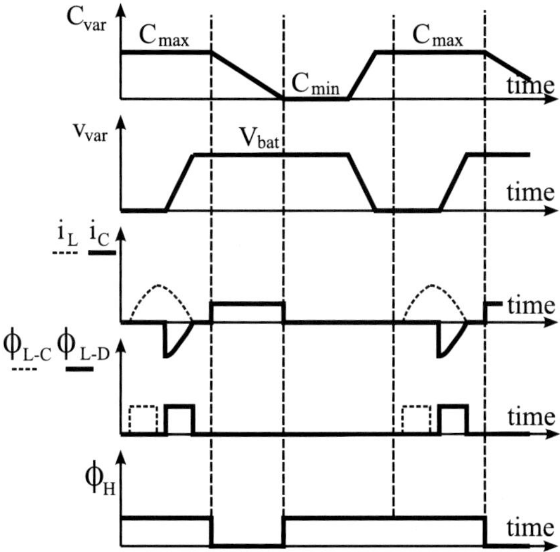

Figure 9.50 Timing waveforms for the electrostatic energy harvesting circuit shown in Fig. 9.49 [36].

9.2.5 *Efficiency Calculation for the Voltage-Constrained Energy Conversion Cycle*

The power losses of the electrostatic energy harvesting generator presented by Torres et al. [36] are analyzed in detail by Rincón-Mora et al. [37]. Due to the power losses present in the voltage-constrained energy conversion cycle of the electrostatic generator, the net energy gained per cycle given by Eq. (9.34) is reduced to

$$\Delta E_{\text{Net}} = \frac{1}{2}\Delta C V_{\text{Bat}}^2 - \sum P_{\text{Losses}} T_{\text{Vib}} \qquad (9.38)$$

where T_{Vib} is the period of the vibrations and $\sum P_{\text{Losses}}$ are the total average power losses during the energy conversion cycle. The power losses of the converter include the losses in MOSFETs and diodes, conduction losses in the inductor and the power consumed by the drivers of the MOSFETs.

9.3 AC–DC Converters for Electrodynamic Transducers

Electrodynamic transducers provide AC power and low output voltages. These voltages are usually increased with a transformer or a voltage multiplier circuit. After the transformer, a rectifier is required for obtaining DC power while the voltage multiplier already provides DC power.

9.3.1 Generic AC–DC Converters

A block diagram of an AC–DC converter for electrodynamic transducers is shown in Fig. 9.51 [38]. This converter employs a transformer X_1 (with 1:10 turns ratio) for increasing the low input voltage v_{gen} delivered by the electrodynamic transducer. Afterward, a half-wave rectifier stores the energy in capacitor C_1. A low-power voltage regulator is required to adapt the voltage V_{C1} on capacitor C_1 to the requirements of the electronic load. The voltage regulator employed in Fig. 9.51 is a synchronous step-down switching converter.

The electrodynamic generator designed by Amirtharajah et al. [39] employs discrete components and provides a power in the order of 400 µW using human walking as input energy. Human walking is characterized with a frequency of 2 Hz and a maximum amplitude of 2 cm, which corresponds to place the electrodynamic generator in a pocket. The vibration source of the

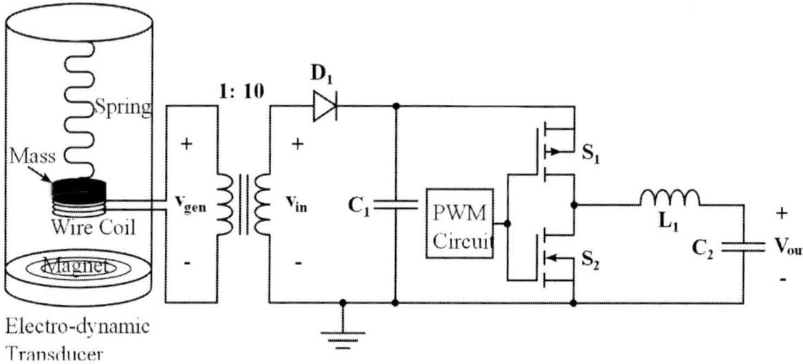

Figure 9.51 Block diagram of the electrodynamic energy harvesting converter based on a transformer [38].

electrodynamic energy harvesting power supply is not considered to be periodic since human motion has often random movements associated.

Yuen et al. [40] designed an AA battery size electrodynamic energy harvesting power supply that includes a voltage multiplier and a large output capacitor. A doubler, a tripler, and a quadrupler multiplier, employing Schottky diodes with a low forward voltage of 230 mV were simulated and their results were analyzed in terms of input power, energy efficiency, and charge time of the output capacitor. The electrodynamic generator includes a startup circuit that connects the storage capacitor to the load once its voltage reaches a certain value. This voltage assures the correct operation of the load. The load powered by the energy harvesting power supply designed by Yuen et al. is a wireless thermometer that transmits the measured data every 20 s when vibrations with an acceleration of 4.63 m/s^2 and an amplitude of 250 μm at 70.5 Hz are applied. The load consumes 27.6 μW.

The operation of the electrodynamic converter shown in Fig. 9.52 is discontinuous. The output capacitor C_{storage} of the voltage multiplier stores the energy provided by the transducer. The startup time of the generator is the time required for charging the capacitor from 0 V to $V_{\text{th}}(\text{H})$, whereas the time between operations is the time necessary for charging the storage capacitor from $V_{\text{th}}(\text{L})$ to $V_{\text{th}}(\text{H})$.

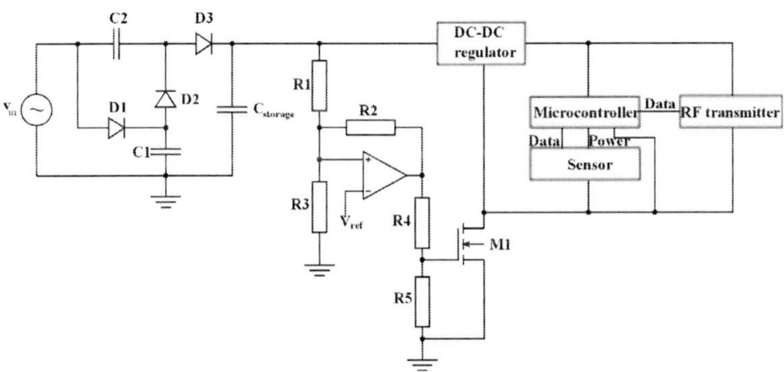

Figure 9.52 Block diagram of the electrodynamic converter based on a voltage multiplier.

James et al. [41] designed two prototypes that employ an electrodynamic transducer. The two prototypes include a conversion circuit that increases, rectifies, and regulates the output voltage delivered by the transducer, a sensor that monitors a physical parameter and a communication interface. The electrodynamic transducer generates 2.5 mW at 0.5 Vrms with a load of 100 Ω when the magnet is displaced 0.4 mm at 102 Hz.

Electrodynamic transducers provide low output voltages that are increased in a different way in each one of the prototypes presented by James et al. The first prototype steps-up the electrodynamic transducer output voltage with a transformer as Amirtharajah et al. [39]. Afterward, a full-wave rectifier is employed. Different core materials (ferrite and iron-based) for the transformer were evaluated in terms of transformer efficiency. The transformers with ferrite core were discarded due to the low operating frequency under consideration (100 Hz). Three different transformers with iron-based cores were compared and the one with a transformer ratio 1:7 gave the best efficiency results and provides an output voltage of 3.5 V.

Schottky and signal diodes were tested in the full-wave rectifier to determine which have the best efficiency. Better results were obtained for the Schottky diodes. The efficiency of the full-wave rectifier decreases with the frequency due to the switching looses. For the voltage regulator, a simple zener regulator is employed since it provided higher efficiencies (80–84%) than commercial DC–DC converters.

The second prototype uses a quadruple voltage multiplier circuit both to step up and to rectify the electrodynamic transducer output voltage like in the design of Yuen et al. [40].

9.3.2 *Dual Polarity Boost Converter*

Mitcheson et al. [33] proposed a dual polarity boost converter for rectifying and increasing the AC voltage obtained from an electromagnetic transducer. In this approach, there is no bridge rectifier and its functionality is replaced by two boost converters where each one is activated during half cycle of the AC signal provided by the transducer. Figure 9.53 shows the dual polarity

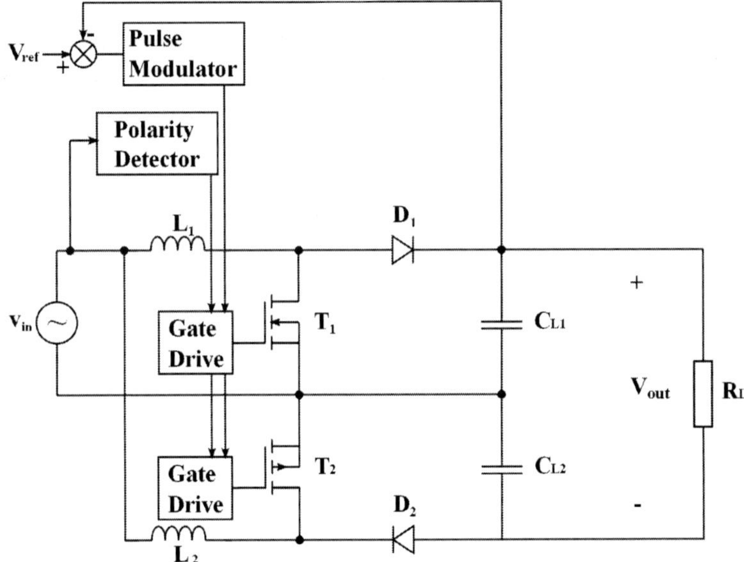

Figure 9.53 Dual polarity boost converter [33].

boost converter where the upper and lower boost converters are employed during the positive and negative half-cycle of the transducer, respectively. A polarity detector circuit is necessary to drive the MOSFETs of the boost converters. Both boost converters are synchronized to avoid the connection of both converters at the same time. The boost converters are operated in discontinuous conduction mode in order to have lower turn-on power losses in the MOSFETs and avoid reverse recovery effects in the diodes. The dual polarity boost converter has an efficiency near 50% for an input power around 50 mW.

9.3.3 *Direct AC–DC Conversion*

9.3.3.1 Physical principles

Dwari et al. [42–45] distinguish between two different topologies for the converters that can be employed for electrodynamic transducers: conventional two-stage converters consisting of a diode bridge rectifier followed by a standard switching DC–DC converter and a direct AC–DC conversion without a diode bridge.

Rectification employing a diode bridge is not feasible because electrodynamic transducers usually provide output voltages in the range of only some hundreds of millivolts. Moreover, the forward voltage drops on the diodes of a bridge rectifier will reduce significantly the efficiency of the converter. Hence, Dwari et al. designed three different topologies of direct AC–DC converters that are introduced in this section.

Dwari et al. [42] expose that the dual polarity boost converter presented by Mitcheson et al. [33] presents a ripple problem at the output that can only be solved employing large output capacitors. Each one of the output capacitors is only charged during half cycle. However, the capacitors charge the load continuously, producing a high ripple in the output voltage of the capacitors. Nevertheless, the employment of high capacitances has two disadvantages: size and slow response of the converter.

9.3.3.2 Electrical circuit of the boost and buck-boost converter

Figure 9.54 displays the proposed direct AC–DC converter of Dwari et al. [42, 45] based on a boost and a buck-boost converter. Inductor L_1, transistor S_1 and diode D_1 are part of the boost converter, whereas the buck-boost converter is composed by inductor L_2, transistor S_2, and diode D_2. Both converters charge the same output capacitor C_L.

The boost converter operates during the positive half cycle of the transducer, whereas the buck-boost converter operates during the negative half cycle. During the positive half-cycle, switch S_2 is open while switch S_1 is operated with its corresponding duty cycle while during the negative half-cycle, switch S_1 is open and S_2 is operated.

The n-channel MOSFETs employed on this circuit have been selected to have a forward voltage drop on the body diode higher than the voltage peak provided by the transducer in order to do not conduct. Diodes D_1 and D_2 are Schottky diodes. Both converters work in discontinuous conduction mode (DCM), which reduces the switching losses and the diode reverse recovery losses of the diodes. Moreover, in DCM, the input voltage and the input current

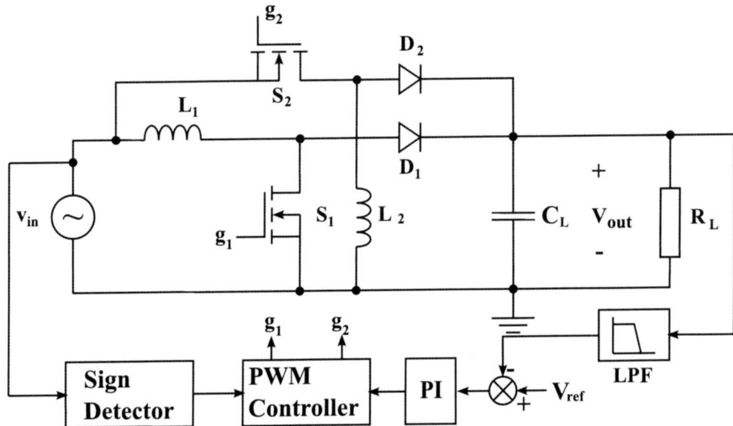

Figure 9.54   Direct AC–DC converter composed by a boost and a buck-boost converter [42, 45].

of the converter for a constant duty cycle are proportional, which means that the current and voltage delivered by the electrodynamic transducer will be in phase; this assures that the maximum power can be extracted from the transducer.

Figure 9.55 shows the four different operation states of this direct AC–DC converter. The first two states correspond to the positive half cycle of the electrodynamic transducer voltage while the two last states correspond to the negative half cycle voltage. Therefore, n-channel MOSFET S_2 is open during the two first states and n-channel MOSFET S_1 during the two last states. During the first state, Fig. 9.55a, n-channel MOSFET S_1 is closed and current i_{L1} flows through inductor L_1 and n-channel MOSFET S_1. During the second state (Fig. 9.55b), n-channel MOSFET S_1 is opened and current i_{L1} flows through inductor L_1 and diode D_1 to charge the output capacitor C_L. During the two first states, the current cannot flow through the body diode of nMOSFET S_2 since its forward voltage drop is higher than the peak voltage of the transducer. In the third state (Fig. 9.55c), n-channel MOSFET S_2 is closed and current i_{L1} flows through inductor L_2 and n-channel MOSFET S_2 while in the fourth state (Fig. 9.55d, n-channel MOSFET S_2 is opened and the current flows through inductor L_2 and diode D_2 to charge the output capacitor C_L.

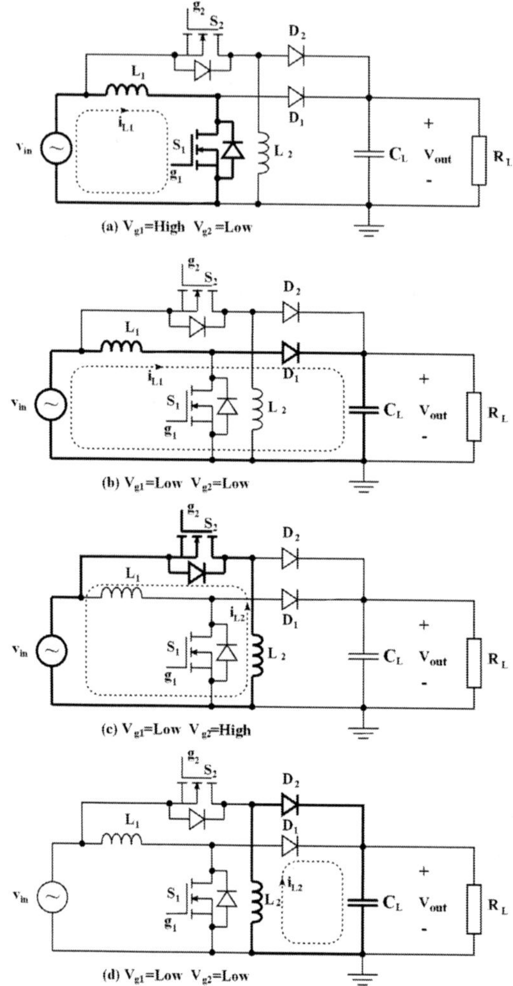

Figure 9.55 Operation states of the direct AC–DC converter composed by a boost and a buck-boost converter [42, 45].

Figure 9.56 shows the circuit for self-starting operation of the boost and buck-boost converter. The circuit proposed by Dwari et al. [45] contains a battery that supplies a voltage V_b to the control circuit through diode D_b while there is no output voltage available. Once the direct AC–DC converter has charged the output capacitor

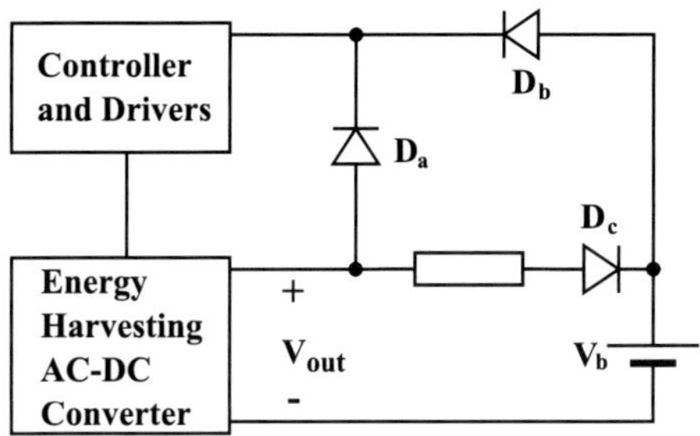

Figure 9.56 Self starting circuit for the direct AC–DC converter composed by a boost and a buck-boost converter using a battery [45].

C_L to its final value, diode D_b is reverse biased and the control circuit is powered by the converter output via diode D_a and the battery is recharged via diode D_c. For ensuring the proper operation of the self-starting circuit, the following condition has to be satisfied:

$$V_b < V_{out} - V_d \tag{9.39}$$

where V_d is the voltage drop of the diodes.

9.3.3.3 Analytical model of the boost and buck-boost converter

Figure 9.57(a) shows the input current waveform of the direct AC–DC converter and Fig. 9.57(b) shows input current i, gate signal V_{g1}, and input voltage v_{in} waveforms during the positive half cycle, where the boost converter operates [45].

The voltage provided by the electrodynamic transducer in a k-th switching cycle is

$$v_{in,k} = V_{ps}\sin\left(2\pi k\frac{T_s}{T_i}\right) \tag{9.40}$$

where T_s is the switching period of the converter and T_i is the period of the transducer voltage signal.

Since the boost converter works in DCM, the peak current flowing through inductor L_1 is

$$i_{L_1,\text{peak},k} = v_{in,k}\frac{D_b T_s}{L_1} \tag{9.41}$$

where $D_b T_s$ is the time that switch S_1 remains closed.

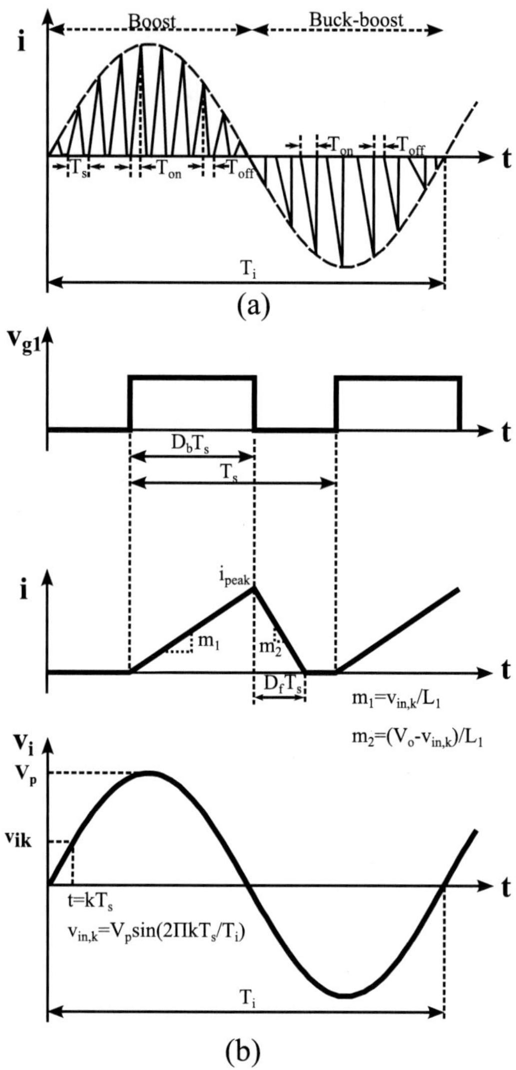

Figure 9.57 (a) Input current waveform of the boost and buck-boost converter; (b) input current waveform, gate signal and input voltage during the positive half cycle [45].

$i_{L_1,\text{peak,k}}$ can also be expressed as:

$$i_{L_1,\text{peak,k}} = (V_\text{o} - v_{\text{in,k}})\frac{D_\text{f}T_\text{s}}{L_1} \tag{9.42}$$

where $D_\text{f}T_\text{s}$ is the time needed by the inductor current to go from its peak value to zero.

The average power extracted from the electrodynamic transducer with the boost converter in one switching cycle is

$$P_\text{kb} = v_{\text{in,k}}i_{L_1,\text{peak,k}}\frac{D_\text{b} + D_\text{f}}{2} \tag{9.43}$$

The average power harvested during the positive half cycle is the summation of all the switching cycles of the boost converter:

$$P_\text{ib} = \frac{2}{N}\sum_{k=1}^{N/2} P_\text{kb} = \frac{2}{N}\sum_{k=1}^{N/2} v_{\text{in,k}}i_{L_1,\text{peak,k}}\frac{D_\text{b} + D_\text{f}}{2} \tag{9.44}$$

where $N = T_i/T_\text{s}$

Replacing in Eq. (9.44) $v_{\text{in,k}}$ for its expression of Eq. (9.40), the following is obtained:

$$P_\text{ib} = \frac{V_\text{p}^2 D_\text{b}^2 T_\text{s}}{4L_1}\beta \tag{9.45}$$

where $\beta = \frac{2}{\pi}\int_\pi^0 \frac{1}{1-\frac{V_\text{p}}{V_\text{o}}\sin\theta}d\theta$ and $\theta = 2\pi t/T_i$.

The output power is equal to the input power times the efficiency, η, of the converter.

$$\frac{V_\text{p}^2 D_\text{b}^2 T_\text{s}}{4L_1}\beta = \frac{V_\text{o}^2}{R_\text{L}}\frac{1}{\eta} \tag{9.46}$$

From the previous expression, the equation for the duty cycle is deduced:

$$D_\text{b} = \frac{2V_\text{o}}{V_\text{p}}\sqrt{\frac{L_1}{R_\text{L}T_\text{s}\eta}\frac{1}{\beta}} \tag{9.47}$$

The duty cycle for the buck-boost converter can be calculated in a similar way.

$$D_\text{c} = \frac{2V_\text{o}}{V_\text{p}}\sqrt{\frac{L_2}{R T_\text{s}\eta}} \tag{9.48}$$

The relation between the duty cycles of the boost converter and the buck-boost converter is given by

$$\frac{D_b}{D_c} = \sqrt{\frac{L_1}{L_2}\frac{1}{\beta}}$$

(9.49)

The value of β approaches to 1 when $V_o >> V_p$. Therefore, it is deduced from Eq. (9.49) that if the boost and buck-boost converters have the same inductances, they can be controlled with the same duty cycle [45].

9.3.3.4 Efficiency of the boost and buck-boost converter

The direct AC–DC boost and buck-boost converter has been simulated and built employing an electromagnetic microgenerator that delivers a sinusoidal peak voltage of 400 mV at 100 Hz. The converter has been tested employing commercial components for providing an output voltage of 3.3 V to a resistive load of 200 Ω. Two inductors with the same value are employed $L_1 = L_2 = 4.7\ \mu H$. The efficiency obtained during simulations was 63% and the control circuit consumed an average power of 2.2 mW [45].

A battery with a voltage value of 3 V was employed for the self-starting operation and the startup time was 4.6 ms.

An efficiency of 61% is calculated from experimental results for the converter for the same input and output conditions employed in the simulations. Table 9.1 shows the estimated power losses that are in good agreement with the measurements [45]. The higher power losses are due to the inductors and the MOSFETs. The conduction loss of the MOSFETs cause a 93% of their total losses and the remaining 7% are caused by switching losses. Thus, the use of MOSFETs with lower on resistance can improve the efficiency of the converter [45].

9.3.3.5 Design optimization of the boost and buck-boost converter

The direct AC–DC converter designed by Dwari et al. can be employed for two different purposes. The duty cycle of the converter can be set to harvest the maximum power provided by the

Table 9.1 Power losses in the direct AC–DC converter based on a boost and buck-boost converter [45]

Component	Estimated power loss (mW)
Boost inductor L_1	4.7
Buck-boost inductor	4.9
Boost N-channel MOSFET S_1	6.2
Buck-boost N-channel MOSFET S_2	6.3
Boost Schottky diode D_1	2.4
Buck-boost Schottky diode D_2	2.5
Output capacitor C_L	0.1
Control circuit	2.2
PCB track resistance and contact resistances	1.6

electrodynamic transducer [46] or to provide a constant output voltage for a certain load [45]. This section calculates the optimum duty cycle of the converter that harvests the maximum power as a function of the mechanical damping of the electrodynamic transducer which varies with the vibration frequency.

Dayal et al. [46] made the approach to the first scenario to extract the maximum power from the transducer. Figure 9.58 shows the internal equivalent circuit of the electrodynamic transducer that has two internal resistors [46]. The expression for the induced

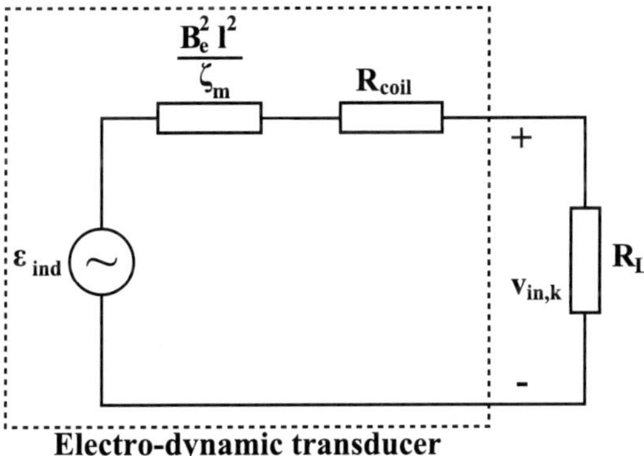

Figure 9.58 Equivalent circuit of the electrodynamic transducer connected to a resistive load [46].

electromotive force in the coil of an electrodynamic transducer is

$$\epsilon_{\text{ind}} = B_e l z'\left(t\right) \tag{9.50}$$

where B_e is the electromagnetic field intensity, l is the length of the wire coil and $z\left(t\right)$ is the magnet vibration amplitude relative to the housing.

If a load is connected directly to the electrodynamic transducer, the power at the load is maximum when it has a value of

$$R_{\text{L,opt}} = R_{\text{coil}} + \frac{B_e^2 l^2}{\zeta_m} \tag{9.51}$$

where R_{coil} is the resistance of the coil of the electrodynamic transducer and ζ_m is the mechanical damping.

For the boost and buck-boost AC–DC converter design, the power delivered by the electrodynamic transducer is calculated in Eq. (9.45). This power is equal to the RMS value of vin_k squared and divided by the equivalent resistance of the direct AC–DC converter.

$$P_i = \frac{V_p^2 D^2 T_s}{4L} \beta = \left(\frac{V_p}{\sqrt{2}}\right)^2 \frac{1}{R_{\text{eq}}} \tag{9.52}$$

Rearranging the terms of the previous equation, it is calculated that the resistance offered by the converter is

$$R_{\text{eq}} = \frac{2L}{D^2 T_s \beta} \tag{9.53}$$

The electrodynamic transducer delivers its maximum power to the direct AC–DC converter when R_{eq} is equal to $R_{\text{L,opt}}$. Thus, the converter harvests the maximum power from the transducer with a duty cycle of

$$D_{\text{opt}} = \sqrt{\frac{2L}{\left(R_{\text{coil}} + \frac{B_e^2 l^2}{\zeta_m}\right) T_s \beta}} \tag{9.54}$$

The mechanical damping changes with the frequency, and therefore a converter that can modify its resistance to match the internal resistance of the transducer can extract the maximum power from it. However, Dayal et al. [46] do not provide any schematic for the control circuit.

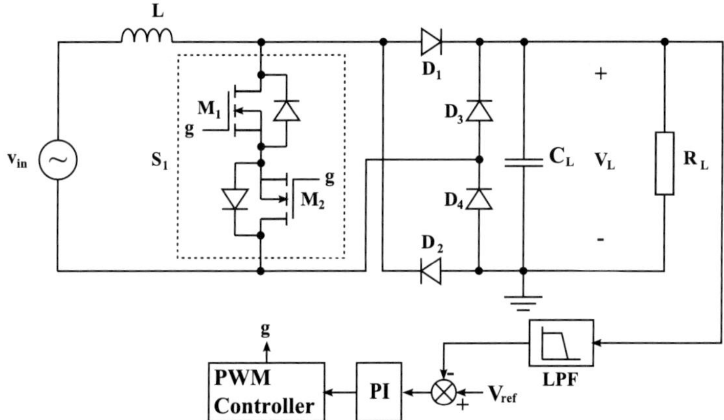

Figure 9.59 Direct AC–DC converter with a secondary side diode topology [43].

9.3.3.6 Electrical circuit of the secondary side diode-based converter

Figure 9.59 shows the direct AC–DC converter with a secondary side diode topology designed by Dwari et al. [43]. The converter has an inductor L, four Schottky diodes (D_1–D_4), a bidirectional switch S_1 composed by two n-channel MOSFETs, a filter capacitor C_L, and a resistive output load R_L.

The topology of the converter is similar to a boost converter followed by a diode bridge ($D_1 - D_4$). The location of the diodes after the boost converter causes that a low current flows through them, and therefore the power losses are lower than if the diode bridge is located after the electrodynamic transducer.

The bidirectional switch S_1 can conduct during the positive and negative half cycles of the transducer voltage. During the positive half cycle, when gate signal g, which is referenced to ground, is high, n-channel MOSFET M_1 conducts in the forward direction while n-channel MOSFET M_2 conducts in the reverse direction. During the negative half cycle, n-channel MOSFET M_2 conducts in the forward direction and n-channel MOSFET M_1 conducts in the reverse direction. When gate signal g is low, due to the connection of the body diodes of the n-channel MOSFETs, no current flows through them.

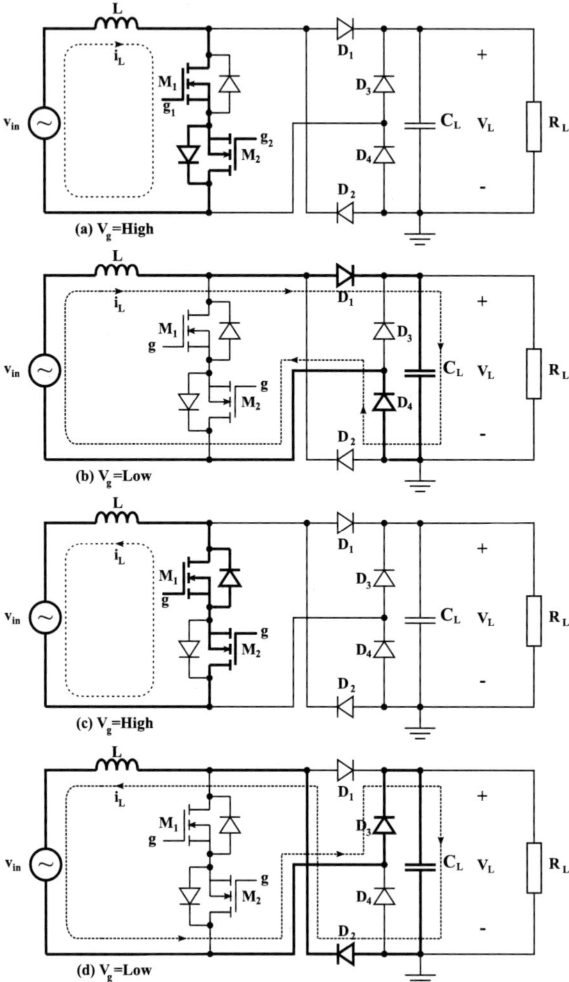

Figure 9.60 Operation states of the direct AC–DC with a secondary side diode topology [43].

Figure 9.60 shows the four operation states of the direct AC–DC converter [43]. During the two first states, the electrodynamic transducer voltage is in the positive half cycle while during the two last states it is in the negative half cycle. During the first state, gate signal g is high and the current flows through inductor L and switch

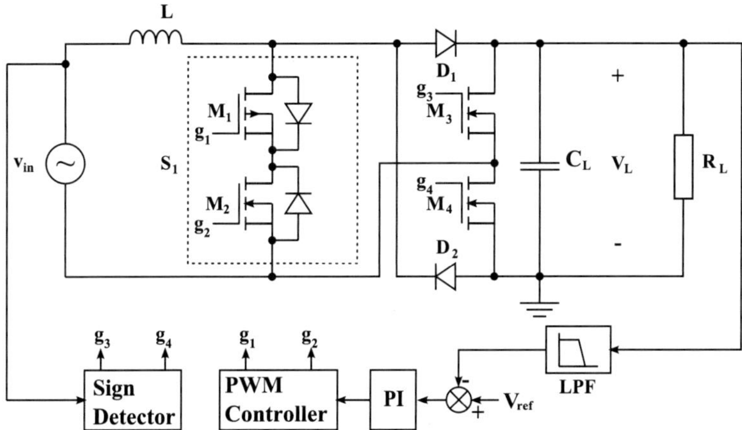

Figure 9.61 Direct AC–DC converter with a secondary side diode topology [44].

S_1. During the second state, gate signal g is low and the MOSFETs do not conduct anymore. Hence, the current flows through inductor L and diodes D_1 and D_4. When the third state takes place, gate signal g is high and the current flows through switch S_1 and inductor L. For the fourth state, gate signal g is low and the current is rectified through diodes D_3 and D_2.

A modified version of this converter with the same functionality: a boost converter and afterward a full bridge rectifier is presented in Fig. 9.61. Dwari et al. [44] substituted in this design diodes D_3 and D_4 for N-channel MOSFETs S_3 and S_4, respectively. Nevertheless, a sign detector circuit is necessary for this new topology in order to control the switching signal for S_3 and S_4. Gate signal g_3 has a high value while gate signal g_4 has a low value during the positive half-wave of the transducer voltage. On a similar way, gate signal g_4 has a high value while the gate signal g_3 has a low value during the negative half-wave of the transducer voltage.

The AC–DC converter in Fig. 9.61 also presents a different topology for the bidirectional switch S_1 with a p-channel MOSFET M_1 and an n-channel MOSFET M_2. Hence, two different gate signals g_1 and g_2 are necessary for the control of M_1 and M_2, respectively.

Table 9.2 Power losses in the direct AC–DC converter with diode based topology [43]

Component	Nominal value	Losses	Estimated power loss (mW)
Boost inductor L_1	4.7 μH	$R_{esr} = 40$ mΩ	7.3
Output capacitor C_1	68 μF	$R_{esr} = 9$ mΩ	—
N-channel MOSFETs M_1 and M_2	20 V, 6 A	$R_{ds,on} = 30$ mΩ at $V_{gs} = 3$ V	11
Schottky diodes D_1, D_2, D_3, and D_4	23 V, 1 A	$V_f = 250$ mV at 15 mA	17.2

9.3.3.7 Efficiency of the secondary side diode-based converter

The efficiency obtained with this converter has been measured with an electrodynamic transducer providing a sinusoidal voltage of 400 mV peak with a resonance frequency at 100 Hz that supplies a load resistance of 200 Ω at 3.3 V. The switching frequency of the converter is 50 kHz with an inductance of 4.7 μH and an output capacitance of 68 μF.

The converter operates in DCM with a duty cycle of 0.7 and a simulated efficiency of 65.6% is obtained. The experimental results for this converter provide a duty cycle of 0.76 and an efficiency of 57.8%. Table 9.2 shows the estimated power loss of various components which are in good agreement with the experimental results obtained.

9.3.3.8 Electrical circuit of the split capacitor converter

A third direct AC–DC converter is displayed in Fig. 9.62. This converter has, as the previous one, just one inductor and the bidirectional switch S_1. In this topology, the diode bridge of the previous converter has been substituted by a half-wave voltage rectifier.

The advantage of this new topology and the secondary side diode topology over the direct AC–DC converter composed by a boost and a buck-boost converter, that was designed by Dwari et al. [42], and the dual polarity boost converter presented by Mitcheson et al. [33] is that employs only one inductor instead of two inductors.

Figure 9.63 shows the four operation states of the split capacitor converter. The first two states correspond to the positive half-wave

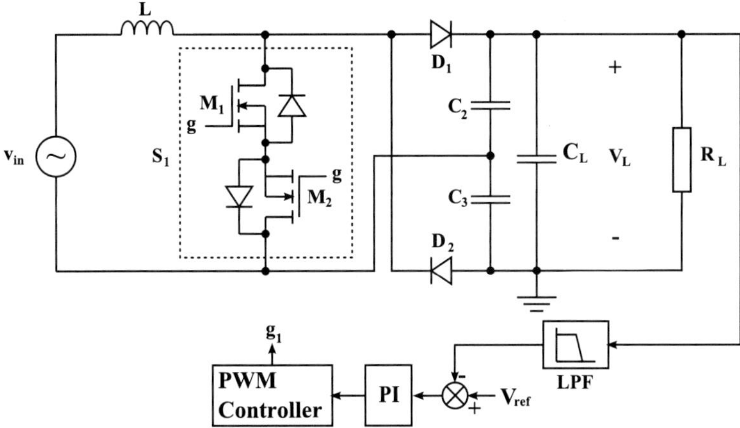

Figure 9.62 Direct AC–DC converter with split capacitor based topology [43].

of the transducer voltage, whereas the two last ones to the negative half-wave. During the first state, gate signal g is high, and therefore the current flows through inductor L and switch S_1. At the second state, gate signal g is low and switch S_1 is turned off. Hence, the current flows through inductor L and D_1, charging capacitors C_2 and C_L, and discharging capacitor C_3. At the third state, gate signal g is high and the current of the transducer flows through switch S_1 and inductor L. During the last state, gate signal g has a low value and the current flows through diode D_2, charging capacitors C_L and C_3, and discharging capacitor C_2.

The parallel connection of capacitors C_2 and C_3 to capacitor C_L is an improvement over the design of Mitcheson et al. [33] since the output voltage ripple is reduced.

9.3.3.9 Efficiency of the split capacitor converter

The efficiency obtained with this converter has been acquired with an electrodynamic transducer providing a sinusoidal voltage of 400 mV peak with a resonance frequency at 100 Hz that supplies a load resistance of 200 Ω at 3.3 V. The switching frequency of the converter is 50 kHz with an inductance of 4.7 μH and an output capacitance of 68 μF.

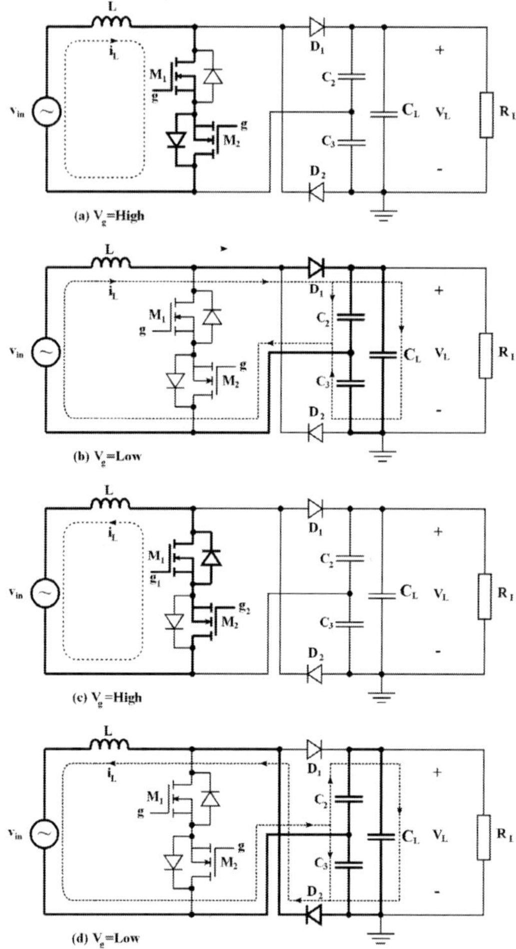

Figure 9.63 Operation states of the direct AC–DC converter with split capacitor based topology [43].

The converter operates in DCM with a duty cycle of 0.68 and a simulated efficiency of 67% is obtained. The experimental results for this converter provide a duty cycle of 0.73 and an efficiency of 60.3 %. Table 9.3 shows the estimated power loss of various components which are in good agreement with the experimental results obtained. The power loss on Schottky diodes includes also the power loss on capacitors C_2 and C_3.

Table 9.3 Power losses in the direct AC–DC converter with split capacitor based topology [43]

Component	Nominal value	Losses	Estimated power loss (mW)
Boost inductor L_1	4.7 μH	$R_{esr} = 40$ mΩ	7.3
Output Capacitor C_1	68 μF	$R_{esr} = 9$ mΩ	—
Split capacitors C_2 and C_3	33 μF	$R_{esr} = 28$ mΩ	—
N-channel MOSFETs M_1 and M_2	20 V, 6 A	$R_{dson} = 30$ mΩ at $V_g = 3$ V	11
Schottky diodes D_1 and D_2	23 V, 1 A	$V_f = 250$ mV at 15 mA	8.6

9.3.3.10 Design optimization of the secondary side diode-based converter and the split capacitor converter

Dwari et al. [44] explore the possibility of using the inductance of the coil present in the electrodynamic transducer as the inductor for the direct AC–DC converters that have only one since the self conductance of the coil can be considered constant over its range of operation.

Dwari et al. [44] simulated the split capacitor converter in conjunction with finite element analysis (FEA) of the electrodynamic transducer. The transducer provided a sinusoidal output voltage with 500 mV peak voltage and a frequency of 100 Hz for a 2 mm peak-to-peak displacement. The FEA simulation confirms that the coil inductance of the electrodynamic transducer was constant over the displacement range, around 13 μH. The converter supplies power to a 200 Ω load with an output voltage of 3.3 V.

9.4 Conclusion

Different topologies of AC–DC converters from several authors for piezoelectric, electrostatic, and electrodynamic transducers have been presented and analyzed. The control circuits required for the operation of such converters have been also discussed. The appearance in the market of ultra-low power operational amplifiers, microcontrollers, and DC–DC converters in the last years have simplified the task to achieve self-powered energy harvesting systems in the power range of few tens of microwatts to some milliwatts that employ mechanical vibrations as ambient source.

The importance of working in the MPP of the transducers has been pointed of and several examples present in the literature have been introduced.

The frequency of the vibrations affects the transducer and the control circuit of the AC–DC converter. The transducers are tuned to resonate at a certain frequency, which is tuned to the vibration applied. Nevertheless, the broadband of the transducers is very narrow and small changes in the frequency of the ambient vibration source can cause a drastic reduction of the generated power. In the literature, there are tunable piezoelectric and electromagnetic transducers that achieve 10 Hz of frequency bandwidth. Moreover, when the control circuit of the AC–DC converters has a limited frequency response, a change in the ambient vibration can affect the behavior of the AC–DC converter.

References

1. J. Han, A. von Jouanne, T. Le, K. Mayaram, and T. S. Fiez, Novel power conditioning circuits for piezoelectric micropower generators, *Applied Power Electronics Conference and Exposition, 2004. APEC '04. Nineteenth Annual IEEE.* **3**, 1541–1546, vol. 3 (2004). doi: 10.1109/APEC.2004. 1296069. URL http://dx.doi.org/10.1109/APEC.2004.1296069.

2. T. Le, J. Han, A. Von Jouanne, K. Mayaram, and T. Fiez, Piezoelectric micropower generation interface circuits, *IEEE J Solid-State Circuits.* **41**(6), 1411–1420 (2006).

3. E. Lefeuvre, A. Badel, C. Richard, L. Petit, and D. Guyomar. Optimization of piezoelectric electrical generators powered by random vibrations. In *Dans Symposium on Design, Test, Integration and Packaging (DTIP) of MEMS/MOEMS.* Citeseer (2006).

4. G. Ottman, H. Hofmann, A. Bhatt, and G. Lesieutre, Adaptive piezoelectric energy harvesting circuit for wireless remote power supply, *IEEE Trans. Power Electron.* **17**(2), 669–676 (September 2002).

5. L. Mateu and F. Moll, Appropriate charge control of the storage capacitor in a piezoelectric energy harvesting device for discontinuous load operation, *Sens. Actuators A.* **132**(1), 302–310 (2006).

6. L. Mateu and F. Moll. Analysis of direct discharge circuit to power autonomous wearable devices using PVDF piezoelectric films. In *Proceedings of International Telecommunications Energy Conference (Intelec)*, pp. 45–50 (September 2005).

7. G. Ottman, H. Hofmann, and G. Lesieutre, Optimized piezoelectric energy harvesting circuit using step-down converter in discontinuous conduction mode, *IEEE Trans. Power Electron.* **18** (2), 696–703 (March 2003).

8. T. Esram and P. Chapman, Comparison of photovoltaic array maximum power point tracking techniques, *IEEE Trans. Energy Conversion.* **22**(2), 439–449 (2007).

9. Lm555 timer. http://www.national.com/ds/LM/LM555.pdf (July 2006). URL http://www.national.com/ds/LM/LM555.pdf.

10. G. Ottman, H. Hofmann, and G. Lesieutre. Optimized piezoelectric energy harvesting circuit using step-down converter in discontinuous conduction mode. In *Power Electronics Specialists Conference, 2002. PESC 02, 2002 IEEE 33rd Annual*, vol. 4, pp. 1988–1994 (June 2002).

11. D. Guyomar, D. Sebald, S. Pruvost, M. Lallart, A. Khodayari, and C. Richard, Energy Harvesting from Ambient Vibrations and Heat, *J. Intell. Mater. Syst. Struct.* **20** (March 2009).

12. Y. Shu, I. Lien, and W. Wu, An improved analysis of the SSHI interface in piezoelectric energy harvesting, *Smart Mater. Struct..* **16**, 2253–2264 (2007).

13. S. Priya and D. Inman, *Energy Harvesting Technologies.* (Springer Publishing Company, Incorporated, 2008).

14. S. Ben-Yaakov and N. Krihely. Resonant rectifier for piezoelectric sources. In *Applied Power Electronics Conference and Exposition, 2005. APEC 2005. Twentieth Annual IEEE*, vol. 1 (2005).

15. M. Lallart, . Lefeuvre, C. Richard, and D. Guyomar, Self-powered circuit for broadband, multimodal piezoelectric vibration control, *Sens. Actuators A.* **143**(2), 377–382 (2008).

16. E. Lefeuvre, A. Badel, C. Richard, and D. Guyomar, Piezoelectric energy harvesting device optimization by synchronous electric charge extraction, *J. Intell. Mater. Syst. Struct.* **16**(10), 865 (2005).

17. Y. Tan, J. Lee, and S. Panda. Maximize piezoelectric energy harvesting using synchronous charge extraction technique for powering autonomous wireless transmitter. In *IEEE International Conference on Sustainable Energy Technologies, 2008. ICSET 2008*, pp. 1123–1128 (2008).

18. J. Brufau-Penella and M. Puig-Vidal, Piezoeletric energy harvesting improvement with complex conjugate impedance matching, *J. Intell. Mater. Syst. Struct.* **20**(5), 597–608 (2009).

19. S. Xu, K. Ngo, T. Nishida, G. Chung, and A. Sharma, Low frequency pulsed resonant converter for energy harvesting, *IEEE Trans. Power Electron.* **22**(1), 63–68 (2007).

20. E. Dallago, D. Miatton, G. Venchi, V. Bottarel, G. Frattini, G. Ricotti, and M. Schipani, Electronic interface for piezoelectric energy scavenging system. pp. 402–405 (2008).

21. H. Xue, Y. Hu, and Q. Wang, Broadband piezoelectric energy harvesting devices using multiple bimorphs with different operating frequencies [Correspondence], *IEEE Transactions on Ultrasonics, Ferroelectrics and Frequency Control.* **55** (9), 2104–2108 (2008).

22. S. Meninger, J. Mur-Miranda, R. Amirtharajah, A. Chandrakasan, and J. Lang, Vibration-to-electric energy conversion, *IEEE Trans. Very Large Scale Integr. Syst.* **9**(1), 64–76, (2001). ISSN 1063–8210.

23. R. Amirtharajah, S. Meninger, J. Mur-Miranda, A. Chandrakasan, and J. Lang. A micropower programmable dsp powered using a mems-based vibration-to-electric energy converter. In *Solid-State Circuits Conference, 2000. Digest of Technical Papers. ISSCC. 2000 IEEE International,* pp. 362–363, 469 (2000).

24. E. Torres and G. Rincón-Mora. Long-lasting, self-sustaining, and energy-harvesting system-in-package (SiP) wireless micro-sensor solution. In *International Conference on Energy, Environment, and Disasters (INCEED), Charlotte, NC* (2005).

25. S. Meninger, J. Mur-Miranda, R. Amirtharajah, A. P. Chandrasakan, and J. H. Lang, Vibration to electric energy conversion, *IEEE Trans. VLSI.* **9**(1) (February 2001).

26. M. Miyazaki, H. Tanaka, T. N. G. Ono, N. Ohkubo, T. Kawahara, and K. Yano. Electric-energy generation using variable-capacitive resonator for power-free LSI: efficiency analysis and fundamental experiment. In *Proceedings of the ISLPED 03,* pp. 193–198 (25–27 August 2003).

27. S. Roundy, P. Wright, and K. Pister. Micro-electrostatic vibration-to-electricity converters. In *Proceedings of ASME International Mechanical Engineering Congress and Exposition IMECE2002,* vol. 220, pp. 17–22 (November 2002).

28. S. Roundy. *Energy scavenging for wireless sensor nodes with a focus on vibration to electricity conversion.* PhD thesis, University of California (2003).

29. T. Sterken, K. Baert, R. Puers, and S. Borghs. Power extraction from ambient vibration. In *Proceedings of the Workshop on Semiconductor Sensors*, pp. 680–683 (November 2002).

30. P. Miao, A. Holmes, E. Yeatman, T. Green, and P. Mitcheson. Micro-machined variable capacitors for power generation. In *Proc. Electrostatics*, **3**, pp. 53–58 (March 2003).

31. B. Stark, P. Mitcheson, P. Miao, T. Green, E. Yeatman, and A. Holmes. Power processing issues for micro-power electrostatic generators. In *Power Electronics Specialists Conference, 2004.PESC 04.2004 IEEE 35th Annual*, vol. 6, pp. 4156–4162 (2004).

32. G. Despesse, T. Jager, J.-J. Chaillout, J.-M. Leger, and S. Basrour. Design and fabrication of a new system for vibration energy harvesting. In *Research in Microelectronics and Electronics, 2005 PhD*, vol. 1, pp. 225–228 vol.1 (25–28, 2005). doi: 10.1109/RME.2005.1543034.

33. P. Mitcheson, T. Green, and E. Yeatman, Power processing circuits for electromagnetic, electrostatic and piezoelectric inertial energy scavengers, *Microsyst. Technol.* **13** (11), 1629–1635 (2007).

34. B. Yen and J. Lang, A variable-capacitance vibration-to-electric energy harvester, *IEEE Trans. Circuits Syst I.* **53** (2), 288–295 (February 2006). ISSN 1549-8328. doi: 10.1109/TCSI.2005.856043.

35. B. Stark, P. Mitcheson, M. Peng, T. Green, E. Yeatman, and A. Holmes, Converter circuit design, semiconductor device selection and analysis of parasitics for micropower electrostatic generators, *IEEE Trans. Power Electron.* **21**(1), 27–37 (2006). ISSN 0885-8993.

36. E. Torres and G. Rincón-Mora. Electrostatic energy harvester and Li-ion charger circuit for micro-scale applications. In *IEEE Midwest Symposium on Circuits and Systems (MWSCAS), San Juan, Puerto Rico* (2006).

37. G. Rincón-Mora and E. Torres, Energy harvesting: A battle against power losses (September 2006).

38. R. Amirtharajah and A. Chandrakasan, Self-powered signal processing using vibration-based power generation, *IEEE J. Solid-State Circuits.* **33** (5), 687–695 (1998). ISSN 0018-9200.

39. R. Amirtharajah and A. Chandrakasan. Self-powered low power signal processing. In *Proceedings of the Symposium on VLSI Circuits Digest of Technical Papers*, pp. 25–26 (June 1997).

40. S. Yuen, J. Lee, W. Li, and P. Leong, An AA-sized vibration-based microgenerator for wireless sensors, *IEEE Pervasive Comput.* **6**(1), 64–72 (January–March 2007).

41. E. James, M. Tudor, S. Beeby, N. Harris, P. Glynne-Jones, J. Ross, and N. White, An investigation of self-powered systems for condition monitoring applications, *Sens. Actuators A.* **110** (1–3), 171–176, (2004).

42. S. Dwari, R. Dayal, and L. Parsa, A novel direct AC/DC converter for efficient low voltage energy harvesting. In Industrial Electronics, 2008. IECON 2008. 34th Annual Conference of IEEE pp. 484–488 (November 2008). ISSN 1553-572X. doi: 10.1109/IECON.2008.4758001.

43. S. Dwari, R. Dayal, L. Parsa, and K. Salama, Efficient direct AC-to-DC converters for vibration-based low voltage energy harvesting. In Industrial Electronics, 2008. IECON 2008. 34th Annual Conference of IEEE pp. 2320–2325 (November 2008). ISSN 1553-572X. doi: 10.1109/ IECON.2008.4758319.

44. S. Dwari and L. Parsa, Low voltage energy harvesting systems using coil inductance of electromagnetic microgenerators. In Applied Power Electronics Conference and Exposition, 2009. APEC 2009. Twenty-Fourth Annual IEEE pp. 1145–1150 (February 2009). ISSN 1048-2334. doi: 10.1109/APEC.2009.4802807.

45. S. Dwari and L. Parsa, An efficient AC-DC step-up converter for low-voltage energy harvesting, *Power Electronics, IEEE Transactions on.* pp. **25**(8), 2188–2199 (2010). ISSN 0885-8993. doi: 10.1109/TPEL.2010. 2044192.

46. R. Dayal, S. Dwari, and L. Parsa, Maximum energy harvesting from vibration-based electromagnetic microgenerator using active damping, *Electron. Lett.* **46**(5), 371–373 (March 2010). ISSN 0013-5194. doi: 10.1049/el.2010.3264.

Chapter 10

Radio Frequency Power Transmission

Josef Bernhard,[a] Tobias Dräger,[a] and Alexander Popugaev[b]

[a]*Fraunhofer Institute for Integrated Circuits IIS,*
Locating and Communication Systems Department,
Nordostpark 93, 90411 Nuremberg, Germany
[b]*Fraunhofer Institute for Integrated Circuits IIS, RF and SatCom Systems Department,*
Am Wolfsmantel 33, 91058 Erlangen, Germany
josef.bernhard@iis.fraunhofer.de, tobias.draeger@iis.fraunhofer.de,
alexander.popugaev@iis.fraunhofer.de

10.1 Introduction

Besides the aforementioned principles of power scavenging from mechanical, thermal, or solar energy source, this chapter deals with energy harvesting from electromagnetic fields. Radio frequency power transmission, or so-called wireless power transmission, is nowadays one of the main research topics among mobile device manufacturers since it is expected to eliminate any cables and connectors for powering or charging devices such as cell phones and laptops. In fact, radio frequency transmission offers the highest energy retrieval among all energy scavenging technologies. Since apart from the sun no natural sources of electromagnetic fields with sufficient energy exist, sources for generating and radiating electromagnetic waves with enough energy to empower an

Handbook of Energy Harvesting Power Supplies and Applications
Edited by Peter Spies, Loreto Mateu, and Markus Pollak
Copyright © 2015 Pan Stanford Publishing Pte. Ltd.
ISBN 978-981-4241-86-1 (Hardcover), 978-981-4303-06-4 (eBook)
www.panstanford.com

electronic device over a certain distance are necessary. The sources are normally intended to be used for power transmission only but can also be sources of radio frequency data transmission systems that offer enough power. The latter is becoming more and more attractive, as the availability of wireless data transmission systems increases. In any case, regulatory constraints regarding the use of frequency bands and maximum radiated radio power have to be considered.

This chapter describes the principles of wireless power transmission based on electromagnetic fields. The two main principles of inductive coupled links, similar to a transformer and the far-field radio transmission will be explained, system concepts will be described, and some hints to create an own wireless power transmission system will be given.

10.2 Physical Principles

10.2.1 *Electromagnetic Field: Generation and Radiation*

A key role in wireless communication is performed by antennas. The *IEEE Standard Definitions of Terms for Antennas* (IEEE Std 145-1983) introduced an "antenna" term, which is described in the following way: "That part of a transmitting or receiving system which is designed to radiate or to receive electromagnetic waves."

The radiating mechanism is depicted in Fig. 10.1. It is well known that in a resonant circuit, which consists of a capacitor and an inductor, energy oscillates between these two elements. Moving apart the capacitor's plates, we modify the closed resonant circuit into an open one to convert it into an antenna.

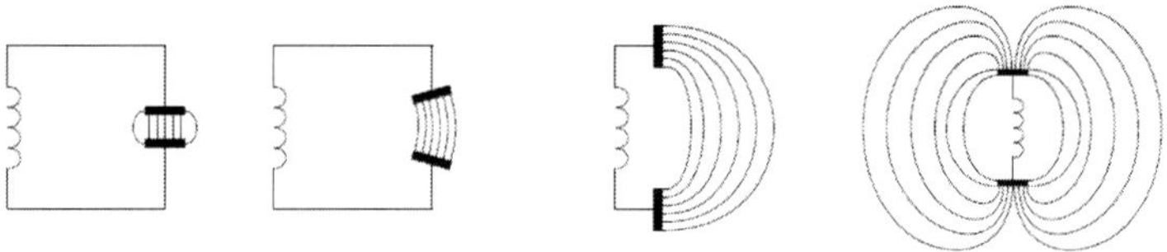

Figure 10.1 Transformation of an oscillating circuit into an antenna.

The capacitor is built in a way that its electric field must spread to the surrounding area. In case of a non-static electric field, the physical link between the electric and the magnetic field will also generate a magnetic field next to this structure. This open resonant circuit behaves as an infinitesimally small electric dipole—the simplest kind of electrical antennas.

The other possibility is to use the coil as a radiating element. The inductivity of the circuit is changed in a way that the magnetic field must spread to the surroundings. Next to this structure, there is a magnetic field of course. If it is non-static, an electric field will also be generated at some distance. This open resonant circuit behaves like an infinitesimally small magnetic dipole—the simplest magnetic antenna [1].

10.2.1.1 Infinitesimally small electric and magnetic dipoles

There are various types of antennas. For the explanation of the physical principles of radio frequency transmission, we confine ourselves to the dipoles. Both antennas are presented in Fig. 10.2. It is convenient to make considerations in a spherical coordinate system.

The electric dipole is described by its length l and effective electrical current i^E. The magnetic dipole is dual to the electric

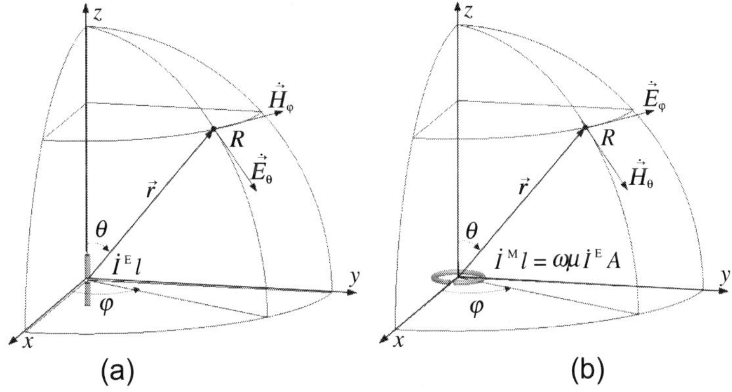

Figure 10.2 Infinitesimally small electric (a) and magnetic (b) dipoles and their dominant field components inside the Fraunhofer zone.

dipole and can be identically described by a fictive magnetic current $\dot{I}^{\mathrm{M}}$. The real model of a magnetic dipole is a small loop with an electrical current $\dot{I}^{\mathrm{E}}$ and an area A:

$$\dot{I}^{\mathrm{M}} l = \omega \mu \dot{I}^{\mathrm{E}} A \tag{10.1}$$

Here and everywhere in this paragraph the currents as well as electric and magnetic field intensities $\vec{E}$ (V/m) and $\vec{H}$ (A/m), respectively, are represented by their complex values that take into account the amplitudes and phases of the oscillation with a constant frequency.

Without a derivation, we give relations for the field components of the two types of antennas [11]:

1. *Electric dipole field components*

Magnetic field intensity:

$$\begin{cases} \dot{H}_\varphi = j\dfrac{\dot{I}^{\mathrm{E}} l k^2}{4\pi}\left[\dfrac{1}{kr} - j\left(\dfrac{1}{kr}\right)^2\right]\sin\theta \cdot e^{-jkr} \\[2mm] \dot{H}_{\mathrm{r}} = \dot{H}_\theta = 0 \end{cases} \tag{10.2}$$

Electric field intensity:

$$\begin{cases} \dot{E}_\theta = j\dfrac{\dot{I}^{\mathrm{E}} l k^3}{4\pi\omega\varepsilon}\left[\dfrac{1}{kr} - j\left(\dfrac{1}{kr}\right)^2 - \left(\dfrac{1}{kr}\right)^3\right]\sin\theta \cdot e^{-jkr} \\[3mm] \dot{E}_{\mathrm{r}} = \dfrac{\dot{I}^{\mathrm{E}} l k^3}{2\pi\omega\varepsilon}\left[\left(\dfrac{1}{kr}\right)^2 - j\left(\dfrac{1}{kr}\right)^3\right]\cos\theta \cdot e^{-jkr} \\[3mm] \dot{E}_\varphi = 0 \end{cases} \tag{10.3}$$

2. *Magnetic dipole field components*

Electric field intensity:

$$\begin{cases} \dot{E}_\varphi = -j\dfrac{\dot{I}^{\mathrm{M}} l k^2}{4\pi}\left[\dfrac{1}{kr} - j\left(\dfrac{1}{kr}\right)^2\right]\sin\theta \cdot e^{-jkr} \\[2mm] \dot{E}_{\mathrm{r}} = \dot{E}_\theta = 0 \end{cases} \tag{10.4}$$

Magnetic field intensity

$$\begin{cases} \dot{H}_\theta = j\dfrac{\dot{I}^{\mathrm{M}} l k^3}{4\pi\omega\mu}\left[\dfrac{1}{kr} - j\left(\dfrac{1}{kr}\right)^2 - \left(\dfrac{1}{kr}\right)^3\right]\sin\theta \cdot e^{-jkr} \\[3mm] \dot{H}_{\mathrm{r}} = \dfrac{\dot{I}^{\mathrm{M}} l k^3}{2\pi\omega\mu}\left[\left(\dfrac{1}{kr}\right)^2 - j\left(\dfrac{1}{kr}\right)^3\right]\cos\theta \cdot e^{-jkr} \\[3mm] \dot{H}_\varphi = 0 \end{cases} \tag{10.5}$$

with $k = \dfrac{2\pi}{\lambda} = \omega \cdot \sqrt{\mu\varepsilon}$

We note here that each of the field expressions presented above contains several terms. However, only some of these terms are responsible for the wave propagation of the fields [2].

10.2.1.2 Antenna field zones

The average power radiated by an antenna can be written as a surface integral:

$$P_{\text{rad}} = \frac{1}{2} \oiint_S \text{Re} \left(\dot{\vec{E}} \times \dot{\vec{H}}^* \right) \cdot ds \tag{10.6}$$

In general, the useful power flow in the radial direction is defined by the complex transversal field components. The analysis of (10.2–10.5) allows to subdivide the field around an antenna into the three principal regions [10]:

- *Near-field region* ($kr \ll 1$): for an electrical dipole only the transversal field components are able to produce radial power, $\dot{E}_\theta$ and $\dot{H}_\varphi$ with the terms $-(1/kr)^3$ and $-j(1/kr)^2$, respectively, they are in time-phase quadrature. Therefore, the magnetic field intensity is negligibly small, thus the field is mainly reactive, and there is no power flow radiated according to Eq. (10.6).
  In a similar way, for a magnetic dipole the power is mostly concentrated in the magnetic field and in this region the electric field can be neglected. The magnetic field distribution in this case is the same as the electrical field distribution near the electrical dipole.
- *Intermediate (Fresnel) region* ($kr \sim 1$): here all of the terms in the field expressions have to be considered because they give a contribution with comparable values. However, only transversal components lead to a radial power flow. Of course, only $\dot{E}_\theta$ and $\dot{H}_\varphi$ lead to a radial part of time-averaged power flow, in case of being in-phase, that is not zero. There are several components and the terms with factors $1/kr$, $(1/kr)^2$, and $(1/kr)^3$, which represent the field distribution, but only components with the factor $1/kr$ lead to power radiation, and that must be marked out.

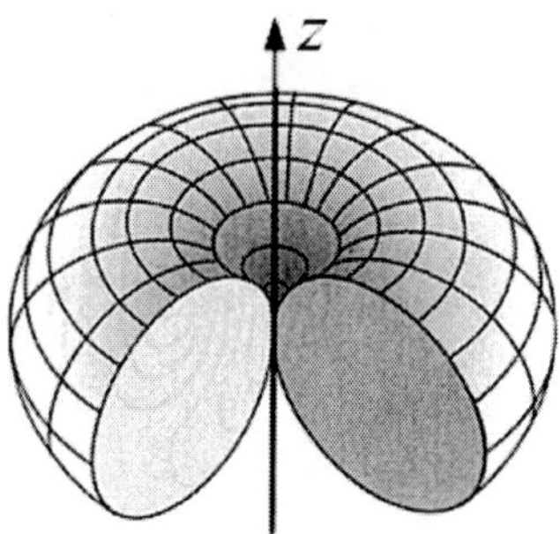

Figure 10.3 Three-dimensional amplitude pattern of an infinitesimally small dipole.

- *Far-field (Fraunhofer) region ($kr >> 1$)*: the dominant field components, $\dot{E}_\theta$ and $\dot{H}_\varphi$, both with the term $1/kr$ are in phase; there is no power stored in the far-field region of space because of the following relation:

$$\mathrm{Im}\left(\vec{\dot{E}} \times \vec{\dot{H}}^*\right) \approx 0 \tag{10.7}$$

All the power flow is radiated and distributed on the sphere with a radius equal to the distance from the antenna to the point of observation. This far-field distribution is generally not uniform and can be depicted as "far-field pattern" like shown in Fig. 10.3 for an infinitesimally small dipole.

The discussion of radiation and power flow depending on the antenna field zone gives the background for three different approaches to achieve radio frequency power transmission that are used today:

- capacitive coupling in the near field
- inductive coupling in the near field (coils)
- microwave transmission in the far field (e.g., half-wave dipoles)

10.2.2 *Frequency Bands: Characteristics and Usage*

Planning systems for radio frequency power transmission in a technical environment is also a question of using the right carrier frequency that allows the best approach to all desired features for the given conditions. Main constraints for the system are the required range and the size of the antenna. The following aspects have to be considered when choosing a frequency band:

- Using low frequencies offers a greater usable near-field zone and thus more power in the proximity of the transmitter antenna. However, it must be proved that the system is not reaching the intermediate zone where different antennas would be needed. Far-field transmission at low frequencies will need a rather long distance to the transmitting antenna.
- The geometrical size of an antenna cannot be arbitrarily reduced without losses in its performance. Due to the longer wavelength, antennas for low frequencies are usually geometrically bigger than for higher frequencies.
- Lower frequency transmissions offer less bandwidth for information transmission.

Table 10.1 gives an overview of the radio frequency spectrum and its main usage.

When designing a wireless power transmission system, the relevant frequency regulations have to be considered. For the most promising applications RFID and other inductive coupled systems, where also power is transmitted without using wires, several frequency bands exist. In these applications passive devices like RFID Transponders are powered by the radiated electromagnetic field of a reading station. Table 10.2 gives an overview of the usable spectrum and its regulations for RFID systems according to the European Regulations Authority [12]. The spectrum is divided into frequency bands for inductively coupled systems—normally below 30 MHz—and UHF and microwave systems above 30 MHz.

For inductively coupled systems frequency bands from 119 to 135 kHz and 13.56 MHz are used, but additional bands below 119 kHz and above 135 kHz are available. The maximum transmitted signal strength is limited and defined as a maximum magnetic field strength measured at a distance of 10 m from the transmitter.

In the UHF spectrum the frequency band between 865 and 867 MHz is used in Europe for RFID applications and allows a maximum

Table 10.1 Use of the radio frequency spectrum

Frequency band	Wavelength	Frequency range	Applications
Extremely low frequency (ELF)	∞ to 100 km	0 kHz to 3 kHz	Acoustics (20 Hz–20 kHz)
Very low frequency (VLF)	100 km to 10 km	3 kHz to 30 kHz	Acoustics, submarines
Low frequency (LF)	10 km to 1 km	30 kHz to 300 kHz	Longwave broadcasting
Medium frequency (MF)	1 km to 100 m	300 kHz to 3 MHz	AM Radio broadcast
High frequency (HF)	100 m to 10 m	3 MHz to 30 MHz	Amateur broadcast, CB Radio (26.6–27.4 MHz), RFID, shortwave broadcast radio
Very high frequency (VHF)	10 m to 1 m	30 MHz to 300 MHz	FM radio broadcast (87.5–108.0 MHz)
Ultra high frequency (UHF)	1 m to 10 cm	300 MHz to 3 GHz	DECT (1880–1900 MHz/1920– 930 MHz),
—	—	—	GSM (824–894 MHz/876–960 MHz/1710–1880 MHz/1850–1990 MHz),
—	—	—	RFID (860–960 MHz),
—	—	—	WLAN (2400–2482 MHz)
Super high frequency (SHF)	10 cm to 1 cm	3 GHz to 30 GHz	WLAN (5.15–5.25 GHz)
Extremely high frequency (EHF)	1 cm to 1 mm	30 GHz to 300 GHz	Radar, satellite HDTV

Table 10.2 Frequency regulation for RFID and other inductive coupled applications (according to [12])

Frequency band	Power/magnetic field	Notes
90–119 kHz	42 dBµA/m	At a distance of 10 m to transmitter
119–135 kHz	66 dBµA/m	At a distance of 10 m to transmitter, level descending 3 dB/octave
135–140 kHz	42 dBµA/m	At a distance of 10 m to transmitter
13.553–13.567 MHz	42 dBµA/m	At a distance of 10 m to transmitter
865.6–867.6 MHz	2 W e.r.p.	—
2446–2454 MHz	500 mW e.i.r.p.	—

equivalent radiated power of 2 W. This energy is sufficient to power a passive device like a UHF identification transponder over a distance of more than 10 meter. Other countries allow a maximum power of 4 W for UHF RFID applications.

At higher frequencies a band between 2.446 to 2.454 GHz within the worldwide 2.4 GHz band for short range devices and wireless data links like WLAN or Bluetooth can be used for RFID applications and allow an equivalent isotropic radiated power of up to 500 mW.

10.2.3 *Basic Concept*

The basic elements (Fig. 10.4) of a wireless power transmission system are a source, generating an electromagnetic field and sink, which is receiving the electromagnetic field and is converting it to a DC-power. Between the source and the sink the electromagnetic wave is passing the transmission channel. It is a medium, normally air, but there could also be solid materials like obstacles in between depending on the distance d of source and sink. Not only the distance of source and sink,

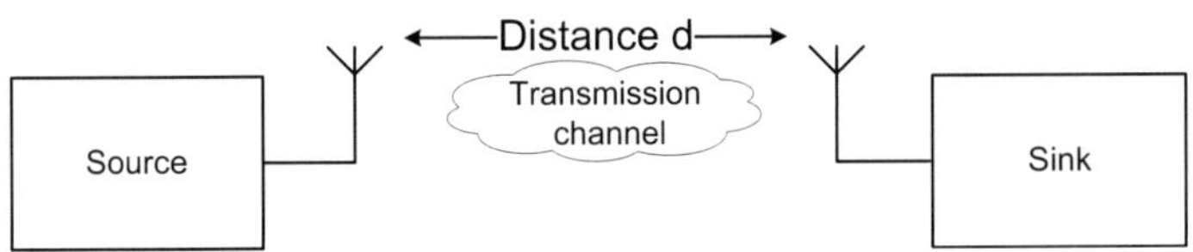

Figure 10.4 Basic elements of a wireless power transmission.

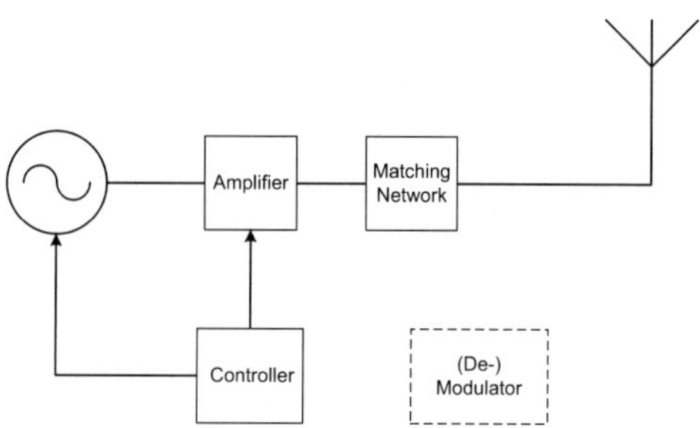

Figure 10.5   Source.

but also the working frequency, physical characteristics (metal-, dielectrical-, permeable materials) of the medium, noise, or unwanted electromagnetic fields should be considered.

The source (Fig. 10.5) consists of a generator for the working frequency and a signal amplifier with a matching circuit coupling the antenna. A controller will take care of the transmission system under different working conditions. In case of additional data transmission a modulator and demodulator is required in addition.

The sink (Fig. 10.6) consists of an antenna with matching circuit followed by a rectifier and a voltage converter. The design of the antenna and the matching toward the receiver circuit has influence on the efficiency of the whole circuit. In general, the amplitudes of the received signals are very low. The conversion of the received RF signals to a DC power with high efficiency is a main challenge when designing the receiver. Solutions for this task are given in other chapters. Finally there will be some sort of load in the sink, which is powered by the received energy.

For lower frequencies with inductive coupling, there is an influence between the primary and secondary coil, which has to be considered. A secondary coil within the coupling area of the primary coil will influence the behavior of the primary coil in the source. The coupling between primary and secondary coil will force to change the matching while the link between both coils is active. According to this effect, a changing load in the sink will be transformed into changes in the source.

In far-field applications, high-gain antennas with a good matching are essential for an efficient energy transmission. A well-known

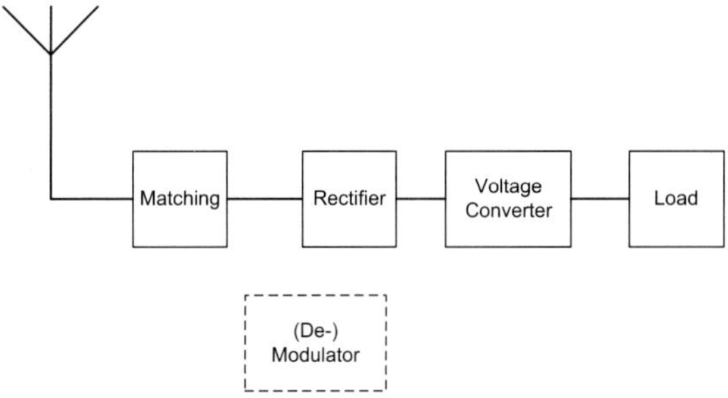

Figure 10.6 Sink.

principle is to build so-called rectennas [5] as a combination of a matched antenna and a voltage converter for low voltages.

10.2.4 *Inductive Coupling*

Within the near-field area of an antenna, both coupling possibilities—capacitive and inductive—are used to transmit power and exchange information. The inductive coupling is the more common principle and can be found in many applications today.

The magnetic field H emitted by a coil as illustrated in Fig. 10.7 can be described by the following equation, where r is the radius of the coil with N windings, I the current through the coil, and x the

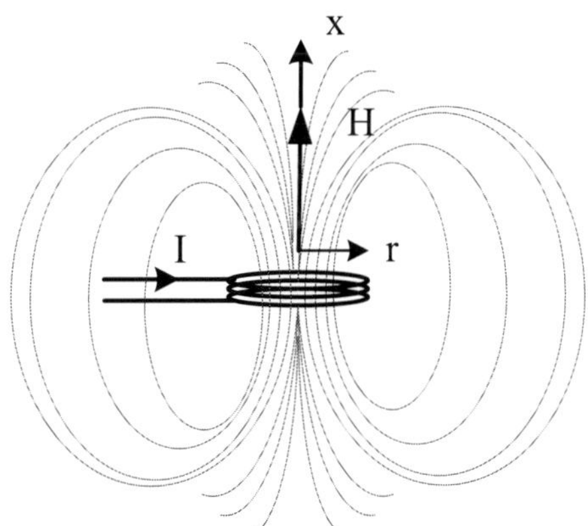

Figure 10.7 Magnetic field of a coil.

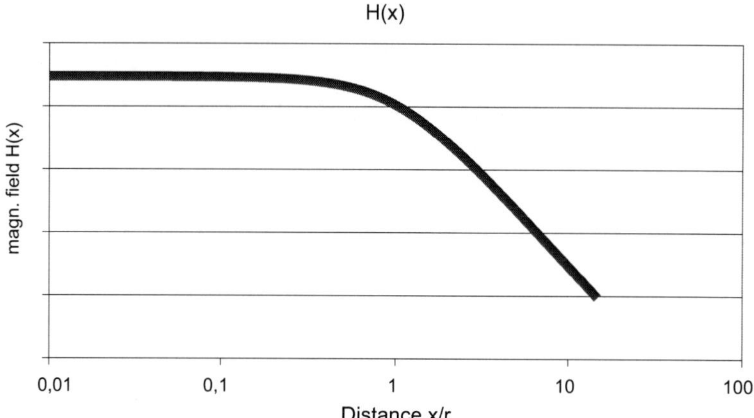

Figure 10.8　Relation between magnetic field and distance.

distance to coil.

$$H(x) = \frac{I \cdot N \cdot r^2}{2\sqrt{(r^2 + x^2)^3}} \tag{10.8}$$

The strength of the field H decreases as the distance x to the coil is increased. A closer look to the equation shows that at small distances x of the coil in relation to the radius the term x^2 can be neglected and the H-field remains almost constant up to a certain distance (Fig. 10.8). At the distance $x = r$, the field strength falls rapidly. In free space, the decay of the field strength is approximately 60 dB per decade in the near field of the coil and 20 dB per decade in the far field [4]. Figure 10.8 shows only the near field area of the field upon the x-axis without specific values for N and I.

The magnetic flux Φ of a coil with a single loop winding, which is the integral of the magnetic field through the area of the coil, can be noted as follows:

$$\Phi = B \cdot A = \mu \cdot H \cdot A \tag{10.9}$$

For a loop with N turns, the total flux Ψ is defined as follows:

$$\Psi = \Sigma\Phi = N \cdot \Phi \tag{10.10}$$

To characterize the behavior of a loop the inductance L is defined with the magnetic field H as

$$L = \frac{\Psi}{I} = \frac{N \cdot \Phi}{I} = \frac{N \cdot \mu \cdot H \cdot A}{I} \tag{10.11}$$

or from its geometry as

$$L = N^2 \cdot \mu \cdot r \cdot \ln\left(\frac{2r}{d}\right) \tag{10.12}$$

with the wire diameter $d \ll 2r$ [4, 7].

To transmit energy in a magnetically coupled system, a second coil L_2 is necessary, which will be located in the vicinity of the first coil. The first coil L_1 (transmitter) is emitting the magnetic field caused by the current I_1. The magnetic flux caused by the current I_1 is passing through the second coil and induces a voltage causing a current flow I_2.

The induced current in the second coil will in turn generate its own magnetic field, which has a reaction to the first coil. It is possible to describe the second coil as an inductance L_2 with N_2, I_2, r_2, A_2, and Ψ_2 in the same manner as it was done for the first coil.

For the energy transmission, the interconnection between the first and the second coil must be known.

Similar to the definition of the (self) inductance L of a conductor loop, the mutual inductance M_{21} is defined as the ratio of the partial flux Ψ_{21} enclosed by the second coil to the current I_1 of the first coil.

$$M_{21} = \frac{\Psi_{21}(I_1)}{I_1} = \oint_{A_2} \frac{B_2(I_1)}{I_1} \cdot dA_2 \tag{10.13}$$

As the whole system is symmetric, a mutual inductance M_{21} and a partial flux Ψ_{21} can also be defined with the relationship

$$M = M_{12} = M_{21} \tag{10.14}$$

In order to have a qualitative prediction about the coupling between two loops the coupling coefficient β ($0 \le \beta \le 1$) of two coils can be calculated with the inductance:

$$\beta = \frac{M}{\sqrt{L_1 \cdot L_2}} \tag{10.15}$$

It has two extreme cases:

- $\beta = 0$ describes two completely decoupled coils. This is possible due to a great distance or a magnetic shielding.
- If both coils are subject to the same magnetic flux , they are totally coupled, $\beta = 1$. The transformer is a technical application of this total coupling, where two or more coils are wound into a highly permeable iron core. The value $\beta = 1$ cannot be reached in real applications due to existing losses.

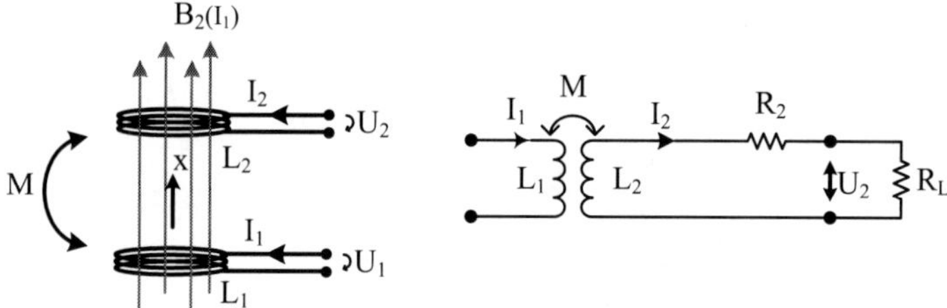

Figure 10.9 (Left) Magnetically coupled loops. (Right) Equivalent circuit diagram for the coupled loops.

The Eq. (10.16) shows which parameters influence the coupling factor for two parallel conductor loops centered on a single x axis ($r_1 \leq r_2$) [8].

$$\beta = \frac{\mu \cdot \pi \cdot N_1 \cdot N_2}{2 \cdot \sqrt{L_1 \cdot L_2}} \cdot \frac{r_1^2 \cdot r_2^2}{\left(\sqrt{r_1^2 + x^2}\right)^3} \tag{10.16}$$

According to Faraday's law, the induced voltage of the second coil can be calculated as

$$u_2 = \oint E_2 \cdot ds = -\frac{d\Psi(t)}{dt} \tag{10.17}$$

For the equivalent circuit in Fig. 10.9, the induced voltage can be calculated:

$$U_2 = \frac{d\Psi_2}{dt} = M\frac{dI_1}{dt} - L_2\frac{dI_2}{dt} - I_2 \cdot R_2 \overset{\omega=2\pi f}{=} j\omega M \cdot I_1 - j\omega L_2 \cdot I_2 - I_2 R_2 \tag{10.18}$$

To improve the efficiency of the equivalent circuit in Fig. 10.9 an additional capacitor C_2 is connected in parallel with the second coil L_2. With this capacity, the circuit is a parallel resonant circuit with a resonant frequency corresponding to the working frequency of the transmission system. It is also possible to add a resonant capacitor to the first coil.

The two coils described above build an energy transmission system with the first coil as a transmitter and the second coil as a receiver (Fig. 10.10) with a load resistor R_L.

The resonant capacitors cancel out the stray inductance in the receiver and the magnetizing inductance in the transmitter. Now, the only remaining loss limit for the power transmission is the winding

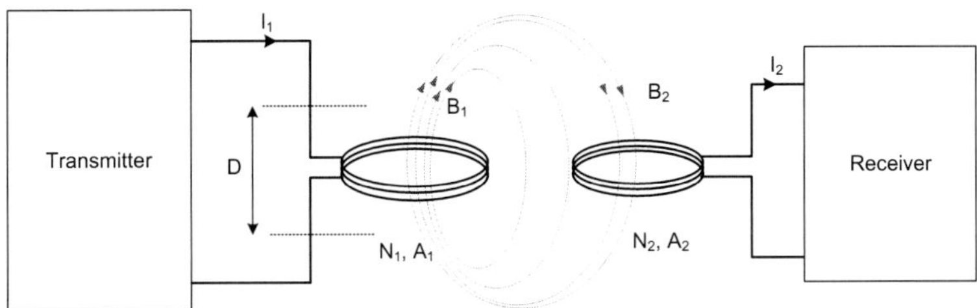

Figure 10.10   Inductive transmission system.

resistances of the coils, whose impedance is one or two orders of magnitude lower than that of the inductances.[a] Therefore, for a given generator source, much more power can be received in resonant transformer systems.

To describe the behavior of a resonant coil the quality factor Q can be calculated. A high Q factor leads to high current in a coil and thus improves the power transmission. However, it also causes a low bandwidth, which has to be considered when thinking about additional data transmission trough the coils.

$$Q \approx \frac{2\pi \cdot f_0 \cdot L_{\text{Coil}}}{R_{\text{Total}}} \tag{10.19}$$

It is calculated from the ratio between the inductive resistance and the resistive loss of the coil. With the quality factor, the bandwidth can be simply calculated [4, 8].

$$B \approx \frac{f_0}{Q} \tag{10.20}$$

The final equivalent circuit of the transmission system is given in Fig. 10.11 [4, 8].

The two coupled coils can be described as an ideal transformer with its mutual inductance M and resistive and reactive losses.

10.2.5 *Far-Field Radio Transmission*

Section 10.2.1 explains the principles of generating and radiating an electromagnetic field using a dipole antenna. For the further discussion of the electromagnetic power transmission, it is assumed

---

[a] By increasing the frequency of the system, additional interrelations like skin effect and proximity effect have to be considered.

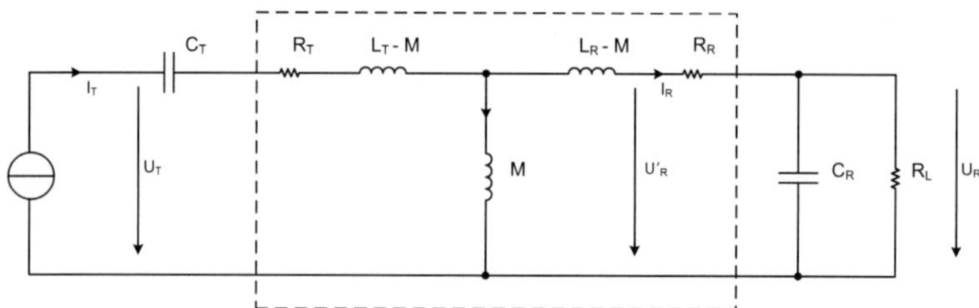

Figure 10.11 Equivalent circuit of an inductive transmission system.

that the receiver of a power transmission system is placed in the far field of the electromagnetic field of a transmitter. As already mentioned the transition from near field to far field depends on the frequency. The higher the frequency the closer is the far-field transition of the antenna.

An electromagnetic wave of an isotropic radiator is propagating spherically into space from its origin and radiates power to the surrounding space. With increasing distance to the source, the power is spread over an increasing sphere surface area. The so-called radiated power density S is defined as the quotient of the effective isotropic radiated power (EIRP) of the transmit antenna and the surface of the sphere [3, 4, 6]:

$$S = \frac{P_{\text{EIRP}}}{4\pi r^2} \tag{10.21}$$

The radiated power density S is the absolute value of the pointing vector $\vec{S}$.

$$\vec{S} = \vec{E} \times \vec{H} \tag{10.22}$$

In the far-field area, the two field vectors are perpendicular.

With this relationship and the characteristic wave impedance in the air, $Z_{\text{F}} = \sqrt{\varepsilon_0 \mu_0} = 377\ \Omega$, it is possible to calculate the field strength at a certain distance r from the radiation source:

$$E = H \cdot Z_{\text{F}} = \sqrt{S \cdot Z_{\text{F}}} = \sqrt{\frac{P_{\text{EIRP}} \cdot Z_{\text{F}}}{4\pi r^2}} \tag{10.23}$$

When travelling through space, the E-field vector of the electromagnetic wave is changing its orientation perpendicular to the direction of the pointing vector. The change can happen in a linear way, vertical or horizontal, or in a circular way, left-hand or right hand.

This effect is called the polarization of the electromagnetic field and is determined by the antenna. An example for a linear polarized antenna is the dipole antenna; an example for circular polarization is the helix antenna. To realize an optimum power transmission transmit and receive antennas with corresponding polarization have to be used [4, 6].

The radiation and reception of electromagnetic waves takes place at any conductor that carries voltage and/or current. An antenna is a conductor that is highly optimized to radiate or receive (or both) a wave within a certain frequency range. There are several characteristics defined to describe the direction-sensitive behavior of an antenna:

- The relative radiation vector $G(\Theta)$ and the main radiation direction $\mathbf{G}_i$ give the connection between directed radiation power and antenna feeding power P_1. Because of this relationship, $\mathbf{G}_i$ is also called the antenna gain:

$$S = \frac{P_{\text{EIRP}}}{4\pi r^2} = \frac{P_1 \cdot \mathbf{G}_i}{4\pi r^2}. \qquad (10.24)$$

- The effective aperture A_{e} is the proportional factor between an incoming plane wave radiated power density S and the received power P_{e}:

$$P_{\text{e}} = S \cdot A_{\text{e}} = \frac{\lambda_0^2}{4\pi} \cdot \mathbf{G}_i \cdot S \qquad (10.25)$$

- The complex input impedance Z_{A} of an antenna is the sum of all resistances seen by the source and has to be matched. It consists of the ohmic loss resistance R_{r} of all current-carrying sections, the radiation resistance R_{V} corresponding to the conversion losses between input power, emitted power and the complex resistance X_{A} describing the frequency dependent losses:

$$Z_{\text{A}} = R_{\text{r}} + R_{\text{V}} + jX_{\text{A}}. \qquad (10.26)$$

Figure 10.12 illustrates some of the above-described values [4, 6].

With these definitions, it is now possible to describe a microwave transmission system mathematically. The Friis formula gives a value for the relationship between the emitted power from the transceiver

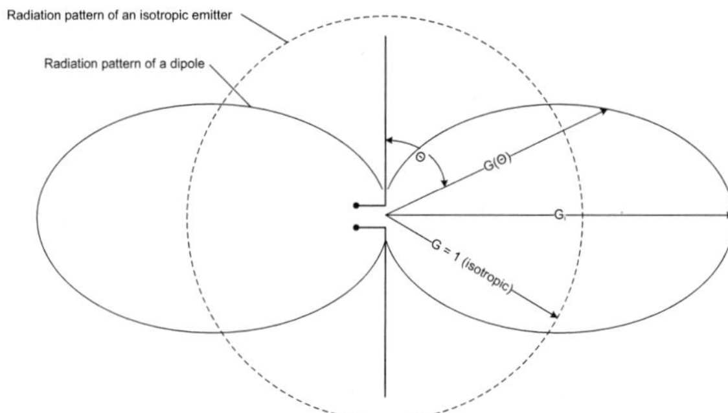

Figure 10.12 Radiation pattern of a dipole compared to an isotropic emitter.

and the finally received RF power from free space, depending on the antenna gains, distance r and the working frequency f:

$$\frac{P_r}{P_t} = G_t G_r \left(\frac{\lambda}{4\pi r} \right)^2 \tag{10.27}$$

Any wireless power transmission within the far-field area of a transmission system will have to deal with these losses proportional to λ^2/r^2. The developer will counteract those with well matched, high gain antennas and by carefully choosing the transmission frequency according to the transmission distance and functionality [4, 5].

10.3 Design Optimization

10.3.1 *Generation and Amplification of High-Frequency Signals*

Crystal oscillators are common sources for sinusoidal signals. They offer high-frequency stability compared to LC oscillators. With the use of frequency dividers, frequency multipliers, and phase-locked loop circuits a wide range of frequencies from one reference frequency can be derived.

Linear (e.g., A and B modes) and switching amplifier circuits (e.g., D and E modes) are used in RF transmitters and are also available as integrated circuits. For energy transmission systems the design of a digital amplifier could be worthwhile. However, with available transistor technologies, switching frequency is still a limiting factor for class D, E, and S amplifier designs in high MHz and GHz applications.

10.3.2 *Antennas and Matching*

10.3.2.1 Low-frequency systems: coil antennas

In Section 10.2.4, the general characteristics of conductor loop antennas are presented. Loop antennas are recommended as most suitable for generating the magnetic (H) field that is required to transfer energy in a magnetically coupled near-field transmission. A loop antenna is a tuned LC circuit for a particular frequency. When the inductive impedance is equal to the capacitive impedance the antenna will be at resonance. From Section 10.2.4, the first conclusion could be that the best antenna for an inductive system is a huge coil with a maximal number of loop turns (maximal inductance L) and a minimal resistance (of the windings) when tuned in resonance.

In a real setup, the coil inductance should not reach too high values because it must still be possible to tune and match the antenna. With the common capacitive matching a high inductance (e.g., 5 μH for 13 MHz systems) will tend to very low capacitor values for resonance and matching. The use of values below 10 picofarad is not practical in circuits as the parasitic influences have about the same dimensions. Even with other matching strategies like a balun, very high inductance together with parasite capacities of the loop (for example, proximity effect) will cause unwanted resonances below the transmitted frequency. Matching and optimization will be difficult or even impossible.

Geometrically huge coils with dimensions reaching about half of the wavelength no longer work only as magnetic antennas. The characteristic behavior toward the magnetic field component begins to disappear.

The quality factor of a coil antenna should be maximized for good power transmission. The antenna has a band-pass characteristic (Fig. 10.13). However, changes in the surrounding or in the load network will influence the quality factor. Especially if the quality factor in a system is very high for best results in a power

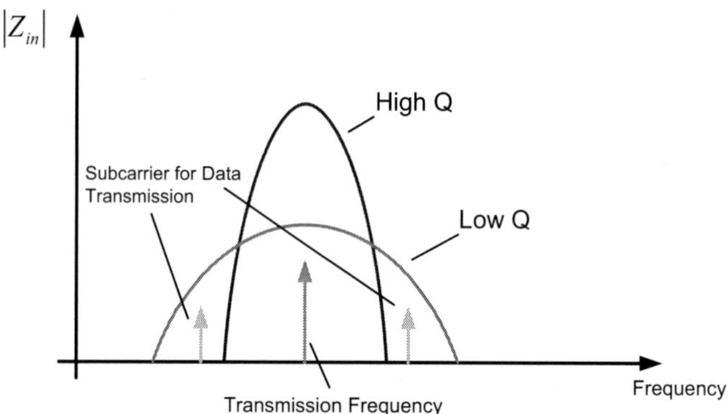

Figure 10.13 Resonant coil antenna band-pass characteristic.

transmission, these influences will put a risk to the whole system to be mistuned and lose efficiency rapidly. For this reason, the tuning and matching of the coils has to be done in presence of the connected network, load, and surrounding material. In case of an additional data transmission, for example, with the modulation of subcarriers, the quality factor has to fit to the required additional bandwidth. A high quality factor leads to a low bandwidth according to Eq. 9.20 [4, 7, 8].

10.3.2.2 High-frequency systems

For high-frequency systems, usually electrical antennas of different types are used like dipole antennas, patch antennas or slot antennas. To transmit and receive electromagnetic waves the antenna must be matched to the free space impedance on the one hand and on the other hand to the impedance of the connected circuit. Without a good matching, the losses at these impedance transitions would be too high.

The matching of an antenna is always a compromise. In general, high-frequency systems use modulated carriers and matching must fit for the full bandwidth needed. Therefore, an exact solution is not possible. The Bode–Fano criterion represents the optimum matching result that can be ideally achieved with a theoretical limit on the minimum magnitude of the reflection coefficient [3–5].

10.3.3 *Voltage Rectification and Stabilization*

In most cases, the output of wireless power receivers should be a DC power for the load circuit. Therefore, the received signal, typically sinusoidal, must be rectified. In many cases, the amplitude of the signal is very low, the use of low threshold diodes and good quality capacitors is recommended. The combination of antenna and rectifier is also called rectenna. Common for voltage reception are full-wave rectifiers, voltage doublers and Greinacher rectifiers. The use of a transformer to generate higher voltage amplitudes can also be found in literature. Circuits for these tasks are well explained in the preceding chapters. However, it should be considered how these circuits influence the impedance behavior of the whole system, concerning the design and matching of the antenna, particularly if the load circuit has a non-linear power consumption [4, 5, 8].

10.4 Efficiency of Wireless Power Transmission

In a wireless power transmission system, the three parts RF-signal generation, air interface and voltage rectifier are the main contributors to the overall wireless transmission system efficiency (Fig. 10.14). The influence of the signal generation ❶ and the rectification/voltage conversion ❸ is obvious and a challenging task for the system and hardware design of the transmission system.

The efficiency η_{Radio} of the wireless radio transmission ❷ between ❶ and ❸ is, of course, influenced by antenna design and matching mainly depends on the distance between transmitter and receiver. The theoretical limits for the power transmission

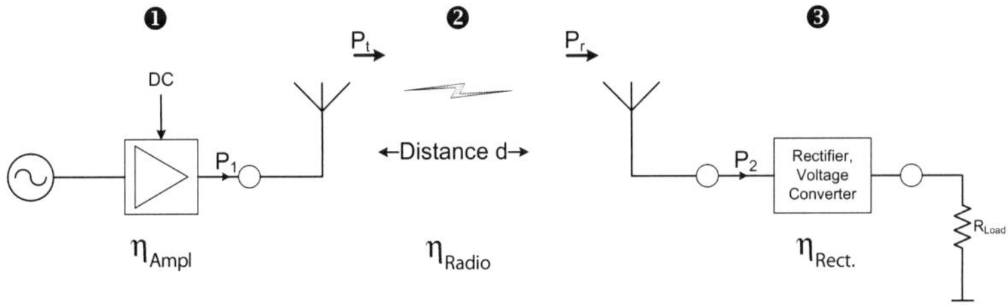

Figure 10.14 Efficiency factors for a transmission system.

efficiency over the air are influenced by the distance, the frequency or wavelength and the transmission principle (near-field coupling, far-field transmission).

10.4.1 *Low-Frequency Transmission Efficiency*

For lower frequencies with inductive near-field coupling the transmission can be described according to [8] as a transmission function for a resonant transformer system with the inductance of the transmitting and receiving coils $L_{1/2}$, the coupling β (Eq. 9.16) between them and the working angular frequency ω_0.

$$\eta = \frac{P_2}{P_1} = \frac{1}{1 + \left[\frac{\omega_0 L_2}{R_L}\left(\frac{1}{\beta^2} - 1\right)\right]^2} - j\frac{\frac{\omega_0 L_2}{R_L}\left(\frac{1}{\beta^2} - 1\right)}{1 + \left[\frac{\omega_0 L_2}{R_L}\left(\frac{1}{\beta^2} - 1\right)\right]^2} \quad (10.28)$$

$$\eta = \left|\frac{P_2}{P_1}\right| = \frac{1}{\sqrt{1 + \left[\frac{\omega_0 L_2}{R_L}\left(\frac{1}{\beta^2} - 1\right)^2\right]}} \quad (10.29)$$

with the coupling

$$\beta = \frac{\mu \cdot \pi \cdot N_1 \cdot N_2}{2 \cdot \sqrt{L_1 \cdot L_2}} \cdot \frac{r_1^2 \cdot r_2^2}{\left(\sqrt{r_1^2 + d^2}\right)^3} \quad (10.30)$$

Figure 10.15 is an example for the efficiency of an inductively coupled transmission at a frequency of 13.56 MHz according to Eq. 9.29. For the calculation, the following values are assumed:

$$L_1 = L_2 = 2.4\,\mu H, \ R_L = 500\,\Omega, \ N_1 = N_2 = 6, \ r_1 = r_2 = 2 \text{ cm}$$

Figure 10.15 shows a quite constant efficiency of 75% over a short distance between the two transmission coils of up to 2 cm. It becomes clear that an effective power transmission is only possible within short distances. In the plot, the red line marks the end of the near-field area for the chosen frequency. Good transmission efficiency will be achieved within a distance at a size of the transmitting and receiving coil. By varying the diameter of the coil, the linear region in the graph can be adapted. By increasing the size of the coils this optimal area for efficient power transmission will also be increased. Although all other chosen parameters, especially the frequency will scale the graph, the principal form will stay as shown [8, 14].

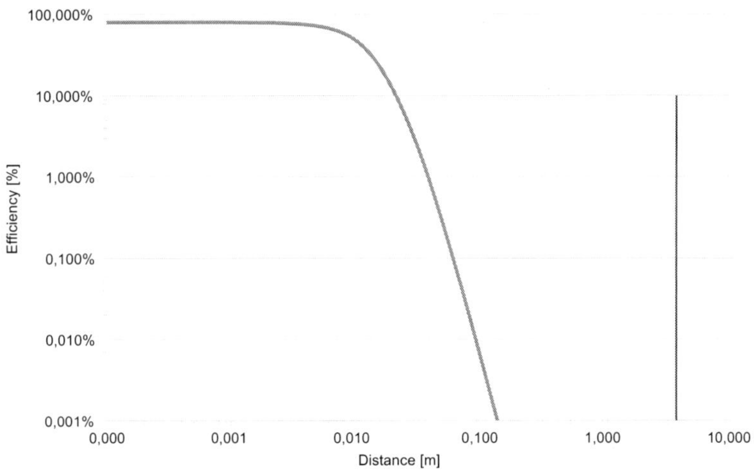

Figure 10.15 Efficiency of low-frequency transmission @ 13.56 MHz.

10.4.2 *High-Frequency Transmission Efficiency*

For higher frequencies with electromagnetic far-field transmission, the efficiency can be described by means of the far-field losses already given in the Friis formula in Eq. 9.27 in its simplest form with the antenna gains:

$$\eta = \frac{P_2}{P_1} = G_2 G_1 \left(\frac{\lambda}{4\pi r} \right)^2 \tag{10.31}$$

The last part in the formula describes the free-space path loss (FSPL), the loss in signal strength for a line-of-sight path through free space depending on the distance d and the frequency f.

Figure 10.16 shows the behavior of a far-field transmission at 868 MHz with a wavelength of 0.35 m, the antenna gain G_1 and G_2 in Eq. 9.31 are set to 1. It has to be noted that the given formula (Eq. 9.31) can only be used in the far-field area of a transmission. Lower distances than about 4λ will cause non-realistic results. In the plot, the red line marks the beginning of the far-field region for the chosen frequency.

The figure shows that only a small amount of energy of less than 0.01% of the transmitted power can be transmitted over a range of 1 m and more. For an RFID system, for example, with a radiated

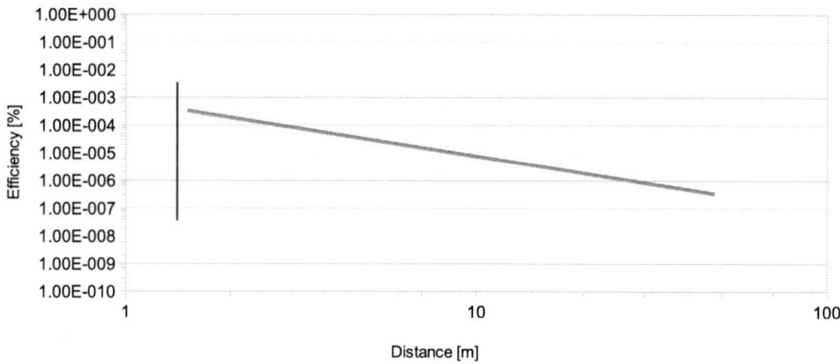

Figure 10.16 Free space path losses for high-frequency transmission @ 868 MHz.

power of 4 W a power of 1 mW or less can be achieved at the receiver side, the RFID transponder.

To increase the efficiency of a far-field transmission, more directive antennas with a very small antenna beam and thus a high antenna gain can be used [6, 7].

10.4.3 *System Efficiency*

Figure 10.17 finally gives an overview to the interrelation between the different parts of a wireless power transmission system and its efficiency:

The efficiency of the free space transmission as it is described in this chapter can vary depending on the distance, frequency, and transmission principle. Other system components of the wireless

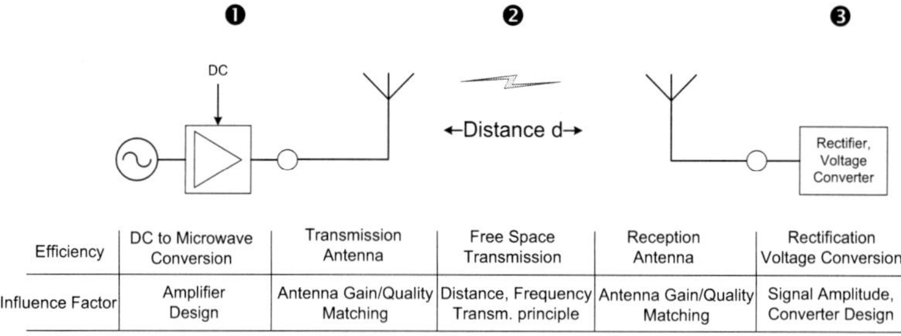

Efficiency	DC to Microwave Conversion	Transmission Antenna	Free Space Transmission	Reception Antenna	Rectification Voltage Conversion
Influence Factor	Amplifier Design	Antenna Gain/Quality Matching	Distance, Frequency Transm. principle	Antenna Gain/Quality Matching	Signal Amplitude, Converter Design

Figure 10.17 Overview on the efficiency influences.

power transmission system such as the signal generation or the rectification will also influence the overall power transmission and lower its efficiency.

For the signal generation and transmission, a high-frequency carrier signal has to be generated from a DC-power source. To create the high transmission power, signal amplifiers are necessary with a non-unity conversion gain. Linear amplifiers according to the literature are reaching efficiencies between 50% for class A and up to 78% for classes B and C, whereas switching amplifiers can reach up to 100% efficiency [13]. These values are theoretical maximum values. Typical values are below and are ranging, for example, from 65% for class B or C amplifiers to 85% efficiency for a switching amplifier.

The transmission and reception antennas of a system have great influence on the system efficiency. With the antenna aperture or the radiation pattern, but also with a good matching the antennas can help to improve the transmission as already described in the earlier sections.

At the receiver side, the incoming power has to be rectified and converted to a fixed, stable output voltage. Depending on the system design and the transmission distance there can be a high receiver signal voltage for close coupling systems and very low signal amplitude when working with long transmission distances at high frequencies. Especially for long-range high-frequency transmissions, the rectification of this low amplitude AC signals to a DC voltage is a challenging task. Due to the low signal voltage, the rectification diodes will not be biased appropriately and cause high losses. This is a serious problem in designing passive UHF-RFID transponder. A typical value for the efficiency of the rectification circuit of a UHF-RFID is about 10–15% [15].

10.5 Example Applications: Passive RFID Systems

As well-known applications of wireless power transmission, passive radio frequency identification (RFID) systems can be mentioned. Among various different types of RFID systems most of the passive systems operate with inductive coupling at the frequencies 125 kHz

Table 10.3   RFID reference values

	LF	HF	UHF
TX frequency	115–135 kHz	13.56 MHZ	865–928 MHz
Typ. operating distance	0–1.5 m	0–1 m	13 m
Antenna principle	Loop antenna	Loop antenna	Mainly dipole antennas
Max. bandwidth	65 kHz	$f_{Tx} \pm 847$ kHz	Max 500 kHz
Modulation uplink	ASK	ASK	ASK
Modulation downlink	ASK, FSK subcarrier	Subcarrier	Backscatter
Typ. data rate	10 kBd	106 kBd	640 kBd

(LF) and 13.56 MHz (HF) and with far-field transmission within the frequency range of 865 to 928 MHz (UHF). These systems consist of active reader devices and one or more passive transponders. The reader transmits a strong carrier signal, which powers the transponder circuits. Beside the power transmission, both devices, reader and transponder, can transmit information to each other by modulating the carrier. The main application for these systems is the wireless identification of objects tagged with a transponder. The bidirectional information transmission makes it possible to either read or write data to the transponder. In some cases, the functionality is expanded to a wireless sensor system.

Table 10.3 presents typical values for passive RFID systems [4].

The main problem in using these passive RFID systems is to deal with the surrounding environment, material properties and field effects that influence the performance of the transmission. Therefore, they must be considered in an application and pointed out to the user who often lacks in theoretical background to evaluate these influences by oneself.

This in an important issue for any wireless power transmission system and has to be considered during the development of the system.

Actual research topics on RFID make efforts to increase operating distance, data rate, received power, and functionality of the transponder devices.

References

1. Hansen, J. E. (1988), *Spherical Near-Field Antenna Measurements*, Peter Peregrinus Ltd., ISBN 0 86341 110 X.

2. Gregson, S. (2007), *Principles of Planar Near-Field Antenna Measurements*, British Library Cataloguing in Publication Data, ISBN 978-0-86341-736-8.

3. Pozar, M. (1998), *Microwave Engineering*, 2nd Ed., John Wiley & Sons, Inc., ISBN 0-471-17096-8.

4. Finkenzeller, K. (2002), *RFID Handbuch*, 3. Auflage. Hansen Verlag München Wien, ISBN 3-446-22071-2.

5. Curty, J. P., Declercq, M., Dehollain, C., Joehl, N. (2007), *Design and Optimization of Passive UHF RFID Systems*, Springer Science+Business Media, ISBN 0-387-35274-0.

6. Stirner, E. (1976), *Antennen Band 1 Grundlagen*, Hüthing Verlag Heidelberg.

7. Rint, C. (1955), *Handbuch für Hochfrequenz- und Elektro-Techniker Band III*, Verlag für Radio-Foto-Kinotechnik GmbH, Berlin.

8. Kolnsberg, S. (2001), *Drahtlose Signal- und Energieübertragung mit Hilfe von Hochfrequenztechnik in CMOS-Systemen*, Dissertation, Gerhard-Mercator-Universität Duisburg Fachbereich Elektrotechnik http://duepublico.uni-duisburg-essen.de.

9. Tietze, U., Schenk, C. (2002), *Halbleiter- Schaltungstechnik*, 12. Auflage, Springer Verlag Berlin Heidelberg New York, ISBN 3-540-42849-6.

10. Balanis, C. A. (1997), *Antenna Theory Analysis and Design*, 2nd ed. New York: John Wiley and Sons.

11. Yu., V. Pimenov, *Linear Macroscopic Electromagnetics*, Intellekt, Moscow, 2008 (in Russian).

12. CEPT/ERC Recommendations 70-03 Version 16, Tromso, October 2009.

13. Nathan O. S O W, RF Power Amplifiers, Classes A Through S—How They Operate, and When to Use Each, 0-7803-3987-8/97 IEEE 1997.

14. Waffenschmidt, E.; Staring, T., "Limitation of inductive power transfer for consumer applications," *Power Electronics and Applications*, 2009. EPE '09. 13th European Conference on, pp. 1, 10, 8–10 Sept. 2009

15. Curty, J.-P., Declercq, M., Dehollain, C., Joehl, N., *Design and Optimization of Passive UHF RFID Systems*, Springer Verlag Berlin Heidelberg New York, 2007, ISBN: 978-0-387-35274-9.

Chapter 11

Electrical Buffer Storage for Energy Harvesting

Robert Hahn[a] and Kai-C. Möller[b]

[a] *Fraunhofer IZM Gustav-Meyer-Allee 25, 13355 Berlin, Germany*
[b] *Fraunhofer ICT, Project Group Electrochemical Energy Storage,*
Parkring 6, 85748 Garching at Munich, Germany
Robert.hahn@izm.fraunhofer.de, kai-c.moeller@isc.fraunhofer.de

11.1 Introduction

Effective intermediate energy storage is required for all energy harvesting concepts, due to the varying availability of ambient energy and varying energy requirements of the device. In most cases, this intermediate storage is best done with the help of a secondary micro battery or with a capacitor or supercapacitor. Secondary batteries have a much higher energy density compared to capacitors and sufficient power pulse capability. On the other hand, capacitors have a higher long-term stability compared to batteries.

Thus, the development of miniaturized power sources, with high volumetric energy density, is of crucial importance for small electronic applications such as small sensor nodes, active smart labels, and MEMS.

Handbook of Energy Harvesting Power Supplies and Applications
Edited by Peter Spies, Loreto Mateu, and Markus Pollak
Copyright © 2015 Pan Stanford Publishing Pte. Ltd.
ISBN 978-981-4241-86-1 (Hardcover), 978-981-4303-06-4 (eBook)
www.panstanford.com

Primary and secondary batteries are well-established power supply technologies for portable applications. For many decades, only two or three chemical systems have dominated the market and by now, a diversification with new applications specified and novel chemical systems is emerging. Still, there is a lot of ongoing development and improvement in battery industry, although the energy density increases by some percent per annum only and not by orders of magnitude. In general, the battery industry has adapted smaller devices with a high power and energy demand to the market requirements within the limits of chemical storage capabilities. A significant research effort is the increase of not only power density of secondary systems but also safety and long-term stability of larger systems.

This chapter will concentrate on secondary batteries and supercapacitors to be used as electrical buffer storage. Nevertheless, one should keep in mind that the power demand of autarkic electronic systems will be further reduced by a factor of 10 to 100 during the next decade. This means that many more applications can be powered with a primary energy source over a life span of 10 to ca. 30 years. Primary batteries for low current drain and a tested durability of up to 30 years are already on the market.

Figure 11.1 gives an overview of energy density of primary and secondary batteries in comparison to fuel cells.

The best, currently available primary batteries are Li-batteries with inorganic electrolytes. At low current drain, the lithium thionyl chloride ($SOCl_2$) battery achieve energy densities up to 1400 Wh/l. For example, the D sized TL-5930 (Tadiran) has a volume of 58 cm^3. A capacity of 19 Ah can be realized at a discharge voltage of ca. 3.6 V, which translates into an energy of 68 Wh and an energy density of 1309 Wh/l. A continuous load of 700 μW can be powered for 10 years with this battery. Hermetically sealed batteries of this chemistry have a lifetime between 10 and 20 years. Today the best rechargeable Li-ion batteries have an energy density 600 Wh/l (18650 cylindrical cells, Panasonic). An average value of 500 Wh/l for the energy density of lithium secondary batteries is shown in Fig. 11.1. It is approximately the same as the best alkaline primary batteries. The energy density of fuel cells, given in Fig. 11.1, is calculated by multiplying the energy density of the fuel with the

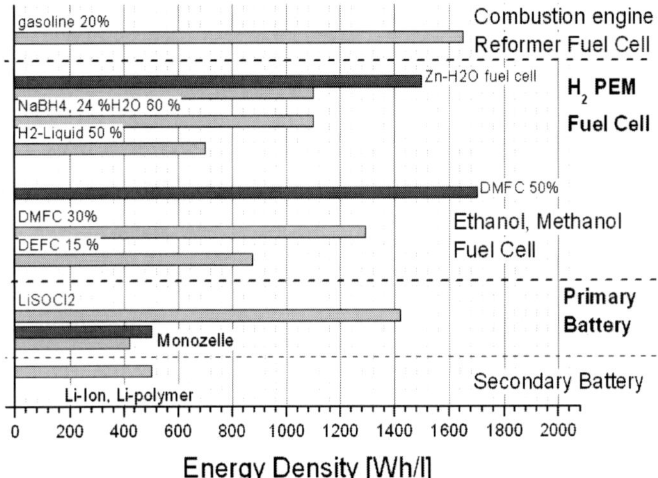

Figure 11.1 Comparison of volumetric energy density of batteries and fuel cells. For fuel cells, the energy density of the fuel is multiplied with the fuel cell efficiency.

efficiency of the fuel cell. The size of the fuel cell is not considered. Therefore, the figures are an upper estimate for long discharge times of several years. In this case, the fuel cell is very small compared to the fuel tank. For example, the best available direct methanol fuel cells have an efficiency of 30% and therefore the electrical energy density is ca. 1300 Wh/l. Only if the efficiency of fuel cells can be further increased, they can keep up with the best primary batteries. It should also be noted that the smallest fuel cells on the market today are much bigger than batteries and miniaturizing them is still a big challenge. The long-term stability of fuel cells over many years is another issue.

The energy density of batteries decreases by reducing the size of the batteries. This is caused by the increasing amount of passive packaging material in proportion to active material in small batteries. There is a significant reduction of energy density at a thickness of below 1 mm because the volume of the packaging becomes dominant. Therefore, only energy densities of power sources of the same size should be compared.

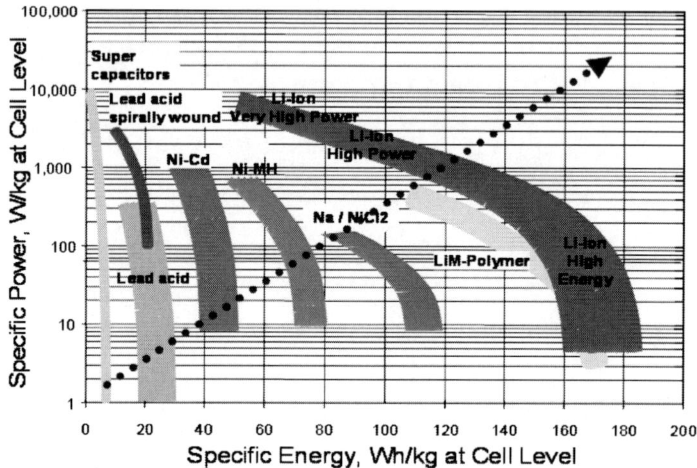

Figure 11.2 Ragone chart, comparison of various secondary batteries and super capacitors in terms of power and energy density. The horizontal axis describes how much energy is available, while the vertical axis shows how quickly that energy can be delivered [1].

The other important parameter of electrochemical storage systems is power density. The energy that can be gained from a primary battery that is designed for several years of use is much lower at high currents or pulsed loads compared to steady low power discharge because of the high internal impedance.

Most secondary batteries have smaller impedance and are capable of high current drain. However, in miniaturized systems, the secondary battery will be designed as small as possible. As a result, a typical load of the system and power pulses (for example, for transmitting the data) may result in a high power density for small secondary batteries. A much higher power density compared to batteries can be achieved with capacitors. The relationship between power and energy density is shown in the so-called Ragone plot (Fig. 11.2). It can be seen that high-power Li secondary batteries are approaching the order of magnitude of power levels of supercapacitors.

In summary, due to the decreasing power demand of the electronic devices in the future more systems can be powered with primary batteries even for service-free time intervals between 10 and 20 years. They will be an alternative to energy harvesting concepts for occasions with a very limited available ambient energy. On the other hand, primary batteries will be displaced by energy harvesting systems for several reasons:

- environmental concerns of primary batteries
- high ambient temperatures where the lifetime of the batteries decreases significantly
- maintenance-free service of the system for more than 10 to 20 years

Energy harvesting systems require electrical buffer storage, which may be secondary batteries or capacitors. The parameters of these buffers like temperature stability and lifetime must be consistent with the requirements of the energy harvesting system.

11.2 Physical Principles

11.2.1 *Secondary Batteries*

Lithium batteries have the highest energy density compared to other chemistries since lithium is the lightest of the metals and also has the highest reduction capability, expressed as the lowest standard reduction potential of all the metals in the electrochemical series, at -3.045 V. Therefore, only lithium secondary batteries will be considered here.

In secondary lithium batteries with organic electrolyte, pure lithium cannot be used as negative electrode material because of dendrite growth and severe safety issues. The deposition of metallic Li during charging is possible for very thin layers only. Instead, graphite is used as intercalation anode with an atomic ratio between Li and C of 1:6. Thus, the energy density of this electrode is reduced by a factor of about ten. On the other hand, the overall energy density is much more determined by the positive than by the negative electrode, as shown in Tables 12.1–12.3 [2].

Therefore, in Li-ion secondary batteries both, the negative and positive electrode consists of so-called host materials into which and from which lithium ions can migrate. The process of lithium

Table 11.1 Comparison of anode materials in Li secondary batteries [2]

	Lithium metal	Amorphous carbon LiC_6	Graphite LiC_6	Lithium alloys	Lithium oxide	Titanate, $Li_4Ti_5O_{12}$
Potential vs. Li/Li$^+$ (mV)	0	100–700	50–300	50–600	50–600	1400–1600
Capacity (mAhg)	3860	ca. 200	372	4000 Si 1000 Sn	<1500,	150–160
Safety	—	+	+	0	+	++
Stability	—	+	+	—	—	++
Cost	+	0	+	++	—	0

Table 11.2 Comparison of cathode materials in Li secondary batteries [2]

	$LiCoO_2$	$LiNiO_2$	$LiMn_2O_4$	$Li(Ni_xCo_yMn_z)O_2$	$LiFePO_4$
Potential vs. Li/Li$^+$ (mV)	3.9	3.8	4.0	3.8–4.0	3.4
Capacity (mAh/g)	150	170	120	130–160	160
Safety	—	—	+	0	++
Stability	—	—	0	0	++
Cost	—	—	+	0	+

Table 11.3 Electrode materials of Li-ion secondary batteries: theoretically achievable energy densities

	Molar mass	No. of intercalated Li atoms	Level of intercalation	Single electrode specific capacity		Specific capacity	Energy density
	[(g/mol])			(As/g)	(mAh/g)	(mAh/g)	(mWh/g)
Anode Li (metal)	6.9410	1	100%	13900.78	3861.33		
Anode $Li_{(0...1)}C_6$	12.0107	6	100%	1338.88	371.91		
Cathode $Li_{(0.4...1)}CoO_2$	97.8730	1	60%	591.49	164.30	114.0	410.3
Anode $Li_{(0...1)}C_6$	12.0107	6	100%	1338.88	371.91		
Cathode $Li_{(0.5...1)}Ni_{0.8}Co_{0.2}O_2$	97.6812	1	50%	493.88	137.19	100.2	360.8
Anode $Li_{(0...1)}C_6$	12.0107	6	100%	1338.88	371.91		
Cathode $Li_{(0.4...1)}Mn_2O_4$	180.8147	1	60%	320.17	88.94	71.8	208.1

Note: Values were calculated according to the atomic mass and Faraday's law. A mean discharge voltage of 3.7 V was assumed for energy density.

ions moving into the electrode is referred to as insertion (or intercalation) and the reverse process, in which lithium ions move out of the electrode, is referred to as extraction (or deintercalation). In electrochemistry, the anode is the electrode where oxidation occurs and the cathode is the electrode at which reduction occurs. When a lithium-based cell is discharging, the lithium ions are extracted from the negative electrode, which becomes the anode and inserted into the positive electrode (cathode). When the cell is charging, the reverse process occurs: lithium ions are extracted from the positive electrode and inserted into the negative (Eqs. 10.1–10.3). For simplification, irrespective of the charge or discharge reactions, the negative electrode and positive electrode are always termed as anode and cathode, respectively. Simultaneously to the reduction and oxidation reactions of the host materials, the insertion and extraction of lithium ions as counter-ions into the host lattice assure the charge balance and electroneutrality. The anode of a conventional Li-ion cell is made from carbon materials such as hard carbons or graphite. The cathode is a transition metal oxide or iron phosphate. The most important characteristics of several anode and cathode materials are plotted in Tables 11.1 and 11.2. Figure 11.3 shows the electrode potentials of the most important anode and cathode materials of Li-ion batteries arranged in the electrochemical series.

The electrolyte of Li-ion batteries consists of a lithium salt in an aprotic, polar organic solvent. Detailed descriptions on Li electrolytes and polymer binders can be found elsewhere [3].

Anode half reaction

$$\mathrm{Li}_x\mathrm{C}_n \leftrightarrow \mathrm{C}_n + x\mathrm{Li} + +xe^- \tag{11.1}$$

Cathode half reaction

$$\mathrm{Li}_{1-x}\mathrm{MO}_2 + xe^- + x\mathrm{Li}+ \leftrightarrow \mathrm{LiMO}_2 \tag{11.2}$$

Total

$$\mathrm{Li}_{1-x}\mathrm{MO}_2 + \mathrm{Li}_x\mathrm{C}_n \leftrightarrow \mathrm{LiMO}_2 + \mathrm{C}_n \tag{11.3}$$

(M: metal)

Li batteries based on graphite anodes and cobalt oxide cathodes are very sensitive to deep discharge and over charge. Therefore,

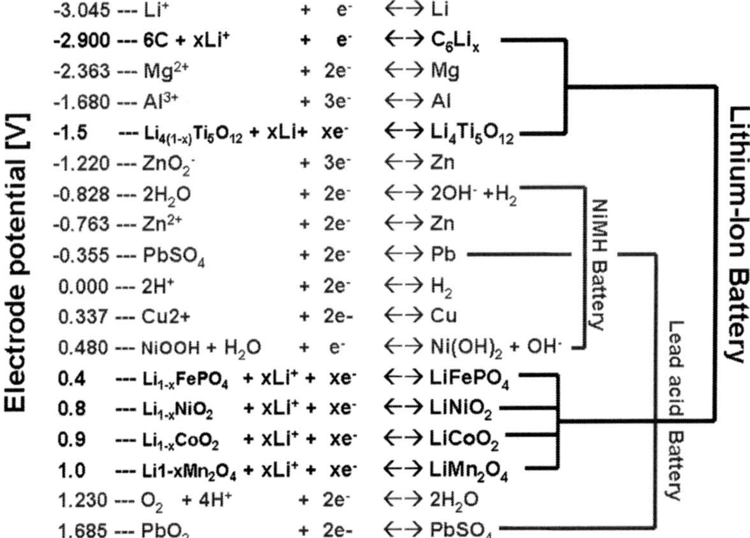

Figure 11.3 Comparison of lithium-ion batteries and NiMH as well as lead acid batteries according to the electrode potential in the electrochemical series.

several safety measures are required. Equations 10.4 and 10.5 show the reactions, respectively.

Overdischarge

$$Li^+ + LiCoO_2 \rightarrow Li_2O + CoO \tag{11.4}$$

Overcharge

$$LiCoO_2 \rightarrow Li^+ + CoO_2 \tag{11.5}$$

The most important lithium-ion battery materials can be characterized as follows:

a. Cathode materials

Lithium Cobalt Oxide, LiCoO₂

Lithium Cobalt oxide is a mature, proven, industry-standard battery technology that provides long cycle life and very high energy density. The cell voltage is typically 3.7 V. Cells using this chemistry are available from a wide range of manufacturers. The use of Cobalt is unfortunately associated with environmental and toxic hazards.

Lithium Manganese Oxide (spinel structure), $LiMn_2O_4$

Lithium manganese oxide provides a higher cell voltage than cobalt-based chemistries at 3.8 to 4 V but the energy density is about 20% less. It also provides additional benefits to lithium-ion chemistry, including lower cost and higher temperature performance. This chemistry is more stable than lithium cobalt technology and thus inherently safer but the trade-off is lower potential energy densities. Manganese, unlike cobalt, is a safe and more environmentally benign cathode material.

Lithium Nickel Oxide, $LiNiO_2$

Lithium nickel oxide-based cells provide up to 30% higher energy density than cobalt but the cell voltage is lower at 3.6 V. They also have the highest exothermic reaction, which could give rise to cooling problems in high-power applications.

Lithium (NCM) Nickel Cobalt Manganese Oxide, $Li(NiCoMn)O_2$

These are mixed tri-metal cathode materials (typically 1:1:1), which combine improved safety with low cost without compromising performance.

Lithium Iron Phosphate, $LiFePO_4$

$LiFePO_4$ with olivine structure belongs to the family of NASICON-type compounds (NASICON—sodium super-ionic conductor) that are known to be fast ionic conductors and which are used as solid electrolytes in electrochemical cells. In $LiFePO_4$ the hexagonal close-packed lattice of oxygen has a two-dimensional channel network that may act as fast diffusion paths for lithium ions. Nanomaterials are currently being investigated in order to further increase the current density, since the $LiFePO_4$ is an electronically non-conducting material. Particle sizes of <200 nm and a specific carbon coating enable higher charge and discharge rates by providing better electronic conductivity pathways and short diffusion lengths for the lithium ions. In addition, $LiFePO_4$ has the highest thermal stability of the so far known materials, which guarantees safe use and stable capacity after numerous work cycles.

Phosphate-based cathodes provide better safety characteristics than those of Lithium-ion technologies made with other cathode materials. Lithium phosphate cells are incombustible in the event of mishandling during charge or discharge; they are more stable under overcharge or short circuit conditions and can withstand high temperatures without decomposing. When abuse does occur, the phosphate based cathode material will not burn and is not prone to thermal runaway. Phosphate chemistry also offers a longer cycle life.

Recent developments have produced a range of new environmentally friendly cathode active materials based on lithiated transition metal phosphates for Lithium-ion battery applications.

Doping with transition metals changes the nature of the active materials and enables the internal impedance of the cell to be reduced.

Phosphates significantly reduce the drawbacks of the cobalt chemistry, particularly the cost, safety, and environmental characteristics. The trade-off is a reduction of ca. 14% in energy density due to the lower voltage of the phosphate cathode.

Due to the superior safety characteristics of phosphates over current lithium-ion cobalt cells, batteries may be designed using larger cell sizes.

Many research activities are under way to further improve the electrode materials and find new material options. For example, $Li(Ni_xCo_yMn_z)O_2$ can be fabricated as a mixture of $LiNiO_2$ and $LiMn_2O_4$ with cobalt oxide or as a solid solution of $LiNiMnCoO_2$ with increased high-temperature features [4].

b. Anode Materials

Carbons

The anodes of most lithium-based secondary cells are based on some form of carbon (graphite or coke) LiC_6. During the formation process of the battery (first charge), the electrolyte directly reacts with the carbon anode and a passivating layer, the so-called solid electrolyte interphase (SEI), is deposited on the anode. This layer is essential for the stability of the cell but it increases the cell internal impedance and reduces the possible charge rates as well as the high and low temperature performance.

Excessive heat can cause the protective SEI barrier layer to break down allowing the anode reaction to restart releasing more heat leading to thermal runaway. The thickness of the SEI layer is not homogeneous and increases with age, increasing the cell internal impedance, reducing its capacity and hence its cycle life.

Lithium Titanate, $Li_4Ti_5O_{12}$

Lithium titanate spinel has been introduced for use as an anode material providing high–power thermally stable cells with improved cycle life. This has the following advantages:

Due to the lower reducing capability Lithium titanate anodes do not react adversely with the commonly used electrolytes in lithium-ion cells hence no SEI layer is formed nor is it needed.

Lower restriction on ion flow hence higher charge and discharge rates are possible as well as better low temperature performance.

The cells have lower internal impedance and can tolerate higher temperatures.

No SEI build-up over time means that a long cycle life is possible.

On the other hand, the nominal cell voltage is reduced significantly to 2.25 V, which means lower energy density.

A comparison of the cell voltage levels at charge and discharge of standard graphite and titanate anodes with cobalt oxide cathodes of cells with equal size is shown in Figs. 11.4a,b. The typical constant current-constant voltage (CC–CV) charge procedure is used. The cell

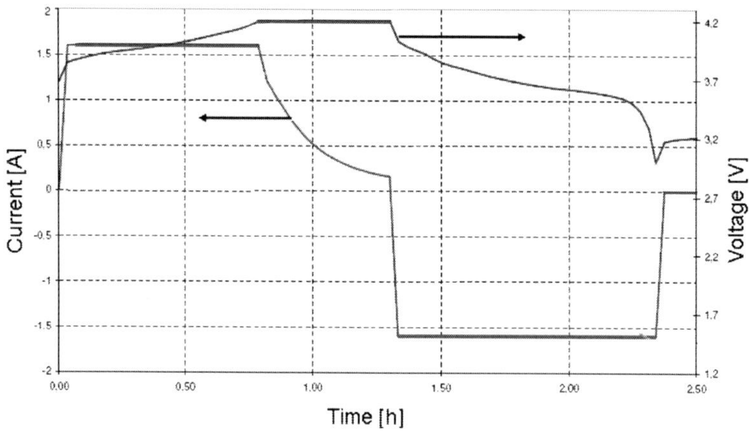

Figure 11.4a Charge and discharge curves of lithium-ion batteries. $LiCoO_2$–Graphite.

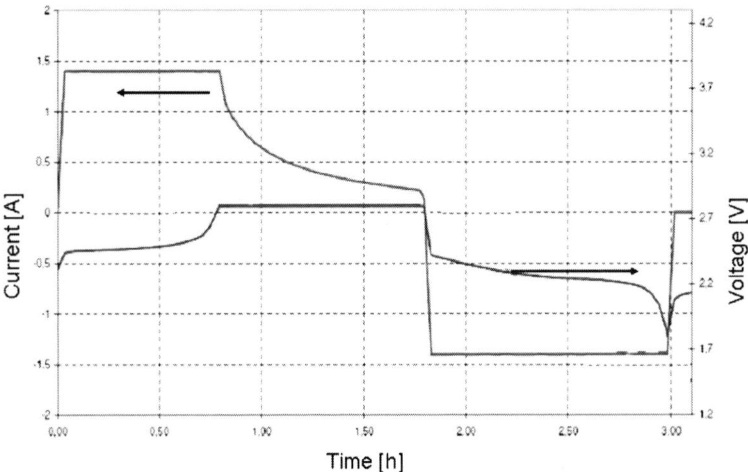

Figure 11.4b Charge and discharge curves of lithium-ion batteries. $LiCoO_2$–Li-Titanate.

is first charged at a constant current until the maximum voltage is reached. Then the voltage is kept constant and charging is continued until the current falls below a threshold value.

11.2.2 *Solid-State Thin-Film Lithium Batteries*

Solid-state batteries are chemical cells that have both solid electrodes and solid electrolytes. Materials are required that have sufficiently high ion conductivity and are insulating towards electrons. Ceramic, glassy, and polymeric materials have been investigated as solid-state electrolytes. Compared to liquid systems the ionic conductivity at room temperature is much lower and the ionic current across the solid-solid interphase is limited. Therefore, applications of solid-state systems were restricted to low power and high-temperature systems.

Low conductivity and interface problems can be tackled if the battery layers are deposited as thin films. If the layer thickness is in the micrometer range, than comparatively high power density can be achieved. Even though the lithium-ion conductivity of the solid electrolyte is 100 × less than for many liquid electrolytes, it is sufficient because just 1 μm-thick films are adequate for a pinhole-free barrier over most thin-film electrodes. Further, the electronic resistivity of the mostly used LiPON is very high, greater than $10^{14}\Omega$ cm.

Other advantages of the thin-film technology for solid-state batteries are as follows:

- The thin-film technology provides clean surfaces and improves the electrode-electrolyte interface contact.
- The deposition of the ionic conductor in vacuum avoids possible moisture problems.
- Thin-film technologies generally give very good adhesion between layers and large areas can also be obtained with these techniques.
- Convenient substrate materials such as silicon wafers can be used.
- Encapsulation of the battery can be achieved by the deposition of an insulating layer on the top of the device.
- Thin-film batteries are easy to miniaturize (by vacuum deposition and lithographic patterning).
- There is no problem with electrolyte leakage. Liquid electrolytes are often highly corrosive. Packaging of small batteries is straightforward since there is no liquid electrolyte.

Solid-state batteries have very long shelf lives, and usually do not have any changes in performance with temperature, such as might be associated with liquid electrolyte freezing or boiling. These are very important characteristics for electrical buffers of energy harvesting systems.

On the other hand, the deposition of the electrolyte layer is a particularly highly sophisticated and time-consuming process that leads to high production costs. Electrical cycling of the solid-state battery leads to cyclic changes of crystallite volume and material tension. Therefore, the total capacity is rather low since layer thickness cannot be increased above a few micrometers. Capacities around ca. 100 μAh/cm^2 are usually obtained with this technology.

At Oak Ridge National Laboratory, the basic process for thin-film batteries has been created [5]. It was shown that addition of nitrogen to the glass structure enhances the chemical and thermal stability of a lithium glass of the ionic conductor. Although the nitrogen partial pressure overwhelms that of the oxygen in the plasma, only a small amount of N replaces the oxygen in the composition, but this nitrogen has a profound effect on the ion conductivity and electrochemical stability. With a N/O ratio as small as 0.1, the ionic conductivity is 1 to 2 μS/cm, which is about 40-fold higher than glassy films of N-free Li_3PO_4.

This battery is a layered structure with alternate layers of lithium metal oxide (in most cases lithium cobalt oxide), lithium phosphate oxynitride (LiPON), and lithium metal. The maximum cell voltage is 4.2 V. All layers are fabricated using vacuum processing.

Meanwhile several materials have been used in thin-film batteries. The best electrodes for rechargeable batteries are those where there is little structural change of the active materials during the charge and discharge reaction to reduce material tensions. The materials should allow for rapid diffusion of lithium ions in solid and good electronic conductivity.

(a) Cathode materials for thin-film batteries

$LiCoO_2$ gives a high capacity (155 mAh/g). It has a layer structure, where lithium and transition metal cations occupy alternate layers of octahedral sites in a distorted cubic close-packed oxygen-ion lattice. The dimensional changes on charge-discharge cycling are small.

$LiMn_2O_4$ has a lower capacity (120 mAh/g) than $LiCoO_2$ but a higher voltage. It has a spinel or tunnel structure, which possesses a three-dimensional space via face sharing octahedral and tetrahedral structures. This provides conducting pathways for the insertion and extraction of lithium ions.

Other intercalation compounds such as TiOS, V_2O_5, and $LiVO_2$ have been used as well.

(b) Anode materials for thin-film batteries

Lithium is the preferred anode material for thin-film solid-state batteries. The specific capacity of lithium metal is 3.86 Ah/g. In order to obtain a reasonable cycle-life, a three- to fivefold excess of lithium is used. Even so, the specific capacity is still much higher than that of the cathode materials.

In the case of lithiated cathodes, insertion compounds can be used as negative electrodes such as Sn_3N_4, Si, Sn, and carbon.

Electrolyte in thin-film batteries

The electrolyte generally used in solid-state thin-film lithium batteries is lithium phosphorus oxynitride (LiPON). LiPON has a good lithium-ion conductivity and excellent stability in contact with metallic lithium.

Battery encapsulation

Lithium is a highly reactive metal. On exposure to moist or dry air, lithium can undergo the following reactions:

$$2Li + H_2O - \frac{1}{2}O_2 \rightarrow 2LiOH \tag{11.6}$$

$$2Li + \frac{1}{2}O_2 \rightarrow Li_2O \tag{11.7}$$

For a lifetime of up to 20 years, only very low levels of water or air diffusion into the package can be tolerated. Since a hermetic package for micro batteries is rather expensive, several protective coatings and encapsulations methods have been developed that allow a high lifetime.

Figure 11.5 shows a schematic cross section of a thin-film battery. The cathode and anode current collectors are deposited by sputtering of the appropriate metals in argon (Ar) atmosphere. Cathode films are deposited by RF magnetron sputtering of sintered targets in order to obtain the desired compound. Some of these materials, mainly the lithiated ones ($LiCoO_2$, $LiMn_2O_4$, etc.), require a thermal treatment after sputtering at high temperatures (300 to 600°C). Annealing the cathode films at temperatures of 300 to 800°C may be used to induce crystallization and grain growth of the desired intercalation compound. Crystallizing the cathode film generally improves the Li chemical diffusivity in the electrode material, and hence the power delivered by the battery, by one to two orders of magnitude. The microstructure is also tailored by the

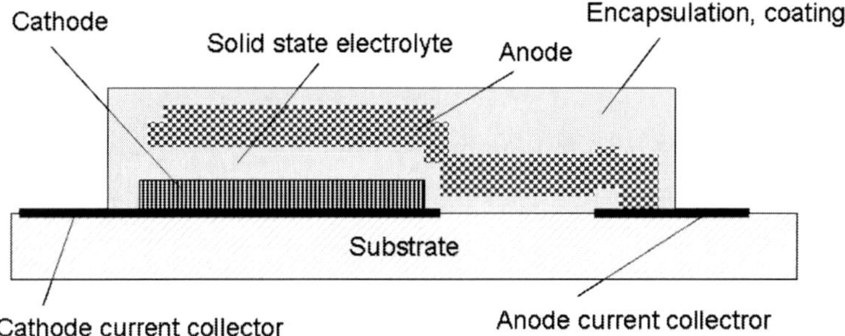

Figure 11.5 Schematic of a deposited thin-film battery.

deposition and heat treatment. To improve the manufacturability of the thin-film batteries, it would be beneficial to eliminate or minimize the temperature or duration of the annealing step. Several efforts have lead to low-temperature fabrication of thin-film batteries on polyimide substrates, but the battery capacity and rate capability are lower than those treated at high temperatures.

A sputtered lithiated glassy electrolyte (Li-ion conducting film and electrically insulator) film covers the cathode and a portion of the substrate up to the anode current collector in order to insulate the substrate from direct contact with the anode. Usually, this electrolyte layer is constituted of LiPON obtained from a Li_3PO_4 target by sputtering in N_2 atmosphere. Lithium is then deposited by thermal evaporation.

In addition, many publications report exploring a variety of other physical and chemical vapor deposition processes, such as pulsed laser deposition, electron cyclotron resonance sputtering, and aerosol spray coating, for one or more components.

Another concept consists in using a lithiated positive electrode and a blocking metallic layer (typically Pt) as current collector for the negative electrode. This concept is easier to be realized and provides electrochemical characteristics similar to the lithium metal battery. Currently this concept does not achieve the cycle life of classical lithium metal batteries, mainly due to lithium protuberances on the Pt layer when the battery is discharged [6].

The microstructures of the thin films may be quite distinct from those of battery electrodes formed from powders. The thin-film cathodes are dense and homogeneous with no added phases such as binders or electrolytes.

11.2.3 *Supercapacitors*

Electrochemical double-layer capacitors (EDLC) are often abbreviated to "supercapacitors," named after the brand name of the first commercialized product by NEC in 1978.

Supercapacitors are using a different principle of charge storage than batteries: While in batteries redox reactions with changes of the oxidation states of the electrochemically active compounds occur, supercapacitors use the electrochemical double layer between a large area electrode and a liquid electrolyte for charge storage. During charging, the electrically charged ions in the electrolyte migrate towards the electrodes of opposite polarity due to the electric field between the charged electrodes created by the applied voltage. Thus, two separate charged layers are produced. The double-layer capacitor is realized by sandwiching two electrode layers and a separator-electrolyte layer. This corresponds to a serial interconnection of two capacitors as shown in Fig. 11.6.

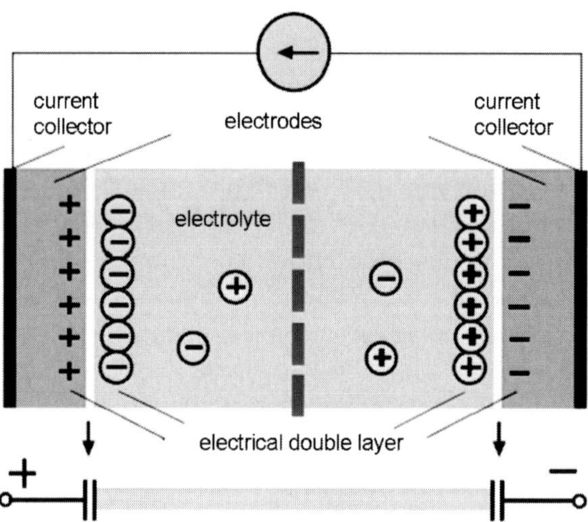

Figure 11.6 Schematic of a double layer capacitor.

According to Eq. 10.8, the capacity is proportional to the surface area S and the permeability ε and indirect proportional to the distance between the electrodes d. The electrode distance is very small since it is the double-layer thickness, which depends on the ion radius and the electrolyte concentration. It is usually in the range of 5 to 10 Å. The active surface S is large due to the porosity of the electrodes. Typical materials have a surface of ca. 2000 m^2/g, which equals a specific capacity between 10 and 20 µF/cm^2. This corresponds to a specific capacity of 100–200 F/g, which could be matched only by a conventional capacitor of much larger dimensions.

$$C = \varepsilon_0 \cdot \varepsilon_r \cdot S/d \qquad (11.8)$$

In electrochemical double-layer capacitors, storage density is improved through the use of a nanoporous material, typically activated charcoal. Activated charcoal is a powder made up of extremely small and very "rough" particles, which in bulk form a low-density volume of particles with holes between them.

The different charge storage mechanisms of batteries and supercapacitors are responsible for their different properties: the redox reactions of the battery electrode materials require solid-state diffusion of (counter) ions into the bulk of the particles, which is a rather slow process. This limits the power density of the batteries ($<<$3 kW/kg). Due to the chemical reactions that are not completely reversible, and also the concomitant volume changes the cycling stability, i.e., the number of charges/discharges, is limited to approximately a few hundred up to 10,000 cycles, depending on the system. Since the bulk of the particles is utilized, the energy densities are high (up to 240 Wh/kg). Supercapacitors are different: Due to the energy storage in the electrochemical double layer only at the surface of the particles, the energy density is quite low (5 Wh/kg), and the power density is high (approximately 20 kW/kg, see also the Ragone chart, Fig. 11.2). Also, supercapacitors have excellent cycling stabilities up to 1,000,000 cycles. A disadvantage compared to batteries is the higher self-discharge, which amounts to about 10% per month, strongly depending on the quality of the manufacturing and the state of charge.

The maximum energy W (in Ws $=$ VAs) of a supercapacitor is calculated as

$$W = \tfrac{1}{2} C \cdot U^2 \tag{11.9}$$

and the maximum power P (in W $=$ VA) as

$$P = \frac{U^2}{4 \cdot R_{ESR}}, \tag{11.10}$$

where R_{ESR} (in Ω $=$ V/A) is the equivalent series resistance, consisting of ohmic conduction and dielectric reversal losses.

The constant current charging (or discharging) of a supercapacitor incorporating identical electrodes gives a linear increase (or decrease) in the cell potential with time, as expected for a conventional capacitor (Fig. 11.7).

As can be seen from the above equations, the maximum voltage is an important set-screw to improve the energy as well as the power of supercapacitors. With aqueous electrolytes, voltages are limited to approximately 1.2 V, the electrolysis of water. Nonaqueous aprotic organic electrolytes allow for potential windows of 2.7 V or even slightly more (lithium-ion batteries use voltages up to 4.2 V). In most cases acetonitrile or propylene carbonate are the solvents of choice. The monitoring of the voltage, similar to a lithium-ion battery management system, is important, since no internal overcharge mechanism exists in the supercapacitors. To achieve higher voltages series-connected individual electric double-layer capacitors have to be matched.

Apart from symmetrical supercapacitors with two identical electrodes with electronic and ionic charge migration the modest charge storage capabilities can be improved by introducing additional Faradaic reactions where in addition to charge migration redox reactions are involved.

Here as so-called pseudo-capacities inorganic systems and systems with conducting polymers as electrodes can be used. Typical inorganic materials are MnO_2 (450 F/g), RuO_2 (up to 800 F/g); organic electrodes are made from polymers such as polyaniline (PA), polypyrrole (PPy), or poly(3-methylthiophene) (200–300 F/g). Higher energy densities can be achieved with these electrochemical systems in comparison to electrostatic double-layer capacitors while the material costs are higher.

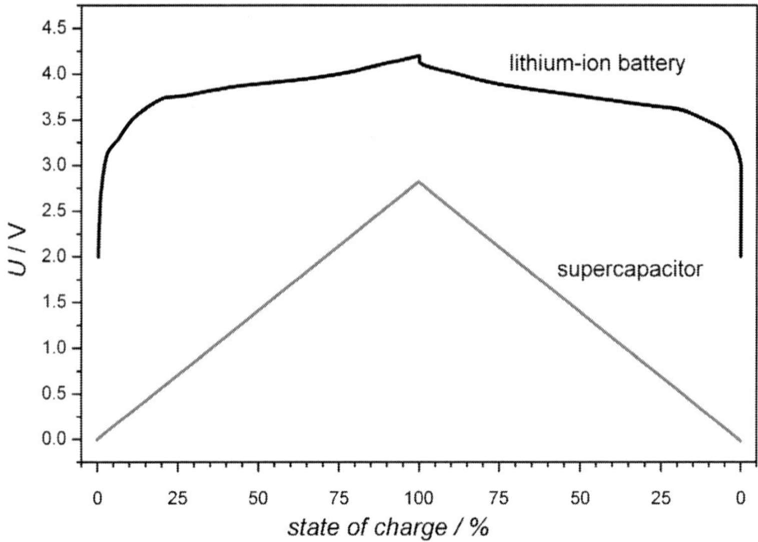

Figure 11.7 Comparison of a constant current charge/discharge cycle of a lithium-ion battery and a supercapacitor.

A third type of supercapacitor is called hybrid capacitor. One electrode uses a double layer at active carbon while the other side is an electrochemical "Faraday" electrode. Higher operation voltages can be achieved with hybrid capacitors. Thus, it is possible to adjust the voltage to a primary or secondary battery cell, and the supercapacitor and the battery can be connected in parallel to improve the pulse power performance without additional electronics [7].

Recent research in electric double-layer capacitors has generally focused on improved materials that offer even higher usable surface areas.

11.3 Realization of Micro Secondary Battery Technology

Energy autarkic systems and their energy harvesting devices can vary widely in terms of load profile, size, and other parameters. Therefore, not one single electrical buffer storage will be used but a variety of systems will be adapted to the individual applications.

Nevertheless, there are some general requirements for secondary batteries as electrical buffers.

- The battery must be rechargeable and should have a high energy density.
- The voltage of the battery should fit to the device voltage.
- The buffer storage should be capable of high pulse discharge rate. This is the case in most systems with transceiver module.
- Small geometrical dimensions and flexible form factors may be necessary for many sensor node applications.
- Minor self-discharge is of crucial importance to safe nearly all of the harvested energy.
- A long life of 10 years and more is of significant importance since the idea of energy harvesting is to have a system that works maintenance-free over decades. A lifetime of 10 years or more is still an issue for most of the rechargeable batteries.
- In some cases, high-temperature stability is required.
- The battery technology must be from high-volume production to reduce the cost.
- The energy storage device must be connected with the system, which can become critical in case of miniaturized systems.
- Encapsulation of the micro batteries is another issue. A near hermetic encapsulation is required for long lifetime. On the other hand, the encapsulation volume should be only a small fraction of the battery. For special applications, the encapsulation must be biocompatible and pressure resistant.
- Safety is a concern of all lithium batteries. On the other hand, safety is less critical in case of very small batteries.

11.3.1 *Coin-Type Cells*

The low-current systems of most used micro batteries are of coin type. In a coin-type battery, the active masses of anode and cathode are filled into a metal cup and a metal cover, respectively, in the form

Table 11.4   Examples secondary coin-type cells

Type	System	Dimension (dm × h) (mm)	Voltage (V)	Capacity (mAh)	Energy density (mWh/cm^3)
ML1220	Li	12.5 × 2	3	16	200
MC 621	Li	6.8 × 2.15	3	3	115
V 6HR	Ni/MH	6.8 × 2.15	1.2	6.2	95
V 40H	Ni/MH	11.5 × 5.35	1.2	43	90
MC 614	Li	6.8 × 1.4	3	1.5	90

Panasonic, Sanyo, Varta.

of powder or pastes in order to provide the two electrical contacts to the outside (Fig. 11.8). Both parts are separated by a separator and joined by means of a polymer sealing ring. Table 11.4 shows the characteristics of several examples coin-type secondary batteries.

In general, the energy density of small coin-type cells of the same chemical system is much lower compared to cylindrical or prismatic cells, due to the volume fraction of the metallic casing. As a comparison, nickel metal hydride (NiMH) cells are shown as well. They offer a higher current capability but at lower energy density compared to lithium cells. NiMH cells will not be suited for most energy harvesting applications since the charge efficiency is much lower compared to lithium and self-discharge is higher.

All coin cells for memory backup can be used only with low currents and are therefore not sufficient for most energy harvesting applications. Nevertheless, they show a low self-discharge rate and withstand overdischarge and overcharge.

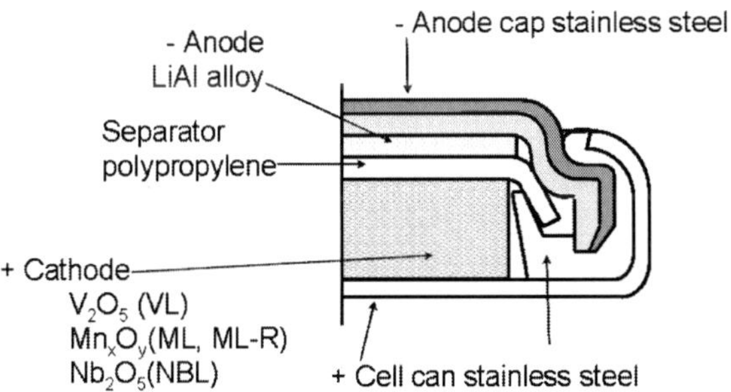

Figure 11.8   Cross section of coin-type batteries (*source*: Panasonic).

Vanadium pentoxide (VL) button cells are mainly for backup of memory data. The self-discharge rate is rather low, only 2% a year. The voltage is high.

Niobium lithium rechargeable batteries (NBL) were developed for memory backup as well but at a lower voltage level consistent to the IC technology.

In manganese compound (ML) lithium secondary batteries, a manganese compound oxide is used for the positive electrode, and a lithium/aluminum alloy for the negative electrode. They can be charged at voltages below 3 V.

Manganese titanium lithium rechargeable batteries (MT) are compact rechargeable batteries that employ a lithium-manganese complex oxide as cathode material, and lithium-titanium oxide as the anode material. These batteries can be used for a wider range of charge and discharge currents. They are developed as main power supply for compact products such as rechargeable watches and can be used as electrical buffer for a variety of energy harvesting devices.

Cobalt titanium lithium rechargeable cells (CTL) are using lithium titanium oxide as anode and lithium cobalt oxide as cathode. They are similar to the manganese titanium system but with a higher voltage (Fig. 11.9). They are developed as main power supply for compact products such as rechargeable watches and can be used as electrical buffer for a variety of energy harvesting devices.

The rechargeable coin cells with lithium titanium oxide as anode differ in terms of cycle life greatly from the cells with LiAl anode. While cells with LiAl anode can be cycled more than 1000 times at 10% depth of discharge (DOD), the can be cycled only ca. 50 times at full discharge. The cells with titanium oxide anode can be cycled 500 times at 100% depth of discharge which make them well suited for a lot of energy autarkic systems where full discharges can occur. Table 11.5 gives an overview of the lithium rechargeable coin cells [8] and characteristics of some battery models are summarized in Table 11.6.

Table 11.5 Overview of rechargeable lithium coin cells

		VL	ML	NBL	MT	CTL
	Cathode	V_2O_5	Li_xMnO_y	Nb_2O_5	Li_xMnO_y	$LiCoO_2$
	Anode	LiAl	LiAl	LiAl	Li_xTiO_y	Li_xTiO_y
Average	Discharge voltage (V)	2.85	2.5	1.5	1.3	2.3
Charge	Voltage (V)	3.25–3.55	2.8–3.2	2.0–2.6	1.5–2.5	
Cutoff	Voltage (V)	2.5	2.0	1.0	1.0	
Self-discharge	Per year (%)	2.0	2.0	2.0	5.0	
Cycles	10% DOD	1000	1000	1000		
Cycles	100 % DOD	ca. 50	ca. 50	ca. 50	500	
Operating	Temperature (°C)	−20 to +60	−20 to +60	−20 to +60	−20 to +60	

Table 11.6 Parameters of manganese titanium lithium coin cells

Model	Nominal voltage (V)	Nominal capacity (mAh)	Diameter (mm)	Thickness (mm)	Weight (g)	Energy density (Wh/l)	Standard load (mA)
MT920	1.5	5.0	9.5	2	0.5	52.9	0.1
MT621	1.5	2.5	6.8	2.1	0.25	49.2	0.05
MT516	1.5	1.8	5.8	1.6	0.15	63.9	0.05

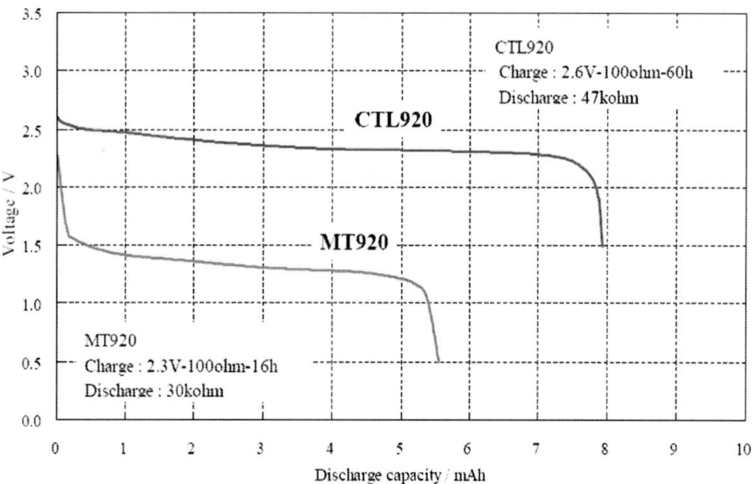

Figure 11.9 Discharge profile of a cobalt and a manganese titanium lithium coin cell.

11.3.2 *Lithium-Ion/Lithium Polymer Batteries*

Battery fabrication

Lithium-based rechargeable batteries are used in very high volumes in applications, such as mobile phones, laptops, cameras, and other consumer electronics products. Lithium-ion polymer batteries use liquid lithium-ion electrochemistry in a matrix of conductive polymers that eliminate the free electrolyte from the cell. The polymer matrix is based on modified PVDF (polyvinylidene fluoride) homopolymer or copolymer. The batteries are packaged as cylindrical cells or in prismatic form. Lithium-ion cylindrical or prismatic cells have a rigid metal case, while polymer cells have a flexible, foil-type (polymer laminate) case, but they still contain organic solvent. The encapsulation with a multilayer foil is called pouch package. The main difference between commercial polymer and lithium-ion cells is that in the latter the rigid case presses the electrodes and the separator onto each other, whereas in polymer cells this external pressure is not required because the electrode sheets and the separator sheets are laminated onto each other. In

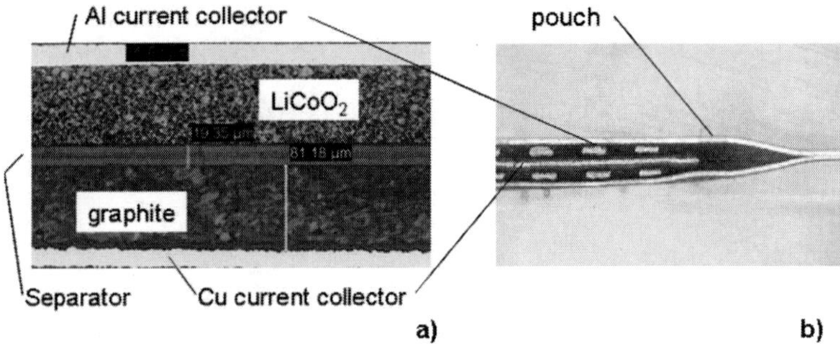

Figure 11.10 Cross-sectional view of lithium polymer batteries, (a) single cell (*source*: Fraunhofer IZM), (b) Bi-cell with central anode (*source*: VARTA).

addition, they are pressed together with help of the atmospheric pressure.

The active masses are fabricated on current collectors as thin foils. Large areas of these foils are stacked together or winded up as a coil. As the active materials are thin, the distance for ionic travel in the electrolyte is small. Therefore, much higher power densities can be achieved compared to button cells where the battery area is small and the electrode thickness is larger. If only one or two battery foils are packaged in a pouch, a very thin lithium battery can be fabricated with a thickness below 1 mm. Figure 11.10 shows cross sectional views of thin lithium polymer batteries.

Miniaturization of lithium polymer batteries down to sizes of ca. 1 cm^2 is not easy, since the pouch package requires a defied sealing edge and sealable current feed through. It was demonstrated that a lithium polymer battery with pouch package can be integrated into a 1 cm^3 sized sensor node. In that case, the size of the electronics module and of the antenna was adapted to the space requirements of the battery as can be seen in Fig. 11.11.

Battery parameters

Figure 11.12 displays the evolution of the energy density of the Panasonic 18650 Li-ion cylindrical cell during the past years. As can be seen, there is a steady increase in energy density. On the other hand, there is a diversification of battery types with application

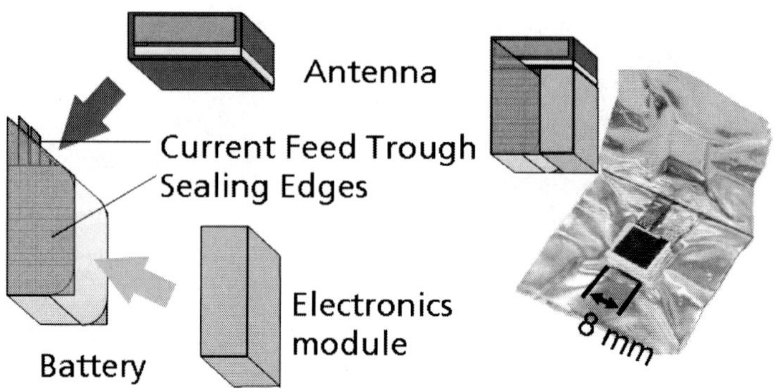

Figure 11.11 Package of lithium polymer battery into as cubic centimeter sensor node (*source*: Fraunhofer IZM).

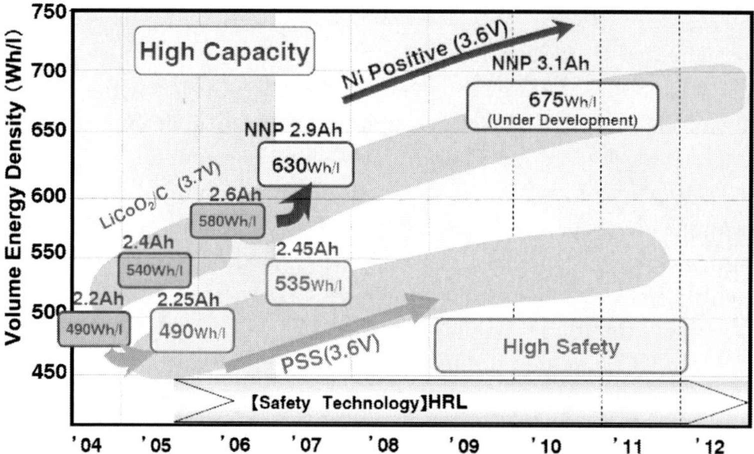

Figure 11.12 Evolution of lithium-ion secondary batteries, cylindrical cells 18650 (*source*: Panasonic).

adapted parameters such as improved safety and high-temperature performance at the cost of lower energy density.

Many research activities are dedicated to lithium secondary batteries for the automotive industry. This will lead to an accelerated introduction of new battery materials and technologies. Micro batteries will benefit of this development as well. Nevertheless, many specifications of automotive batteries are different compared

Table 11.7 Comparison of lithium-ion battery specifications for automotive and micro batteries for energy autarkic systems

	Automotive Li-batteries	Lithium micro batteries
Specific energy (Wh/kg)	Very important (40–60)	Not so important
Power density (kW/kg)	Important (5–10)	Important (pulse load)
Self-discharge	Medium	Should be very low
Reliability, lifetime	10 years	5 to 30 years
Cycles	1,000 (5000: 2012)	100 to 20,000
Low-temp. performance	Important	For several applications
Safety	Very important	Medium to low
Cost €/Wh	<1 (0.3: 2014)	10 to 10,000)

to micro batteries for energy autarkic systems as shown in Table 11.7. The battery weight is most important for automotive batteries since fuel consumption is directly proportional to the weight. For micro systems the battery volume is more important than the weight and only if miniaturization is a prime concern. No difference is in power density since the high current loads during acceleration and recuperation for the automotive battery may result in the same power density compared to the pulse load of a sensor node of energy autarkic microsystems. Lifetime is important in both cases. A change of batteries after ca. 10 years may be more realistic for automobiles than for energy autarkic systems where service-free operation for 10 to 30 years is required.

United States Advanced Battery Council (USABC) in the Freedom-CAR research initiative, e.g., demand a calendar life-time of 15 years for 42 V battery systems and hybrid electrical vehicles (HEVs), and 10 years for electrical vehicles (EVs). In terms of cycle life, a lifetime of up to 1000 cycles at 80% DOD is requested.

The number of charge/discharge cycles of micro systems varies greatly for different applications. Safety is a prime concern for large lithium batteries while it is much relaxed if the battery has the size below 1 cm^2.

According to Table 11.7 the most important parameters of secondary batteries for energy autarkic systems are lifetime and self-discharge. Both are strongly dependent on temperature. The maximum storage life is achieved at low temperatures and ca. 40%

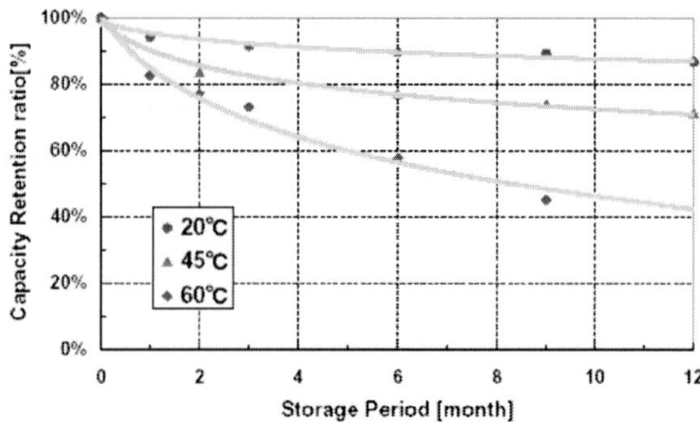

Figure 11.13 Capacity loss of CGR18650C cells All cells were stored full-charged, charge: CCCV 4.2 V 0.7 C, 100 mA cutoff. Discharge: CC, 1 C, 3.0 V cutoff (*source*: Panasonic).

charge. After a full charge, a Li-ion battery will typically lose about 5% capacity in the first 24 h, then approximately 3% per month at 20°C. This is the recoverable capacity loss. Figure 11.13 shows the capacity loss over time of 18650 cylindrical lithium-ion cells as function of temperature.

Permanent capacity loss refers to losses that are not recoverable by charging. A multitude of degradation mechanisms are responsible for permanent losses [9]. They are mainly due to the number of full charge/discharge cycles, battery voltage and temperature. Battery ageing, increasing cell impedance, power fading, and capacity decay origin from multiple and complex mechanisms.

Material parameters, as well as storage and cycling conditions, have an impact on battery life-time and performance. Depending on the cell chemistry, both high and low state of charge may deteriorate performance and shorten battery life. At high temperatures, the decay is accelerated, but low temperatures, especially during charging, also can have a negative impact.

The more time the battery remains at 4.2 V or 100% charge level (or 3.6 V for Li-ion phosphate), the faster the capacity loss occurs.

Lithium polymer batteries were tested as buffer for micro solar modules [10]. This is a typical situation where the cells are held at maximum voltage for long time periods. It was shown that cycle life can be increased significantly if the maximum voltage is reduced

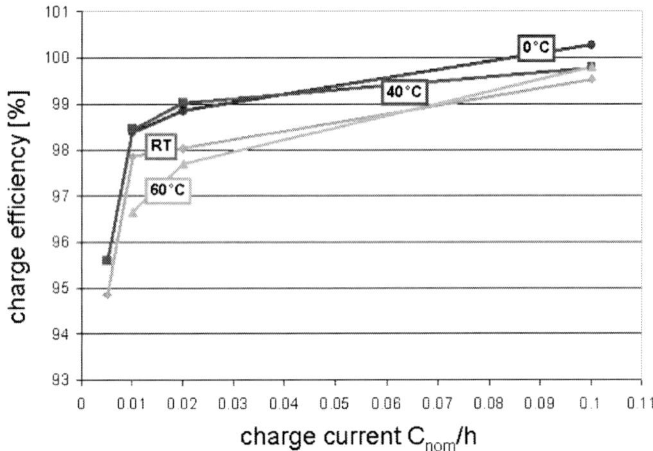

Figure 11.14 Charge efficiency of lithium polymer batteries as function of charge current and temperature. The lowest current corresponds to 200 h charge [9].

from 4.2 to 4.0 V or 3.9 V. More than 1000 full cycles were achieved with a capacity reduction of ca. 0.0125% per cycle.

Another issue in energy autarkic systems is low charge and discharge currents. While most specifications are made for 5 h charge or discharge (0.2 C) up to ca. 200 h can occur in energy autarkic systems. In that case, the charge efficiency of lithium-ion batteries can decrease to ca. 95% as shown in Fig. 11.14. Higher ambient temperatures further decreased the charge efficiency. Nevertheless, even at low currents, lithium-ion batteries show a much better charge efficiency than other battery chemistries.

The smallest lithium polymer batteries that are fabricated on a large scale are for Bluetooth modules. Table 11.8 gives an overview of typical samples.

11.3.3 *Solid-State Thin-Film Batteries*

-Parameters of solid-state thin-film batteries

Thin-film batteries with several different cathode and anode materials have been cycled to hundreds and many thousands of deep cycles with little loss of capacity. This is attributed to the

Table 11.8 Examples of small lithium polymer batteries (Guangzhou Markyn Battery Co., Ltd)

Model	Nominal voltage (V)	Nominal capacity (mAh)	Depth (mm)	Length (mm)	Width (mm)	Energy density (Wh/l)	Weight (g)	Specific energy (Wh/kg)	Discharge current (mA)
300910	3.7	15	3	9	10	206	2.25	24.7	120
041225	3.7	80	4.2	12.5	26	217	1.4	211.4	120
051235	3.7	130	4.7	12.5	36	227	2.7	178.1	195
062030	3.7	270	5.7	20.5	31	276	5.8	172.2	405

stability of the Lipon electrolyte film, the ability of the thin-film materials to accommodate the volume changes associated with the charge-discharge reactions, and the uniformity of the current and charge distribution in the thin-film structure. With cycling, the batteries gradually become more resistive with aging rates that depend on the particular electrode materials, the thickness of the films, and the temperature and voltage range during operation of the battery. As all-solid-state devices, thin-film batteries can operate over a wider temperature range than most lithium-ion batteries. Results of reasonable cycling performance are reported for -40 and $150°C$ [11]. When cycled at high temperatures, battery degradation increases and may be due to gradual microstructure or phase changes in the electrodes or at the interfaces. For example, aging for $LiCoO_2$ cathodes cycled above $75°C$ is associated with a lattice transformation. There are no reactions reported where safety becomes a concern. The self-discharge rates of most thin-film batteries with a LiPON electrolyte are negligible. Thin-film batteries as fabricated at Oak Ridge National Laboratory can be expected to hold a full charge for years [12]. The Ragone plot in Fig. 11.15 shows results for the power and energy densities that are obtained for thin-film batteries over a wide range of constant current discharge conditions. The values are normalized by the active area of the battery for ease in assessing the dimensions required for a particular application. The batteries all have a 3 μm-thick metallic lithium anode and the designated cathode. The cathode film thickness in micrometers is indicated beside each curve. This determines the capacity of the cell. Reference points giving the specific and

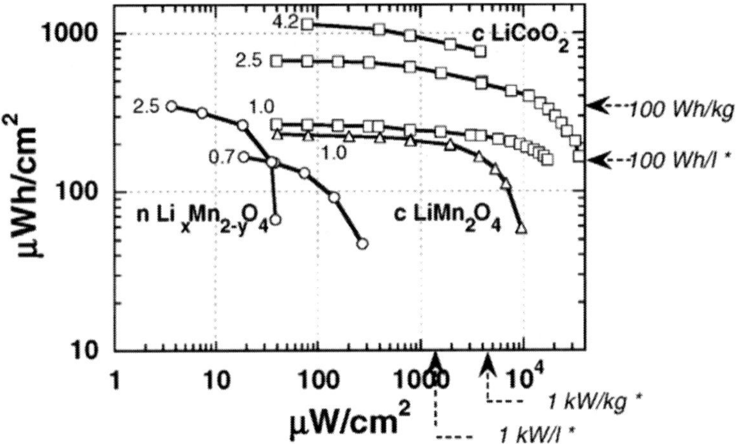

Figure 11.15 Energy and power density normalized to active battery area of solid-state thin-film batteries with different cathode materials and crystallinity (nanocrystalline (n)). Values indicating the cathode thickness.* All layers including 7 μm passivation but not the substrate [12].

volumetric energy, and power density are marked at the margins. These estimates include the volume and mass for all the active battery components, including the current collectors, electrodes and electrolyte, as well as a 7 μm-thick parylene and Ti protective coating. The substrate and the battery container have not been included and this obviously has a large impact on the battery specifications for the final product.

As can be seen from Fig. 11.16, the Li - LiCoO$_2$ battery discharged between 4.2 and 3.0 V give the best energy and power density. With a 4 μm-thick LiCoO$_2$ cathode, batteries can provide energy of 1 mWh/cm^2 at a discharge power of 1 mW/cm^2. The highest power density was achieved with 2.5 μm-thick LiCoO$_2$ films.

Thin-film batteries with LiCoO$_2$ cathodes can be discharged at very high continuous and pulsed current densities. Similarly, they can be rapidly charged at high currents. On the contrary, thin-film batteries can be charged at trickle currents of $<$1 μA/cm^2, such as might be produced by an energy-scavenging device. For application of thin-film batteries as energy storage for energy harvesting and -scavenging devices, the energy efficiency is important. Energy

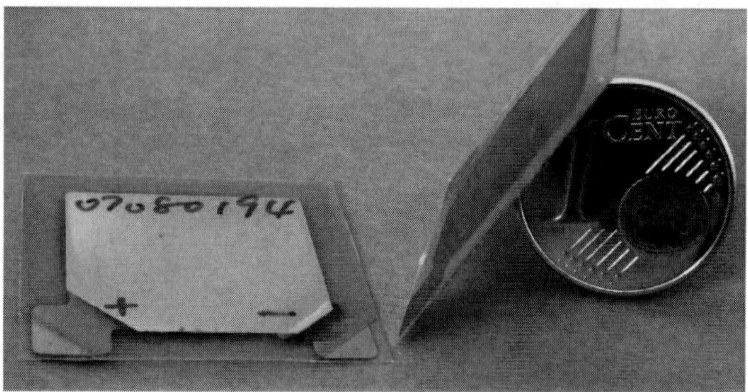

Figure 11.16 Example of thin-film battery with thin glass package (photo: Fraunhofer IZM) [11].

dissipated in the battery charge and discharge is evident as a voltage gap between the charge and discharge curve of a battery cycle. For batteries with highly crystalline and reversible electrode materials such as $LiCoO_2$ and $LiMn_2O_4$ cathodes with Li anodes, the battery cycling efficiency at low currents exceeds 96%. For batteries with amorphous cathodes, the energy efficiency is lower even for low current densities because of the hysteresis in the lithium insertion reaction for these materials. Efficiencies may be only 80%.

Examples of solid-state thin-film batteries

Meanwhile several manufacturers are fabricating and supplying thin-film batteries [11, 13–15]. While all systems are based on the $LiCoO_2$, Li, LiPON system, there are differences on details in the manufacturing process, the substrate, and the packaging. For example, flexible metal and polyimide substrates are used which allow the fabrication of flexible batteries [14]. Polyimide substrates require processing temperatures (annealing) below ca. 350°C. Currently thin-film batteries are the best solution of energy buffers for energy harvesting devices since their long-term and cycle stability, self-discharge and temperature range is much better compared to Li-ion polymer and coin-type cells. On the other hand, the energy density (including package) is much lower and the costs are high.

A thin-film battery that is packaged between two glass substrates is shown in Fig. 11.16. A battery product line that is especially dedicated to energy harvesting takes advantage of the same Li–LiCoO$_2$ thin-film technology [13]. Stacked package are available to increase the energy density. Therefore, various capacities are available for the same footprint. The parameters of both commercial systems are summarized in Table 11.9.

The batteries can be charged at rather high currents. The charging time required to obtain 95% of the rated capacity is 4 min at the first cycle and increases to 6 min at the 1000th cycle [13]. Since the batteries are tolerant to charge current, only the voltage (4.2 V) has to be controlled. The current will be limited due to the internal resistance of the battery. As can be seen in Fig. 11.17, in that case the maximum current is ca. 200 µA for a 50 µAh battery [15].

The thin-film batteries perform better at elevated temperature due to lower internal resistance. When operated at 100°C, charge and discharge can be performed at higher rate and with higher capacity. Table 11.10 shows typical discharge capacities of a 0.1 mAh thin-film battery discharged at room temperature, 60°C, and 100°C. It can even be operated at temperatures as high as 170°C; however, the capacity drops much faster during cycling. The capacity nearly doubles at 100°C compared to 25°C. Note that the capacity of a lithium polymer battery would degrade after weeks at 60°C and that it is irreversibly damaged at 100°C.

Thin-film batteries can be operated at temperature as low as −40°C, however, with lower charge and discharge rates. Table 11.11 shows discharge values of a 0.9 mAh battery discharged at 30, 0, and −40°C. The discharge current was 0.5 mA at 30 and 0°C, and 0.01 mA at −40°C. At 0°C it takes about 80 min to reach 95% of the rated capacity. When charged at −40°C, the battery can be charged to 0.6 mAh in 25 h.

The influence of discharge rate on the capacity of a thin-film battery is shown in Table 11.12. Approximately half of the capacity of a 0.9 mAh thin-film battery can be achieved at a continuous discharge at 10 mA compared to the capacity at 0.5 mA. The average voltage is in that case 3.5 V. That means that 35 mW can be continuously delivered from such a small battery which is sufficient for most transceivers in wireless sensor nodes.

Table 11.9 Parameters of commercial solid-state thin-film batteries

Model:	NX0201 (Front End Technology [11])		
System:	Li-LiCoO$_2$ LiPON electrolyte		
Size:	25 × 25 × 0.15	42 × 25 × 0.4 mm^3	
Capacity:	0.7 mAh	5 mAh	
Self-discharge:	5%/year		
Weight:	190 mg		
Max. operation			
Temperature:	150°C		
Cycles:	3500 (70%)		
Energy density:	13 Wh/kg, 26 Wh/l (including full package)		
Maximum current:	10 C (20 C pulse)		
	10 mA	50 mA	

Model:	MEC (Micro Energy Cell, Infinite Power Solutions [13])		
		MEC125	MEC101
System:	Li-LiCoO$_2$ LiPON electrolyte		
Thickness:	170 μm		
Size:		12.7 × 12.7	25.4 × 25.4 mm^2
Capacity:		0.1/0.2 mAh	0.5/0.7/1.0 mAh
Weight:			450 mg
Operation			
Temperature:	−40°C to +85°C		
Cycles:	10000		
Self-discharge:	2%/year		
Energy density:	(including full package)	14/27 Wh/l	17/24/34 Wh/l

Maximum current:	70 C		7.0/14 mA	40 mA
Model:	Excellatron battery [14]			
System:	Li-LiCoO$_2$ LiFON electrolyte			
Size:	customized sizes, integration in RFID tags			
Capacity:	0.135 mAh/cm^2 per layer			
Operation				
Temperature:	$<150°$C			
Cycles:	40000			
Self-discharge:	$<1\%$ / year			
Maximum current:	50 C at 25°C, 300 C at 150°C			

Model:	Enerchip, SMD-package (Cymbet [15])			
			CBC012	CBC050
System:	Li-LiCoO$_2$ LiPON electrolyte			
Size:			5 × 5	8 × 8 mm^2
Capacity:			0.012 mAh	0.05 mAh
Operation				
Temperature:	$-20°$C - $+70°$C			
Cycles:	at 25°C, 1000 at 50% DOD. 5000 at 10% DOD			
Self-discharge:	8%/year recoverable, 2.5%/year non-recoverable			
Charge time:	to 80% SOC		30 min	50 min
Max. pulse current:	0.1 mA		0.3 mA	

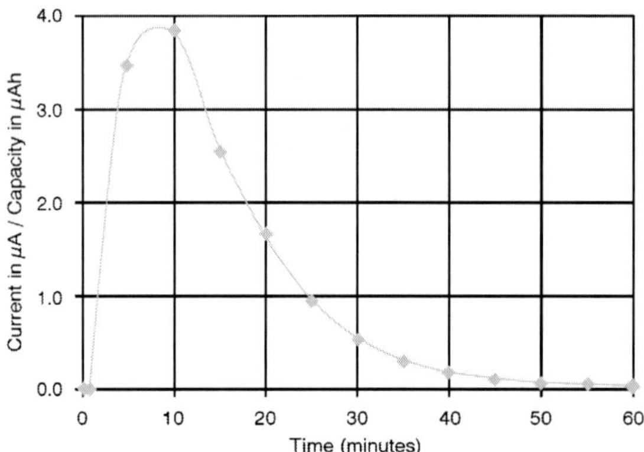

Figure 11.17 Charge current of a thin-film battery connected to a 4.1 V stabilized voltage source. The current is normalized to the rated capacity of the battery [15].

Table 11.10 Capacity of 0.1 mAh thin-film battery as function temperature [11]

Temperature (°C)	Capacity (mAh)
25	0.11
60	0.16
100	0.195

Table 11.11 Capacity of 0.9 mAh thin-film battery as function temperature [11]

Temperature (°C)	Current (mA)	Capacity (mAh)
30	0.5	0.9
0	0.5	0.76
−40	0.01	0.72

The non-recoverable capacity loss of thin-film batteries is specified between 1 and 2.5% per year. Thus, a lifetime of more than 10 years should be easily obtainable.

An important point regarding interconnection and packaging is that thin-film batteries in the fully discharged state can survive

Table 11.12 Capacity of 0.9 mAh thin-film battery as function discharge current [11]

Current (mA)	Voltage (V)	Capacity (mAh)
0.5	3.9	0.9
3.0	3.8	0.7
10	3.5	0.46

repeated solder reflow bonding procedures at 265°C without degradation of properties. This is not possible with Li-Polymer and most coin cells. One type of thin-film batteries is already available in form of SMD packages (16 pin QFN, Fig. 11.18) [15]. The same thin-film batteries are combined with power management logic in 20 pin SMd package (DFN). Thus, a complete μW power uninterruptable power source is available. Currently two capacities are used: 12 mAh and 50 μAh. The range of the input voltage is 2.5 to 5 V. An internal charge pump with an external capacitor generates a precise voltage of 4.1 V for battery charging.

11.3.4 *Other Micro Batteries*

Two primary batteries are included here since they are very small and therefore well suited for integration into micro systems. In both cases, secondary battery counterparts of these batteries are under development.

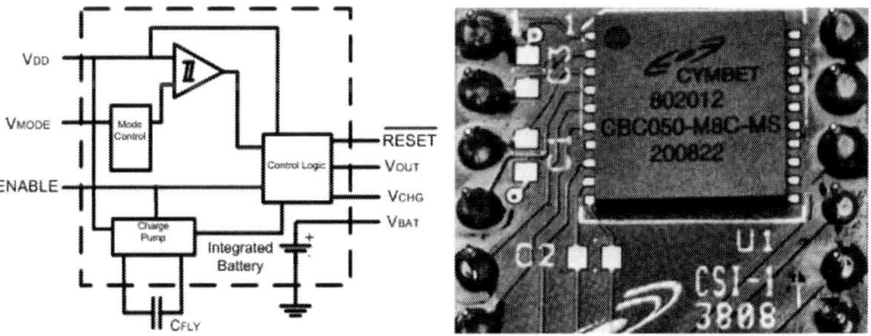

Figure 11.18 Functional block diagram of thin-film battery integrated control circuit (left) and SMD packaged thin-film battery (right) [15].

Table 11.13 Other commercial micro batteries

Battery	Medical [16]	Chip card, LFP25 [17]
Supplier	Eagle Picher	Varta Microbattery
System	Li MnO$_2$ primary	Li MnO$_2$ primary
Dimensions (mm)	6.73 × 2.37 diameter	29 × 22 × 0.4
Package	Welded Ti, glass	Metal foil, polymer seal
Capacity (mAh)	2.7 (at 30 μA)	25
Self-discharge	2%/year	—
Weight (g)	0.09	0.65
Volume (cm^3)	0.03	0.25
Cycles	—	—
Maximum temperature	—	—
Energy density (Wh/kg)	77	115
Energy density (Wh/l)	233	300

An implantable-grade micro battery was developed by downsizing the technology for laser welded metal casings and glass feed through technology [16]. The battery is so small (2.37 mm diameter) that it can be deployed via a minimally-invasive catheter. Since this battery has a cylindrical format, integration and interconnection is not straightforward. The cost of this battery is rather high, which prevents utilization in the mass market of energy harvesting devices. Due to the real hermetic packaging, reliability and long-term stability should be rather high.

A well-established technology is that for primary chip card batteries [17]. In that case, the metal foil current collectors of the electrodes are acting as battery housing. Both current collectors are laminated together at the edges with help of a polymer seal such that a reliable housing results. Special epoxy resin for cold lamination of flat cell LFP25 in the smart card battery leads to excellent behavior for the hard bending test. They are proven for >1000 bendings in cards after ISO bending test.

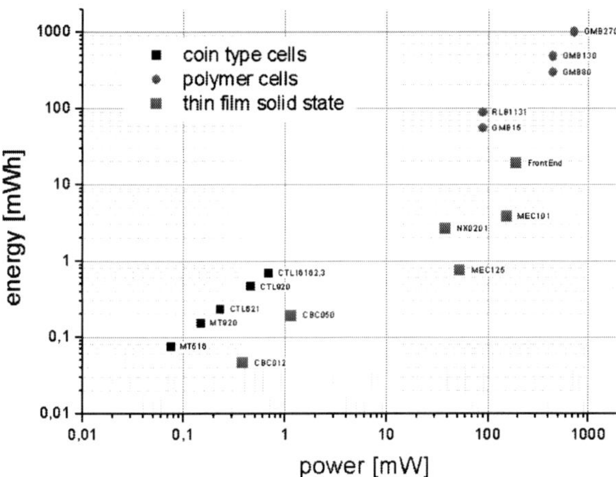

Figure 11.19 Overview of lithium secondary micro batteries. Comparison of maximum power and stored energy of coin polymer and thin solid-state cells.

11.3.5 *Summary*

Energy and power of the battery samples shown in Tables 11.6, 11.8, and 11.9 are plotted in Fig. 11.19. The power was calculated as product of maximum current and nominal voltage. The coin cells have the lowest power density while some thin-film batteries are capable of a current delivery equal to lithium polymer cells, which are much bigger. Figure 11.20 compares the energy and power density as function of battery size.

Thin-film batteries have the highest power density and the lowest energy density compared to the other battery technologies. Since they will be used not as the primary energy source but as buffer storage, power density is much more important than energy density. Different models of thin-film batteries vary much more in terms of energy and power density compared to the other technologies. This may due to different packaging concepts and the possible stacking of batteries in one package. All types (coin type, thin film and lithium–polymer are available at a volume of ca. 0.3 cm^2, in that case the smallest polymer and the largest thin-film and coin batteries.

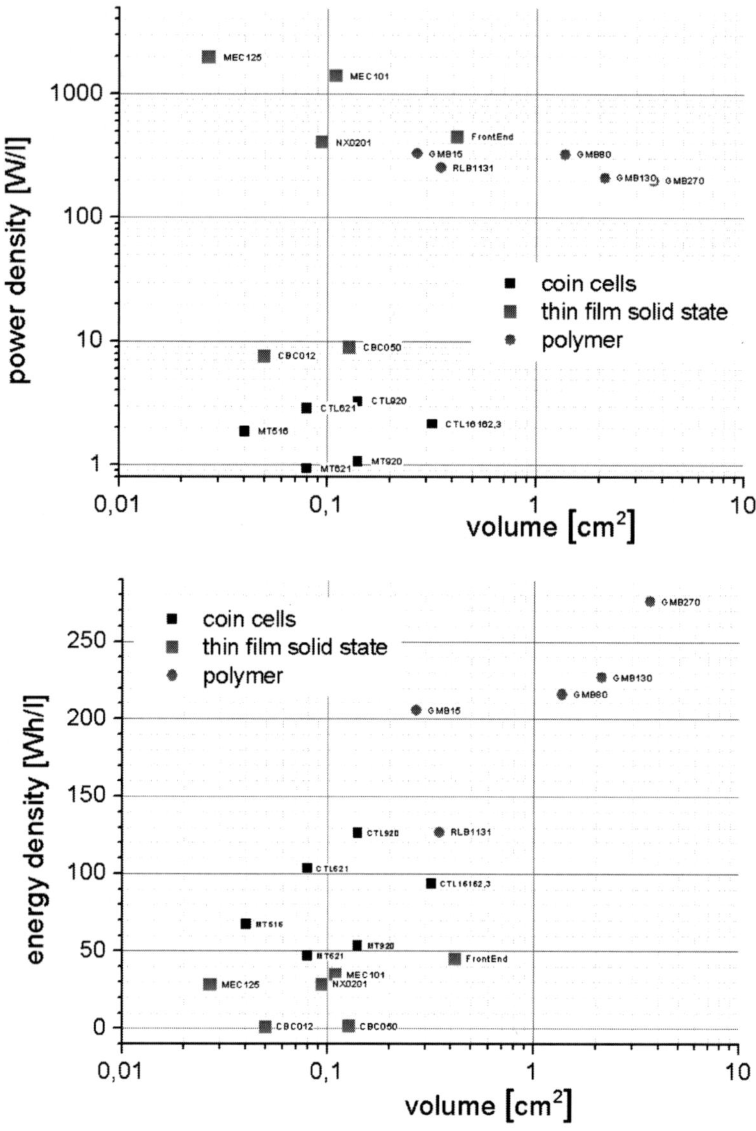

Figure 11.20 Overview of power (top) and energy density (down) of the considered secondary batteries as function of size.

Thin-film batteries will be the first choice in case of high ambient temperatures and if miniaturization is of importance. If there is enough space, then for power levels below 1 mW coin-type batteries could be considered. For higher power levels, small lithium polymer cells may be an alternative. Higher degradation of the polymer cells over time may be compensated by the more than 10 fold higher capacity compared to thin-film batteries. The depth of discharge will be much smaller for a lithium-polymer battery of higher capacity than for a thin-film battery, and thus the polymer battery may be cycled for some thousand times as well. On the other hand, the harvester must deliver enough energy to compensate for the higher self-discharge and degradation of the polymer battery.

Last but not least, battery costs will dictate the battery choice. Currently coin-type cells have the lowest cost.

11.4 Battery Dynamic Behavior and Equivalent Circuits

Most energy harvesting devices are highly dynamic. In most of the cases, it is the main purpose of the energy storage device to adjust the different and varying dynamics of energy input and energy drain. For example, the energy input can be relatively stable in case of thermo electric generators or very erratic and in form of short time pulses for some piezoelectric generators. Similarly, the current drain of the system can vary over several orders of magnitude from low current stand-by to short-term high current pulses in the microseconds to seconds range in the transceiving mode. Therefore, in most cases the buffer battery not be applied in a steady, constant current but in a pulsed, dynamic mode and the dynamic behavior of the battery must be taken into consideration. The battery dynamics is caused by many different physical, electrochemical, and long-term effects that influence the time response over many orders of magnitude in different ways. Table 11.14 gives an overview of typical time ranges of different dynamic effects of batteries.

Since the time ranges of pulsed charge and discharge are in the range between microseconds and seconds, the electric, the double layer, and part of the mass transport properties of the batteries are important for understanding and optimizing the system. In this time

Table 11.14 overview of typical time ranges of batteries

Effects	Time range (s)	Time range
Electromagnetic, electric	10^{-6}–10^{-3}	Microseconds
Electric double layer	10^{-3}–10^{2}	Milliseconds to seconds
Mass transport	10^{0}–10^{5}	Seconds to hours
Cycling and state of charge effects	10^{3}–10^{6}	Hours to months
Ageing	10^{7}–10^{9}	Months to years

interval, equivalent circuits are best suited to describe the relevant performance parameters. Influences on the larger time scale (state of charge and aging effects) are described in most cases with numerical models that can be implemented in a microcontroller or higher system levels.

The short time relevant effects can be described as follows:

Mass transport effects

In lithium secondary batteries, the diffusion of lithium ions is largely responsible for the mass transport. It occurs in the free electrolyte and the separator as the ions have to be transported from one electrode to the other. Diffusion also occurs within the porous electrode. Here it is limited by geometrical constrains which influence the dynamic behavior as well. Another important step is the diffusion of the ions through the solid electrolyte interphase on the negative electrode and inside the active mass particles. In solid-state thin-film batteries, the situation is slightly different since there is no liquid electrolyte and no SEI. Here diffusion is basically solid state inside the grains and through grain boundaries.

Double-layer effects

The charge zone between the electrode and the electrolyte was already mentioned in Section 11.2.3, where it is used for supercapacitors. Since the surface of the porous electrodes is large, the double-layer capacitance plays an important role in batteries as well. As the double-layer capacitor is on the electrode surface, it occurs in parallel to the electrochemical charge transfer reaction. The charge transfer over-potential is described by the charge transfer resistor

R_{CT} and the double-layer capacitor by the capacitor C_{DL}. This is shown in Fig. 11.21b. It is important to know that C_{DL} and R_{CT} are not constant elements. They are impacted by the state of charge, the temperature, the battery age, and the current.

The current that flows through the battery is divided at the phase boundary into a part that flows in the charge transfer reaction and a part that flows into the double-layer capacitor. As the capacitor can store only a limited charge amount, it is mainly charged in the first moment of a charge pulse. After a short time, the whole current flows through the charge transfer reaction.

When the charge pulse is finished and the battery goes into a rest phase or phase with a smaller charge current, the double-layer capacitor is discharged and the charge amount flows into the charge transfer reaction. This means that the elements $R_{CT}//C_{DL}$ form a low-pass filter for the charge transfer reaction. The double-layer capacitor can only carry alternative currents with a "high frequency," which results in filtering for the charge transfer reaction. As the two electrodes of a battery are not equal, the dynamic characteristics of both electrodes can also be different.

Thus, pulse currents are filtered as well and the charge transfer reaction sees the average current. This effect is the main reason why losses due to high current pulses are not as high as if one would only consider the internal resistance of the battery.

Equivalent circuits and electrical battery models

The simplest equivalent circuit is shown in Fig. 11.21a. It includes only a serial resistor and therefore does not represent dynamic effects. Usually here the open-circuit voltage E_o and the serial resistor R_s are a nonlinear function of the state-of-charge. R_s can also be changed for charge and discharge currents. The ohmic resistance R_s is the sum of the electrolyte resistance, the resistance of the current collector, the active mass, and the transition resistance between the current collector and active mass. The voltage at the ohmic resistance immediately follows the battery current according to Ohm's law.

Figure 11.21b includes the charge transfer resistance and the double-layer capacity as discussed above.

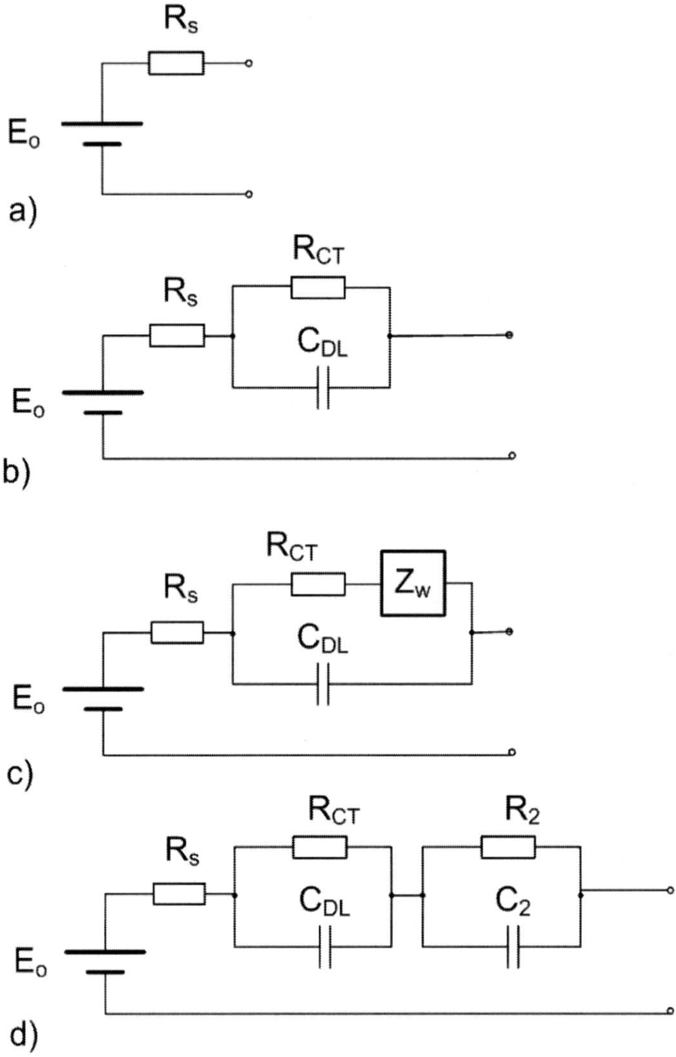

Figure 11.21 Battery equivalent circuits.

Mass transport or diffusion processes are difficult to describe with RLC elements. Therefore, in most cases they are shown as impedance elements, the so-called Warburg impedance Z_W as shown in Fig. 11.21c. Another way to represent diffusion mechanisms is to use a transmission line model that is a chain of RC elements or with help of an additional RC element as shown in Fig. 11.21d. On the other hand, the two RC elements in Fig. 11.21d can also be interpreted as elements from the anode and cathode side respectively. Impedance spectra or pulse response curves can be fitted with two circuits elements as shown in Figs. 11.21c and 11.21d and yield sufficient results in many cases.

Electrochemical Impedance spectroscopy

Electrochemical Impedance spectroscopy is an interesting technology for analyzing the dynamic behavior of batteries.

The Nyquist plot that shows the complex impedance in a single curve over the entire frequency range is in most cases used to show the impedance. As the characteristic of batteries is mainly capacitive, the sign of the imaginary axes is reversed to bring the curve into the upper part of the diagram. The diagram allows the effects caused by mass transport, the electrochemical double layer and the electrical effects to be separated. Figure 11.22 demonstrates how the different regions of the Nyquist plot can be interpreted.

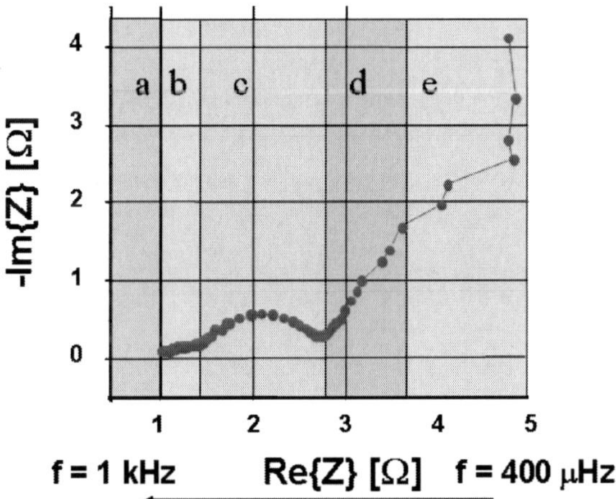

Figure 11.22 Nyquist plot of a 50 mAh lithium polymer battery and its qualitative interpretation. (a) Ohmic conductance; (b) SEI effects (c) effects by charge transfer and electrochemical double layer (d, e) mass transport effects.

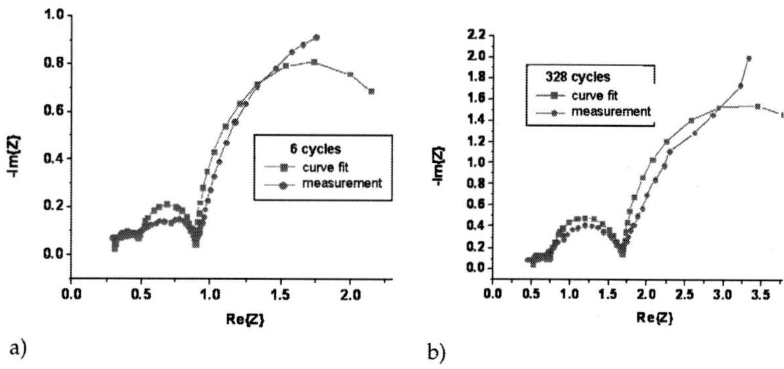

Figure 11.23   Nyquist plot of a 100 mA lithium-polymer battery after 6 and 328 full cycles measurement between 800 µHz (right) and 1 kHz (left) fit according to equivalent circuit of Fig. 11.21d.

Figure 11.23 shows a Nyquist plot for a new and an aged Li-ion polymer cell. This 100 mAh cell was cycled 328 times and the capacity has decreased to 65%. The aged cell has a much higher impedance in comparison to the new cell. The change in impedance can be seen over the whole frequency range. In other investigations, where the battery ageing was limited to 80% of the initial capacity, the change in the Nyquist plot was more significant at low frequencies and nearly unchanged at high frequencies. The measurement curve was fitted with an equivalent circuit according to Fig. 11.21d with an additional third RC element. As an example the resulting values for the circuit elements are shown in Table 11.15. Analysis shows that this relatively small battery has a doublelayer capacity of ca. 5 F.

Variations of ambient temperature and the self-heating at high pulse power have to be taken into consideration. The temperature is

Table 11.15   Resulting circuit elements for curve fitting of the measurements shown in Fig. 11.23

Cycles	R_s (Ohm)	R_1 (Ohm)	C_1 (F)	R_2 (Ohm)	C_2 (F)	R_{CT} (Ohm)	C_{DL} (F)
6	0.32	0.17	0.009	0.4	0.17	1.6	6.8
328	0.51	0.23	0.016	0.9	0.24	3.1	4.5

a key influencing factor in the diffusion coefficient. The limitations of the diffusion of the ions result in locally changed ion concentrations. From the electrical perspective, the diffusion results in an over-potential caused by reduced or increased ion concentration at the location of charge transfer. As a reservoir for ions exists in the free electrolyte and the porous structure of the electrodes, the diffusion shows a dynamic characteristic. In particular, the electrolyte conductivity increases with increasing temperature, which results in a reduction of transport losses. This must be represented with temperature dependent circuit elements.

11.5 Outlook

Today thin-film solid-state batteries are best adapted to the parameters of energy harvesting devices but the costs are even in mass production too high for many applications. The required parameters cannot be addressed with state-of-the-art lithium-polymer materials as well. New materials based on carbon free anodes, ceramic electrolytes and long-term stable cathodes are necessary. Nanomaterials can help to achieve these goals.

At the moment, lithium titanate ($Li_4Ti_5O_{12}$) is a promising alternative material for the negative electrode. This material has a better cycle stability than conventional mixed graphite anodes. The stability is much higher since there is no solid electrolyte interphase. Tin-based anodes are another alternative to graphite. The cycle life and rate capability may be improved with help of nanostructured Sn-based anodes.

Optimized active materials, such as nanoparticles of $Li_4Ti_5O_{12}$ as anode material with olivine structures ($LiMPO_4$) as cathode intercalation matrix, can be used to reduce the bulk diffusion and, hence, increase the ability to operate at high current peaks.

Additional performance improvements may be realized through the use of 3D architectures. This is especially convenient for microscale batteries since lithography and technologies adapted from Si and MEMS processing can be used. The use of the vertical dimension enables the battery to have a small areal footprint. Power density can be improved due to short ion diffusion length and high surface area.

Special crystal growth and vacuum deposition technologies, which will never be scalable for large batteries can be used to increase the performance of micro batteries. This may be carbon nanotubes or nano-sized silicon columns for improved conductivity and Li intercalation [19].

Thus, the most efficient use of micro batteries is in a fully integrated device. When physically integrated, the materials needed to support and protect the battery should serve a dual purpose as an active or protective component of the device. Examples include thin-film batteries deposited onto the back of a ceramic integrated circuit chip carrier and thin-film batteries deposited on a substrate shared with thin-film solar cells. Full integration of the battery and device requires coordination and compatibility of the fabrication processing which favors an all solid-state battery because of the high-temperature stability.

A three-dimensional battery assembly process [18] was developed whereby battery materials are filled into high aspect ratio holes of a substrate such as glass or silicon. Capacity up to 2 mAh/cm^2 and volumetric energy density of approximately 80 Wh/l were achieved.

Substrate integration of micro batteries into silicon chips is an important practical issue [20]. One approach is to assemble battery laminates into cavities of silicon wafers. Battery laminates from Li-polymer mass production were used for this technology. Thus, costs are relatively low. A near hermetic package was achieved with help of glass lids and UV curable epoxy as sealing. Thus, the silicon chip acts as housing. In contrast to thin-film batteries, this technology results in higher energy densities. The active masses have a thickness between 100 and 300 μm. On the other hand, the temperature stability of these batteries is the same as for conventional lithium-polymer batteries.

In mass production, usually liquid electrolyte is dispensed before encapsulation is performed. The liquid electrolyte is not easy to handle and will degrade the sealing surfaces of a micro battery. In addition, the high vapor pressure of the battery prevents the use of vacuum technology for encapsulation. Therefore, a gel-type electrolyte would be required that is compatible with the materials of the secondary battery. All materials and processes have to be optimized for ultra low water content and ultra low humidity and gas permeation [21].

References

1. Wayne Pitt, Energy storage at the heart of WSN IDTEchEx Energy Harvesting and storage Europe 2012, 15. May 2012, Berlin, Germany

2. Jossen, A., Weydanz, W. (2006) *Moderne Akkumulatoren,* 1.Auflage, U Books-Verlag, ISBN:3-937536-01-9.

3. Wakihara, M., Yamamoto, O. (1998) *Lithium Ion Batteries*, Wiley-VCH, Weinheim, pp. 156–180.

4. Ohzukua, T. Brodd, R. J. (2007) An overview of positive-electrode materials for advanced lithium-ion batteries, *J. Power Sources*, **174**, 449–456.

5. Bates, J. B., Dudney, N. J., Neudecker, B., Ueda, A., Evans, C. D. (2000) Thin-film lithium and lithium-ion batteries, *Solid State Ionics,* **135**, 33–45.

6. Salot, R., Martin, S., Oukassi, S., Bedjaoui, M., Ubrig, J. (2009) Micro-battery technology overview and associated multilayer encapsulation process, *Appl. Surf. Sci.,* **256**, S54–S57.

7. Chung, K. I., Lee, J.-S., Ko, Y.-O. (2005) Electrical analysis of $Li/SOCl_2$ cell connected with electrochemical capacitor, *J. Power Sources*, **140**(2), 376–380.

8. http://www.panasonic.com/industrial/includes/pdf/Panasonic_Lithium_Rechargeable.pdf

9. Vetter, J., Novák, P., Wagner, M. R., Veit, C., Möller, K.-C., Besenhard, J. O., Winter, M., Wohlfahrt-Mehrens, M., Vogler, C., Hammouche, A. (2005) Ageing mechanisms in lithium-ion batteries, *J. Power Sources,* **147**, 269–281.

10. R. Hahn, Power Supply for Wearable Applications Based on 3D Solar Module Technology, Energy Harvesting and Storage Europe, May15–16 2012 Berlin, Germany

11. Jeffrey Arias (2007) Thin-film solid-state rechargeable lithium battery, *Proceedings of the NanoPowerForum 2007*, June 4–6, San Jose, CA.

12. Dudney, N. J. (2005) Solid state thin-film rechargeable batteries, *Mater. Sci. Eng. B,* **116**, 245–249.

13. Bradow, T. (2009) Solid state micro energy cells uniquely enable energy harvesting, *Industrial Embedded Systems*, May 19, 2009, www.infinitepowersolutions.com.

14. Johnson, L. (2008) High power density thin film batteries for RFID tags, Excellatron Solid State, LLC, *Proceedings of the NanoPowerForum 2008*, June 2–4, Costa Mesa, CA.

15. Cantrell, T. (2009) Battery in a chip technology, *Circ. Cellar,* **228**, 62–69, www.cymbet.com.

16. Brand, C. (2007) A micro battery for low power applications", *Proceedings NanoPower Forum* **2007**, San Jose, CA, June 4–6, 2007

17. Eddie Shaviv (2007), Thin-film lithium polymer batteries, *Proceedings NanoPower Forum*, San Jose, CA, June 4–6, 2007.

18. Nathan, M., Golodnitsky, D., Yufit, V., Strauss, E., Ripenbein, T., Shechtman, I., Menkin, S., Peled, E. (2005) Three dimensional thin film microbatteries for autonomous MEMS, *J. Microelectromechan. Syst.,* **14**(5), 879–885.

19. Chan, C. K., Peng, H., Liu, G. (2007) High-performance lithium battery anodes using silicon nanowires, *Nat. Nanotechnol.,* **3**, 31–35.

20. Hahn (2006) Battery, especially a microbattery, and the production thereof using wafer-level technology, WO2005036689 (A3), WO2005036689 (A2), EP1673834 (A3), EP1673834 (A0), DE10346310 (A1).

21. Marquardt, K., Hahn, R., Blechert, M., Lehmann, M., Töpper, M. Reichl, H. (2009) Development of near hermetic silicon/glass cavities for packaging of integrated lithium micro batteries, *J. Microsyst. Technol.,* Springer 2009, 0946-7076 (Print) 1432–1858 (Online).

Chapter 12

Applications of Energy Harvesting Power Supplies

Peter Spies

Fraunhofer Institute for Integrated Circuits IIS,
Nordostpark 93, 90411 Nuremberg, Germany
peter.spies@iis.fraunhofer.de

This chapter deals with typical applications of energy harvesting power supplies. The different fields of applications are described and related system architectures and devices are introduced. Furthermore, suitable converter principles and types for the different scenarios are discussed. Also, examples from research and development are given. Finally, a comparison of the properties of the ambient energy sources in the different fields of applications is made.

Energy harvesting makes use of ambient energy sources to power small electronic devices such as wireless sensors, micro-controllers and displays. Typical examples of these environmental power sources are light from the sun or any artificial source, vibrations from vehicles or machines or heat from motors or the human body. Energy transducers such as solar cells, thermogenerators and piezo materials convert this ambient energy into electrical energy. The goal of energy harvesting is to replace batteries and wires used

Handbook of Energy Harvesting Power Supplies and Applications
Edited by Peter Spies, Loreto Mateu, and Markus Pollak
Copyright © 2015 Pan Stanford Publishing Pte. Ltd.
ISBN 978-981-4241-86-1 (Hardcover), 978-981-4303-06-4 (eBook)
www.panstanford.com

for power supply or at least, to extend recharge intervals of energy storage elements.

A first large application field is the building automation sector with self-powered light switches. Upcoming applications are condition-monitoring systems for large industrial plants or structural health monitoring in huge buildings. Most often, wireless sensors or sensor networks are powered with ambient energy sources. Another promising market is the consumer area with bags, apparel and the like which exhibit integrated energy transducers in form of solar cells to recharge consumer products like mobile phones or audio players.

This chapter introduces applications of energy harvesting power supplies. Since energy harvesting is a relatively new technology, these applications have not yet achieved mass-market success. Some of them are already available in form of devices, products or services, and others are just in the state of development or demonstration.

The following paragraphs will introduce the system architectures of these applications and explain which electrical consumers are employed in terms of application device and what their tasks are. It will be discussed which energy harvesting converter principles are relevant in the application environment and which are the most promising converter types here. Also, the main challenges to cope with in the different applications will be highlighted. A general overview of estimated energy to be harvested in different application environments is summarized in Table 12.1. "Human" refers to applications using energy from or at the human body. "Industrial" stands for the environment of machines, engines or plants.

All electronic devices exhibit some kind of intelligence per-formed by microcontrollers. Since this functionality of controlling actions and processing data is the minimal requirement of most electronic devices, energy harvesting must at least power these building blocks. Table 12.2 compares the power consumption of different state-of-the-art microcontrollers in certain operation modes. Since manufacturers name the operation modes not in a unique way, it is not easy to compare the power requirements of the devices directly. However, the difference between full operation and several sleep modes is shown. Other functions besides these

Table 12.1 Energy harvesting estimates (Raja, 2009)

Energy source	Harvested power
Vibration/motion	
Human	4 μW/cm^2
Industrial	100 μW/cm^2
Temperature difference	
Human	25 μW/cm^2
Industrial	1–10 μW/cm^2
Light	
Human	10 μW/cm^2
Industrial	10 mW/cm^2

Table 12.2 Comparison of the power consumption of several microcontrollers in different operational modes

Texas Instruments		Energy Micro AS		Microchip Technology	
MSP430F5437	3V, 1 MHz	EFM32 G890F128	3V, 1 MHz	PIC24F16KA102	3.3 V, 1 MHz
Active mode:	1110 μW	EM0	660 μW	Standard operating	1204 μW
Low-power mode 0	258 μW	EM1	309 μW	Idle mode	26.4 μW
Low-power mode 2	24 μW	EM2	2.7 μW	Sleep mode	6.6 μW
Low-power crystal mode 3	7.8 μW	EM3	1.77 μW	Deep sleep mode	0.07 μW
Low-power VLO mode 3	5.4 μW	EM4	0.06 μW		
Low-power mode 4	5.07 μW				

data processing tasks are measuring and transmitting data like physical parameters or processed values. The power consumption of such actions depends strongly on the application since they differ significantly in sensor principle, data rate, transmission distance and duty cycle.

Table 12.3 gives an overview of typical power requirements of state-of-the-art electronic devices and products. Energy harvesting covers the upper rows of this table, whereas by technology development more and more applications from the lower rows become feasible. Table 12.4 indicates an important detail of wireless

Table 12.3 Power consumption of typical electronic devices and products (Harrop, 2009)

Application	Power requirement
Standby	10 nW
32 kHz quartz oscillator	100 nW
Electronic watch or calculator	1 μW
RFID tag	10 μW
Hearing aid	100 μW
FM receiver	1 mW
Bluetooth transceiver	10 mW
Palm, MP3	100 mW
GSM	1 W
μP laptop	10 W
μP desktop	100 W

Table 12.4 Typical active times and power consumptions (Cepnik, 2011)

Application	Switch-on time (ms)	Power requirement (mW)
Radio (868 MHz, 4 dBm)	1	57
Microprocessor	Permanent	1
Energy management	Permanent	0.1
Radio + microprocessor, standby	Permanent	0.003
Mobile phone (without user intervention)	Permanent	5
IPod shuffle, music play back	Permanent	75

communications systems which enables energy harvesting in a lot of cases. Wireless communication systems always work in burst mode, sending or receiving data only in very short time frames. This is indicated by the switch-on time of 1 ms of the radio transceiver in the table. The rest of the time, the components of the wireless transceiver are in standby or totally powered off. Typical active times or burst length are 1 ms or shorter. This leads to a low average power consumption, which makes self-powered devices with energy harvesting possible.

Energy harvesting collects permanently small amounts of ambient energy in a storage device like a rechargeable battery or a capacitor. After a given time this storage device is able to supply the large power pulse required from wireless transceivers working in burst mode. This fact also makes clear why some kind of energy storage element is always required in energy harvesting systems.

12.1 Building Automation

Building automation systems are networks of electronic devices to control different functionalities in a building. They manage lights, heaters, air conditioning systems, doors, valves, security systems and the like. They also monitor the status of the building and send messages to the building's engineering staff or a central control computer. The advantages of building automation are reduced energy and maintenance costs, improved security and comfort. Typical energy savings with the help of building automation systems are around 30%. These energy savings appear mostly by controlling consumers like lights, heaters or air conditioning systems depending on the actual requirement. Examples are occupancy sensors turning off the lights or window sensors controlling heating and ventilation (Daintree Networks, 2012).

Energy harvesting in building automation generates a lot of benefits compared to power supplies by wires or batteries. Most important, it saves cost and energy. The cost savings appear in different categories. Energy harvesting saves installation costs, because no drilling and wiring is needed to provide power lines. Typical reduction in cabling can be around 70%. In this way, it also preserves the building envelope and enables the installation on difficult surfaces like stone, brick or glass in, e.g. greenhouses or atriums (see Fig. 12.1). It also reduces operational costs, because no energy is needed for the devices powered by energy harvesting. Additionally, it saves maintenance costs, because no batteries have to be replaced or recharged. Moreover, using ambient energy sources provides ecological compatibility, eliminating batteries and the use of cable material (copper, plastics, etc.). Finally, it reduces potential displacement costs, because the devices are wireless and can easily be moved from one place to another. By energy harvesting, a lot of money for installation work and time, energy, operation,

Figure 12.1 Battery-less EnOcean radio switch module, assembled within plastic frame by PEHA, mounted on a glass wall (EnOcean).

service and maintenance can be saved achieving a return on invest of the energy harvesting device in a very short time (Distech1).

12.1.1 *System Architecture and Application Devices*

Besides wired bus systems like RS232, Ethernet or optical fibre, wireless networks are used for the communication of different devices of the building automation system. Examples are wireless, battery-less sensors grouped around a controller with wireless receiver on-board, forming a star subnetwork. The controller processes the measurement information from the sensors as an input to the application control scheme. The controller acts as a communication hub and relays data to a higher level network like LonWORKS, EIB, BACnet, or EC-Net control network (Distech2).

Like in most other applications of sensor systems, wiring costs and labour time are huge and thus, wireless sensors and sensor networks are the preferred choice in building automation systems.

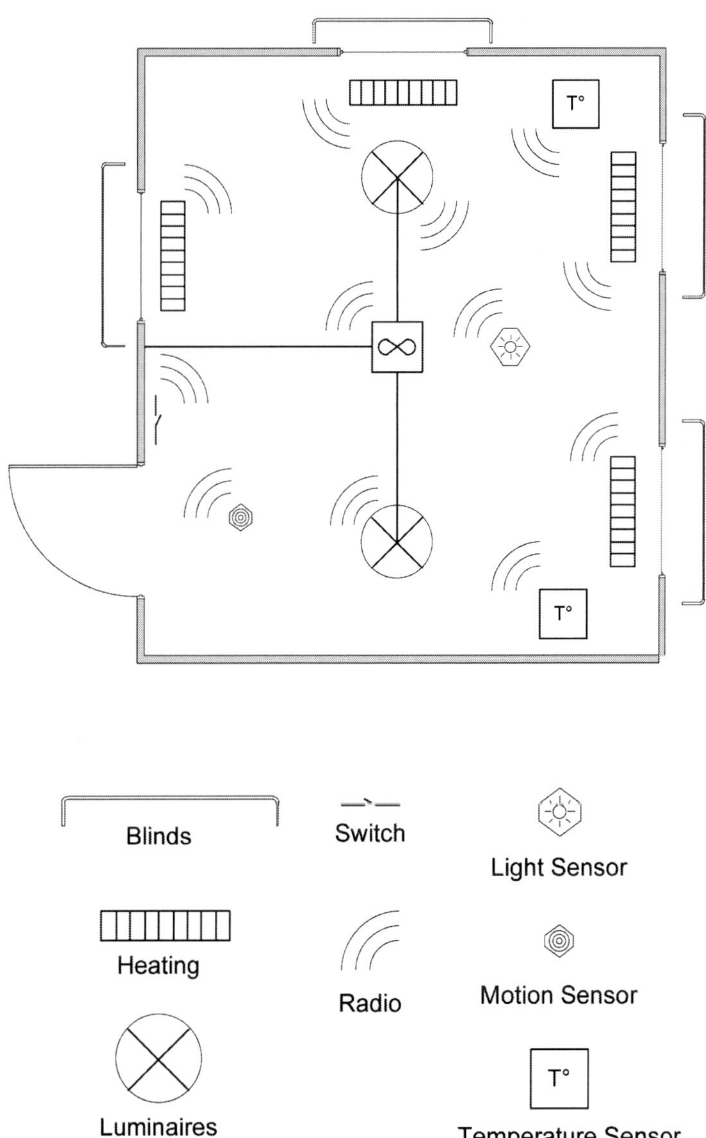

Figure 12.2   System architecture of a building automation system.

Especially in new and retrofitting projects, wireless transceivers are easy to install. So, the devices powered by ambient energy sources in building automation systems are wireless transceivers or transmitters and low-power sensors like in most other applications.

Typical sensors used in building automation are temperature sensors, humidity sensors, door and window sensors or pressure sensors. Moreover, light sensors, occupancy or motion sensors are useful for controlling building functions (see Fig. 12.2). In the case of security systems, smoke detectors, gas sensors, e.g. for CO_2, or glass break sensors are common. Actuators are switches or valves, speed controls of fans, dimmers for lighting, controls for sunblinds, for air handler, water pumps and the like. Other application devices are sensors for monitoring the position of flaps and valves in heating, ventilation and air conditioning systems (Kreitmair et al., 2009; Malux).

The big advantage for all these sensors is the slow variation of the physical quantities to monitor. Thus, the measurements and transmissions of these data need to be implemented only with a very small duty cycle, leading to a low overall power consumption of sensors and transmitters. Typical wake-up times of such systems to carry out a reading of a sensor are 1, 10 or 100 sec. Also, the sensor itself makes a first data processing and rating, sending an alarm or warning only when certain thresholds, e.g. light levels, temperatures or CO_2 values are exceeded. So, even lower power consumption is achieved by reducing the amount of data sent or the number of transmissions. Most of the time, these sensors are in sleep mode. Implementation of timers, which are the only component active in sleep mode, with a power consumption of only a few nano Watts enable the use of energy harvesting in these applications (Strba, 2009).

Radio transmitters are used to broadcast the signals from the sensors, either just for information or for generating a certain activity, like dimming the lights, moving the sunblind or controlling a valve or a fan speed. Thus, the transmitters are used directly as radio switchgear to generate signals to open and close doors, gates and barriers (Malux).

Especially the transmitters are optimized for the application in these building automation systems with energy harvesting power supplies. The optimization goal is always a minimum power consumption to enable energy harvesting with lowest hardware effort and minimized costs. Typical transmission power is, for example, 10 mW, the frequency band is most often the license-free SRD (short range device) band at 868 MHz or similar. Transmission telegrams are just, e.g. 60 bits and require energy of, e.g. 50 μWs for several transmissions. Incoherent ASK is often employed as energy-efficient modulation scheme, since only the "1" (high) bits are transmitted. Typical data rates are 120 kBit/sec. State-of-the-art transmission ranges are 30 m indoors and 300 m outdoors (free-field) (Schmidt and Heiden).

Transmitters send, e.g. a 32-bit identification number and the useable information. Most often, only unidirectional (simplex) connections with just an uplink to transmit information from the wireless transmitter to a base station or receiver are implemented. A duplex connection, with receive and transmit functionality would require more power for the additional receiver and would thus challenge the energy harvesting power supply even more. To achieve a high reliability without a bidirectional (duplex) connection, multi-transmission is used. This means the information or control signal is sent several times (e.g. three times) to ensure at least one correct reception at the base station or receiver. Powering an additional receiver in the wireless sensor unit to get an acknowledge signal from the base station would require in most cases more power than repeating the transmission several times. Of course, this depends on the transmission range and thus on the required transmission power to cover that transmission range. The power consumption in these transmitters is, above a certain transmission range significantly influenced by the power for transmission, especially the power amplifier (Malux).

In the free-field with no shadowing or obstacles for the radio transmission the mean propagation loss is calculated as

$$\frac{P_E}{P_S} = \frac{G_E \cdot G_S}{(4\pi df/c)^2} = \frac{G_E \cdot G_S}{(4\pi[d/\lambda])^2}$$

This formula is called Friis' transmission equation. Here, d is the distance between the wireless transmitter and base-station, f the frequency, c the speed of light as propagation speed and λ the wave length. P_S is the transmit power, P_E the receive power, G_S the transmit and G_E the receive gain of the antennas employed.

The receive power is mostly dependent on the distance to the transmit antenna. Any metallic obstruction should be avoided. Elevator shafts, electric risers and metal enclosures, for example, are major obstacles for wireless transmission in building automation.

The base station or receiver is often powered by the power grid and so there is no big need to optimize its operation towards lower power consumption. Thus, being permanently powered to wait for signals from the wireless unit with energy harvesting power supply is not a problem. Especially, when these receivers control a light switch or door opener, there is a mains power supply available.

More care must be taken in the operation of the whole transmission system when the receiver is also powered by energy harvesting, e.g. in wireless sensor networks with access to the power grid only for a small number of sensor nodes. Here, duplex connections with well-thought-out protocols, synchronization and duty cycles can be implemented.

Moreover, care must be taken in the radio communication protocol when several or even worse a huge number of transmitters are used in the same or neighbouring environment. This is typical for the case of buildings. Using very short messages, e.g. just several milliseconds, enables the operation of a large number of senders within the same radio cell. The error rate caused by collisions remains extremely low in this scenario. Statistically viewed, the transmission reliability is still greater than 99.99% in the case of 100 radio sensors that transmit once every minute. High interference protection is achieved by redundant asynchronous transmissions in random intervals. This means that even large office buildings or industrial facilities can be equipped with a large number of wireless, battery-free sensors (Distech1; Distech2).

12.1.2 Converters

In building automation, light as well as motion and heat can be used to power electronic devices. Depending on the given location and the locally available ambient energy sources, the appropriate generator type has to be chosen.

Switches are often operated by the hand and thus, can use motion in conjunction with, e.g. electrodynamic (inductive) converters to generate electrical energy. In buildings, there is no special need to make the transducer or harvester small or tiny, so electrodynamic generators with their long lifetime are very well suited. Moreover, pull-wire switches are used, e.g. at garage doors, where the housing of the switch and the transmitter are mounted on the wall, unreachable for the user. Lifecycles of more than one million switching cycles are achieved with these systems. Lifetime is an issue here, but the frequency is low when powered by human action (Malux).

Furthermore, solar cells are used in buildings in combination with a supercap or a rechargeable battery to back-up night operation (see Fig. 12.3). If necessary, primary or pre-charged secondary batteries can be used as a back-up precaution, but here aging and self-discharge of the battery has to be considered. Battery lifetime can vary depending on chemistry, the load profile and temperature variations (Distech1).

Sometimes, only several hours of intensive sunlight are sufficient to fully charge an energy storage element and secure an operation of a certain device for a lot of hours without illumination. As additional safety feature, a presence signal can be transmitted to signal the proper operation of the system. Typical brightness in homes is between 100 and 500 lx. Most health and safety workplace standards require a minimum illumination of 500 lx in office workplaces (see Table 12.5). The minimum illumination time required under daylight will be about 30% shorter than with fluorescent light of the same brightness. When illuminating the sensor with direct artificial light such as spotlights, the angle of incidence relative to the solar cell should not be too steep. Applications at windows are, of course, particularly suited for solar cells. Power output of solar cells at different light intensities can be found in the related chapter of this book.

Thermal gradients are a further option to use in building automation. Especially at heaters, air conditioning systems or hot water ducts, thermogenerators can be used to power sensors or actuators. Due to the high room temperature of typically around $20°C$, depending on the temperature of the heat source, heat sinks

Table 12.5 Typical illumination intensities (Kreitmair, 2009)

School	
Blackboard	500–1000 lx
General class room	300–500 lx
Office building	
PC workplace	200–500 lx
Conference room	300–700 lx
Corridor	50–100 lx
Hotel	
Reception	300–700 lx
Restaurant	150–300 lx
Staircase	50–150 lx

Figure 12.3 Room sensor for temperature and humidity and operation panel from Thermokon.

have to be used to maintain a sufficient thermal gradient. When varying temperature differences are present, maximum power point trackers (MPPTs) are helpful to match the power management to generate and care for maximum power output of the TEG. Furthermore, changing directions of the thermal gradients could require additional circuitry like low-voltage rectifiers. Examples for such an application scenario are, e.g. heating pipes, which are hot in winter and cold in summer.

Vibrations can also be used in buildings to gain electrical energy. The problem is that only small vibrations at small frequencies are available, which would require large generators with large seismic masses. Typical values are 0.01 g and a frequency range between

1 and 10 Hertz. Fortunately, in the case of buildings, large dimension generators are acceptable and thus it is possible to harvest small vibrations with huge transducers to arrive at a sufficient power level for the target application. However, these large generators would damp small vibrations in some applications, like in the case of windows, steel enclosures and the like. Problems arise in special buildings like bridges due to the change in vibration frequencies between summer and winter or day and night, which could require broadband or self-tuning transducers. These are subject to present research and investigation.

12.2 Condition Monitoring

The idea of condition monitoring is to measure physical parameters like temperature, vibration, speed or position to care for security and efficiency. Condition monitoring can be used for all kinds of machinery like motors, pumps, fans or compressors. By analysing the physical parameters, abuse or defects can be detected and further damage is avoided by appropriate maintenance or repair. Condition monitoring can determine possible failures, required maintenance and repair times. It is the most effective means to ensure efficient operation, help prevent unexpected failures and reduce repair costs and downtime (Discenzo et al. 2006; Kafka, 2011).

The benefits of condition monitoring are improved maintenance planning, scheduling and increased safety and environmental compliance. Moreover, asset reliability, availability, higher throughput and quality are the positive effects of condition monitoring. On the other hand, reduced maintenance costs, energy consumption, spare parts inventory and catastrophic asset failures and downtime are achieved. With help of condition monitoring, only parts that actually need replacing are replaced. Parts will be used for longer times and the number of spare parts in stock or produced in advance is reduced. Most important, with on-side monitoring, parts which are wearing out can be replaced before a serious defect occurs and operation is stopped totally (GE Energy, 2008).

12.2.1 *System Architecture*

Condition monitoring employs different kinds of sensors to measure the operational parameters of the system under surveillance. All these data have to be collected and analysed by microprocessors. Finally, if predefined thresholds are exceeded, warnings are reported to the responsible engineer via displays and the like. Moreover, the dynamic behaviour is evaluated and deviations from the usual behaviour are detected.

Most common is the offline data collection with handheld devices carried out by service personal, e.g. once per month. Alternatively, all the employed sensors will be connected and powered by wires, producing a huge amount of installation and material costs. Certain condition-monitoring systems provide an Ethernet connection to collect all data and pass it to a central computer. For example, in many industrial applications the cost of wiring may be US$40 per foot or more and often exceeds the cost of the remote sensor (Discenzo et al., 2006; GE Energy, 2008; Kafka, 2011).

More feasible are wireless sensors networks, which transmit the data by radio signals, eliminating the need for the installation of wires. Galvanic isolation is required in some applications as well, e.g. in high power transmission systems, thus wireless data transmission is best suited for these cases (see Fig. 12.4).

Although the wiring costs for communications and power cables are eliminated with wireless devices, these costs are replaced with the continuous cost of batteries and maintenance. This includes manpower, logistics, environmentally conscious disposal, protection from potential leakage, and potential equipment downtime due to unexpected battery failure (Discenzo et al., 2006).

With energy harvesting, sensors, radios, actuators, processors and displays can be powered without batteries or at least without the need for recharging or replacing them. Furthermore, installation in inaccessible, remote or dangerous areas and embedding them into machinery becomes possible (Discenzo et al., 2006).

Besides the transmission of the measurement values like acceleration and temperature, displays directly at the sensors can be applied to enable monitoring of the interesting parameters at the location of occurrence.

The need to measure physical parameters like vibration or temperature at different locations of a machine or equipment further facilitates the use of wireless networks. Depending on the equipment to monitor, different duty cycles in terms of data acquisition and transmission are required because of different operational modes, speeds, stresses and strains. Thus, a certain programmability of the sensor nodes to realize different duty cycles is necessary. This will also affect the power requirement and thus the size and system cost of the energy harvesting power supply significantly.

The self-powered sensor node performs local processing and transmits analytic results or raw sampled data to a central node for database storage and more extensive diagnostics and prognostics. The decision about raw data transmission or pre-signal processing in the sensor node is depending on the available energy in the node and the required energy for these tasks compared to the energy for the transmission of data. Sometimes it is more power-efficient to spend more energy for data processing in the node and thus reduce the time and so the power for data transmission. On the other hand, it can be more efficient to transmit everything and save the energy for data processing. This is most often depending on the distance to cover, the data rate or bandwidth and the environment like walls, machines or other metallic obstacles, which influence the transmission power significantly. Anyway, some programmable threshold values are monitored in the sensor node and just alarms will be sent when these are exceeded.

The interesting parameters to monitor in condition-monitoring systems do not change fast. Defects build up during hours or days; thus the duty cycle in terms of measuring and analysing the parameters is more in the range of hours and days then seconds and minutes. Condition monitoring is done in the form of snapshots sampling the interesting data only once in a while. This has significant influence on the power consumption and the design of the energy harvesting power supply.

Strict power management can be applied to keep the power consumption at a minimum. This is usually done by turning off components presently not required. Thus, sensors are powered only for the small periods of the measurement, including a certain

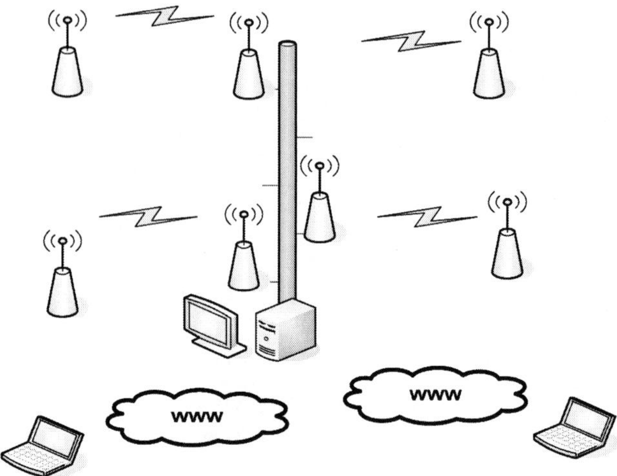

Figure 12.4 System architecture of condition-monitoring systems.

transient time to power on. Also, the transceiver and especially the power amplifier are switched on only during transmission of small data bursts like in cellular communication networks.

There are certain applications where a real-time location (RTL) functionality is helpful to determine and transmit the exact or relative position of a certain sensor. Furthermore, RTL can ease the installation process, because the point of measurement is provided by the sensor itself and need not to be recorded during the mounting of the sensors. The technologies for RTL will be discussed in a later section of this chapter.

12.2.2 *Application Devices*

The amplitude and frequency of vibrations can provide important information about any defects in, e.g. bearing surfaces or unbalanced shafts of any machine or engine. The frequency of vibration is determined by rotational speed and the construction of the shaft and the bearings. In case of a smooth running, the vibration amplitude is low. As soon as any defects in bearing surfaces, unbalanced or misaligned shafts occur, the amplitude is increased. By monitoring the change of the amplitudes, upcoming failures and problems can be predicted by a trend increase. Thus, application devices are most often accelerometers. Maintenance is then planned when convenient and prior to a failure actually occurring, which can be called 'predictive maintenance' (Mars and Parker, 2008).

Another important parameter in condition monitoring is the temperature and thus temperature sensors are applied frequently. Unusual heating of machinery can be a hint of an abnormal behaviour or any defects. Detecting moisture can be an indication of seal leak and thus a prediction of a dysfunction or upcoming error. Pressure of any liquids in the machinery is a further physical quantity to monitor to detect potential faults or the like. Thus, pressure and moisture sensors are often used for condition monitoring. More sophisticated approaches use corrosion, tribology or ultrasonic sensors and oil samples as input (Discenzo et al., 2006; Kafka, 2011).

Because some of the application areas are outdoor locations, the device needs an appropriate housing designed for rough or hazardous environments with waterproofed capability. Examples are chemical or petrochemical facilities. This has to be considered in the design, especially when deciding about the best suited energy transducer like solar cells or piezoelectric generators.

In a gas plant from Shell in Nyhamna, Norway, British company Perpetuum has successfully installed a self-powered wireless sensor network. The sensors report temperature and overall vibration every 5 min and a full vibration frequency spectrum every 6 h. The whole system is designed to work for 20 years. The sensors and radios are powered from the vibrations of the machinery, employing Perpetuum's inductive vibration transducers. The installation showed that a much greater number of monitoring points, e.g. in hazardous areas, can be regularly monitored and so a potential system breakdown can be indicated in advance. Key component according to the customer was the battery-free power supply, because batteries do not work well in outdoors environments. Further important properties were easy-to-install feature, low cost, low maintenance, retrofitting capability and that it is an open industry standard (Mars and Parker, 2008).

In the Yorkshire Water plant in Blackburn Meadows, Perpetuum and Nanotron installed a condition-monitoring wireless sensor network, which is powered by vibration harvesters. The wireless technology for data acquisition is well suited for covering large areas in this application. Also, in this case, the energy harvesting power supply rules out other kinds of supply because it is reliable and robust, has a low maintenance and can be installed quick and easy. Further big points were the accurate and flexible measurement of the whole wireless system and the simple integration into existing infrastructure (Mars and Parker, 2008; Smit and Albers, 2008).

12.2.3 *Converters*

Condition-monitoring systems can use different kinds of ambient energy sources. However, some of them are better suited to fulfil the requirements in these applications.

Solar cells are a good choice, but dirt, snow and ice require a certain maintenance, which is sometimes not accepted here. Furthermore, an energy storage element for night operation is required. Also, the degradation of solar cells has to be considered in long-life applications. Finally, shadowing from large obstacles like buildings and trees should be taken into account.

Most popular are kinetic transducers to harvest the vibrations of the different machineries. Anyway, energy storage elements have to be considered to power the electronic consumer during periods of inactivity of the machineries. Because vibration is often an important parameter to measure in condition-monitoring systems, it is obvious to use the vibrations also for power supply. Typical vibrations of machinery are in the range between 50 and 100 Hz with amplitudes around 100 mg (Kafka, 2011).

A vibration generator from Perpetuum in such an application works over a relatively large vibration and frequency range. It produces 1 mW peak with 25 mg acceleration in a 2 Hz bandwidth. They claimed it typically provides more than 0.3 mW on 95% of possible machines (see Figs. 12.5 and 12.6).

Critical and a challenge in condition-monitoring systems is the fact that resonant energy harvesting transducers have to be tuned to the exciting application frequency, which is often variable due to speed of the machinery, environmental conditions or lifetime. This problem might be solved by optimizing the harvester to work in a

Figure 12.5 Vibration harvester Perpetuum PMG FSM.

wider bandwidth, which on the other hand increases the damping of the device and so decreases the power output. An option would be several transducers electrically connected in parallel, which are tuned to adjacent frequencies to provide a broadband harvester.

Heat is also available in the presence of large machineries or plants. Moreover, heat sinks to cool one side of the thermoelectric generator are not a problem in condition-monitoring systems, since enough space is available in most cases. Thus, neither the transducers nor the sensors have to be small in these environments. The required energy storage only poses a challenge when they have to operate at extreme temperatures like below -20 or above $+80°$C.

Micropelt GmbH, a German specialist in thin-film thermo-electrics, has developed the so-called TE-Power CORE (see Fig. 12.7). This is a universal power supply using thermal gradients as low as 10 K to produce electrical energy. It consists of a specialized SMD-housing for a chip-sized TEG between two plates of aluminium. With an on-board dc-dc converter, a predefined dc-voltage is provided in a range of 1.9 and 4.5 V. Depending on the input thermal gradient, the module delivers between 150 µW and 10 mW. An integrated magnet enables an easy mounting and the heat sink can simply

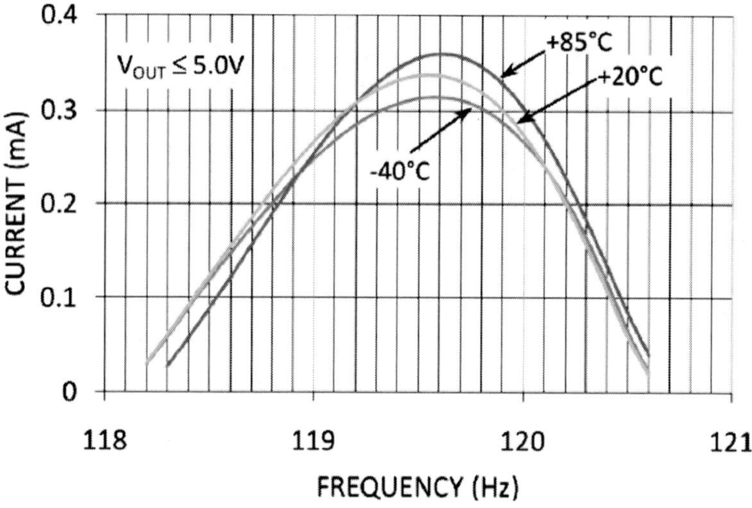

Figure 12.6 Output power of Perpetuum PMG FSM at 0.025 mg acceleration.

be replaced for optimization. Typical applications for this module are, besides heaters or air conditioning mainly machines or engines (Markt&Technik, 2011).

12.3 Structural Health Monitoring

Structural health monitoring (SHM) is the process of detecting damage in components and constructions. It is the static counterpart to condition monitoring where moving or vibrating systems are observed. The goal of SHM is to improve the safety and reliability of aerospace, civil and mechanical infrastructure by predicting and detecting damage before it reaches a critical state. Especially extreme events like earthquakes and typhoons, weight of snow and storm or just the aging of the materials and the environmental deterioration causes serious concerns on the integrity of any structures. These are closely related to the public safety; thus, knowledge of the structures health, load-bearing capacity, and remaining life is the primary goal of any strategy of SHM (Cho et al., 2008; Sazonova et al., 2004).

Figure 12.7   Micropelt's TE-power CORE.

To achieve this goal, technology is being developed to replace qualitative visual inspection and time-based maintenance procedures with more quantifiable and automated damage assessment processes. These processes are implemented using both hardware and software with the intent of achieving more cost-effective condition-based maintenance (Park et al., 2008).

An obstacle in SHM systems, especially for large structures is the intensive costs of hardware, installation and maintenance. To guarantee a reliable communication and thus system functionality, often coaxial cables are used between the sensors and the central or control unit for data evaluation and post processing. These cables exhibit large costs, arriving at prices of US$500 per sensor channel. Due to the fact, that the costs of the SHM grows faster than linear with the number of sensors, 350 sensing channels on the Tsing Ma Suspension Bridge in Hong Kong were estimated to over US$8 million. SHM methods in vehicles like aeroplanes or ships often necessitate the complete immobilization, which produces large economic losses (Celebi, 2002; Cho et al., 2008; Farrar, 2001).

A promising approach in SHM are 'smart materials', which are defined by an automatic adaption of their own properties as a reaction to external influences. In other words, smart materials couple two forms of energy, like electric and mechanical energy in the case of piezoelectric materials. Thus, these materials can be used as sensor or as actuator and exhibit a large application range in SHM (Musiani et al., 2007).

12.3.1 *System Architecture*

Structures which are able to sense operating temperatures, pressures, and strains could reduce the weight and costs of composite materials. This would result in improved condition based maintenance. Weak point would be the connectors and cables for passing the sensor information outside the structure. These are always subject to fatigue and breakage. Wireless transmission with extremely miniaturized electronics could solve this problem halfway, leaving the question of power supply. Here, energy harvesting in the structure would provide the final solution (Arms et al., 2007, 2008).

In SHM, several requirements exist which facilitate the deployment of wireless sensor networks. To detect and locate defects in structures, several sensors in conjunction with an actuator, which generates a test signal, can be employed. The large number of individual measurement points and thus large number of sensor nodes, densely deployed in possibly random configurations in the sensing environment demand the use of wireless sensor nodes. Moreover, the capability for self-organization and near-neighbour awareness is required so that information exchange between an individual node and a user may be achieved via point-to-point hopping protocols. Finally, the cooperation between sensor nodes is a must, where local processing capability is used to perform data fusion or other computational duties and then transmit only required or partially processed data onward. All these circumstances make WSNs useful in structural health monitoring (Park et al., 2008).

Besides the pure measurement of physical parameters, it is often vital to know at which point of the structure these parameters are measured. Thus, RTL feature, providing information about the measurement spot with a related accuracy is of interest.

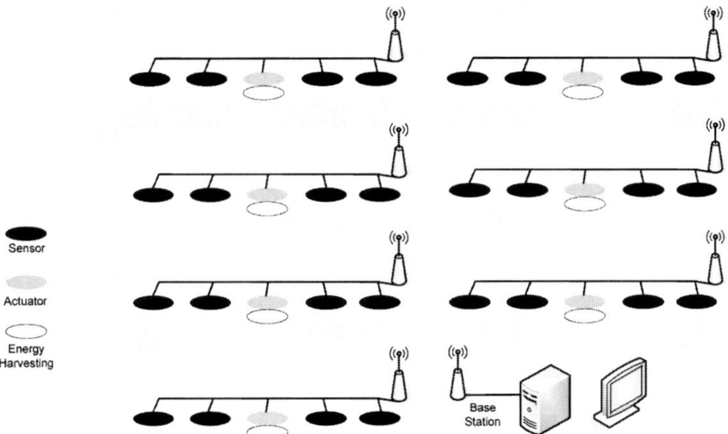

Figure 12.8 Architecture of a wireless, active SHM system (Park et al., 2008).

Technologies for localization will be explained in one of the following chapters.

Figure 12.8 shows the architecture of a wireless, active SHM system with energy harvesting. In each floor of the building, a wired sensor network and an actuator is powered by an energy harvesting transducer. Since only one harvester is employed, a wired connection for powering all sensors is required. However, each sensor could as well have its own harvester, increasing the costs because of several harvesters but saving cable connections and thus cost for these.

A wireless link is used between the floors of the building, which controls the actuation and measurement actions and collects data for further analysis and assessment.

12.3.2 *Application Devices*

The goal of any SHM sensor network system is to make the sensor reading as directly correlated with, and as sensitive to damage as possible. The most common measurements currently employed in

SHM are acceleration, strain, Lamb wave and electrical impedance (Park et al., 2008).

Due to commercial availability and relative maturity, acceleration is the easiest and thus most common form of measurement. It can be implemented with piezoelectric, piezoresistive, capacitive or fibre-optic sensors. The sensor principles work passive; thus, they do not require any energy. Anyway, all principles of measuring acceleration need appropriate charge-amplification, ADCs, signal processing and demultiplexing. These procedures require significant energy, resulting in high power consumption. Micro-electromechanical (MEMS) accelerometers for SHM applications are under investigation and development, but have only little actual meaning up to now (Park et al., 2008).

Strain is a non-dimensional measure of an object's deformation resulting from an applied stress. It is defined as the displacement per unit length of the object. To measure strain, the fact that some conductors change their properties like resistance, capacitance or inductance is employed. Most often due to the simplicity of the measurement, the resistive approach is used. A resistive network formed by conductors, embedded in an elastic layer is directly bonded on the object to measure. Options for measuring strain are piezoelectric patches or fibre-optics. Sensor signal processing is used to achieve sufficient signal strength directly at the sensor. As these measurement principles require similar signal processing, the energy consumption is in the same range of that of acceleration measurements (Park et al., 2008).

Besides the pure passive measurement of physical quantities, the active excitation with the help of actuators can be used. Here, after this excitation the response of the structure is measured by the appropriate sensor. Thus, also actuators can be system components in SHM applications. Piezoelectric materials can be used as a sensor and as an actuator, using the inverse piezoelectric effect. This effect describes the deformation of the piezo patch when an electrical voltage is applied. With these actuators, local motion can be introduced to a structure and the response can be measured for SHM purposes (Park et al., 2008).

Regarding these active measurements, there are two different approaches which are the impedance-based technique and the

Lamb wave technique. The impedance-based technique uses high-frequency vibrations to investigate the local area of a structure for variation in structural impedance. With the help of piezoelectric materials whose electrical impedance is directly related to the structure mechanical impedance, changing parameters like resonant frequencies will indicate possible damages (Musiani et al., 2007).

The Lamb wave method is often used in plate-like structures and employs in the first step some kind of actuator integrated in the structure. This actuator, e.g. a PZT material, is powered by an alternating voltage, defining the shape, amplitude and frequency of a test wave. Because of the coupling between electrical and mechanical domains, this electrically charged material generates a mechanical wave, called Lamb wave in the structure. The second step of this method is the sensing phase, where an appropriate sensor measures the response of the structure to the Lamb wave. Of course, for sensing purpose a PZT material can be used as well. The electrical signal received from the sensor compared to the exciting signal is a measure of the integrity of the structures. Cracks, delaminations or other defects will alter the response of the structure (Musiani et al., 2007).

12.3.3 *Converters*

State-of-the-art sensors for SHM are powered by batteries, which have a limited shelf-life in best cases up to 5 to 10 years. This generates a lot of maintenance effort for replacing or recharging the batteries and making it a major obstacle for the wide acceptance of SHM. Especially, the monitoring of arbitrary-shaped parts like rotor blades, wind turbine or hip-sockets requires that the sensors occupy small volume so that they do not affect the aerodynamic and hydrodynamic integrity of the structure. This makes batteries or large energy storage devices impractical. Thus, using ambient energy sources to power the sensors and sensor nodes is a promising solution in SHM (Huang et al., 2010).

Mechanical structures to monitor buildings, bridges or large vehicles often exhibit large dimensions making the use of vibration generators as well as solar cells feasible. Solar cells in SHM have the same downsides as in other applications, whereas operation by night can be buffered by batteries. Of course, these will add more weight to the system, making it difficult in lightweight applications. Dirt, ice and snow are even more serious in SHM systems, recommending other energy harvesting principles.

Since the vibration in form of acceleration and frequency is an important parameter to evaluate the health of a mechanical structure, it is obvious to use vibrations also for powering electronic circuits. Several principles like electrodynamic and piezoelectric systems are feasible here, where the piezo materials have superior performance due to their low weight. Depending on the way of coupling, piezos could easily be integrated into the monitoring structure and be used as sensor and as energy transducer in one device.

However, the limiting factor is the form factor. It can be shown, that the maximum power P_d which can be harvested from a piezoelectric element using strain under certain conditions is

$$P_d = \frac{2 \cdot \pi \cdot f \cdot Y_E^2 \cdot d_{31}^2 \cdot V}{\varepsilon_0} \Delta\varepsilon,$$

where f is the frequency of loading, V is the volume of the piezoelectric element, d_{31} is a piezoelectric constant, Y_E is the short circuit elastic modulus, ε_0 is the electrical permittivity and $\Delta\varepsilon$ is the strain in the piezoelectric element. With this equation, the maximum power from two different piezoelectric materials (PZT, PVDF) as a function of the feature size can be compared, shown in Fig. 12.9. The volume scales with the cube of the feature size. Lead zirconate titanate (PZT) is brittle and suitable for civil SHM, whereas polyvinylidene fluoride (PVDF) is flexible, bio-compatible and suited for in-vivo SHM. The calculations in Fig. 12.9 are based on a 1000 $\mu\varepsilon$ mechanical loading at a frequency of 1 Hz. From these graphs, it is obvious that from a harvester with dimensions of less than 1 cm^3 the maximum power is less than 10 μW. Additional losses due to imperfect matching or voltage conversion can lead to power budgets of 1 μW (Huang et al., 2010).

Of minor significance are thermoelectric generators in SHM, since sufficient thermal gradients in mechanical structures are rare except if there are significant heat sources available like machines or hot fluids in pipes or tubes.

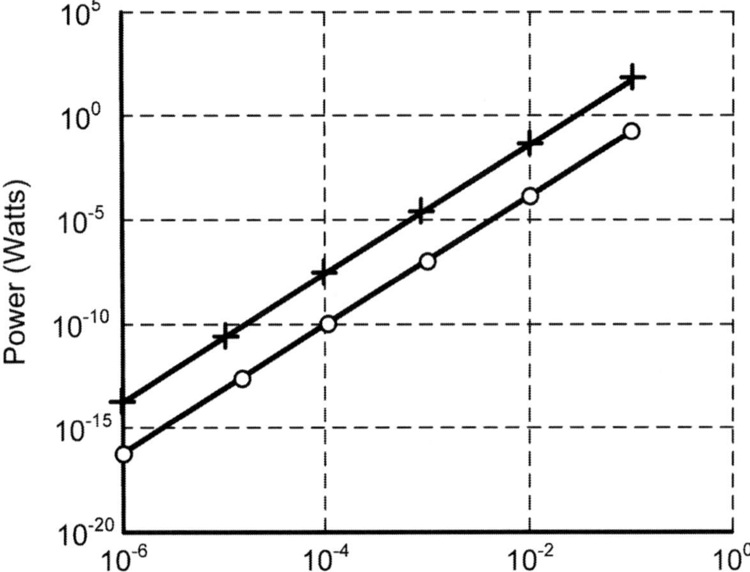

Figure 12.9 Maximum power from two different piezoelectric materials (PVDF; PZT) with a mechanical loading of 1000 $\mu\varepsilon$ at 1 Hz $\Delta\varepsilon$ (Huang et al., 2010).

In the German Project PiezoEN, funded by the Federal Ministry of Education and Research, a self-powered wireless sensor was implemented to measure and transmit the vibration of motorway bridges. The piezoelectric generator was built by 28 piezo patches from German company Invent and is producing around 0.3 mW of electrical power. The resonance frequency of the bridge was at 2 Hz and the acceleration peaks were up to 8 mg (Fig. 12.10).

12.4 Transport

The transport sector includes the traffic of people, goods or nowadays also data between different points of locations. The first popular and very successful example for energy harvesting in the transport sector was the bicycle dynamo as a form of electrodynamic generator a long time ago. The transportation area offers a lot of other applications for energy harvesting, because of its mobility and

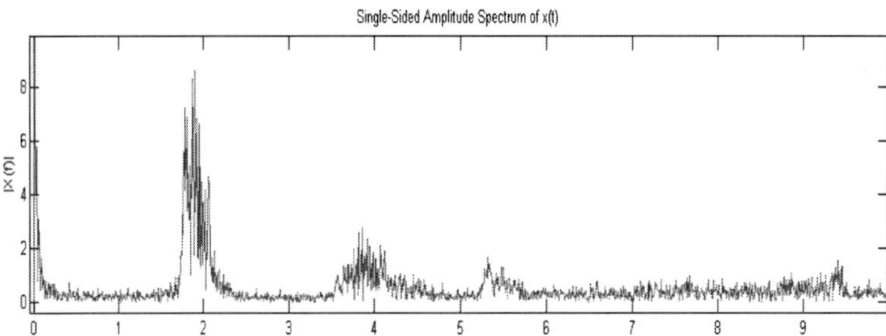

Figure 12.10   Vibration spectrum of a motorway bridge (y-axis in mg).

dynamic behaviour. Due to the affinity with condition monitoring, structural health monitoring and logistics applications, only the technology of tyre pressure monitoring and the field of aeronautics will be explained here. Some of the already-discussed applications in latter sections of this chapter are also found in the transport area. These are, for example, condition monitoring for machines or structural health monitoring in aeroplanes.

12.4.1 *Tyre Pressure Monitoring*

Tyre pressure monitoring (TPM) is a means to increase safety and efficiency of all kinds of vehicles using inflated tyres by monitoring the pressure. Tyre pressure monitoring systems (TPMS) were first adopted in 1986 in a Porsche 959. The technology was further used in top range luxury vehicles like AUDI A6, Mercedes Benz S-Class and BMW7 Series to increase safety and maintenance economy. Since 2000, first mid-size passenger cars were equipped with this feature as well (TMPS, 2012).

In 1990, more than 100 deaths from rollovers following a tyre tread-separation led Firestone to a large recall. This pushed the National highway traffic safety administration (NHTSA) of the Clinton administration to publish the TREAD Act in the USA. This act mandates the use of a suitable TPM technology in order to alert drivers of a severe under-inflation condition of their tyres. It affects all light motor vehicles (under 10,000 pounds) sold after September 1, 2007. In Europe, TPMS will be supported by regulation starting 2012 for new cars. Besides personal cars, TPMS have also potential application fields in airplanes, commercial trucks, buses, recreational and off-highway vehicles and motorbikes (Just-Auto, 2012; TMPS, 2012).

System Architecture and Applications Devices

Indirect Systems

Available are direct and indirect TPM systems. Indirect TPMS do not use pressure sensors. These systems measure the air pressure in the wheel by monitoring individual wheel rotational speeds. Indirect TPMS take use of the fact that an under-inflated tyre has a slightly smaller diameter than a correctly inflated tyre. Therefore, the tyre has to rotate at a higher angular velocity to cover the same distance as a correctly inflated tyre. Other developments can also detect simultaneous under-inflation in up to all four tyres using vibration analysis of individual wheels or analysis of load shift effects during acceleration or cornering. These systems need additional suspension sensors, making it more complex and expensive. Alternatively, the spectrum of the wheel speed sensor can be used and the computation can be carried out by the microprocessor in the ABS or ESC units (Just-Auto, 2012; TMPS, 2012).

A big disadvantage of indirect TPMS is that inflation differences below 0.5 bar cannot be detected, but such differences will increase the fuel consumption. A further disadvantage of indirect TPMS is that the driver must calibrate the system at an optimum inflation status (Just-Auto, 2012; TMPS, 2012).

Direct Systems

Direct-sensor systems use physical pressure sensors inside each tyre and send that pressure data together with a temperature measurement value from the tyre to the vehicle's control unit. These systems can identify simultaneous underinflation in all four tyres in any combination with a precision below 0.2 bar, as well as for the spare wheel. These systems are specifically designed to cope with ambient and road-to-tyre friction-based temperature changes, both

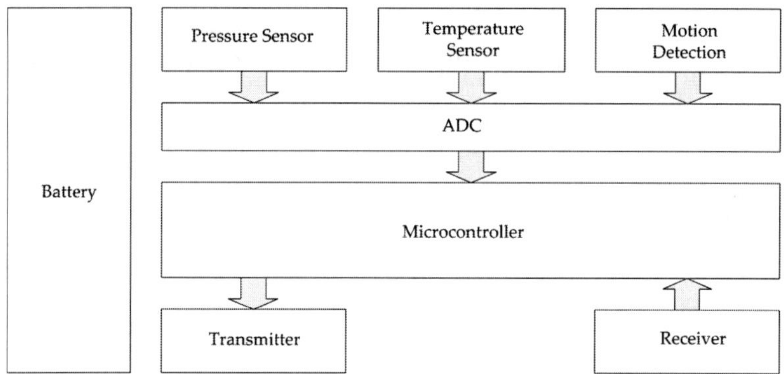

Figure 12.11   Block diagram of a TPMS (Loehndorf et al., 2007).

of which heat up the tyre and increase the pressure (Just-Auto, 2012; TMPS, 2012).

The TPM systems are either mounted on the well bed of the rim or attached to the bottom end of the tyre (Matsuzaki and Todoroki, 2008).

Besides a pressure and a temperature sensor, these systems use an accelerometer to detect when the car is parking or driving. Depending on that information, the system control can determine how often the measurements and transmissions are made.

To convert the analogue measurement values to a digital format for transmission, an analogue-to-digital converter and a multiplexer are needed. A microcontroller manages all measurement as well as communication and power management actions (see Fig. 12.11).

Communication

In order to transfer the pressure and further data from the wheel to the vehicle control unit radio-frequency (RF) communication or electromagnetic coupling are used. Due to the rotation of the wheel, no wired communication may be employed. The power supply of the sensor in the tyre poses a challenge in direct TPM systems. Here batteries exhibit the same problems as in many other applications as well, like limited lifetime, insufficient temperature range, production of waste, additional weight and additional costs for the device itself and for its replacement or recharge (Just-Auto, 2012; TMPS, 2012).

State-of-the-art systems use lithium batteries for power supply. To fulfil the OEM requirement of 10-year lifetime, these batteries take about 30–40% of the system's volume. In addition to size and cost, the batteries' weight is a major drawback for further integration of the system in the wheel. Typical transmission intervals are 30 sec when driving, with larger intervals to save energy when the car is parking. State-of-the-art data rates implemented today are 10 kBit/sec. Although the data transmission causes a significant part of the power consumption in TPM systems, the biggest part is the standby energy consumption due to the long standby periods and the small duty cycle. Typical values for power consumption nowadays are 200–250 µWs for one telegram, which can be optimized by improvement of transmission level and increase of date rate down to about 10–15 µWs (Loehndorf et al., 2007).

Converters

Due to the high amount of vibration energy, TPMS are a potential application field for vibration energy harvesting. The required small volume and low price facilitate the use of self-powered, battery-less and maintenance free systems. Best suited for the use in tyres are the piezoelectric and the electro-static principles, mainly because of their small implementation size and weight and related integration costs. Besides the energy transducer, a sophisticated ac–dc converter and an energy storage element are required to supply the pulses for the transmission bursts of the communication module of the TPMS.

Tyre pressure monitoring systems are subject to three acceleration directions, whereas the two main ones are the radial and the tangential. The vibration spectrum of a tyre exhibits components between 5 Hz and 1 kHz and is influenced by the type of tyres, road conditions and driving speed (see Fig. 12.12). At lower frequencies, there are some discrete spectral lines and harmonics, and at the higher part, the spectrum smears out. So using energy harvesting in TPMS large bandwidth systems are far better suited than resonant systems (Loehndorf et al., 2007).

The achievable output power of vibration transducers will increase with velocity and the weight of the seismic mass in the

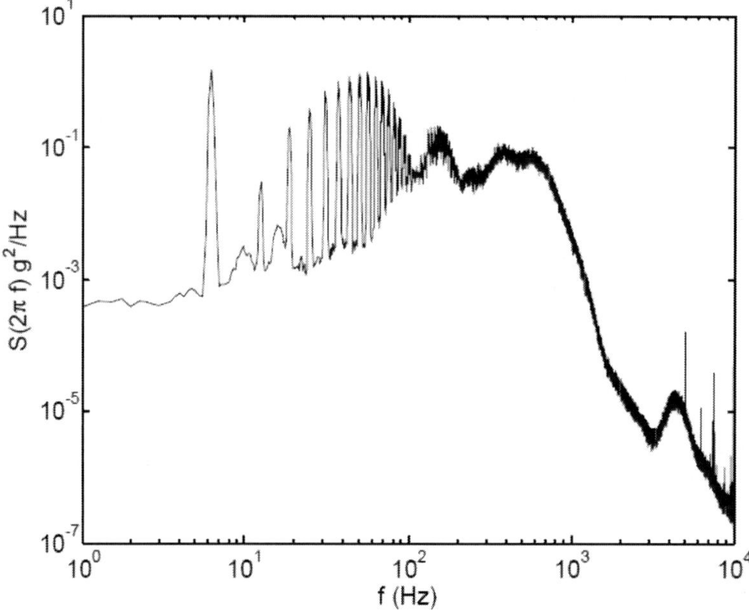

Figure 12.12 Spectrum of a tyre (Loehndorf et al., 2007).

transducer. For the design of an optimal energy harvesting power supply for TPM systems, the maximum acceleration peaks of up to 5000 g have to be considered, to achieve the lifetime and stability required in the automotive environment.

12.4.2 *Aeronautics*

Objectives of energy harvesting are reduction of installation costs and weight, improvement of modularity and to enable rapid introduction of new features and fast reconfiguration. These objectives are especially interesting in large buildings and due to a lot of similarities, in large vehicles like ships or airplanes. In planes, there are several application areas for self-powered systems like condition or structural health monitoring, flight tests and passenger comfort, which can be implemented with energy harvesting (Mitchell, 2007).

In structural health monitoring, strain and corrosion are of special interest. Corrosion costs the aircraft industry about 2.2

billion dollar per year, caused among others by unnecessary and costly aircraft downtime. Current prevention programs rely on scheduled, invasive visual inspections. Inspections are vital in aeronautics, since failures in the air have dramatic consequences, risking health or even life of passengers. These inspections are usually done on the ground, leading to financial losses because the plane is not used during these inspection times. Furthermore, a lot of inaccessible areas have to be covered, increasing the inspection time and thus the financial losses (Mitchell, 2007).

A concept change from scheduled maintenance to on-condition maintenance would require integrated sensors instead of ground test equipment, possibility of remote sensing of inaccessible and distant areas and sensor networks for complex airframe structure monitoring. The benefits would be maintenance and cabin crew support, decreased MRO (maintenance, repair and overhaul) effort and increased in-service time. The technical solution for this on-condition maintenance could be wireless sensor networks, because they are easy to install, without cables or special preparations and they provide intelligent network capabilities.

In the aeronautic application, an operation in harsh environment, e.g. between $-55°C$ and $+85°C$, at humidity, ice and acceleration is vital. Furthermore, lightweight and long lifetime comparable to that of the aircraft is required. All these specifications demand a self-powered energy supply which can be provided with energy harvesting technologies. The power requirements are comparable to that of other applications of sensor networks, with a required electrical output power in the range of a few milli Watts and the capability to provide peak currents for transmission bursts.

Another aeronautic vehicle well suited for energy harvesting is the helicopter. To increase the lifetime of a helicopter, dynamic component removal, refurbishment and replacement must be optimized. By tracking loading histories, components are monitored and individual maintenance of actual usage is possible. Tracking damage on the rotating structural components of helicopters can be implemented by wireless strain gauges. One of those critical components of the helicopter is the control rod or "pitch link". The pitch link is responsible for controlling the rotors' angle. The loads on these pitch links vary strongly with the present flight regime with much higher loads during manoeuvres compared to straight or level flight. Thus, this pitch link is a good indicator for the stress of the vehicle and thus will report important data for condition based maintenance (Arms, et al., 2008).

Recently, energy harvesting also starts to be adapted for unmanned aerial vehicles (UAVs), used in many military operations. A critical issue is the endurance of these increasingly smaller aircrafts. The limited size of the fuel system gives rise to several optimization goals and the rechargeable battery consumes much of the total mass (Anton and Inman, 2008).

System Architecture and Application Devices

Due to the large dimensions of the airplanes, wired sensors would pose an immense installation and cost effort. Thus, wireless sensors with self-powered energy supply are the optimum solution.

During flight tests, strain, temperature and acceleration are important measurement values. Regarding the passenger comfort, the traditional system architectures are complex, heavy and expensive (Mitchell, 2007).

Especially powering sensors like strain gauges, accelerometers and thermocouples and transmitting their data is helpful in structural health monitoring and reporting (SHMR) in aircrafts. Often these data are combined with GPS position, velocity and precise timing information to reference the sensor data to certain flight procedures. Additionally, inertial sensors report the orientation of the vehicle.

Converters

There are different possible energy transducers possible in aircrafts. The employment of solar cells is depending on the light source and thus the installation place. They could work indoors in the cabin as well as outdoors. Anyway, an energy storage is needed because light is not always available all day and night.

Vibration energy harvesting is a further energy source in aircrafts. Significant vibration amplitudes in the range of several tens of mg are available in the frequency range of 30 to 60

Hz. First generator implementations achieved up to 330 µW of electrical power. Problem here is that aircraft vibrations are broadband vibrations and these vibrations are not welcomed there, so engineers try to suppress them. In helicopters, the employment of kinetic transducers is more promising, since higher amplitudes are present (Arms et al., 2008; Kluge, 2011).

Thermogenerators are a third option for powering wireless sensors in aircrafts, although adopters have to cope with the good isolation of the plane in possible applications. Another issue is the required heat sink for maintaining large thermal gradients (Becker, 2008).

12.5 Logistics

Logistics deals with the flow of goods between different locations. It includes the information, the security and the management of the goods. Tasks to fulfil from the technical system are the tracking and tracing of assets to provide the user or customer with the exact position. Additionally, important parameters of the goods will be collected. Typical examples are temperature measurements in frozen cargoes or shock sensors for fragile goods, which act like a quality control. Other parameters to monitor are air quality or humidity, e.g. for plants or animals. Reporting history or status of goods is another issue in logistics. Examples are containers, trucks or pallets which can carry all different kinds of goods. Special cases are dangerous goods like chemicals or toxic waste which has to be handled with a certain care. Another example for a technical system in logistics is inventory control, where the actual number and location of certain assets are monitored. This could be done, for example, by access control of buildings or identification of vehicles which deliver goods and so on (Havinga, 2010).

12.5.1 *System Architecture and Application Devices*

Due to the mobility of assets like containers or pallets, wired communications and power supplies are not feasible. On the other hand, batteries exhibit the same downsides like additional weight and costs, limited lifetime and required maintenance as in other applications.

The system architecture has to be chosen depending on the area where goods and assets are monitored and their specific requirements. Wireless sensor networks work well in restricted areas where each asset is equipped with one or several sensors. With the help of sensor networks, also a localization of the single sensors and thus goods is possible.

Typical examples for logistic applications well suited for wireless sensor networks are buses, trucks or trains in depots or car parks. Other areas are supermarkets or shops, where sensor networks monitor the freshness of food with temperature sensors. These networks have an Internet connection to pass information to an Internet portal with alert functions. Additionally, applications of tracking diary herds or other animals are discussed (Havinga, 2010; Thurman, 2010).

Besides monitoring physical parameters like temperature, humidity or mechanical stress or just identifying physical objects, the exact or rough estimation of the location of an object is important in logistics. Requirements might be accuracies below 3 m for trailer location in a yard (Thurman, 2010).

Within wireless sensor networks, real time locating services (RTLS) are a means to determine the relative position of an object equipped with a wireless sensor. These RTLs often work with the receive signal strength indication (RSSI). This is a measure of the received signal strength from a neighbour sensor node which is available in most RF chips. Due to the fact that the signal strength is a function of the distance, it is possible to calculate the relative position of a sensor with the help of several RSSI measurements from different neighbours by triangularization (see Fig. 12.13). Alternatively, RF fingerprinting can be employed for localization. In this case, a measurement of the RF signal strength from different transmitters in various locations is stored in the sensor node. During localization, the sensor node compares present measurements with the RF measurement maps in its memory and is so able to estimate its own position. Another technique for localization employs the TOF (time-of-flight) of the RF signal, taking into account that the time of the signal travel is also proportional to the distance between

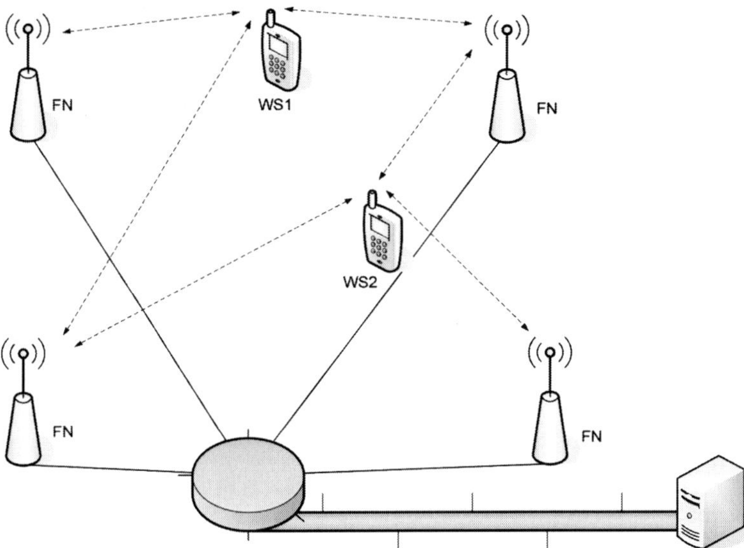

Figure 12.13 Architecture of a WSN with fixed nodes (FN) for real-time localization (WS: Wireless Sensor).

transmitter and receiver. Similar to satellite navigation systems, with at least three signals and by triangularization it is possible to calculate the position relative to the transmitters. A typical accuracy is 3 to 10 m or approximately between 20 and 50% of the distance to the transmitter (Havinga, 2010).

Outside a restricted area, cellular or satellite communication systems are able to transmit data of any purpose. Besides the data transmission via GSM or UMTS, a localization with the help of a cellular network is also possible. There are several principles to calculate the position of a communication terminal which exhibit different accuracies. Of course, most powerful localization systems are satellite based like the Global Positioning System (GPS) or the German counterpart GALILEO. They work nearly all over the world and enable a tracing and monitoring of goods and assets independent of the location with an accuracy of several meters.

Further application devices depend on the functionality or task of the system to fulfil. These might be all kinds of sensors measuring certain parameters of the goods like temperature, humidity, vibration, shock, position and the like.

12.5.2 *Converters*

The choice of converters depends strongly on the available space and the employed ambient energy source. Due to the mobility of the goods to monitor, a huge portion of vibration is available in logistic applications. Trains or trucks exhibit vibrations between 1 and 100 Hz with amplitudes in the higher milli Gs, depending on velocity, type of ground or rail and mounting location at the vehicle. Regarding the generator type, price and lifetime issues as well as required power and available board space has to be discussed. Challenges in mobile applications are variable amplitudes and especially fluctuant frequency range. Here, broadband generators which work efficiently in a certain band of frequencies are required. An example is the acceleration spectrum of a railway train, which exhibits several peaks of mechanical energy and where the energy is distributed over a certain band of frequencies (Fig. 12.14).

Solar cells are most common in trucks, since large surfaces directed to the sun are available. If dirt, snow or degradation is not an obstacle, these are well suited for the power supply of sensors, sensor nodes or tracking systems. Furthermore, for small goods like food or any kind of package for small products, thin-film solar cells are feasible.

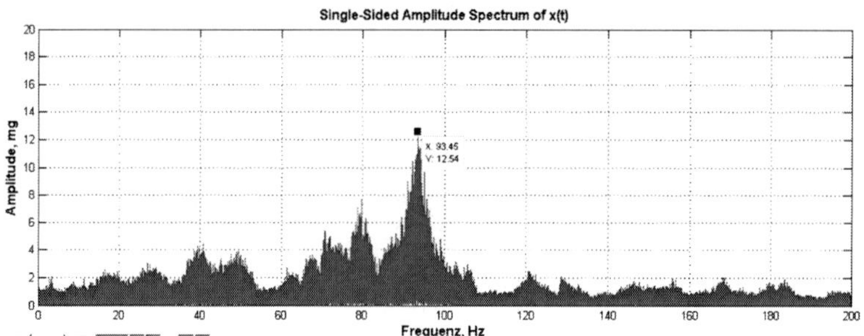

Figure 12.14   Acceleration spectrum of a railway train.

For a minor range of scenarios, also thermal gradients can be used for power supply. These are frozen goods, refrigerated containers or any kind of air-conditioned transport. Here, thin-film generator types are preferable.

12.6 Consumer Electronics

Consumer electronics are devices for everyday use, usually for entertainment, communication and office applications. Due to the large quantities of primary and secondary batteries in consumer electronics, like watches, headsets, toys, mobile phones, cameras or remote controls, this sector provides a huge potential for energy harvesting. However, in all the present applications, the energy harvesting power supply has to compete with the battery, especially its price, but also with its weight and volume, which are critical parameters of consumer devices.

Since the technologies of energy harvesting have not yet reached the mass fabrication and mass market, the material and device costs are presently often too high to enter the consumer market right now. Thus, in this chapter, mainly promising prototypes and demonstrators are reported.

12.6.1 *System Architecture*

In most consumer applications which have the potential to use energy harvesting, the goal is to recharge the battery already in use. In this way, recharging at the power grid becomes unnecessary, the battery size, thus cost and volume can be shrunken or it can be replaced by a capacitor with significantly smaller energy capacity. The system architecture here is mostly straightforward where the energy harvester just recharges the battery. For this purpose, the typical power management blocks like rectifier, dc-dc regulator and charge regulator with an optional protection and limiting circuit are employed (see Fig. 12.15). Depending on the application requirements and scenarios, a battery or optionally a capacitor is used as energy storage. For storing energy for a longer duration, longer periods without any input energy or for

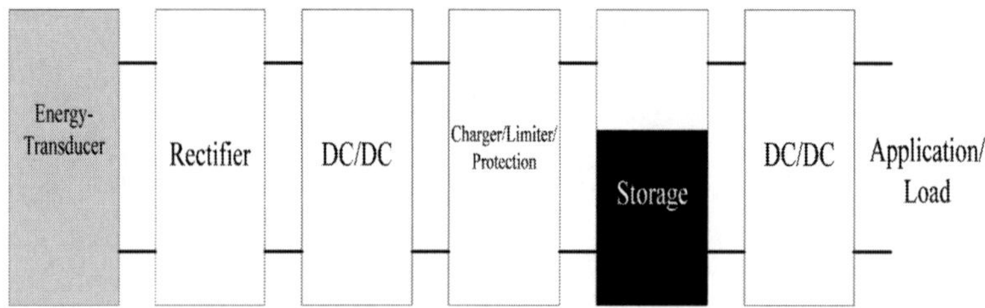

Figure 12.15 System architecture of an energy harvesting power supply in consumer applications.

larger power requirements, batteries are used. If smaller actions are powered or very frequent input energy is available, a capacitor can be used for storage. Some implementations provide battery monitoring functionalities to indicate the state-of-charge (SOC) of the battery and thus show the user when the maximum energy is harvested.

12.6.2 *Application Devices*

Generic applications in the consumer field are chargers which allow recharging different devices and thus provide a universal connectivity to such devices. Besides these chargers, a lot of specialized energy harvesting systems are developed to power electronics in different applications and scenarios. The following paragraphs describe some of them.

Information

One of the most convenient parts of the human body to harvest thermal energy are the wrist joints. So it was obvious to use that energy for wrist watches. The first commercially available wristwatch was produced by Seiko Instruments Inc. in 1998. It had more than 1000 miniaturized thermocouples stacked together and placed inside to use the heat of the human arm to power its electronics with a thermal gradient of just about 5 K. A small lithium battery was used as energy buffer. For periods when the watch was not worn and has no energy, the hands moved to a given position and just a low-power timer was running to count the time. When the watch was powered again due to sufficient thermal gradient, the hands are able to update the time because of the internal timer.

Entertainment

First applications of piezoelectric transducers in dance clubs are realized to use the mechanical energy during dancing. Designed by Dutch company Sustainable Dance Club, the floor deflects half an inch as people dance and it compresses a piezoelectric material under it. The floor produces between 2 and 20 W of power in a given area, depending on the impact of the patrons feed. For now, only the LED light on the floor are powered from it, but in the future, with newer technology more gain is expected. In London, Surya, another eco-nightclub, uses this technology.

American company Powerleap is developing flooring systems which harvest energy from pedestrian and vehicular traffic for different kinds of applications. Their target application fields start with powering small electronic devices like wireless transmitters to track consumer data, create interactive environments and control lighting and HVACs. One of the benefits here will be a substantial energy saving in buildings, compared to the wireless sensors and transceivers from the building automation application. More power intensive applications include displays for interactive media or way-finding and light bulbs. Finally, a goal of these transducers integrated in floors is to back up the power grid. Their generators are designed for high traffic areas like train platforms, sport stadiums, city sidewalks, doorways and the like. Typical power output of these floor generators can range up to 1 mJ per step (Redmond, 2011).

The fashion retailer Inditex places the Powerleap mats in 1500 Zara stores worldwide and achieves a saving of 20% on their energy bill. Furthermore, they collected new information on in-store shopping patterns (Redmond, 2011).

In the entertainment sector, a big consumer of batteries and thus a potential candidate for energy harvesting are remote controls. Because of direct interaction with the human body, energy from the human body can be used for powering the electronics inside these devices. SoundPower Corporation and NEC Electronics Corporation presented a prototype of a battery-free remote control.

A piezoelectric generator from SoundPower for power supply and a microcontroller with integrated RF technology from NEC are used to transmit data patterns to control home appliances. By pressing any button on the remote control, the piezoelectric material provides electrical energy to power the electronic circuits inside the device. The prototype was ready in 2010 and the actual end product was predicted for the consumer market in 2011 (Greendiary, 2012).

The French start-up Arveni developed a 12 key remote control, which was introduced in 2009. The bidirectional radio is IEEE805.15.4 compliant and was designed by their partner Wytek, also a French start-up.

Apparel and accessory

Apparel and accessory provide a great opportunity to integrate energy harvesting technologies to use energy from the human body or the environment. The most promising and most advanced examples are solar cells in backpacks, jackets or shoulder bags. The solar cells use the sunlight to generate electrical energy to power electronic devices like mobile phones, PDAs, audio players and the like.

German company Solarc was one of the pioneers of such products when they introduced a shoulder bag with flexible solar modules in cooperation with Artbag24. This shoulder bag has flexible solar modules with dimensions of 200 times 100 mm and a thickness of 1.5 mm which provide up to 1 W electrical power to any device connected inside. In 2010, Samsonite also announced the launch of solar consumer products. They integrate flexible solar cells from Ascent Solar Technologies, Inc., into their carrying case solutions.

Italian company Zegna commercializes in cooperation with German companies Solarc and Interactive Wear AG the jacket SOLAR JKT with solar cells on the collar to recharge mobile phones or iPods. The solar modules of 9 cm by 5.5 cm provide about 0.5 W under full sun. The output voltage of the whole system is selectable and a lithium-ion battery is used as energy buffer.

Sports

In sports applications, there are several electronic devices which are usually powered by batteries and thus exhibit opportunities

Table 12.6 Power consumption of typical sports devices in different operational modes (Ravise, 2011)

	Mode	Time of use
Watches, mini computers	Normal: μW	Normal: 24 h
	Sport: μW– mW	Sport: min–h
	Lighting: mW	Lighting: sec–min
Watches accessories	mW	min–h
Small lighting devices (Hiking lamp, classic lamp)	1–10 mW	min–h
Large lighting devices (bike lighting, camping)	> 100 mW	min–h
MP3, radio	5–50 mW	min–h
MP4, speaker, walkie-talkie	> 100 mW	min–h

for energy harvesting. Alike other user scenarios, the power consumption differs depending on operational mode, like active, standby or lighting mode (see Table 12.6). Most popular are sports watches with a power requirement of some micro Watts in normal operation mode. Certain products are already equipped with energy harvesting power supplies. Examples are watches with solar cells, camping tents with large solar modules or the bike dynamo as one of the first energy harvesting products (Ravise, 2011).

Aside from powering sports devices or related accessory with energy harvesting technologies, the energy spent during sports activities could be used for other purposes. The gym Green Microgym in Portland, Oregon, uses the machines like stationary bikes to harvest the energy during the workout. For now, only a small fraction of the energy consumption of the building is supplied with such machines.

12.6.3 *Converters*

For the application at the human body, several converter principles are feasible. Solar cells as a mature technology use light to produce electrical power. Since the human seldom operates at total or piecewise darkness, light is always present at the human body. Solar cells are often integrated in textiles like jackets or accessories like bags. Several promising implementations are available, especially flexible thin-film modules adding neither weight nor volume to the garment.

Table 12.7 Energy from the human body (Jansen and Stevels, 1999)

	Mechanical	Electrical	Thermal	Chemical
Muscles (active)	X			
Movement (passive)	X			
Skin potential		X		
Perspiration				X
Body heat			X	

However, light is not energy directly from the human body like thermal or mechanical energy, which are more related to physiological actions (Table 12.7) and can be employed to power electronic devices.

Heat dissipates from the human body in the form of respiration, skin convection, sweat and skin radiation. Due to practical reasons, the skin surface is ideal to place thermoelectric generators to harvest parts of that thermal energy (see Fig. 12.17). Depending on the size and the thermal gradient and thus the room or ambient temperature, several micro Watts can be harvested per unit area (see Fig. 12.18). However, this energy is limited by the Carnot efficiency. As an example, the Carnot efficiency is calculated with 37°C body temperature in a 20°C room:

$$\frac{T_{\text{human}} - T_{\text{environment}}}{T_{\text{human}}} = \frac{310K - 293K}{310K} = 5.5\%$$

In a warmer environment, the Carnot efficiency drops even further. On the other hand, in a cold environment, it does not perform better because when the skin surface detects cold air, the blood vessels constrict rapidly and reduce the skin temperature. Given that today's thermogenerators do not approach Carnot efficiency, this calculation is somewhat optimistic (Jia and Liu, 2009).

While sitting, the human energy expenditure is about 116 W. Due to several obvious limitations, this energy cannot be used fully to harvest electrical energy. Because of the fact, that a part of this energy is expended by vaporization, the maximum power available for the use with TEGs drops to 2.4–4.8 W, (Fig. 12.16). Practical reasons prohibit the use of all the skin surface of the human body to harvest thermal energy. Furthermore, by using some of the body

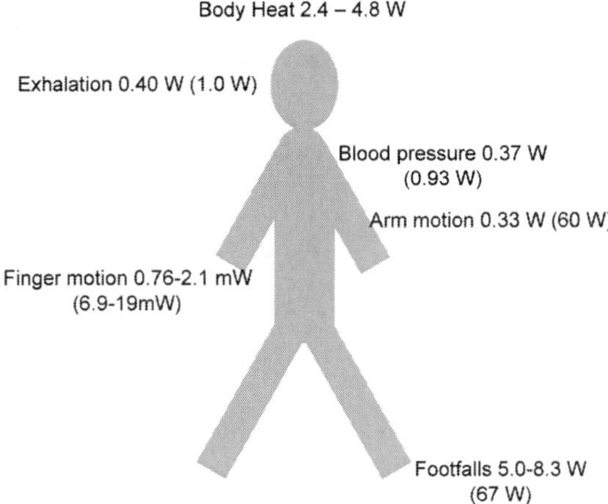

Figure 12.16 Possible power to be harvested at the human body (total power for each action in parentheses) (Paradiso and Joseph, 2004).

Figure 12.17 Different kinds of thermoelectrical generators.

heat at a given place with a TEG, the location would cool down, because the body would restrict blood flow to that area. When the skin surface is cooled down, a constriction of the blood vessels in the skin allows the skin temperature to approach the temperature of the interface. These circumstances would diminish gain of energy with a Carnot engine even further (Paradiso and Joseph, 2004).

Due to practical reasons, only some places at the human body remain to be used for thermal energy harvesting. Especially the fact that a thermal gradient is needed instead of only a heat source limits

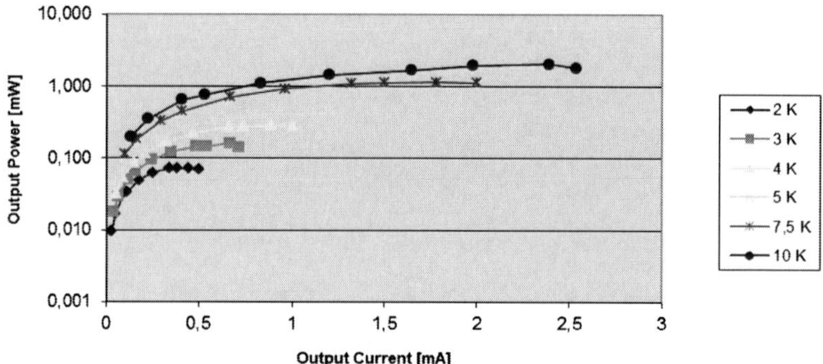

Figure 12.18 Output power of thin-film TEG at low thermal gradients, typical for the application at the human.

the location to the areas which are exposed directly to the ambient air without being covered by textiles or shoes. A good example is the neck, which is approximately 1/15 of the surface area of the core region. The core region parts are kept warm by the human body at all times. Assumptions predict a maximum of between 0.2 and 0.32 W of energy available at the total neck. Another convenient spot to use TEGs for catching energy from the human body is the head. It is about three times of that of the neck and could provide between 0.6 and 0.96 W of power with optimal conversion (Paradiso and Joseph, 2004).

Mechanical energy is another form of energy to be harvested at the human body. Here it must also be distinguished which activity is done.

Joint rotation at the elbow or at the knee will provide much mechanical energy. Anyway, adding the extra load of the generator to the body will increase the required energy from the human, which might not be acceptable to the user nor is it the idea of energy harvesting. A better idea would be to use the energy during the braking phase of walk. This energy is usually converted to heat in the muscle and useless.

Another mechanism is the linear displacement of body weight. The work of this movement can be calculated as follows:

$$W = \int_{s_1}^{s_2} F_s(s)\,ds,$$

where F_s is the component of the force along the direction of the displacement, d_s is the differential displacement vector, s_1 is the initial location and s_2 the final location. Recovering the energy from walking is a promising approach due to the high amount of energy involved. A 68 kg man, walking at 3.5 mph, or 2 steps per second uses 280 kcal/h or 324 W of power.

Considering to use just the fall of the heel through 5 cm shows that

$$(68\text{kg}) \cdot \left(\frac{9.8\,\text{m}}{\text{sec}^2}\right) \cdot (0.05\,\text{m}) \cdot \left(\frac{2\,\text{steps}}{\text{sec}}\right) = 67\ \text{W}$$

of power is available. This is, of course, the maximum number and trying to use the full stroke of 5 cm would generate additional load to the walker, which is not the goal of energy harvesting. Padded running shoes deflect about 1 cm while walking. This could be used as an upper limit which results in about 13 W (Paradiso and Joseph, 2004).

For using mechanical energy at the human body, different generator principles are feasible. The most common piezoelectric materials used for energy harvesting are PZT and PVDF. Because PZT is very hard and brittle, it is not well suited for the integration into textiles and apparel where some flexibility is required. PVDF has a much lower coupling constant than PZT, but it is very flexible, easy to handle and to shape, has a good stability over time and exhibits no depolarization in the presence of high electrical fields. Anyway, the application in textiles is limited by the efficiency degradation of the materials, which is, of course, subject of present research and material optimization. Another problem are the low frequencies available at the human body, which are far below any feasible PZT generator implementation (Paradiso and Joseph, 2004).

Table 12.8 gives an overview of the upper bound of electrical energy from daily activities using piezo generators and related losses (González et al.).

The use of inductive generators, perhaps in conjunction with pulley systems, is difficult at the human body, because of the weight of these systems. On the other hand, these generator types provide relative large amounts of power, which is related to their weight.

Table 12.8 Upper bound of electrical energy from daily activities using piezo generators and related losses (González et al.)

Activity	Mechanical power losses (%)	Electromechanical power losses (%)	Electrical power losses (%)	Daily activity (%)	Electrical power available (mW)
Upper limbs	50	11.2	10	16.6	24.6
Breath	10	11.2	10	100	74.8
Walk	75	50	10	16.6	1.265

12.7 Conclusions

Energy harvesting is adopted in a lot of promising application fields. Conventional ways of power supplies like batteries or wires are replaced by energy harvesting, saving resources like material, manpower or expenses for maintenance or retrofit. Moreover, new applications become feasible with energy harvesting like permanent monitoring systems for machines, plants or the human body.

Present implementation scenarios differ significantly concerning the properties of the employed ambient energy source and the requirements of the electrical consumers. Not just the type of energy is different, but especially its parameters like amplitude and frequency, thermal gradient or light intensity and the variation over time. Regarding the power requirements of the application devices, these changes are significantly depending on the sensor reading frequency or transmission and control duty cycle resulting in longer latency or response times. Furthermore, the transmission distance (coverage) has an influence on the power consumption.

For the present application fields of energy harvesting a rough estimate of the properties of different energy sources is possible. Table 12.9 (together with Tables 12.10 and 12.11) compares these properties from the different applications discussed in this chapter.

According to the expression 'The exception proves the rule' there might exist a lot of special cases where this table does not apply. Anyway, it is a first estimate for the engineer seeking for ambient energy sources in a given application scenario.

The majority of the introduced and discussed applications in this chapter are not yet mass applications. They are niche market scenarios or pilot and test installations.

Due to the small device quantities required in these first applications, the costs for energy harvesting generators and required materials are still high because a mass fabrication is not yet established. So, only markets with low cost pressure are suitable for such initial applications.

Furthermore, competing with batteries is still a problem, since the costs of batteries are much smaller compared to energy

Table 12.9   Properties of ambient energies in different applications

Parameter	Thermal gradient	Vibration amplitude	Vibration frequency	Light level	Data rate	Coverage
Building automation	L/M/H	L	L	L/M/H	L	M
Condition monitoring	H	M/H	M/H	L/M	L/M	M
Structural health monitoring	L	L	L	L/M	M	M/H
Transport	L	L/M	M	L	H	H
Logistics	L	L/M	M	L	H	H
Consumer	L	L/M	M	M	—	L

Notes: See Tables 12.10 and 12.11 for explanation of L, M and H.

Table 12.10 Rating/Quantification of ambient energies

Thermal gradient	
Low	0–$5°C$
Medium	5–$20°C$
High	$> 20°C$
Vibration amplitude	
Low	<100 mg
Medium	100 mg – 1 g
High	> 1 g
Vibration frequency	
Low	1–10 Hz
Medium	10–50 Hz
High	> 50 Hz
Light level	
Low	< 300 lx
Medium	300–500 lx
High	> 500 lx

Table 12.11 Rating/quantification of coverage

Low	Centimetres
Medium	Metres
High	Kilometres

harvesting transducers in a comparable power range. The economic benefits of energy harvesting become obvious when the costs for battery replacement or recharging have to be considered as well. Thus, energy harvesting is the preferred solution in applications where one single battery life of a non-rechargeable battery or one charge cycle is not sufficient.

The picture is different with applications or environments where batteries are not really feasible. These are, for example, environments with harsh conditions like extreme low or high temperatures or the employment of electronics in inaccessible or remote locations. Here, energy harvesting can enable new applications and needs not compete with batteries; thus there is no price pressure.

With the help of first applications from lead users, the maturity of the technology is demonstrated and more scenarios for energy harvesting become obvious. This will increase the requested device quantity and accelerate mass production with the expected fall of the material and generator cost.

Large quantities of energy harvesting devices are presently only sold in the building automation sector and in the consumer market. In the building sector, EnOcean as the pioneer of wireless, self-powered technology started several years ago and improved and optimized its technology over the past years.

In the consumer market, mainly solar cells are used for energy harvesting. These devices entered the market many years ago and already achieved an optimized state for mass production. Major research and development activities enhanced the solar cells mainly for power grid supply a popular technology.

Considering these two successful mass-market applications of energy harvesting, it is anticipated that the same will happen with kinetic transducers and thermo-generators in these and other applications like condition monitoring or structural health monitoring. Thus, it should be just a matter of time that energy harvesting will enter more mass-market sectors and enable innovative and promising self-powered microsystems.

References

Anton, S. R., Inman, D. J., Energy harvesting for unmanned aerial vehicles, Center for Intelligent Material Systems and Structures, Virginia Polytechnic Institute and State University, Blacksburg, VA 24060, SPIE, 2008.

Arms, S. W., Hamel, M. J., Townsend, C. P., *Multi-Channel Structural Health Monitoring Network, Powered and Interrogated Using Electromagnetic Fields*, Microstrain, Inc. 310 Hurrican Lane, Unit 4, Williston, Vermont 05495, 2007.

Arms, S. W., Townsend, C. P., Churchill, D. L., Galbreath, J. H. Corneau, B., Ketcham, R. P., Phan, N., *Energy Harvesting, Wireless Structural Health Monitoring and Reporting System*, Microstrain, Inc., 310 Hurrican Lane, VT, USA, Branch Head, Rotary Wing/Patrol Aircraft, NAVAIR Structures,

Naval Air Systems Command, Lexington Park, MD, 2nd Asia-Pacific Workshop on SHM, Melbourne, 2–4 December 2008.

Becker, T., EADS Innovation works: energy scavenging for wireless sensor, networks in aircraft, *3rd Fraunhofer Symposium Micro Energy Technology*, Nuremberg, 4 December 2008.

Celebi, M., *Seismic Instrumentation of Buildings (With Emphasis of Federal Buildings)*, Technical Report No. 0-7460-68170, United States Geological Survey, Menlo Park, CA, USA, 2002.

Cepnik, C., Energie aus Vibration, Elektronik 14/2011.

Cho, S., Yun, C.-B., Lynch, J. P., Zimmerman, A. T., Spencer Jr., B. F., Nagayama, T. Smart wireless sensor technology for structural health monitoring of civil structures, *Steel Structures* 8 (2008) 267–275.

Daintree Networks. The Value of Wireless Lighting Control. Last accessed March 2012 [Online]. Available http://www.daintree.net/downloads/ whitepapers/smart-lighting.pdf.

Discenzo, F. M., Pump, D. C., Loparo, K. A. *Condition Monitoring Using Self-Powered Wireless Sensors*, Rockwell Automation, Inc., Mayfield Heights, Ohio, Case Western Reserve University, Cleveland, Ohio, Sound and Vibration, May 2006.

Distech1. Distech Controls. *Wireless Resource Guide—Building open control products*, 05DI-DSWLSEN-21, www.distech-controls.com. Last accessed August 2014 [Online]. Available http://www.distech-controls. com/Wireless/Open-to-Wireless.html

Distech2. Distech Controls. *Wireless Battery-less Solution for Open Building Automation Systems*, www.distech-controls.com. Last accessed August 2014 [Online]. Available http://www.hvacc.net/pdf/distech/ Distech_CD/autorun/Brochures/wireless.pdf

Farrar, C. R., *Historical Overview of Structural Health Monitoring*. Lecture Notes on Structural Health Monitoring Using Statistical Pattern Recognition, Los Alamos Dynamics, Los Alamos, NM, USA, 2001.

GE Energy. *Condition Monitoring for Essential Assets*, General Electric Company, GEA-13979C (11/08), 2008.

González, J. , Rubio, A., Moll, F. *Human Powered Piezoelectric Batteries to Supply Power to Wearable Electronic Devices*, Electronic Engineering Department, Universitat Politècnica de Catalunya, C/ Jordi Girona, 1–3, Modul C4—Campus Nord, 08034 Barcelona, Spain. March, 2002.

Greendiary. *NEC's Concept Piezoelectric Remote Control Runs without Batteries*, Desh Raj Sharma, November 2009, Last accessed March 2012 [Online]. Available http://www.greendiary.com/entry/nec-s-concept-piezoelectric-remote-control-runs-without-batteries/.

Harrop, P. *An Introduction to Energy Harvesting*, IDTechEx, www.IDTechEx.com, 2009.

Havinga P. *Towards the Real Internet of Things, Ambient Systems*, IDTechex Energy Harvesting & Storage Boston, November 2010.

Huang, C., Lajnef, N., Chakrabartty, S. Infrasonic energy harvesting for embedded structural health monitoring micro-sensors, Michigan State University, East Lansing, MI, USA, in *Proc. SPIE Sensors and Smart Structures Technologies for Civil, Mechanical, and Aerospace Systems*, San Diego, USA, Mar 2010.

Jansen, A. I., Stevels, A. L. N., Human power, a sustainable option for electronics, Delft University of Technology, Faculty of Design, Engineering and Production, *IEEE International Symposium on Electronics and the Environment*, May 11–13, 1999, Boston, USA.

Jia, D., Liu, J. *Human Power-Based Energy Harvesting Strategies for Mobile Electronic Devices*, Higher Education Press and Springer-Verlag, 2009.

Just-Auto. Last accessed March 2012 [Online]. Available www.just-auto.com

Kafka, T., *Condition Monitoring on Machines Using Wireless Systems and Energy Harvesters*, GE Energy Germany GmbH, Neu-Isenburg, Germany, IDTechex Energy Harvesting & Storage Munich, June 2011.

Kluge, M., Autarke flexible Monitoringeinheiten zur Überwachung technischer Systeme – AMETYST, Öffentliches Statusmeeting EAS/AVS, Berlin, 2011.

Kreitmair, M., Micro Energy Harvesting, Demonstration, use cases standardization, *MRS Symposium Z: Energy Harvesting—From Fundamentals to Devices*, Boston, MA, USA, December 2009.

Loehndorf, M., Kvisteroy, T., Westby, E., Halvorsen, E. *Evaluation of Energy Harvesting Concepts for Tire Pressure Monitoring Systems*, Infineon Technologies AG, 81726 Munich, Germany, Infineon Technologies SensoNor AS, Horten Norway, Vestfold University College, Norway, PowerMEMS 2007.

Malux. *Industrial Building Automation, safe Switchgear for Demanding and Critical Applications*, Porvoo, Finland, www.malux.fi. Last accessed August 2014 [Online]. Available http://www.nhp.com.au/files/editor_upload/File/SLP/Steute/Brochures/Steute_Wireless-Product-Overview.pdf

Markt&Technik. Thermoharvesting-Gleichstromquelle als Batteriealternative, Markt&Technik, Nr. 46, 11.11.2011.

Mars, P., Parker, J., *Vibrational Energy Harvesting Case Study*, Darnell nanoPower Forum, June 2008.

Matsuzaki, R., Todoroki, A. Wireless monitoring of automobile tires for intelligent tires, *Sensors*, 8, 8123–8138, DOI: 10.3390/s8128123, ISSN, 1424–8220, www.mdpi.com/journal/sensors.

Mitchell, B. J., *Product Development, Boeing Commercial Airplanes: Energy Harvesting Applications Architectures at Boeing Commercial Airplanes*, Darnell NanoPower Forum, San Jose, USA, June 2007.

Musiani, D., Lin, K., Rosing, T. S. *Active Sensing Platform for Wireless Structural Health Monitoring*, Department of CSE, UCSD, La Jolla, Ca 92093, IPSN'07, Cambridge, Massachusetts, USA, 2007.

Paradiso, A., Joseph, T. S. *Human Generated Power for Mobile Electronics*, GVU Center, College of Computing, Georgia Tech, Atlanta, GA 30332-0280, Responsive Environments Group, Media Laboratory, MIT, Cambridge, MA 02139, 2004.

Park, G., Farrar, C. R., Todd, M. D., Hodgkiss, W., Rosing, T. Energy harvesting for structural health monitoring sensor networks, LA-UR-07-0365, *ASCE Journal of Infrastructure Systems*, 14(1), 64–79, 2008.

Raja, M., Micro-energy harvesting systems scavenge milliwatts for ULP devices, *Energize*, p. 44, Texas Instruments, June 2009.

Ravise, A. *Harvesting Energy for Electronic Sport Products, Oxylane Decathlon*, IDTechex Energy Harvesting & Storage Munich, June 2011.

Redmond, E., *Fully Integrated Energy Harvesting & Data Tracking Flooring System*, Powerleap Inc., IDTechex Energy Harvesting & Storage USA, Boston, MA, USA, November 2011.

Sazonova, E., Janoyan, K., Jha, R. *Sensor Network Application Framework for Autonomous Structural Health Monitoring of Bridges*, Clarkson University, 8 Clarkson Ave, Potsdam, NY, 13699, SMT 2004.

Strba, A., Embedded systems with limited power resources, EnOcean GmbH, Germany, January 2009.

Smit, G., Albers, J. *Case Study: High Performance Wireless Sensing Network in a Challenging Environment Using Perpetuum Energy Harvesters and Nanotrons NanoNET*, Darnell nanoPower Forum, June 2008.

Thurman, A. *Seamless Geolocation in a Wireless Sensor Network*, Omnisense Ltd., IDTechex Energy Harvesting & Storage USA, Boston, MA, USA, November 2010.

TMPS. Tire Pressure Monitoring Systems—Tire tech information/General Information. Last accessed February 2012 [Online]. Available http://www.tirerack.com/tires/tiretech/techpage.jsp?techid= 44

Index